Lecture Notes in Networks and Systems

Volume 147

Series Editor

Janusz Kacprzyk, Systems Research Institute, Polish Academy of Sciences, Warsaw, Poland

Advisory Editors

Fernando Gomide, Department of Computer Engineering and Automation—DCA, School of Electrical and Computer Engineering—FEEC, University of Campinas—UNICAMP, São Paulo, Brazil

Okyay Kaynak, Department of Electrical and Electronic Engineering, Bogazici University, Istanbul, Turkey

Derong Liu, Department of Electrical and Computer Engineering, University of Illinois at Chicago, Chicago, USA; Institute of Automation, Chinese Academy of Sciences, Beijing, China

Witold Pedrycz, Department of Electrical and Computer Engineering, University of Alberta, Alberta, Canada; Systems Research Institute, Polish Academy of Sciences, Warsaw, Poland

Marios M. Polycarpou, Department of Electrical and Computer Engineering, KIOS Research Center for Intelligent Systems and Networks, University of Cyprus, Nicosia, Cyprus

Imre J. Rudas, Óbuda University, Budapest, Hungary

Jun Wang, Department of Computer Science, City University of Hong Kong, Kowloon, Hong Kong

AF876250

The series "Lecture Notes in Networks and Systems" publishes the latest developments in Networks and Systems—quickly, informally and with high quality. Original research reported in proceedings and post-proceedings represents the core of LNNS.

Volumes published in LNNS embrace all aspects and subfields of, as well as new challenges in, Networks and Systems.

The series contains proceedings and edited volumes in systems and networks, spanning the areas of Cyber-Physical Systems, Autonomous Systems, Sensor Networks, Control Systems, Energy Systems, Automotive Systems, Biological Systems, Vehicular Networking and Connected Vehicles, Aerospace Systems, Automation, Manufacturing, Smart Grids, Nonlinear Systems, Power Systems, Robotics, Social Systems, Economic Systems and other. Of particular value to both the contributors and the readership are the short publication timeframe and the world-wide distribution and exposure which enable both a wide and rapid dissemination of research output.

The series covers the theory, applications, and perspectives on the state of the art and future developments relevant to systems and networks, decision making, control, complex processes and related areas, as embedded in the fields of interdisciplinary and applied sciences, engineering, computer science, physics, economics, social, and life sciences, as well as the paradigms and methodologies behind them.

Indexed by SCOPUS, INSPEC, WTI Frankfurt eG, zbMATH, SCImago.

All books published in the series are submitted for consideration in Web of Science.

More information about this series at http://www.springer.com/series/15179

Nikhil Ranjan Das · Santu Sarkar
Editors

Computers and Devices for Communication

Proceedings of CODEC 2019

Springer

Editors
Nikhil Ranjan Das
Institute of Radio Physics and Electronics
University of Calcutta
Kolkata, India

Santu Sarkar
Institute of Radio Physics and Electronics
University of Calcutta
Kolkata, India

ISSN 2367-3370 ISSN 2367-3389 (electronic)
Lecture Notes in Networks and Systems
ISBN 978-981-15-8365-0 ISBN 978-981-15-8366-7 (eBook)
https://doi.org/10.1007/978-981-15-8366-7

© Springer Nature Singapore Pte Ltd. 2021
This work is subject to copyright. All rights are reserved by the Publisher, whether the whole or part of the material is concerned, specifically the rights of translation, reprinting, reuse of illustrations, recitation, broadcasting, reproduction on microfilms or in any other physical way, and transmission or information storage and retrieval, electronic adaptation, computer software, or by similar or dissimilar methodology now known or hereafter developed.
The use of general descriptive names, registered names, trademarks, service marks, etc. in this publication does not imply, even in the absence of a specific statement, that such names are exempt from the relevant protective laws and regulations and therefore free for general use.
The publisher, the authors and the editors are safe to assume that the advice and information in this book are believed to be true and accurate at the date of publication. Neither the publisher nor the authors or the editors give a warranty, expressed or implied, with respect to the material contained herein or for any errors or omissions that may have been made. The publisher remains neutral with regard to jurisdictional claims in published maps and institutional affiliations.

This Springer imprint is published by the registered company Springer Nature Singapore Pte Ltd.
The registered company address is: 152 Beach Road, #21-01/04 Gateway East, Singapore 189721, Singapore

Preface

The world is witnessing a tremendous progress in science and technology through significant transformation in industrial manufacturing, which has resulted in the emergence of the Fourth Industrial Revolution (Industry 4.0). Now, we talk about materials, structures and devices with prefix 'smart', and we wish to see ourselves in smart home, in smart cities. This vision is being fulfilled due to the advances in high-speed communication, optimized computation, artificial intelligence, machine-to-machine communication, material science, novel devices, data science and several other areas of science and technology.

To discuss such advances and recent trends in industry and research, the Seventh International Conference on Computers and Devices for Communication (CODEC) was organized on 19–20 December in Kolkata, India, by bringing together researchers from different parts of the world to exchange their views. CODEC was initiated in 1998 to begin the golden jubilee year celebration of the Institute of Radio Physics and Electronics (IRPE), a department of the University of Calcutta. The IRPE (estd.1949) is the first university department in India to begin teaching and research in electronics, devices, communication, control and all areas covered by radio science. CODEC is organized by the IRPE, and since 2014, it has become a triennial event, though the seventh CODEC was held after a 4-year gap due to some unavoidable reasons. All CODECs were participated by international delegates from different parts of the world and at different times were technically sponsored by several IEEE societies (EDS, PS, APS, MTTS), IET, SPIE, etc. The deliberations from experts were in the form of keynote address, plenary talks, invited lectures and contributory research papers.

Selected contributory research papers of CODEC-19, after peer review (minimum two reviewers) and plagiarism checks, have been considered for publication in Springer Lecture Notes in Networks and Systems (LNNS). In this book, each paper has appeared as a chapter. The chapters cover almost all important areas of electronics and communication and are grouped into four broad categories as follows:

- Computer Applications and Control (CAC)
- Communication and Space Science (CSS)

- Microwave and Lightwave Technology (MLT)
- Nanoscale Materials and Devices (NMD)

It was a hard task to arrange peer reviews and plagiarism checks for all the contributory papers to comply with the publication requirement of Springer LNNS. The editors are thankful to their colleagues and staff in the department, who made a meticulous effort to complete the task towards the final selection of papers for publication. The editors are grateful to the reviewers for their time and efforts towards critical review of the submitted contributory research papers. Full credit goes to them for their evaluation to ensure the high quality and standard of the papers for the Springer Lecture Notes. It is a pleasure to acknowledge the help received from the Springer publication/production team, including G. Ayyasamy, S. Ravivarman, P. Thirumani, to name a few, and especially, Mr. Aninda Bose, Senior Editor of Springer Nature (Research Publishing).

Kolkata, India

Nikhil Ranjan Das
Santu Sarkar

Committees

ORGANIZING COMMITTEE

Chief Patron

Prof. Sonali Chakravarti Banerjee, Honb'le Vice-Chancellor, University of Calcutta, Kolkata, India

Patron

Prof. Gopa Sen Head of the Dept., Institute of Radio Physics and Electronics, University of Calcutta, Kolkata, India

Chairman

Prof. Abhirup Das Barman, IRPE, University of Calcutta

Convener

Dr. Jawad Y. Siddiqui, IRPE, University of Calcutta

Jt. Conveners

Dr. Anisha Halder Roy, IRPE, University of Calcutta
Dr. Suchismita Tewari, IRPE, University of Calcutta

Technical Programme Chairs

Prof. G. Ghosh, IRPE, University of Calcutta
Prof. Animesh Maitra, IRPE, University of Calcutta
Prof. Partha P. Goswami, IRPE, University of Calcutta
Dr. Anirban Bhattarcharyya, IRPE, University of Calcutta
Prof. Abhijit Biswas, IRPE, University of Calcutta

Finance Chair

Dr. Bratati Mukhopadhyay, IRPE, University of Calcutta

Publication Chairs

Prof. Nikhil Ranjan Das, IRPE, University of Calcutta
Dr. Santu Sarkar, IRPE, University of Calcutta

Registration Chairs

Dr. Sumitra Ghosh, IRPE, University of Calcutta
Dr. Arpita Das, IRPE, University of Calcutta
Ms. Shampa Guin, IRPE, University of Calcutta

Proceedings Chair

Dr. Kaushik Mandal, IRPE, University of Calcutta
Mr. Sumit Dasgupta, IRPE, University of Calcutta
Dr. Prasun Ghosal, IIEST, Shibpur

Local Arrangements Chairs

Mr. Souvik Majumder, IRPE, University of Calcutta
Dr. Pulak Mondal, IRPE, University of Calcutta
Dr. Santanu Mandal, IRPE, University of Calcutta

Publicity Chairs

Dr. Soumya Pandit, IRPE, University of Calcutta
Dr. Chinmoy Saha, IIST, Thiruvananthapuram
Dr. Somak Bhattacharyya, IIT, BHU

Sponsorship Chair

Prof. Ashik Paul, IRPE, University of Calcutta

INTERNATIONAL ADVISORY COMMITTEE

Prof. C. Jagadish, Australian National University (Australia), Chair
Prof. Guo-En Chang, National Chung-Cheng University (Taiwan)
Prof. Basabi Chakraborty, Iwate Prefectural University , Japan
Prof. M. Jamal Deen, McMaster University (Canada), Co-Chair
Prof. Eddy Simoen, IMEC, Leuven, Belgium
Prof. Tapan K. Sarkar, Syracuse University, USA
Prof. G. Cuniberti, TU Dresden (Germany)

Prof. Antonella Bogoni, CNIT Pisa, Italy
Prof. Goutam Chakraborty, Iwate Prefectural University , Japan
Prof. B. M. A. Rahaman, City University of London (UK)
Prof. Jayanta Sarma, University of Bath, UK
Prof. Partha Pratim Pande, Washington State University, USA
Prof. Sanjit K. Mitra, University of California Santa Barbara, USA
Prof. Samar Saha, Santa Clara University, USA
Prof. Subhasish Chakraborty, The University of Manchester, UK

NATIONAL ADVISORY COMMITTEE

Prof. Shankar Pal, ISI, Kolkata, Chair
Prof. Asis Kumar Chattopadhyay, Pro-Vice-Chancellor for Academic Affairs, University of Calcutta
Prof. Mahua Bhattacharya, Indian Institute of Information Technology & Management, Gwalior
Prof. R. Muralidharan, IISC, Bangalore, Co-Chair
Prof. S. C Dutta Roy, Ex-IIT, Delhi
Prof. Bhabani P. Sinha, ISI, Kolkata
Prof. Bhargab B Bhattacharya, Ex-ISI, Kolkata
Prof. Ranjan Gangopadhyay, Ex-Professor (Emeritus); IIT, Kharagpur
Prof. Partha P. Sahu, Tezpur Central University, Tezpur
Prof. Debasish Datta, IIT, Kharagpur
Prof. P. K. Basu, Institute of Radio Physics and Electronics, Kolkata
Prof. P. K. Saha, Institute of Radio Physics and Electronics, Kolkata
Prof. J. B. Roy, Institute of Radio Physics and Electronics, Kolkata

Contents

About the Editors

Dr. Nikhil Ranjan Das is a Professor in the Department of Radio Physics and Electronics, University of Calcutta. He served as the Head of the Department, and the Dean of Engineering & Technology, and the Director of the Centre for Research in Nanoscience and Nanotechnology, the University of Calcutta. He was a Postdoctoral Fellow and later a Postdoctoral Research Associate at McMaster University, Canada, from 1999 to 2002. He was a Senior Guest Scientist in Trieste (Italy), and Visiting Professor at McMaster University (Canada), the University of Sheffield (UK), and Pohang University of Science and Technology (POSTECH, South Korea). He was also a Visiting Scientist at the Technical University Dresden, Germany, in 2018 as a recipient of the INSA-DFG award. His research focuses on design, simulation, and optimization of optoelectronic and photonic devices, including semiconductor nanostructures. He has been a reviewer for several IEEE, IET, and other journals. He is a Fellow of IETE and IE(I) and senior member of the IEEE. He was the founder and Chairman of the IEEE Photonics Society Calcutta Chapter, and has been the Branch Counselor of IEEE Calcutta University Student Branch. He completed international collaborative programs such as UKIERI with the University of Sheffield and was the key person from the University of Calcutta for Erasmus Mundus 'LEADERS' program.

Dr. Santu Sarkar received his B.Sc. (Hons.) degree in Physics from the University of Calcutta, in 1995, and B.Tech., M.Tech., and Ph.D. degrees in Radio Physics and Electronics from the University of Calcutta, in 1998, 2000, and 2011, respectively. From 2000 to 2004, he was a Lecturer at Asansol Engineering College, University of Technology, West Bengal. From 2004 to 2016, he was an Associate Professor and Head of the Electronics and Communication Engineering Department, Academy of Technology. He is currently an Assistant Professor in the Department of Radio Physics and Electronics, University of Calcutta.

Computer Applications and Control (CAC)

IEEE 754 Floating Point Pipelined Multiplier with Karatsuba for Mitigations of Area and Power

Mohammed Abdul Raheem(✉) and Mohammed Abdul Rahman Shareef

ECE Department, Muffakham Jah College Engineering and Technology, Hyderabad, India
abdulraheem.mj@gmail.com

Abstract. The proposed architecture implements IEEE 754 floating point pipelined multiplier merge single and double precision using Karatsuba. This paper is presented in order to reduce power and area expenditures to fast the process of the adder and reduce delay. To achieve, the design Verilog language is used and targeted on Xilinx Virtex-5 (XC5VLX155FF1760-3) and Cadence on TSMC-180 nm CMOS technology. The architecture reduces the processing block as it uses exception block as replacement, whereas in the Karatsuba multiplier, the adders are replaced with different adder to obtain reduce results of area and power. Karatsuba is used in placing the mantissa in the multiplier, and in the Karatsuba architecture, the previous design is replaced in order to reduce the numbers of operations in block processing; here, left shifting is in the 27 * 27 bits multiplier. By implementation of various adders, better results are obtained when compared to the previous work in terms of DSP 6 (4%), LUTs 1367 (1%), frequency 168.741 MHz, dynamic power 123.61 mW.

Keywords: Xilinx Virtex-5 · Cadence · Verilog · Karatsuba · Floating point pipelined multiplier

1 Introduction

In recent years, there are fabulous developments in computation of results in field of adders, subtractors, multipliers, divisions and other operations. IEEE 754 plays all the function in it as floating point multiplier. There are different types of precisions; for instance, two types of precisions are used in this as single precision and double precision.

2 System-level Architecture

We purposed an exception processing block replacing three processing blocks of multiplicand processing, exponent processing and sign processing, and there are two blocks in stage of 1–5 where 27 * 27 bits multipliers are show in Fig. 1.

© Springer Nature Singapore Pte Ltd. 2021
N. R. Das and S. Sarkar (eds.), *Computers and Devices for Communication*,
Lecture Notes in Networks and Systems 147,
https://doi.org/10.1007/978-981-15-8366-7_1

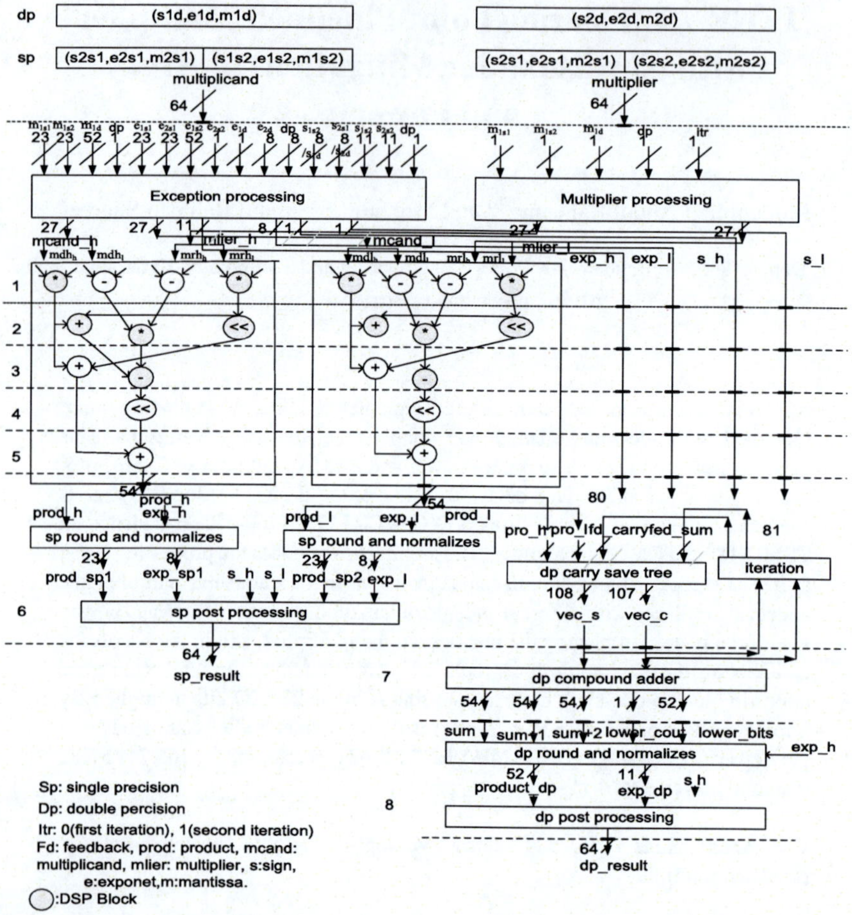

Fig. 1 Data path of the proposed single- and double precision floating point

3 Simulation Results Obtained in Xilinx ISE Software

Simulation results of IEEE 754 floating point pipeline multiplier merged single precision and double precision using Karatsuba with various adder is showed in Figs. 2, 3 and 4.

3.1 Carry Propagate Adder

Inputs: A_{in} [7:0] = 36; B_{in} [7:0] = 129; C_{in} = 1; Width [31:0] = 8.

Output: SUM [7:0] = 166; Carry out = 0.

3.2 Kogge–Stone Adder

Inputs: Clk = 1; dp = 0; itr = x; Multiplicand [63:0] = e2ca4ec5f4007ae8; Multiplier [63:0] = de8e28bd2e584…; **Outputs**: SP Result [63:0] = 90a7fa5380000000.

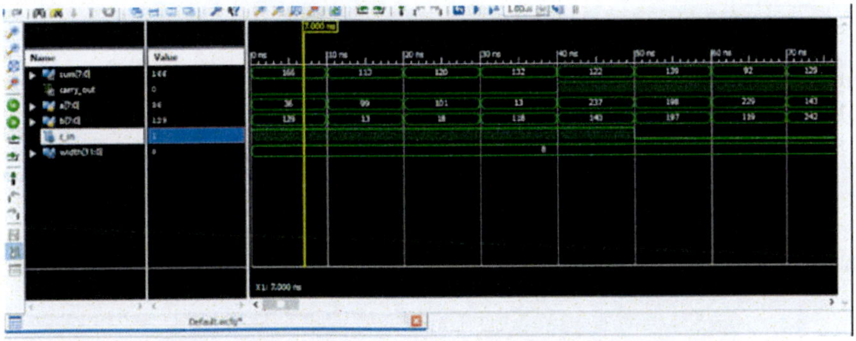

Fig. 2 Simulation result of carry propagated adder

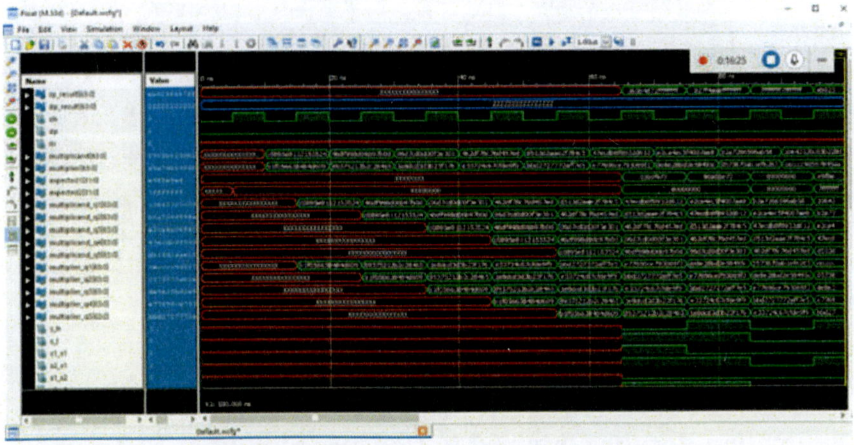

Fig. 3 Simulation result of Kogge–Stone adder

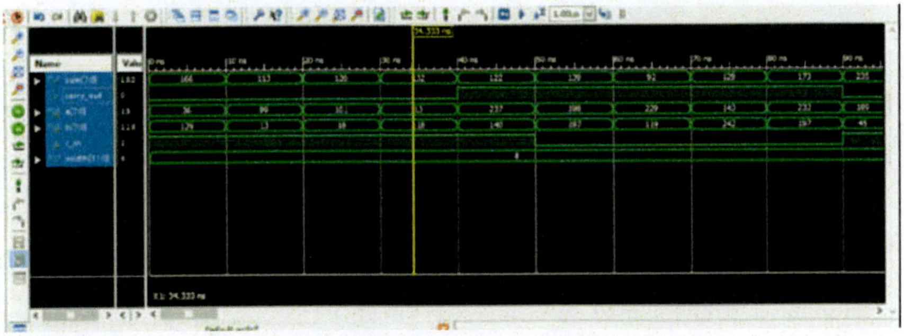

Fig. 4 Simulation result of carry look ahead adder

3.3 Carry Look Ahead Adder

Inputs: A_{in} [7:0] = 13; B_{in} [7:0] = 118; C_{in} = 1; Width [31:0] = 8.

Outputs: SUM [7:0] = 132; Carry out = 0.

4 Schematic of Floating Point Multiplier in Cadence

Figure 5 is the structure view of complier RC 14.28 this is implemented in Cadence in TSMC-180 nm CMOS technology.

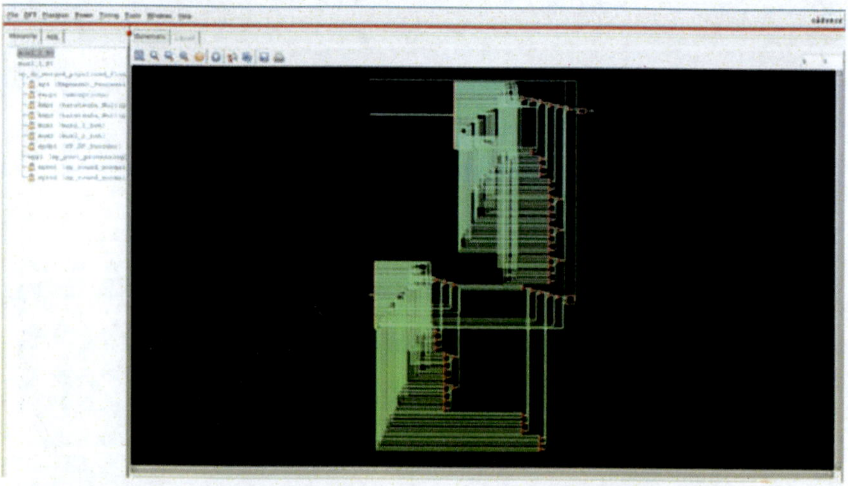

Fig. 5 Schematic of floating point multiplier

5 Simulation Results Obtained in Cadence Simulator Software

See Fig. 6.

a Dynamic power = 17,928,133.149 nW; **b** Area = 17,226.

6 Comparison Results

See Table 1.

7 Conclusion

The multiplier is implemented in Xilinx on Virtex-5 (XC5VLX155FF1760-3) and Cadence, for high-performance Kogge–Stone adder is best in terms as it obtained better results as less power as per the desired specifications by implementation of the floating point multiplier. The adder is implemented successfully with Karatsuba in Xilinx and Cadence, with less area and power consumptions achieved as compare to previous work [1]. This work can be further increased in by using Cadence with different multipliers [2], Alteria-7 [3].

(a) Dynamic Power Report of adder.

(b) Area Report of adder

Fig. 6 **a** Dynamic power report and **b** area report for adder

Table 1 Comparison results of various adders

Device	DSP	LUT's	Frequency (MHz)	Dynamic power (mW)
Virtex-5 adder	6 (4%)	1565 (1%)	156.33	168.7
Virtex-5 carry look ahead adder	6 (4%)	1759 (1%)	125.142	170.37
Virtex-5 carry propagate adder	6 (4%)	1434 (1%)	117.542	170.10
Virtex-5 Kogge–Stone adder	6 (4%)	1367 (1%)	168.741	123.61
Virtex-5 [1]	6 (4%)	1117 (1%)	348.19	247.70

References

1. Zhang, H., Chen, D., Ko, S.-B.: Area- and power-efficient iterative single/double-precision merged floating-point multiplier on FPGA. IET Comput. Digit. Tech. **11**(4), 7 (2017)
2. Deepak, P., Sharma, R.K.: Design of low power reconfigurable floating point multiplier. IET computers and digital techniques. In: 2016 Conference on Advances in Signal Processing (CASP), Cummins College of Engineering for Women, Pune, 9–11 June 2016
3. Zhang, H., Chen, D., Ko, S.-B.: High performance and energy efficient single precision and double-precision merged floating-point adder on FPGA. IET Comput. Digit. Tech. **12**(1) (2018)

Check for updates

Digital Fault Detection Techniques: A Review

Vivekananda Mukherjee, Pradip Kumar Ghosh, Manabendra Maiti, and Judhajit Sanyal(✉)

Department of Electronics and Communication Engineering, Techno International New Town, 1/1, Service Rd, DG Block, Action Area I, Newtown, Kolkata, West Bengal 700156, India
vivek.ananda@rediffmail.com, pkghosh_ece@ieee.org, manab1972@yahoo.com, sanyaljudhajit@gmail.com

Abstract. The testing and detection of faults in digital circuits has become extremely relevant to both industry as well as academia in recent years. Effective and cheap testing procedures and algorithms help to decrease the manufacturing costs for digital circuits to a significant extent. The present work discusses state-of-the-art digital fault detection techniques. Specific emphasis is laid on a relatively new class of techniques that employ machine learning techniques to classify and detect faults in digital circuits and systems, especially focusing on Bayesian networks and support vector machines (SVMs). The work is concluded by the proposal of a machine learning-based fault probability estimation technique and brief deliberations on the enhancements that can be made in future to the technique.

Keywords: Testing · Digital faults · Machine learning · Bayesian network · Support vector machines

1 Introduction

Digital systems currently define the way the majority of humans interact and remain connected to each other as well as the natural world. The advent of digital circuits has brought about new possibilities in terms of computation, control and design of electronic systems. Digital circuits have revolutionized many domains of engineering in direct or indirect ways. However, certain critical issues still remain in the field of digital circuits and systems. One such issue is the testing of digital circuits for the possibility of occurrence of faults. Since a fault in a critical digital circuit or system can have far-reaching effects resulting in malfunctioning, researchers have proposed various methods for detection of digital faults. However, due to the ever-increasing complexity of digital systems, new methods need to be regularly investigated to keep up with the increase in number and complexity of digital circuits.

Researchers have outlined extensive testing procedures and principles for the testing of digital circuits. Many tools and algorithms have been designed to ensure the functional integrity of digital circuits and systems. However, new tools are constantly being proposed and investigated by scholars for the possible applicability to the domain of

© Springer Nature Singapore Pte Ltd. 2021
N. R. Das and S. Sarkar (eds.), *Computers and Devices for Communication*,
Lecture Notes in Networks and Systems 147,
https://doi.org/10.1007/978-981-15-8366-7_2

digital circuit and system testing. Consequently, some of the important state-of-the-art techniques are reviewed in this work.

The paper is organized in the following manner. Section 2 contains the survey of research papers outlining novel techniques for determining digital faults. This section lays special emphasis on machine learning-based techniques, which have gained in prominence over the recent few years. Section 3 elucidates the proposed machine learning-based scheme for detection and classification of digital faults. Section 4 concludes the paper with a brief discussion on the avenues of research expected to stem from the present work.

2 Literature Survey

Digital circuits have been tested using various techniques since their inception. Parameters such as the number of states have been studied for identifying fluctuations in their numbers and estimating fault coverage [1]. The tests used by researchers have at times been singular; however, in recent years, adaptive testing has gained prominence due to the highly accurate results available using such testing procedures [2]. Binary decision diagrams have also been used by some scholars to identify conditional digital faults [3]. Among the classes of digital faults, delay faults can cause massive depreciation in circuit performance. Such delay faults have been studied by certain researchers in order to reduce their testing time [4].

Automated design flows have been put forward by authors in a bid to apply triple modular redundancy in such digital circuits [5]. The patterns generated for testing a digital circuit have also been validated by some researchers using automatic test pattern generator (ATPG) [6]. Research has also been carried out to reduce testing time related to test pattern generation [7, 8]. Reseeding of the linear feedback shift register (LFSR) has been employed by the scholars to that end [7].

Extensive research has also been done in the domain of detection and elimination of hidden faults [9]. The evolution of hidden faults has been examined in conjunction with their proposed solutions [10]. Injection of faults has been employed for the purpose of diagnosis of digital faults in FPGA-based systems [11]. Power supply glitch-based faults have also been investigated, and appropriate solutions based on fast clock modulation have been proved to be applicable in such cases [12]. Other state-of-the-art approaches to digital fault detection include quantum-based test generation methods [13]. Nowadays, the utilization of novel digital VLSI circuits is increasing exponentially, and the design trends are to decrease the node capacitance and power supply requirements. This results in huge susceptibility to transient faults to increase in the nano-range digital CMOS designs [14].

In recent times, machine learning has gained importance in the domain of digital fault testing, diagnosis and mitigation. Classification and recommendation-based algorithms involving neural networks (NNs) and support vector machines (SVMs) have been studied and developed for digital fault localization in object-oriented paradigms [15]. Digital fault identification and mitigation has been carried out for vehicular systems connected by wireless networks through the application of NNs [16].

Among the machine learning techniques popularly employed, Bayesian techniques have been used frequently by researchers in recent years due to their comparative simplicity, as well as their accuracy. Bayesian networks have been extensively examined by some researchers in the context of their applicability to the testing and fault identification of digital systems of varying specifications [17]. The detection of data faults in wireless sensor networks has, for example, been accomplished through a distributed Bayesian network technique, which allows for rapid detection and classification of faults [18]. Minimal risk Bayesian decision theory has been used in conjunction with other techniques to diagnose faults [19]. Principal component analysis (PCA) has been employed by some researchers to determine faults even in situations with missing information [20].

The present work therefore proposes a Bayesian method of prediction of probability of occurrence of a fault in any particular type of digital circuit or system, as discussed in the following section.

3 Proposed Technique

Multiple parameters may influence the occurrence of faults in digital circuits. These include but are not limited to physical parameters such as aging of the circuit, power supply noise, sudden drops in voltage at certain points in the circuit and crosstalk effect between connecting wires [1]. These factors may or may not cause faults in digital circuits at any given instance of their occurrence. Consequently, it is possible to design a probabilistic Bayesian model for estimation of the probability of occurrence of a fault in a digital circuit in the presence of these parameters. The only assumption used to formulate such a model is the fact that the parameters used are independent of each other.

A frequency table containing the instances where faults occur or do not occur in the presence of each of the above-mentioned parameters is constructed. The instances are represented by suitable variables. The number of times faults do not occur despite aging of circuit is represented by A_{nf}, and the number of times faults occur in an aged circuit is represented as A_{of}. In a similar manner, where fault does not occur despite power supply noise being present, S_{nf} represents such instances, with corresponding number of faulty instances being represented by S_{of}. The corresponding variables for voltage drops are V_{nf} and V_{of}, respectively, and for crosstalk, C_{nf} and C_{of}, respectively. Table 1 is presented as follows.

From Table 1, the corresponding probabilities can be found. The overall probability of a fault not occurring in presence of these parameters is represented by P_{FF} and the corresponding probability of occurrence of a fault by P_{TF}. The total number of instances that each parameter is present in circuits is determined by the sum of the number of times faults occur and do not occur in circuits, for each parameter. The corresponding number of instances for aging, power supply noise, sudden voltage drops and connector crosstalk is represented by A, S, V and C, respectively.

$$A_{\mathrm{nf}} + A_{\mathrm{of}} = A \tag{1}$$

$$S_{\mathrm{nf}} + S_{\mathrm{of}} = S \tag{2}$$

Table 1 Digital faults occurring in the presence of various parameters

Parameter	Fault not occurring	Fault occurring
Aging of circuit	A_{nf}	A_{of}
Power supply noise	S_{nf}	S_{of}
Sudden voltage drops	V_{nf}	V_{of}
Connector crosstalk	C_{nf}	C_{of}

$$V_{nf} + V_{of} = V \tag{3}$$

$$C_{nf} + C_{of} = C \tag{4}$$

$$P_{FF} = (A_{nf} + S_{nf} + V_{nf} + C_{nf})/(A + S + V + C) \tag{5}$$

$$P_{TF} = (A_{of} + S_{of} + V_{of} + C_{of})/(A + S + V + C) \tag{6}$$

The total probabilities of occurrence of faults for each of the parameters are also determined through the following equations. The probability that an aging circuit will experience a fault is given by OF_A, while the probability of no occurrence of a fault of an aging circuit is given by NF_A. Similarly, for power supply noise, the variables are OF_S and NF_S, while for sudden voltage drops, they are OF_V and NF_V, respectively, and for crosstalk, the parameters are OF_C and NF_C. The corresponding equations are given as follows.

$$OF_A = A_{of}/(A + S + V + C) \tag{7}$$

$$NF_A = A_{nf}/(A + S + V + C) \tag{8}$$

$$OF_S = S_{of}/(A + S + V + C) \tag{9}$$

$$NF_S = S_{nf}/(A + S + V + C) \tag{10}$$

$$OF_V = V_{of}/(A + S + V + C) \tag{11}$$

$$NF_V = V_{nf}/(A + S + V + C) \tag{12}$$

$$OF_C = C_{of}/(A + S + V + C) \tag{13}$$

$$\mathrm{NF_C} = C_{\mathrm{nf}}/(A + S + V + C) \tag{14}$$

Now, the following Bayes' theorem, Prob(x|y) is the probability of event x occurring provided event y has occurred, Prob(y|x) is the probability of event y occurring provided event x has occurred, Prob(y) is the total probability of event y occurring, and Prob(x) is consequently the total probability of event x occurring. By Bayes rule, the equation governing these probabilities is given by

$$\mathrm{Prob}(x|y) = \mathrm{Prob}(y|x) * \mathrm{Prob}(x)/\mathrm{Prob}(y) \tag{15}$$

Using the previous Eqs. (1)–(14) with Eq. (15), we can determine the probability of a fault occurring in the presence of any parameter. As an example, the numbers of faults occurring or not occurring at various instances in the presence of the parameters are simulated. The simulation results are shown in Table 2.

Table 2 Digital fault simulated to occur in presence of parameters

Parameter	Fault not occurring	Fault occurring
Aging of circuit	13	112
Power supply noise	66	57
Sudden voltage drops	58	43
Connector crosstalk	80	113

Using Eqs. (1)–(15), the probabilities of faults occurring due to each of the factors can be calculated, as shown in Table 3.

Table 3 Probabilistic estimates for occurrence of digital faults

Parameter	Probability of fault not occurring	Probability of fault occurring
Aging of circuit	0.02	0.21
Power supply noise	0.12	0.1
Sudden voltage drops	0.11	0.08
Connector crosstalk	0.15	0.21
Total probabilities	0.4	0.6

It can be seen from Table 3 that the proposed Bayes estimation model can estimate individual fault occurrence probabilities as well as joint probabilities, for example, the probability of a fault occurring in the presence of both power supply noise as well as connector crosstalk is $(0.1 + 0.21) = 0.31$. Hence, the Bayes estimation technique proposed in this work can help to predict fault probabilities for all combinations of parameters considered in the model.

4 Conclusion

The machine learning model proposed in the present work is able to estimate the probability of occurrence of faults in digital circuits with respect to the presence of multiple parameters, using Bayesian estimation. The model can be further improved by allowing nonlinear relationships between the model parameters as well as incorporation of more parameters to further increase the accuracy of prediction using the model. Techniques such as SVM and neural network-based estimation can also be combined to increase the effectiveness of the proposed estimation technique.

Acknowledgements. The authors acknowledge the Department of Electronics and Communication Engineering, Techno International New Town, for providing the necessary support and resources.

References

1. Vinarskii, E., Laputenko, A., López, J., Kushik, N.: Testing digital circuits: studying the increment of the number of states and estimating the fault coverage. In: 2018 19th International Conference of Young Specialists on Micro/Nanotechnologies and Electron Devices (EDM), Erlagol, pp. 6403–6408 (2018)
2. Stratigopoulos, H., Streitwieser, C.: Adaptive test with test escape estimation for mixed-signal ICs. IEEE Trans. Comput. Aided Des. Integr. Circuits Syst. **37**(10), 2125–2138 (2018)
3. Jürimägi, L., Ubar, R.: Conditional fault collapsing in digital circuits with shared structurally synthesized BDDs. In: 2018 16th Biennial Baltic Electronics Conference (BEC), Tallinn, pp. 1–4 (2018)
4. Hirai, S., Yotsuyanagi, H., Hashizume, M.: Test time reduction on testing delay faults in 3D ICs using boundary scan design. In: 2018 IEEE 27th Asian Test Symposium (ATS), Hefei, pp. 7–12 (2018)
5. Benites, L.A.C., Kastensmidt, F.L.: Automated design flow for applying triple modular redundancy (TMR) in complex digital circuits. In: 2018 IEEE 19th Latin-American Test Symposium (LATS), Sao Paulo, pp. 1–4 (2018)
6. Boutobza, S., Popa, S., Costa, A.: An automatic testbench generator for test patterns validation. In: 2018 IEEE East-West Design & Test Symposium (EWDTS), Kazan, pp. 1–11 (2018)
7. Dilip, P.S., Somanathan, G.R., Bhakthavatchalu, R.: Reseeding LFSR for test pattern generation. In: 2019 International Conference on Communication and Signal Processing (ICCSP), Chennai, India, pp. 0921–0925 (2019)
8. Bozzoli, L., Sterpone, L.: MATS**: an on-line testing approach for reconfigurable embedded memories. In: 2018 IEEE International Symposium on Defect and Fault Tolerance in VLSI and Nanotechnology Systems (DFT), Chicago, IL, pp. 1–6 (2018)
9. Drozd, O., Kuznietsov, M., Martynyuk, O., Drozd, M.: A method of the hidden faults elimination in FPGA projets for the critical applications. In: 2018 IEEE 9th International Conference on Dependable Systems, Services and Technologies (DESSERT), Kiev, pp. 218–221 (2018)

10. Drozd, A., Kuznietsov, M., Antoshchuk, S., Martynyuk, A., Drozd, M., Sulima, J.: Evolution of a problem of the hidden faults in the digital components of safety-related systems. In: 2018 IEEE East-West Design & Test Symposium (EWDTS), Kazan, pp. 1–5 (2018)
11. Nemikumar Mandaogade, N., Ingole, P.V.: Development of fault diagnosis system using fault injection for discrete architectures. In: 2018 Fourth International Conference on Computing Communication Control and Automation (ICCUBEA), Pune, India, pp. 1–6 (2018)
12. Singh, A., Kar, M., Chawla, N., Mukhopadhyay, S.: Mitigating power supply glitch based fault attacks with fast all-digital clock modulation circuit. In: 2019 Design, Automation & Test in Europe Conference & Exhibition (DATE), Florence, Italy, pp. 19–24 (2019)
13. Hahanov, V., Gharibi, W., Litvinova, E., Chumachenko, S.: Qubit-driven fault simulation. In: 2019 IEEE Latin American Test Symposium (LATS), Santiago, Chile, pp. 1–7 (2019)
14. Singar, S., Ghosh, P.K.: Fault-free D-latch configurations for low power applications. J. Nanoelectron. Optoelectron. **12**, 1–7 (2017)
15. Ascari, L.C., Araki, L.Y., Pozo, A.R.T., Vergilio, S.R.: Exploring machine learning techniques for fault localization. In: 2009 10th Latin American Test Workshop, Buzios, Rio de Janeiro, pp. 1–6 (2009)
16. Sargolzaei, A., Crane, C.D., Abbaspour, A., Noei, S.: A machine learning approach for fault detection in vehicular cyber-physical systems. In: 2016 15th IEEE International Conference on Machine Learning and Applications (ICMLA), Anaheim, CA, pp. 636–640 (2016)
17. Cai, B., Huang, L., Xie, M.: Bayesian networks in fault diagnosis. IEEE Trans. Ind. Inform. **13**(5), 2227–2240 (2017)
18. Yuan, H., Zhao, X., Yu, L.: A distributed Bayesian algorithm for data fault detection in wireless sensor networks. In: 2015 International Conference on Information Networking (ICOIN), Cambodia, pp. 63–68 (2015)
19. Zheng, Y., Mao, S., Liu, S., Wong, D.S., Wang, Y.: Normalized relative RBC-based minimum risk Bayesian decision approach for fault diagnosis of industrial process. IEEE Trans. Ind. Electron. **63**(12), 7723–7732 (2016)
20. Atoui, M.A., Verron, S., Kobi, A.: PCA in a Bayesian framework for fault detection. In: 11th IEEE International Conference on Control & Automation (ICCA), Taichung, pp. 354–359 (2014)

Machine Learning-Based Rain Attenuation Prediction Model

Md Anoarul Islam, Manabendra Maiti, Pradip Kumar Ghosh, and Judhajit Sanyal(✉)

Department of Electronics and Communication Engineering, Techno International New Town, Block—DG 1/1, Action Area 1, New Town, Rajarhat, Kolkata 700156, India
mdanoarul@gmail.com, manab1972@yahoo.com, pkghosh_ece@ieee.org, sanyaljudhajit@gmail.com

Abstract. Rain attenuation is an important factor affecting wireless communication systems throughout the world. Rain attenuation and rain rate data are collected using multichannel radiometer and laser precipitation monitor at a tropical location. The dataset obtained is used to initially propose an empirical model for prediction of rain attenuation from rain rate data. An alternative model using linear spline regression-based machine learning is also used to predict rain attenuation. The machine learning-based model is found to be more accurate by an appreciable degree compared to the empirical model proposed in the previous instance.

Keywords: Rain attenuation · Multichannel radiometer · Empirical model · Linear spline · Regression · Machine learning

1 Introduction

Attenuation of communication signals due to rainfall is a significant parameter that affects terrestrial and satellite communication links worldwide. Different methods of measurement and estimation of attenuation are employed by researchers at present. The effect of rain attenuation is especially felt in the tropics due to the concentration of the major amount of global rainfall in that region of the world [1]. In fact, majority of the efforts have been made during the last decades to characterize the statistical and dynamical aspects of rainfall rate and subsequent attenuation at the desired frequencies in the earth-space path. The physical approach to predict rain attenuation is to calculate the path attenuation through an integral of all the individual increments of rain attenuation caused by the drops encountered along the path [2, 3]. However, since the rain cannot be accurately described along the aforementioned path, all proposed models are empirical in nature.

The present paper outlines a machine learning-based estimation approach to predict rain attenuation at multiple frequencies. The paper is divided into the following sections. The succeeding Sect. 2 outlines some of the relevant research in the domain of estimation of rain attenuation through empirical models. Section 3 briefly discusses the instruments used for measurement of rain rate. Section 4 presents the empirical model initially

© Springer Nature Singapore Pte Ltd. 2021
N. R. Das and S. Sarkar (eds.), *Computers and Devices for Communication*,
Lecture Notes in Networks and Systems 147,
https://doi.org/10.1007/978-981-15-8366-7_3

proposed in the paper and then introduces the machine learning-based linear spline regression model which gives a more accurate estimate than the previously introduced curve fitting model. Accordingly, results obtained using both models are presented in comparative fashion in this section. Section 5 then concludes the paper with a brief discussion on future avenues of research.

2 Literature Survey

Extensive rain attenuation data collection has been carried out for temperate regions, where the trends displayed by such datasets do to occur in other regions with comparatively more rainfall, such as tropical regions [4]. Generally, in these regions, rain is mostly of stratiform structure, which is 'light' with relative large rain cell diameters [5]. But in tropics, rain is at times from convective rain cells with relatively small diameter which often results in heavy down pour for a short period [6]. However, different types of rain show different types of spatial structure, and hence, the vertical path is affected differently in the presence of stratiform rain and convective rain [7, 8]. Rain attenuation at high frequencies for 5G applications has also been investigated in recent years [8]. Previously proposed methods of calculation of rain attenuation have been shown to be less reliable under certain rainy conditions [9]. Specific rain attenuation is especially variable, as can be seen in recent works by researchers [10]. It has also been observed and reported by certain scholars that rain rate variations may cause loss-of-lock in communication and navigational systems [11].

As a consequence to the importance of rain attenuation measurement, machine learning models have been extensively used by researchers in recent years [12, 13]. Machine learning tools such as neural network-based models have become popular among researchers in recent years due to their accuracy [14].

The present paper therefore proposes a machine learning-based approach for predicting rain attenuation and investigates the efficacy of such an approach through comparison with other estimation models as well as actual values of the variable.

3 Measuring Instruments

Radiometer and laser precipitation monitor are used to collect data. The features of the radiometer and laser precipitation monitor are outlined in brief.

The radiometer used is a radiometrics profiling radiometer. This radiometer (MP-3000A) series incorporates two radio frequency (RF) subsystems in the same cabinet. The RF subsystems share the same antenna pointing system. The selected frequency bands are the water vapour absorption band (22–30 GHz) and the oxygen absorption band (50–60 GHz). The radiometer is controlled by radiometrics proprietary software and pre-installed control computer. The control computer is connected directly to the radiometer via supplied RS-422 cable. The operating code provides a single use graphical interface that allows the selection of user-defined observation procedures and automated calibration procedure. Real-time observations and calibration data are displayed in graphical format. Now, depending upon the choice of operating codes, it begins logging data to level '0' file (raw sensor data in volts), level '1' file (brightness temperature) and others along with calibration files. The radiometer antenna has its beam with and side lobes 4.9–6.3°; −24 dB (for 22–30 GHz) and 2.4–2.5°; −27 dB (for 50–60 GHz). It has the integration time 0.01–2.5 s and pre-detection channel bandwidth 300 MHz, and depending on the integration time, it possesses the resolution of 0.1–1.0 K. Here, in our study, we have chosen the resolution as 1.0 K.

The laser precipitation monitor of make Thies Clima is used to measure the rainfall intensity even though although it has the capability to measure particle sizes down to 0.16 mm diameter. The instrument has the following characteristics.

The optical laser operates at a wavelength of 785 nm with a maximum power of 0.5 nW. Measuring area is 46 cm^2. The minimum intensity that can be registered by the instrument is 0.0005 mm/h which corresponds to a drizzle, while maximum intensity measurable is 250 mm/h. The integration time for the device is 10 s.

The two instruments mentioned above are used to collect the data on which the estimation techniques are applied. Measurements of attenuation corresponding to rain rate are taken for the frequencies 22.234, 22.5, 23.034, 23.834, 25, 26.234, 28 and 30 GHz. Approximately, 7500 data points are obtained for each of the above-mentioned frequencies, for both attenuation as well as rain rate. Now, estimation models is proposed to estimate rain attenuation from rain rate.

4 Proposed Models and Results

The attenuation and rain rate data obtained through measurement are initially used to propose a power law model for estimation of attenuation from rain rate. The nature of the proposed model is observable in Eq. (1). The rain rate is represented as r, and the corresponding rain attenuation is represented as A.

$$A = X * r^Y \tag{1}$$

The values of X and Y for different frequencies obtained through curve fitting for minimum mean square error (MMSE) fit considering the power law model are shown in Table 1.

The corresponding values of mean error and root mean square error (RMSE) obtained through comparison with actual readings at different frequencies using the power law model are shown in Table 2.

The variation of mean error and RMSE over the given set of frequencies is shown in Fig. 1.

To better the estimation achieved by this model, a machine learning-based adaptive spline model is proposed, which calculates estimates by splitting up the datasets into subsets of 1500 points each and running a linear regression spline estimation on the subsets.

The spline model is piecewise linear in nature, and hence, the equations are constructed using constants P, which is the slope of the equation, and Q, which is the intercept constant, as shown in Eq. (2).

$$A = P * r + Q \tag{2}$$

Table 1 Values of X and Y parameters at different frequencies for power law model

Frequency (GHz)	X	Y
22.234	2.5259	0.5345
22.5	2.3576	0.561
23.034	2.9708	0.4939
23.834	3.6545	0.4269
25	3.2588	0.4757
26.234	2.84	0.5243
28	3.69	0.45
30	4.31	0.4171

Table 2 Mean error and RMSE at different frequencies for power law model

Frequency (GHz)	Mean error	RMSE
22.234	0.25	1.85
22.5	0.38	1.88
23.034	−0.11	1.91
23.834	−0.79	1.91
25	−0.8	1.94
26.234	−0.67	1.93
28	−1.29	1.98
30	−1.64	2.07

The sets of mean error values and RMSEs for all frequencies obtained using the linear spline regression model are shown in Table 3 (Fig. 2).

The RMSE is a better measure of accuracy compared to mean error, and therefore, the RMSEs for both techniques are compared to determine the relative accuracy of the two models. The comparison is graphically represented in Fig. 3.

The estimation values obtained by using the linear spline model are seen to be much more accurate compared to values obtained using the power law model. This clearly shows the advantage of using linear spline for estimation of rain attenuation from rain rate.

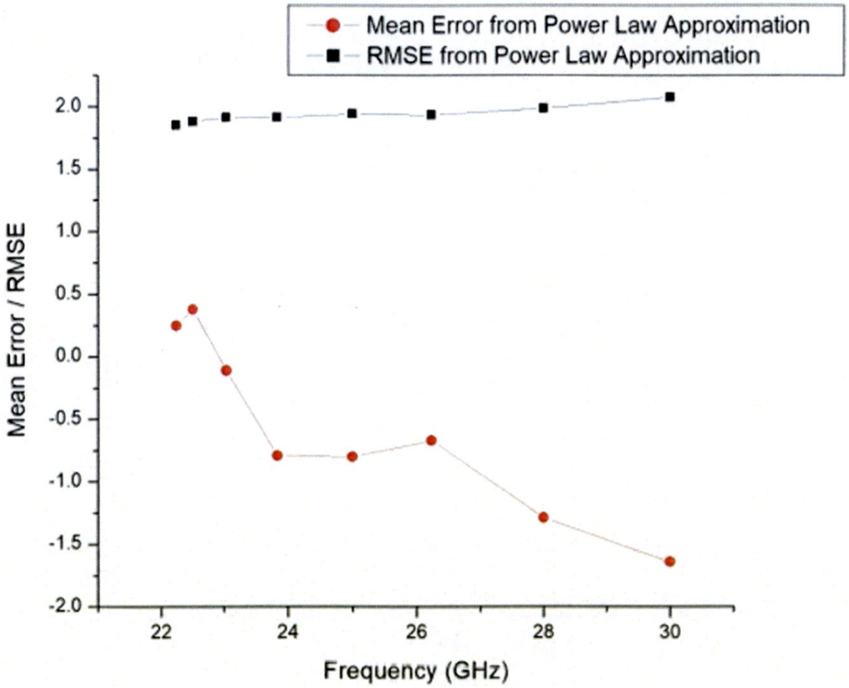

Fig. 1 Mean error and RMSE from power law estimation model

Table 3 Mean error and RMSE at different frequencies for linear spline model

Frequency (GHz)	Mean error	RMSE
22.234	−0.15	1.09
22.5	0	1.04
23.034	0	1.03
23.834	0	1.06
25	0	1.12
26.234	0	1.18
28	0	1.25
30	0	1.36

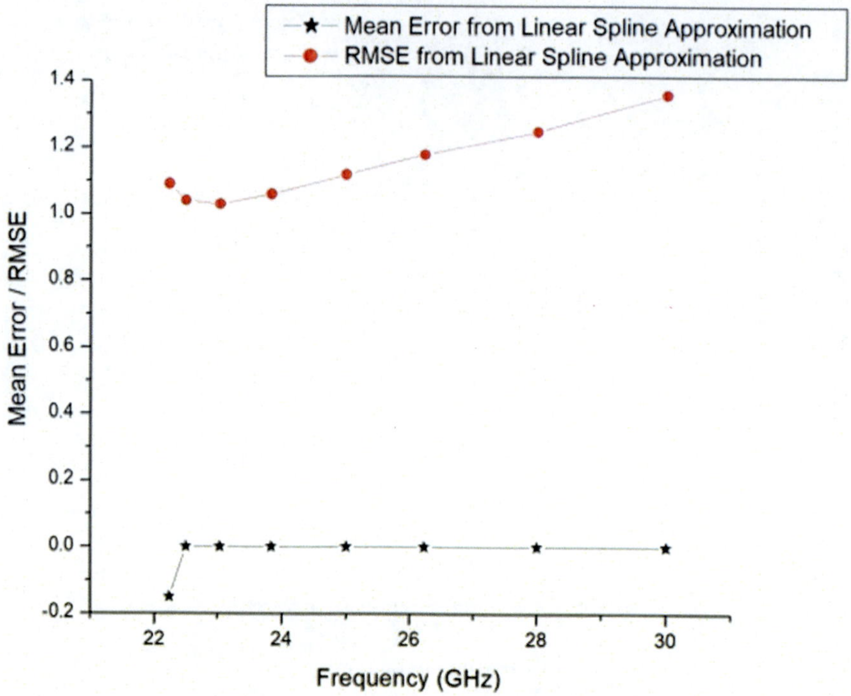

Fig. 2 Mean error and RMSE from linear spline estimation model

5 Conclusion

The machine learning technique used in the current paper has better accuracy compared to the empirical power law model proposed earlier. However, the linear spline estimation used in this paper can be further improved by nonlinear equations for more accurate prediction of nonlinear data points which otherwise may be regarded to be outliers by the linear spline estimation model. Also, classification of data into subsets may be done using other machine learning algorithms such as decision trees or support vector machine (SVM) models. Thus, the authors would endeavour to investigate combinations of machine learning models to improve accuracy. Another avenue of research would be estimation based on neural networks.

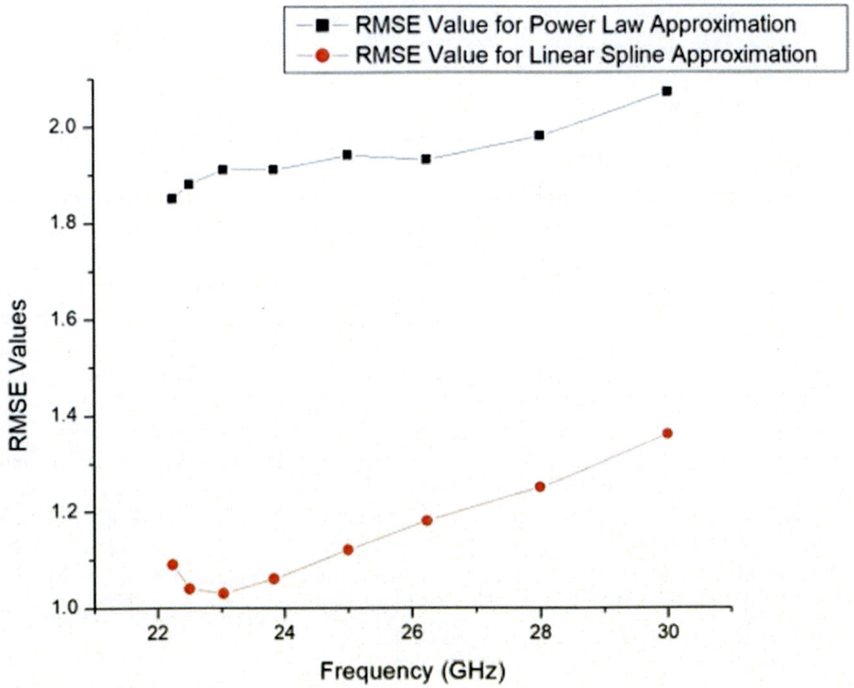

Fig. 3 Comparison of RMSE values obtained by power law and linear spline estimation models

Acknowledgements. The authors acknowledge the Department of Electronics and Communication Engineering, Techno International New Town, for providing the necessary support and resources.

References

1. Verma, A.K., Jha, K.K.: Rain drop size distribution model for Indian climate. Indian J. Radio Space Phys. **25**, 15–21 (1996)
2. Matricciani, E.: Service oriented statistics of interruption time due to rain fall in earth space communication systems. IEEE Trans. Antennas Propag. **52**(8), 2083–2090 (2004)
3. Kumar, V., Ramachandran, V.: Rain attenuation measurement at 11.6 GHz in Suva, Fiji. Electron. Lett. **40**(2), 1429–1431 (2004)
4. Singh, M.S.J., Hassan, S.I.S., Ain, M.F., Igarashi, K., Tanaka, K., Iida, M.: Rain attenuation model for South East Asia countries. IET Electron. Lett. **43**(2), 75–77 (2007)
5. Mandeep, J.S., Allnutt, J.E.: Rain attenuation predictions at ku-band in South East Asia countries. Prog. Electromagn. Res. **76**, 65–74 (2007)
6. Ramchandran, V., Kumar, V.: Modified rain attenuation model for tropical regions for ku-band signals. Int. J. Satell. Commun. Netw. **25**, 53–67 (2007)
7. Allnutt, J.E.: Slant path attenuation and space diversity results using 11.6 GHz radiometer. Proc. IEE **123**, 1197–1200 (1976)

8. Shayea, I., Rahman, T.A., Azmi, M.H., Islam, M.R.: Real Measurement study for rain rate and rain attenuation conducted over 26 GHz microwave 5G link system in Malaysia. IEEE Access **6**, 19044–19064 (2018)
9. Regonesi, E., Luini, L., Riva, C.: Limitations of the ITU-R P.838-3 model for rain specific attenuation. In: 2019 13th European Conference on Antennas and Propagation (EuCAP), Krakow, Poland, pp. 1–4 (2019)
10. Verma, A.K., Nandan, D.R., Verma, M.A.: Rain drop size distribution and variability of specific rain attenuation for Indian climate. In: 2019 URSI Asia-Pacific Radio Science Conference (AP-RASC), New Delhi, India, pp. 1–4 (2019)
11. Brost, G., Magde, K.: On the processing of loss-of-lock attenuation and rain rate statistics. In: 2019 13th European Conference on Antennas and Propagation (EuCAP), Krakow, Poland, pp. 1–3 (2019)
12. Ahuna, M.N., Afullo, T.J., Alonge, A.A.: Rain attenuation prediction using artificial neural network for dynamic rain fade mitigation. SAIEE Afr. Res. J. **110**(1), 11–18 (2019)
13. Yang, H., He, C., Song, W., Zhu, H.: Using artificial neural network approach to predict rain attenuation on earth-space path. In: IEEE Antennas and Propagation Society International Symposium. Transmitting Waves of Progress to the Next Millennium. 2000 Digest. Held in conjunction with: USNC/URSI National Radio Science Meeting (C), Salt Lake City, UT, vol. 2, pp. 1058–1061 (2000)
14. Sujimol, M.R., Acharya, R., Shahana, K.: Prediction and estimation of rain attenuation of Ka-band signals. In: 2019 URSI Asia-Pacific Radio Science Conference (AP-RASC), New Delhi, India, p. 1 (2019)

IoT-Based Real-Time Remote ECG Monitoring System

Samik Basu(✉), Anwesha Sengupta, Anindita Das, Mahasweta Ghosh, and Soma Barman (Mandal)

Institute of Radio Physics and Electronics, University of Calcutta, 92, A.P.C. Road, Kolkata 700009, India
samikbasu2010@gmail.com, anweshasengupta72@yahoo.com, dasanindita.konnagar@gmail.com, mahasweta94g@gmail.com, barmanmandal@gmail.com

Abstract. One of the most promising applications of Internet of things (IoT) is to interconnect the available medical resources and offer smart, reliable, and effective healthcare service, to meet the demands of growing population and increasing medical expenses. In remote places where people do not have easy and fast access to healthcare facilities, IoT enabled system can easily cater to such issues. IoT-based remote healthcare service allows patients to take quality medical advice from any corner of the world. Electrocardiogram (ECG) is a common diagnostic tool used to detect heart problems. This paper presents an IoT-based real-time ECG monitoring system design for remote patients. AD8232 sensor is used to acquire ECG signal which is interfaced with notch filter, for removal of power-line frequencies and is then sent to Arduino Uno for further processing. This signal is further processed in MATLAB platform, for removal of baseline wander and then uploaded to the cloud, which is then accessed via an external web page.

Keywords: ECG · AD8232 ECG sensor · Arduino Uno · Notch filter · Baseline wander · IoT · Cloud · Remote health monitoring

1 Introduction

Within the period between 1990 and 2016, the death rate due to cardiovascular diseases (CVD) in India increased by almost 34% [1]. In 2016, there was an estimated 62.5 million premature loss of life due to CVD in India. Regular visits to a doctor's chamber are a challenge for the elderly people. Also, in rural areas, there is acute shortage of health providers and infrastructure. India has 1 million doctors of modern medicine to treat its 1.3 billion population. Just 1.1 lakh out of them work in the public health sector, to whom India's 900 million rural population turns for treatment [2].

The modern technology, IoT, used in health monitoring of a patient wirelessly is a major development in medical arena. Electrocardiogram (ECG) is a diagnostic tool, used to monitor the electrical activity of our heart. For detection of heart diseases remotely,

© Springer Nature Singapore Pte Ltd. 2021
N. R. Das and S. Sarkar (eds.), *Computers and Devices for Communication*,
Lecture Notes in Networks and Systems 147,
https://doi.org/10.1007/978-981-15-8366-7_4

we used IoT in ECG monitoring system. A person can regularly monitor their health using this system. Doctor, sitting at any part of the world, can check the monitored data and accordingly can provide quick services and necessary instructions to the patient. Hence, the time spend by people in travelling to hospitals and waiting in long queues at hospitals can be saved.

Mishra and Chakraborty proposed a smart healthcare application using IoT system based on AD8232 sensor interfaced with Arduino Uno and further connected to cloud using Esp8266 Wi-Fi module [3]. Singh and Jasuja proposed an IoT-based remote ECG monitoring system, where the ECG signal is collected by AD8232 sensor [4]. After the required processing by Arduino Uno and Raspberry Pi 2, the data is sent to the cloud IBM Bluemix for further analysis. Deshpande and Kulkarni proposed an implementation of ECG monitoring system based on Cypress Wireless Internet Connectivity for Embedded Devices (WICED) in IoT platform [5].

Our proposed IoT-based ECG monitoring system consists of the AD8232 ECG sensor module to capture the ECG signal from the body. This signal is first filtered by our designed notch filter for removal of power-line frequencies and then processed by the Arduino Uno. The measured real-time ECG data is displayed on both Arduino's serial monitor and serial plotter. This signal is further filtered by Symlet wavelet transform in MATLAB environment for removal of baseline wander and uploaded to the cloud server via Wi-Fi. The data stored in the cloud can be then accessed from our designed external Website, for remote access. The main novelty of the system lies in the hardware design of notch filter and the use of Webhostapp Internet hosting service to create our own cloud server in healthcare application.

2 Methodology

Figure 1 shows the block representation of the proposed IoT-based ECG monitoring system. It consists of a sensing unit, a processing unit, and an ECG monitoring unit.

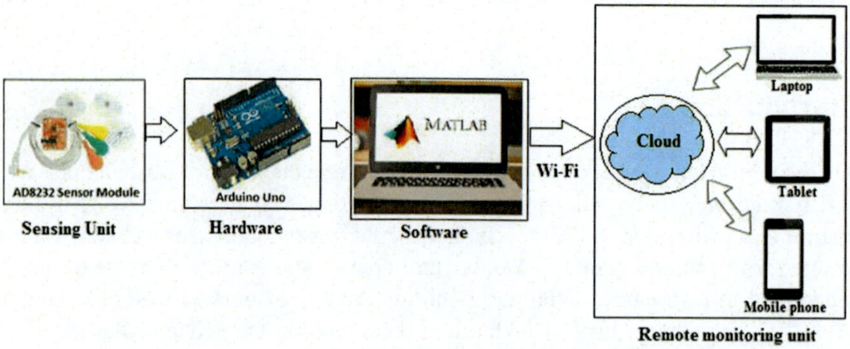

Fig. 1 Block representation of our proposed ECG monitoring system

The *sensing unit* consists of the *AD8232 sensor module* [6]. It is an integrated signal conditioning block for ECG and other bio-potential measurement applications. The *processing unit* consists of

Hardware Component. Arduino Uno [7].

Software Component. MATLAB environment [8].

The patients can self-monitor their ECG image and communicate to a remote doctor via IoT platform.

The experimental set-up of our IoT-based real-time remote ECG monitoring system and the placement of the ECG electrodes on the body are shown in Fig. 2a, b, respectively.

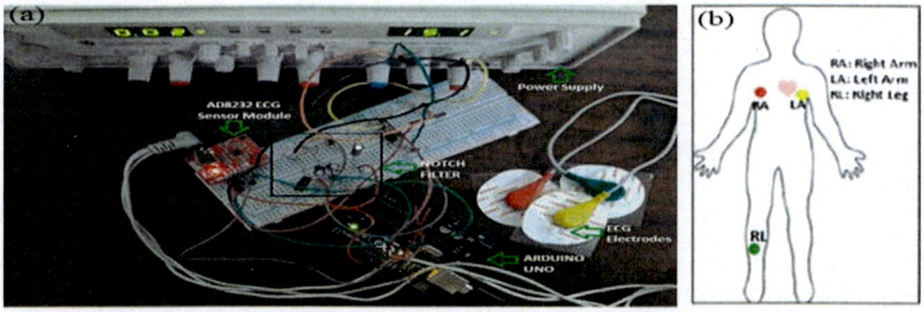

Fig. 2 **a** Experimental set-up of ECG monitoring system. **b** Placement of ECG electrodes on our body

The Arduino UNO is connected to the PC via an USB cable, and the power supply is switched on. When the red LED of the AD8232 ECG sensor module turns on, our system is in operation.

The noise (50 Hz frequency) from electric power system is a major source of noise during the recording of ECG. ECG signal has frequency range from 0.5 to 80 Hz, and power-line frequency of 50 Hz lies within that signal. This corrupts the ECG readings which affect the diagnosis of diseases. Hence, a notch filter is designed to eliminate the power-line frequencies [9, 10] as shown in Fig. 3.

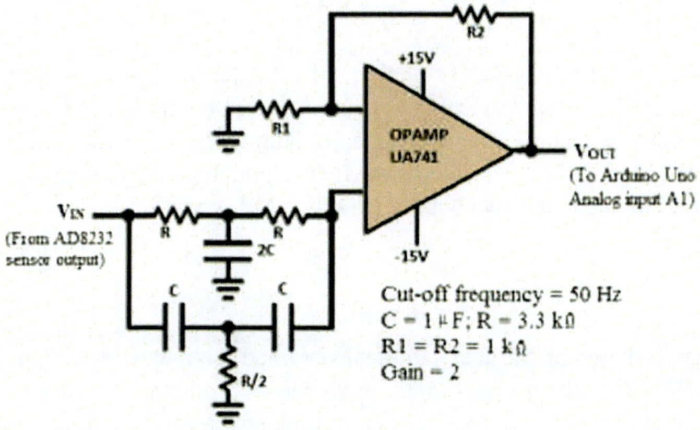

Fig. 3 Design of notch filter

The Arduino Uno receives powerline interference free signal, and processed ECG voltage is displayed on serial monitor at every instant of time. The real-time ECG signal, measured directly from the volunteer, is visible on Arduino serial plotter during the run-time [11].

The CoolTerm application [12], a freeware serial terminal software, is used to save the discrete data that appeared on the serial monitor of Arduino Uno in a text file, titled as "ecg.txt". This "ecg.txt" file (containing ECG data) is used in MATLAB environment for further filtering to remove baseline wander (BLW) which is a major noise source in ECG signal. BLW is a low frequency noise (0.05–3 Hz), characterized by the varying electrical isoline of the ECG signal, arising mainly due to movement during breathing, patient movements, poor contact between electrode cables and ECG recording equipment, inadequate skin preparation where the electrode is placed, and dirty electrodes. BLW severely limits the usefulness of ECG records. Hence, the elimination of this kind of noise is necessary for a better clinical evaluation.

Symlet, a type of discrete wavelet transform (DWT), is used to remove BLW from ECG signal [13] as Symlet wavelets nearly resemble the ECG wave [14]. Figure 4a shows a Symlet6 wavelet, where 6 represent the order, while Fig. 4b shows a typical ECG signal. Figure 5a, b shows the BLW removal algorithm developed in MATLAB R2015a platform and the algorithm for uploading the ECG data to the cloud server, respectively.

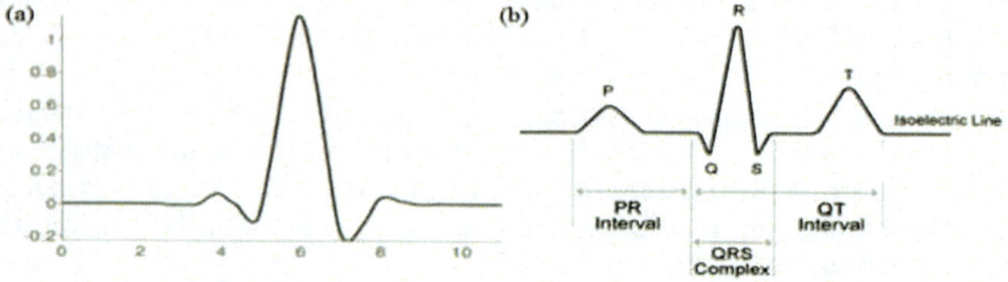

Fig. 4 **a** Scaling function of Symlet6 wavelet. **b** A typical ECG signal

After removing BLW, the ECG signal is uploaded to the cloud server. Webhostapp [15], a free Internet hosting service, is thus used to create our cloud server https://rpe-iot-ecg.000webhostapp.com. The image file containing the ECG signal of a person is sent to this cloud server. We can observe the ECG signal on this Website, and the signal can be accessed by the respective doctor, in any part of the world.

3 Result and Discussion

An ECG signal (1 out of 50 data samples) obtained from the sensor directly has a lot of noise (Fig. 6a), and our designed notch filter eliminates such noises (Fig. 6b). Figure 7a, b show, respectively, the ECG signal obtained after BLW removal in MATLAB environment and the signal observed from the remote end on our developed Website https://iotlabecg.000webhostapp.com. Both the uploading site (https://rpe-iot-ecg.000

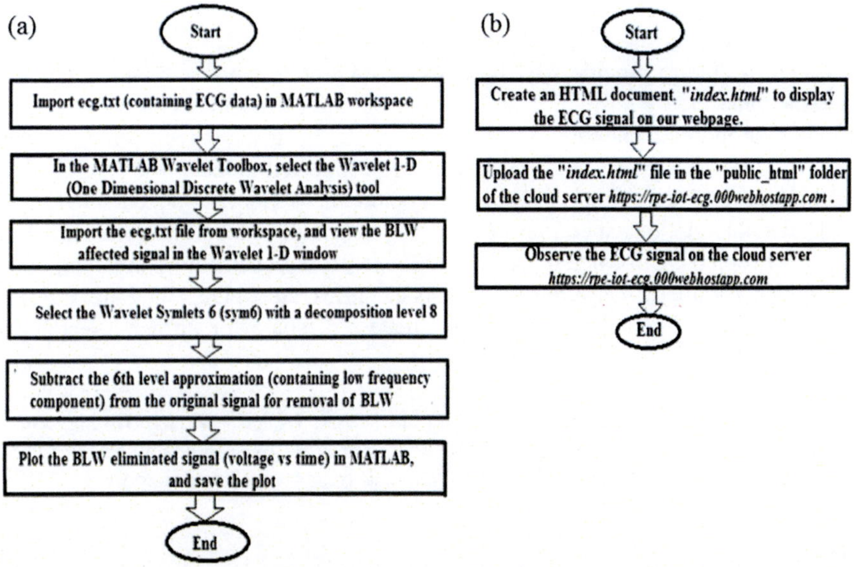

Fig. 5 Flowchart for **a** removal of baseline wander (BLW) from ECG signal in MATLAB. **b** Uploading ECG signal to cloud server

webhostapp.com) and the display site (https://iotlabecg.000webhostapp.com) for the ECG data can be viewed by entering the required address in the address bar.

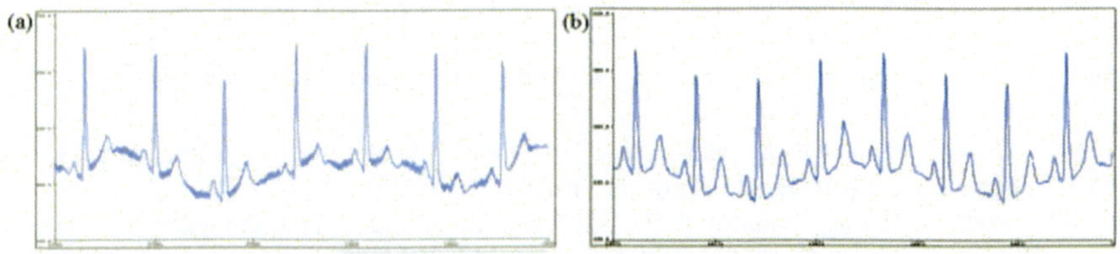

Fig. 6 **a** Noisy ECG signal (without using notch filter). **b** ECG signal (using notch filter)

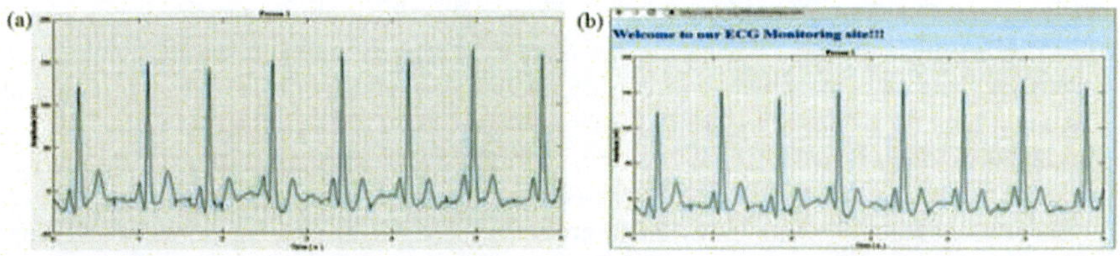

Fig. 7 **a** ECG signal observed after baseline wander removal. **b** ECG signal observed from cloud server https://rpe-iot-ecg.000webhostapp.com

4 Conclusion

Our system is able to provide 24/7 ECG monitoring of patients, and hence, this will reduce the unnecessary hospital visits and transportation cost and consequently result in a reduction in treatment cost. Any patient can take medical advice from any corner of the world because physicians and patient are connected globally via Internet. The MBBS doctor to patient ratio in India is far below the WHO recognized standard till date. Under this circumstance, the proposed system will reduce the huge burden of patients at hospitals. Our system has further scope of improvement by including more physiological parameter measurement sensors like heart rate, SpO2, blood pressure, etc. These uploaded ECG data can be used to create a database for further research and analysis.

Acknowledgements. The authors would like to thank UGC UPE II, "Modern Biology Group B: Signal Processing Group", University of Calcutta for providing research facilitiesand SMDP-C2SD, University of Calcutta,funded by MeitY, Govt. of India for technical support.

References

1. Dey, S.: Heart disease deaths rise in India by 34% in 26 years. The Times of India. https://timesofindia.indiatimes.com/india/heart-disease-deaths-rise-in-india-by-34-in-15-years/articleshow/64924601.cms
2. Sharma, S.: India's public health system in crisis: too many patients, not enough doctors. Hindustan Times. https://www.hindustantimes.com/india-news/public-health-system-in-crisis-too-many-patients-not-enough-doctors/story-39XAtFSWGfO0e4qRKcd8fO.html
3. Mishra, A., Chakraborty, B.: AD8232 based smart healthcare system using internet of things (IoT). Int. J. Eng. Res. Technol. (IJERT) **7**(4), 13–16 (2018)
4. Singh, P., Jasuja, A.: IoT based low-cost distant patient ECG monitoring system. In: International Conference on Computing, Communication and Automation, pp. 1330–1334 (2017)
5. Deshpande, U.U., Kulkarni, M.A.: IoT based real time ECG monitoring system using Cypress WICED. Int. J. Adv. Res. Electr. Electron. Instrum. Eng. **6**(2), 710–720 (2017)
6. Single-Lead, Heart Rate Monitor Front End Data Sheet AD8232. Analog Devices. https://www.analog.com/media/en/technical-documentation/data-sheets/AD8232.pdf
7. Arduino Uno. https://www.farnell.com/datasheets/1682209.pdf
8. MATLAB-Wikipedia. https://en.wikipedia.org/wiki/MATLAB
9. Band Stop Filters are Called Reject Filters. Electronics Tutorials. https://www.electronics-tutorials.ws/filter/band-stop-filter.html
10. Pavani, P., Adithya, N.V.R., Krishna, K.H., Kumar, M.V.S.: Removal of power line interference in ECG signal by adaptive LMS, NLMS and RLS algorithms. Int. J. Adv. Res. Comput. Commun. Eng. **6**(4), 800–806 (2017)
11. Verma, S.: Arduino Based ECG & Heartbeat Monitoring Healthcare System. Instructables. https://www.instructables.com/id/ECG-Monitoring-System-by-Using-Arduino-or-AD8232/
12. Visualize Data From Sensors Using Arduino + Coolterm. Instructables. https://www.instructables.com/id/Visualize-data-from-sensors-using-Arduino-coolte/
13. Ara, I., Hossain, Md.N., Mahbub, S.M.Y.: Baseline drift removal and de-noising of the ECG signal using wavelet transform. Int. J. Comput. Appl. **95**(16), 15–17 (2014)
14. Yadav, T., Mehra, R.: Denoising ECG signal using Daubechies and Symlet wavelet transform techniques. Int. J. Adv. Res. Comput. Commun. Eng. **5**(9), 438–443 (2016)
15. Webhostapp. https://in.000webhost.com/

Detection of Coronal Holes in Solar Disk Image Using Fast Fuzzy C-Means Algorithm

Sanmoy Bandyopadhyay(✉), Saurabh Das, and Abhirup Datta

Discipline of Astronomy, Astrophysics and Space Engineering, Indian Institute of Technology Indore, Indore, India
sanmoy1985@rediffmail.com, das.saurabh01@gmail.com, abhirup.datta@iiti.ac.in

Abstract. Detection of coronal holes (CHs) in the solar disk is one of the important tasks in the domain of astronomy. In this work, a technique based on image segmentation has been used for the detection of the CHs in the solar disk image. Here in the work, the task of image segmentation has been carried out using fast fuzzy c-means algorithm (FFCM). In the work, modified classical fuzzy c-means clustering technique has been induced for the segmentation of solar-disk image and detection of CHs in the image. For the purpose, the histogram of image intensities has been taken into account in order to faster the clustering and segmentation process. The output of the proposed algorithm has been compared visually with other existing technique of CHs detection in order to validate the capability of the proposed technique.

Keywords: Solar-disk image · Coronal holes · Image segmentation · Fuzzy C-means · Fast fuzzy C-means · Hough transformation · Solar limbs

1 Introduction

In the domain of astronomical studies nowadays various researches are being gone on the study of the different character of Sun. It is the only star, which is nearer to the planet Earth. As such it is a vital source of energy for the planet and plays a vital role in maintaining the space environment [1]. The Sun is responsible for many important phenomena in space environment starting from flares to coronal mass ejection to geomagnetic storm [1].

Among various areas of sun-oriented characteristic considers, one of the critical areas is the discovery and consideration of the solar coronal holes. CHs is the most source of sun-oriented wind and is responsible for the geomagnetic storm. This incorporates a colossal effect on the space environment as-well-as on the soil surface. The sun-powered wind, as-well-as geomagnetic storm, straightforwardly influence the counterfeit satellites arranged on the circle. This comes about in breaking down of the satellites. In addition, the life span of satellites gets diminish due to the effect of a geomagnetic storm. It intensely influences the communication satellites situated within the ionospheric

© Springer Nature Singapore Pte Ltd. 2021
N. R. Das and S. Sarkar (eds.), *Computers and Devices for Communication*,
Lecture Notes in Networks and Systems 147,
https://doi.org/10.1007/978-981-15-8366-7_5

locale, which lead to a serious breakdown within the communication framework. Nearby, sun-powered wind and geomagnetic storm causes hinder within the electric supply by influencing the control lattices. As such it ended up vital to identify the measure and the area of CHs in the solar surface in for legitimate forecast of solar wind and geomagnetic storm. So that congruous safeguard can be taken in order to overcome the ill-effect arise due to geomagnetic storm and solar wind.

In the domain, there exists some image segmentation as-well-as machine learning-based techniques for extraction of CHs from solar disk image. In this paper, we have introduced a fast fuzzy c-mean clustering algorithm (FFCM) for extraction of CHs region from the solar disk image. The proposed algorithm basically works in two steps. In the first step, CHs regions are extracted using the FFCM, and in the second step, the unwanted solar limb regions are removed using the circular Hough transform. This has been done in order to extract only the CHs regions and to remove the unwanted region with a view to reduce the misclassification of the regions.

The organization of the rest of the paper goes as follows; the related work on CHs detection using the existing image segmentation technique has been discussed in Sect. 2. In Sect. 3, the proposed algorithm has been elaborated. Discussion on the results has been done in Sect. 4. In Sect. 5, conclusion on overall work has been drawn.

2 Related Work

Till late, 90s human supervision-based procedures such as hand-drawn maps had been broadly utilized for the assurance of CHs in sun-based disk [2, 3]. In any case, with the progression of image handling strategies, presently a days few robotized image segmentation-based object extraction strategies have been utilized for the reason of recognizing CHs in sun-based disk images. Local intensity thresholding has been utilized by Krista and Gallagher for the location of CHs in sun-based disk image of wavelength 195 Å [3]. Yield from this sort of thresholding-based segmentation depends on the legitimate determination of window measure. Rotter et al. [4] had connected histogram-based thresholding strategy for extraction of CHs from 193 Å. The precision of CHs discovery depends on the correct finding of edge esteem display in nearby minima. The strategy can appear its confinement for the case of histogram cover. Lowder et al. [5], as-well-as Kirk et al. [6] had moreover utilized the thresholding-based division procedure for CHs location. But fundamental concern related to the outcome from this procedure is the right dispersion of the histogram, which frequently shifts with changes in the wavelength of sun-oriented image and condition of imaging [7]. Region-growing-based image segmentation procedure had to been actualized by Caplan et al. [8] for mapping of CHs gotten from multi-instrument images. In any case, the victory of region-growing-based strategy depends on the legitimate choice of the seed focuses on images. Ciecholewski [9] utilized watershed change procedure for robotized coronal division in sun-powered disk images. But, the strategy is having an impediment that it requires tremendous variety in color of the image, as-well-as the strategy at some point, comes about in oversegment the images. Spatial possibilistic clustering algorithm (SPoC) has been connected for different locale location in sun-oriented disk image of different wavelength [10]. But the strategy is having scientific complexity for utilizing three distinctive algorithms. As of late, convolution neural network (CNN)-based strategy has been utilized for the sun-powered image division reason [11]. For division exactness, the neural organize require preparing with a colossal set of sun-oriented disk image. Chan-Vese active contour model, active contour without edge (ACWE) had too been connected for extraction of CHs in images by Boucheron et al. [7]. The strategy requires a legitimate area of the CHs in the account for initialization of the form.

3 Proposed Work

With a view to overcome the limitation of the existing methods of CHs detection in the solar-disk image and to reduce the computational time of extraction of CHs regions from the solar-disk image in this work, a novel technique of FFCM coupling with circular Hough transform has been introduced. In the first stage of the work, FFCM has been introduced in order to segment the image into CHs and non-CHs regions. The FFCM-based segmentation has been carried out by taking image global statistic in the account; that is the histogram image pixels intensities have been considered in place of raw image data [12]. The cost function for classical fuzzy c-mean (FCM) algorithm is given by Eq. (1) [13];

$$J = \sum_{j=1}^{N} \sum_{i=1}^{c} \mu_{ij}^{m} x_j - \vartheta_i^2. \tag{1}$$

where μ_{ij} is the membership function of the jth pixel intensity x_j, in the ith cluster. The term N denotes the total number of pixels in the image, c indicates total cluster number, and m is the fuzziness constant. The term ϑ is given by

$$\vartheta_i = \frac{\sum_{j=1}^{N} \mu_{ij}^{m} x_j}{\sum_{j=1}^{N} \mu_{ij}^{m}}; \tag{2}$$

where the term μ_{ij} is initially denoted by

$$\mu_{ij} = \sum_{k=1}^{c} \left(\frac{x_j - \vartheta_i}{x_j - \vartheta_k} \right)^2. \tag{3}$$

Now for the FFCM segmentation purpose, the raw data term x_j is replaced by the frequency of occurrence of the intensity value, h_l. Here h indicates the histogram of the gray-level intensity l. Thus, the overall cost function for FFCM is given by

$$J = \sum_{l=0}^{255} \sum_{i=1}^{c} \mu_{il}^{m} h_l - \vartheta_i^2. \tag{4}$$

In the next stage of the work, circular Hough transform developed by Duda and Hart [14] has been applied in order to eliminate the effect of solar limb present in the disk image. The presence of solar limbs leads to two major issues. Firstly, it leads to a probability that portion other than CHs regions to be classified as CHs due to variation in the intensity level of the solar limb. Secondly, the background in the solar disk images are having the same color intensities as that of CHs region. So, background and CHs region, both, have a chance to get classified as the region of interest (RoI). The application of the circular Hough transform generates a mask only on the solar-disk region and thus eliminating the solar limb from the image.

4 Results and Discussions

The incorporated algorithm of FFCM, for the detection of CHs regions, has been applied and tested on a few benchmark datasets. The datasets are basically solar-disk images consisting of three colors channel, namely red, green, and blue, collected for the different time span. The datasets have been collected from Solar Dynamics Observatory[1]. The solar-disk images used for the experimental purpose and corresponding results of CHs extraction obtained using the proposed method and other existing techniques has been shown in Fig. 1. In Fig. 1, the proposed technique has been visually compared with ACWE method [7], CNN-based technique [11] of CHs segmentation, along with CHs extraction performed by CHIMERA[2]. In the work, due to lack of proper ground-truth data, the output obtained using CHIMERA has been considered as benchmark ground-truth data for visual comparison purpose [7, 11]. From Fig. 1, it can be seen easily stated that the output obtained using the proposed technique generates comparable result as CHIMERA-based technique. While the output using the other two methods has under segment the RoI. From images in the first row of Fig. 1, it can be noted that ACWE method fails to detect the non-coronal whole region like other compared method. At the same time CNN-based technique misclassified CHs region as non-coronal hole region.

The elimination of solar limbs has been handled by introducing the circular Hough transform, and the corresponding outcome have been shown in Fig. 2. From the second row of Fig. 2, it can be seen easily that without application of solar-limb correction technique lead to the misclassification of CHs region. The non-coronal holes regions are also has been considered as the CHs. While in the third column of Fig. 2, it can be noticed easily the application of solar-limb removal technique by means of circular Hough transform has eliminated the problem of misclassification and has detected only the CHs region, which can also be established by the final outcome of the proposed technique shown in the last column of Fig. 2.

After implementation, it can be seen that the proposed technique has detected a few smaller non-CHs regions present inside the solar-disk image which the other existing method has unable to localize. This due to the virtue of fuzzy-based technique which has the capability to classify the RoI with vague boundary. The class memberships value for CHs and non-CHs regions and corresponding class assignment has been shown in Fig. 3.

---

[1] https://sdo.gsfc.nasa.gov/.

[2] https://www.solarmonitor.org/chimera.php.

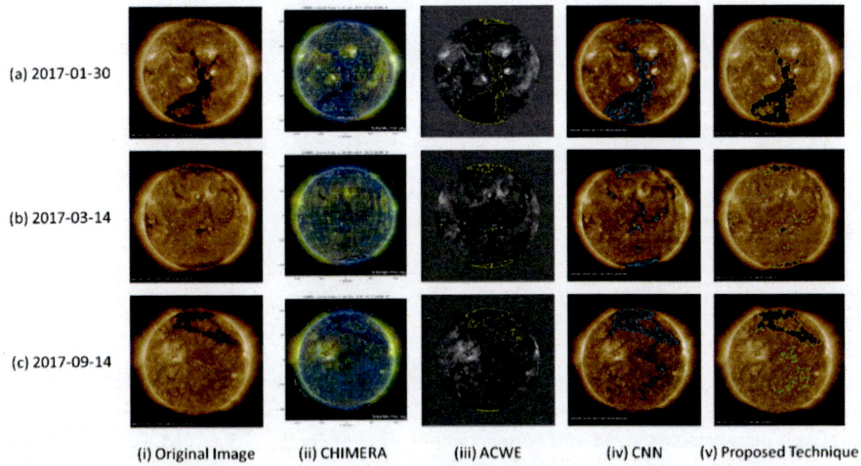

Fig. 1 Pictorial summarization of the CHs extraction results obtained using the proposed method and existing state-of-the-art methods, for the solar-disk image of a different time period

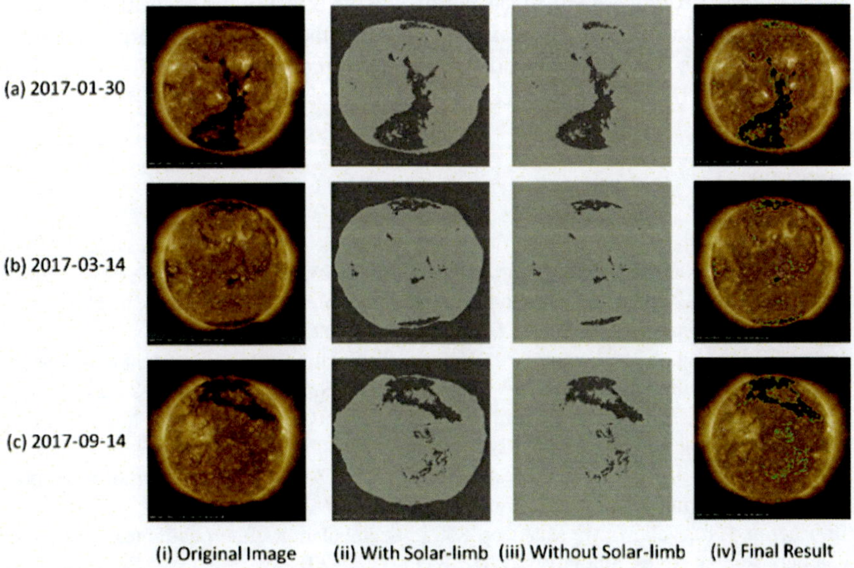

Fig. 2 Pictorial summarization of the CHs segmentation obtained with solar-limb and without solar-limb. The last column shows the final results obtained by eliminating solar-limb

5 Conclusion

In this work, fast fuzzy c-mean technique has been introduced for CHs region detection in a solar-disk image. Due to the capability of fuzzy c-mean technique the method perfectly segmenting out the CHs region even in the presence of an ambiguous boundary, from the solar-disk image. The use of color intensities histogram in place of individual pixels

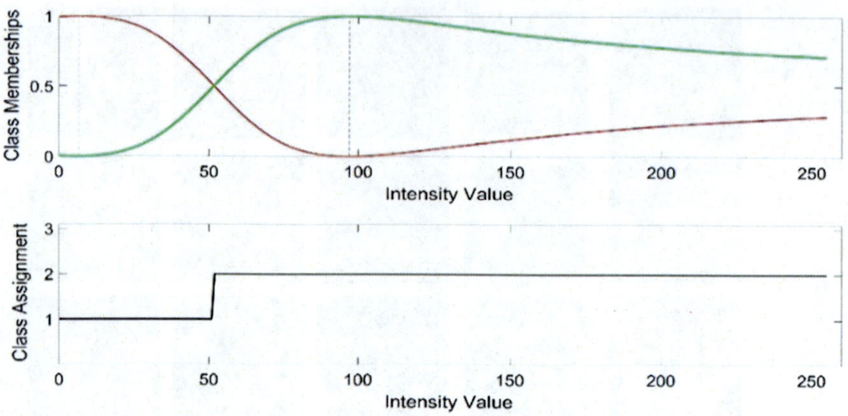

Fig. 3 Class memberships value and corresponding class assignment obtained using the FFCM technique for CHs detection

intensities leads fasten the operation of classification of CHs region in the image. Also, the use of circular Hough transform has eliminated the effect of the solar limb, which in turn, has overcome the problem of misinterpretation of the RoI. In the future, it would be in aim to incorporate the proper ground-truth image of the CHs, for quantitative analysis of the performance of the proposed technique.

References

1. Gombosi, T.I.: Physics of the Space Environment. Cambridge University Press (1998)
2. Harvey, K.L.: Polar coronal holes during cycles 22 and 23. Sol. Phys. **211**(1), 31–52 (2002)
3. Krista, L.D., Gallagher, P.T.: Automated coronal hole detection using local intensity thresholding techniques. Sol. Phys. **256**(1), 87–100 (2009)
4. Rotter, T., Veronig, A.M., Temmer, M., Vršnak, B.: Relation between coronal hole areas on the sun and the solar wind parameters at 1 AU. Sol. Phys. **281**(2), 793–813 (2012)
5. Lowder, C., Qiu, J., Leamon, R., Liu, Y.: Measurements of EUV coronal holes and open magnetic flux. Astrophys. J. **783**(2), 142 (2014)
6. Kirk, M.S., Pesnell, W.D., Young, C.A., Webber, S.H.: Automated detection of EUV polar coronal holes during solar cycle 23. Sol. Phys. **257**(1), 99–112 (2009)
7. Boucheron, L.E., Valluri, M., McAteer, R.T.J.: Segmentation of coronal holes using active contours without edges. Sol. Phys. **291**(8), 2353–2372 (2016)
8. Caplan, R.M., Downs, C., Linker, J.A.: Synchronic coronal hole mapping using multi-instrument EUV images: data preparation and detection method. Astrophys. J. **823**(1), 53 (2016)
9. Ciecholewski, M.: Automated coronal hole segmentation from solar EUV images using the watershed transform. J. Vis. Commun. Image Represent. **33**, 203–218 (2015)
10. Verbeeck, C., Delouille, V., Mampaey, B., De Visscher, R.: The SPoCA-suite: software for extraction, characterization, and tracking of active regions and coronal holes on EUV images. Astron. Astrophys. **561**, A29 (2014)
11. Illarionov, E.A., Tlatov, A.G.: Segmentation of coronal holes in solar disc images with a convolutional neural network. Mon. Not. R. Astron. Soc. **481**(4), 5014–5021 (2018)

12. Semechko, A.: Fast fuzzy c-means image segmentation. https://github.com/AntonSemechko/Fast-Fuzzy-C-Means-Segmentation/ (2019)
13. Bezdek, J.C., Ehrlich, R., Full, W.: FCM: the fuzzy c-means clustering algorithm. Comput. Geosci. **10**(2–3), 191–203 (1984)
14. Duda, R.O., Hart, P.E.: Use of the Hough transformation to detect lines and curves in pictures. Commun. ACM **15**(1), 11–15 (1972)

Automatic Detection and Classification of Enhanced Brain Tumor Using Machine Learning Algorithm

Poulomi Das[1(✉)] and Arpita Das[2]

[1] OmDayal Group of Institutions, MAKAUT, Howrah, India
das.poulomi214@gmail.com
[2] Department of Radio Physics and Electronics, University of Calcutta, Kolkata, India

Abstract. The early detection and proper treatment of brain tumor are essential to prevent permanent damage of brain. Present study proposes an automatic and effective approach to detect brain lesion in early stage that refers to the process of automated contrast enhancement of magnetic resonance (MR) brain images by incorporating simple power law transformation followed by segmentation and identification of the region of interest (ROI) using fuzzy c-means clustering technique and then finally classification of ROI into benignancy/malignancy classes by capturing six significant morphological feature selection. Finally, benignancy/malignancy of masses is examined and assessed by using well-known receiver operating characteristic method of ANN classifier based on significant feature selection. The result of the proposed method is enterprising with very low computational time and accuracy of 90.8%.

Keywords: Automated power law transformation · Region of interest · Benignancy/malignancy

1 Introduction

Automation in medical science has become one of the most essential and challenging tasks in today's medical scenario. Present study proposes an effective approach to analyze the magnetic resonance (MR) brain images for detecting the lesion characteristics in early stage. For years, cancer ranks one of the large threats to human life. People with brain tumor may experience long suffering symptoms of seizures, headaches, personality changes, vision problems, speech difficulties, etc., which are very common to other diseases and lead to difficulty in proper diagnosis.

Patients having stage II tumors (cancerous) in brain need regular monitoring by magnetic resonance imaging (MRI) or computed tomography (CT) scan by every 6 months. However, from the previous studies, we have found that appearance of tumors may also be obscured by the surrounding soft tissues, the presence of cerebral cortex, ventricles, etc., which may lead to serious inaccuracies regarding the location and exact boundaries of tumors. Poor visibility of MRI may misguide the radiologists to predict the prognosis

© Springer Nature Singapore Pte Ltd. 2021
N. R. Das and S. Sarkar (eds.), *Computers and Devices for Communication*,
Lecture Notes in Networks and Systems 147,
https://doi.org/10.1007/978-981-15-8366-7_6

of the diseases. Various filtering technique [1], many local and global intensity-based contrast enhancement methods have also been used for contrast enhancement of medical images. After contrast enhancement of medical images, the goal of the proposed approach is to segment the suspicious region and choose those significant features of segmented lesion that allow the classifier to categorize the masses into benign/malignant stages. K-means clustering algorithm, fuzzy set theory based FCM algorithm [2] and Genetic algorithm based search technique are also very useful for segmenting the suspicious regions. Rangayyan et al. proposed changes of shape patterns from benign to malignant lesion expresses the prognosis of cancer stages [3]. Based on feature selection, classifiers are to be modeled to classify the benignancy/malignancy of lesions. Accuracy and efficiency of classifiers mostly depend on the choice of significant features.

Present study enhances the MRI of brain images by applying γ-based power law transformation to improve the visual quality. Then, we segment the suspicious region using fuzzy c-means clustering techniques and emphasize on significant shape describing feature extraction which are able to convey the precise information about benignancy/malignancy. Following this feature selection approaches, classification result with conventional multilayer neural network is able to produce satisfactory detection accuracy.

2 Methodology

In this section, detail discussion of proposed approach has been given. A typical overview of the proposed methodology for tumor detection is given in Fig. 1.

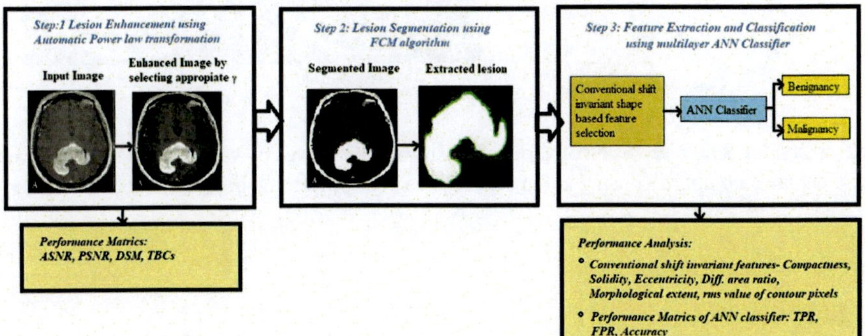

Fig. 1 Schematic diagram of the proposed methodology to determine benignancy/malignancy

Proposed approach exploits automatic power law transformation technique to achieve the best enhancement results. Followed by this automated enhancement procedure, fuzzy c-means clustering is applied for segmentation of suspicious regions (ROI). In addition, an optimal gray level thresholding method is also used. Using significant morphological feature, artificial neural network classifier is designed to classify benignancy/malignancy of lesions. Each of the processing blocks of Fig. 1 is discussed in the following section.

2.1 Lesion Enhancement

In the proposed method, nonlinear spatial filter (median filter) [1] is applied to denoise the input image and enhance the lesion region by automated power law transformation technique by employing some efficient contrast analyzing metrics like average signal-to-noise ratio (ASNR), peak signal-to-noise ratio (PSNR), distribution separation measure (DSM), target to background contrast ratio (TBC) [1] in an iterative manner to achieve the best γ value for the best enhancement results. Power law transformation technique is represented by the following equation [1].

$$s = cr^{\gamma} \tag{1}$$

where r and s are the intensity variation of input and output images, respectively, c is a constant generally kept at 1, and the value of γ is varied to achieve different level of enhancement (Figs. 2 and 3).

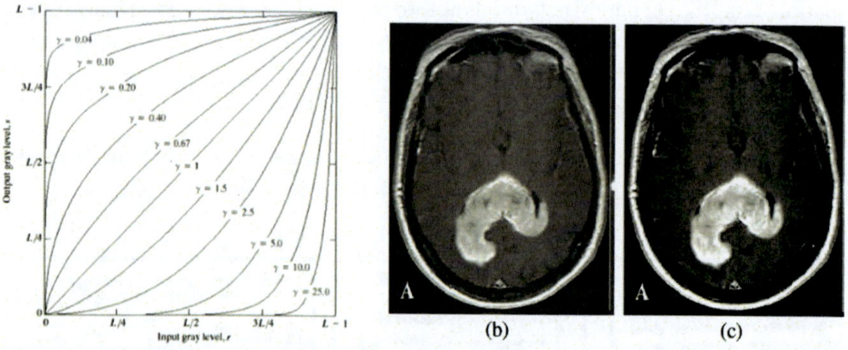

Fig. 2 **a** Graphical representation of various transformation curves by varying γ; **b** original brain MRI; **c** enhanced brain MRI using automated approach of power law transformation

2.2 Segmentation

Followed by lesion enhancement operation, suspected ROI must be extracted by applying fuzzy c-means clustering method [2]. FCM is a method of splitting a collection of n data, $X = \{x_1, \ldots x_j, \ldots x_n\}$ into several number of clusters or groups, called fuzzy groups, and finds the cluster center in each group such that a cost function to measure dissimilarity is minimized. In the present study, five number of clusters are used for segmenting the five distinct brightness regions (cerebrospinal fluid, brain tissues, gray matter, white matter, tumor) of enhanced brain MRI. Following FCM approach, morphological opening by reconstruction is applied to remove the unwanted glitches appeared in MRI due to heterogeneous nature of lesions. Now, based on the segmented boundaries, feature of benignancy/malignancy of tumors is determined accordingly (Fig. 4).

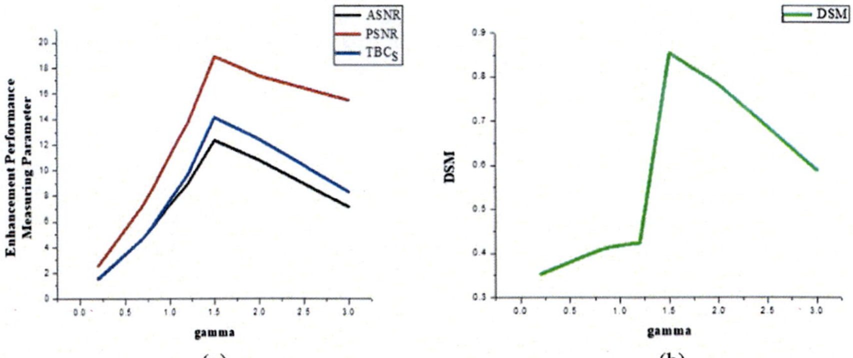

Fig. 3 Graphical representation of ASNR, PSNR, DSM, TBC with variation of γ; **a** highest value of ASNR, PSNR, TBC at $\gamma = 1.5$ shows the best enhancement result; **b** similarly highest value of DSM at $\gamma = 1.5$ indicates the best enhancement achievement

Fig. 4 **a** Original brain MRI; **b** enhanced by automated power law transformation; **c** segmentation by fuzzy c-means clustering algorithm; **d** segmented boundary of lesion, inclusion of CSF and white matter; **e** extracted lesion by morphological opening

2.3 Feature Analysis

From the reviews, we have found that change of shape patterns from benign to malignant expresses the prognosis of cancer stages. Following shape descriptors which are relatively insensitive to the orientation and size of the lesion used for classification.

$$\text{Compactness} = \frac{P^2}{4\pi A} \tag{2}$$

$$\text{Solidity} = \frac{A}{CA} \tag{3}$$

$$\text{Eccentricity} = \frac{W}{L} \tag{4}$$

$$\text{Difference area ratio} = \frac{|CA - A|}{A} \tag{5}$$

$$\text{Morphological extent} = \frac{A}{\text{Smallest bounding box area}} \tag{6}$$

$$\text{r.m.s value of contour pixels} = \sqrt{\frac{\sum_{j=0}^{N-1} \left[f_x(j) - f_x(j+1)\right]^2 + \left[f_y(j) - f_y(j+1)\right]^2}{N}} \quad (7)$$

where, P, A, CA are the perimeter, area and convex area of the object, respectively; W and L are the width and length of smallest bounding box, $f_x(j)$ and $f_y(j)$ are jth contour pixel for x and y coordinates, respectively; N is the total number of pixels in the contour (Fig. 5).

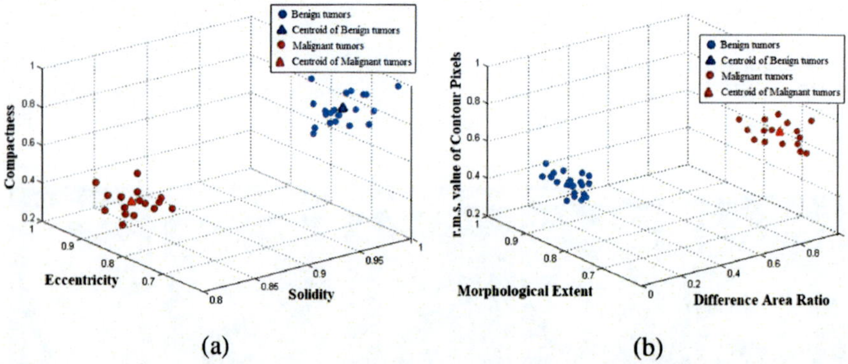

Fig. 5 **a** 3D feature distribution of solidity, eccentricity, compactness; **b** 3D scattered distribution of difference area ratio, morphological extent and r.m.s. value of contour pixels

2.4 Classification

In the present study, a conventional 3-layer feed-forward neural network (NN) structure is designed for computation of benignancy/malignancy of tumor. Initially, NN is trained with well-defined samples, and hence subsequently, test dataset (a set of features with unknown classes) are examined to evaluate the discrimination accuracy. Effectiveness of the proposed approach is assessed using several parameters such as sensitivity, specificity and overall classification accuracy. In the present study, five-fold cross validation is developed for producing five different sets of training and test data. Initially, brain MRIs are randomly divided into five number of groups, and each group consists of same number of data. The classifier is trained by the data of any four groups, and remaining one group is used for testing purposes. This procedure is repeated five times, and each of the group is used once for testing. In step of classification, performances of the classifier are evaluated, and finally, the average performance indices are obtained.

3 Experimental Results

Here, we have examined the accuracy of the proposed model by evaluating 78 numbers of test samples and then assessing the outcomes with https://www.med.havard.edu/aan

lib/home.html. Some conventional performance indices such as sensitivity, specificity, overall classification accuracy are applied to estimate the average effectiveness of the proposed scheme.

From the summary of Table 1, we have achieved encouraging value of accuracy and sensitivity for threshold value of 0.55. For this particular threshold value, proposed model shows the highest accuracy elsewhere it is degraded. ROC assessment of the proposed model is shown in Fig. 6.

Table 1 Summary of performance indices for a threshold value to achieve highest accuracy of ANN classifier

Feature combination	Compactness, solidity, eccentricity, difference area ratio, extent, r.m.s. value of contour pixel
Confidence threshold value	0.55
$\mathrm{TPR} = \frac{\mathrm{TP}}{\mathrm{P}}$ (Sensitivity)	0.901
$\mathrm{FPR} = \frac{\mathrm{FP}}{\mathrm{N}}$ (1 − Sensitivity)	0.052
$\mathrm{Accuracy} = \frac{\mathrm{TP}+\mathrm{TN}}{\mathrm{P}+\mathrm{N}}$	0.908

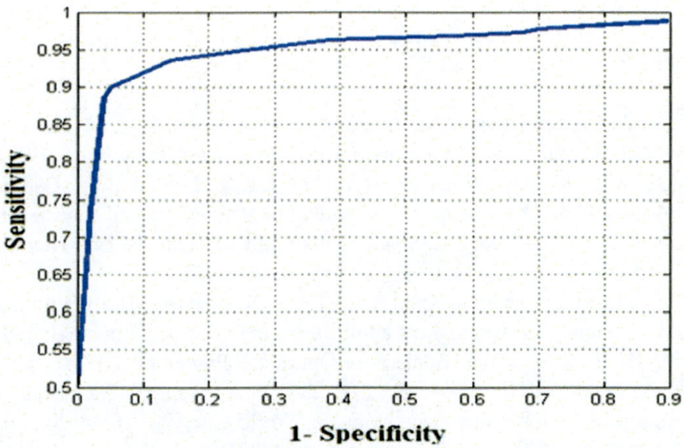

Fig. 6 Graphical representation of sensitivity versus 1 − specificity (ROC)

In the following section, a comparison of experimental results with some other scientific research works using different approaches has been executed to compare the overall accuracy of the proposed model. Table 2 shows this summary.

4 Conclusion

Present study proposes an automatic and effective approach of early detection of brain lesion from MRI. This work focuses on comprehensive contrast enhancement of brain

Table 2 Accuracy comparison among the proposed and state-of-the-art methods

Method	Tustison et al. [4]	Paul et al. [5]	Abiwinanda et al. [6]	Proposed
Accuracy	83	90.26	84.18	90.80

lesion by automated enhancement technique followed by segmentation of ROI using FCM technique and then finally classification of benignancy/malignancy of tumors by capturing the significant morphological features and ANN classifier. Proposed approach describes the 3D scatter distributions of different features for benign/malignant groups. The ROC assessment shows that accuracy reaches its optimized value 90.8% when sensitivity is 90.1%.

Acknowledgements. This work is partially supported by the Center of Excellence (CoE), funded by the World Bank, MHRD India. We would also like to thank to Dr. S. K. Sharma of EKO X-ray and Imagining Institute, Kolkata, for providing us valuable comments.

References

1. Gonzalez, R.C., Woods, R.E.: Digital Image Processing, 2nd edn. Pearson Education (2003)
2. Adhikari, S.K., Sing, J.K., Basu, D.K., Nasipuri, M.: Conditional spatial fuzzy C-means clustering algorithm for segmentation of MRI images. J. Appl. Soft Comput. **34**, 758–769 (2015)
3. Rangayyan, R.M., El-Faramawy, N.M., Desautels, J.E.L., Alim, O.A.: Measures of acutance and shape for classification of breast tumors. J. IEEE Trans. Med. Imaging **16**, 799–810 (1997)
4. Tustison, N.J., Shrinidhi, K.L., Wintermark, M., Durst, C.R., Kandel, B.M., Gee, J.C., Grossman, M.C., Avants, B.B.: Optimal symmetric multimodal templates and concatenated random forests for supervised brain tumor segmentation (simplified) with ANTsR. Neuroinformatics **13**, 209–225 (2015). https://doi.org/10.1007/s12021-014-9245-2
5. Paul, J.S., Plassard, A.J., Landman, B.A., Fabbri, D.: Deep learning for brain tumor classification, medical imaging. In: Biomedical applications in molecular, structural, and functional imaging, p. 1013710. International Society for Optics and Photonics (2017)
6. Abiwinanda, N., Hanif, M., Hesaputra, S.T., Handayani, A., Mengko, T.R.: Brain tumor classification using convolutional neural network. In: World Congress on Medical Physics and Biomedical Engineering 2018, pp. 183–189. Springer (2019)

An Artificially Intelligent Fusion Approach for Prognosis of Alzheimer's Disease

Suranjana Mukherjee(✉) and Arpita Das

Department of Radio Physics and Electronics, University of Calcutta, Kolkata, India
mala35b@gmail.com

Abstract. In the present study, we have merged functional and anatomical information to study the prognosis of Alzheimer's disease (AD). To restore all possible directional information of the source images, the shift-invariant non-subsampled shearlet transform (NSST)-based decomposition technique is employed. Then, principle component analysis (PCA) and position-based maximum energy selection scheme are applied for lower subband and higher subbands, respectively, to combine multimodal information in a single frame. Finally, fused images have been generated by inverse NSST. Efficiency of the proposed fusion scheme has been reflected in the experimental results.

Keywords: Multimodal imaging · WT · NSCT · NSST

1 Introduction

In medical diagnosis, multimodal images have an important role because a single imaging procedure cannot provide all the information alone. For example, magnetic resonance imaging (MRI) provides the anatomical information like soft tissue regions. On the other hand, from positron emitted tomography (PET), functional information like blood flow, food activity and metabolism of affected organ is obtained [1, 2]. Hence, the combination from different modalities in one composite frame saves valuable time of a doctor in diagnostic procedure.

So far, many researchers have proposed various fusion schemes for multimodal medical images. Few years ago, wavelet transform (WT) has been used widely [2–4]. But, WT approaches are able to capture the information of only three directions namely, horizontal, vertical and diagonal.

Thus, to overcome those fundamental shortcomings, non-subsampled contourlet transform (NSCT) and NSST are proposed [5, 6]. In these approaches, source images and subimages are of same size. Hence, they are named so. In NSST, source images are decomposed into a low frequency subband and several numbers of trapezoidal high frequency subbands called shearlets by employing the shift-invariant shearing filter. In comparison with NSCT, shearlets do not have any restriction for shearing [6]. Moreover, NSST is computationally efficient, as the inverse NSST only requires a summation

© Springer Nature Singapore Pte Ltd. 2021
N. R. Das and S. Sarkar (eds.), *Computers and Devices for Communication*,
Lecture Notes in Networks and Systems 147,
https://doi.org/10.1007/978-981-15-8366-7_7

of shearing filters rather than inverse directional filter bank of NSCT. Hence, we have preferred it in our work.

After the multiscale decomposition, fusion rule plays a crucial role in combining anatomical and functional information of higher and lower frequency subbands. Many researchers had combined low and high frequency subbands using simple averaging and max rule [4, 7, 8]. But, there is a significant reduction of contrast, and some significant information of the source images may be degraded in the fused image. Hence, we have employed PCA and position-based maximum energy selection rule for lower and higher subbands, respectively.

2 Proposed Fusion Method

For NSST analysis, 'maxflat' is used as pyramidal filter, and images are decomposed to three levels. We can avoid the pseudo-Gibbs phenomenon by using non-subsampled pyramidal filter. Basic framework of the proposed fusion scheme is described in Fig. 1.

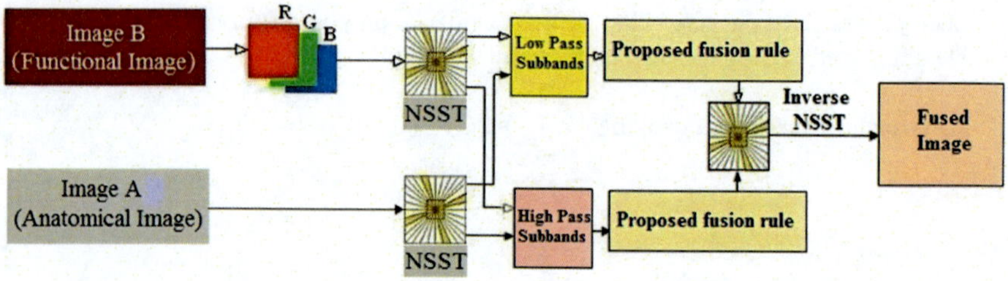

Fig. 1 Brief framework of the proposed approach

Figure 1 describes the basic steps of the proposed fusion scheme as follows.

1. Extract Red (R), Green (G) and Blue (B) components of the functional image-B.
2. Decompose the anatomical image-A and each R, G, B component of image-B by employing three levels NSST with non-subsampled Laplacian pyramidal filter bank.
3. Combine the high and low frequency coefficients according to the proposed fusion rules.
4. Reconstruct the fused image by applying inverse NSST.

2.1 Non-Subsampled Shearlet Transform (NSST)

NSST overcomes the lack of directionality of conventional WT. Two basic steps of NSST are partitioning and directional localization [6]. The shift invariance in the multiscale partitioning is achieved by the non-subsampled Laplacian pyramidal filter scheme, in which Gibbs phenomenon is suppressed properly as downsamplers are replaced with convolutions, and as a result, edge/color information of source images is preserved in the fused images. In the directional localization, the frequency plane is decomposed into a low frequency subband and several numbers of trapezoidal high frequency subbands utilizing shift-invariant shearing filter. For j level of decomposition, a pair of trapezoids known as shearlets of size $2^{2j} \times 2^{j}$ (approx) is produced which are oriented along the line slope $l \times 2^{-j}$, where l is any integer value and it is described in Fig. 2. Inverse NSST is obtained by the summation of shearing filter.

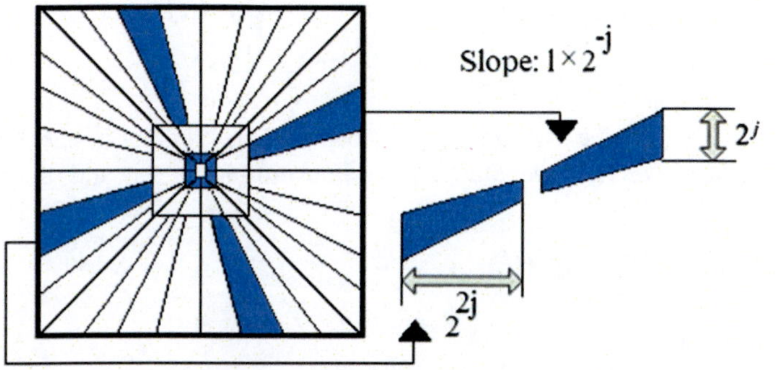

Fig. 2 Representation of shearlets

2.2 Proposed Fusion Rule

After multiscale decomposition, appropriate fusion rules are imposed on the low and high frequency subband images. To combine anatomical and functional information of different frequency subbands, *PCA* and *position-based absolute maximum energy selection* rule have been designed as follows [2].

2.2.1 Combination of Low Frequency Subbands

In the present study, we have employed *PCA*-based weighted averaging approach to integrate overall gross features of lower subband coefficients [2]. *PCA* is a linear mapping technique to adjust the contribution of the source images by analyzing the principal axis containing maximum eigenvalue which is associated with the principal eigenvector. Hence, the computation is also less complex in determining accurate weight factor of two lower subband coefficients, and it is described as follows

$$C_F = q_A * C_{A,k} + q_B * C_{B,k} \tag{1}$$

where $C_{A,k}$ and $C_{B,k}$ are the low frequency subbands of image A and B with kth decomposition level.

Estimation of q_A and q_B.

1. Calculate covariance (covar) of $C_{A,k}$ and $C_{B,k}$.
2. Calculate eigenvectors (e) of covar in which eigen values are arranged in descending order.
3. Calculate q_A and q_B as follows. $q_A = \frac{e_1}{e_1+e_2}$ and $q_B = \frac{e_2}{e_1+e_2}$.

2.2.2 Combination of High Frequency Subbands

Position-based maximum energy selection rule is also very simple and computationally efficient [2]. Position-based selection of information is performed using the following decision mapping

$$\begin{aligned} D_{F,k,l}(i,j) &= D_{A,k} \quad \text{for } D_{A,k} \geq D_{B,k}(i,j) \\ &= D_{B,k} \quad \text{for } D_{A,k} < D_{B,k}(i,j) \end{aligned} \tag{2}$$

where $D_{A,k}$ and $D_{B,k}$ are the high frequency subbands (shearlets) of image A and B with kth decomposition level; (i, j) denotes any pixel position in ith row and jth column.

3 Experimental Evaluation

3.1 Experimental Setting

In the present study, we have produced some of the experimental results using our proposed fusion scheme and compared those results with discrete wavelet transform (DWT), stationary wavelet transform (SWT) and NSCT. To obtain a fair comparison, we have decomposed to three levels for every case.

3.2 Dataset Preparation

The entire fusion scheme has been implemented either on different sagittal slices of brain MR-T2 and PET-FDG images of a 70-year old man, who is suffering from mild AD as described by Harvard University, on January2, 2019, 11:30 a.m. [9]. MR-T2 and PET-FDG images reveal the widened hemispheric sulci and abnormalities in regional cerebral metabolism as well as blood flow, respectively.

3.3 Subjective Analysis

DWT and SWT techniques are suffering from poor contrast and lack of visual clarity. The fusion scheme using DWT suffers from Gibbs phenomenon severely. But in SWT, it is suppressed moderately as down and up sampling is avoided by inserting zeros in between the filter coefficients. Since SWT captures information of only three directions like DWT, some of the directional information of source images is lost in the fused images. But in NSCT and NSST, directional reformations are restored well.

3.4 Objective Analysis

Apart from subjective analysis, some quantitative measurements like *entropy standard deviation (STD)* and *average gradient (AG)* are conducted to evaluate the information content, contrast and sharpness of texture distribution of various fusion schemes [2, 10].

Those parametric details are depicted in Table 1, and the efficiency of the proposed fusion method is proved.

Table 1 Quantitative evaluation of different fusion schemes

Image set	Evaluation parameter	DWT	SWT	NSCT	Proposed
Set 1 of Fig. 3	Entropy	4.4270	4.0531	4.6167	4.6811
	STD	0.1937	0.201	0.2075	0.2112
	AG	0.0175	0.0169	0.0174	0.0176
Set 2 of Fig. 3	Entropy	4.6619	4.0611	4.8061	4.8505
	STD	0.2148	0.2218	0.2808	0.2926
	AG	0.0172	0.0163	0.0184	0.0188
Set 3 of Fig. 3	Entropy	4.7037	4.0713	4.8237	4.8881
	STD	0.1964	0.2038	0.2872	0.2907
	AG	0.0188	0.017	0.0288	0.0294

In DWT, entropy is large with respect to SWT, as there is incorporation of artifacts during up and down sampling process. In SWT, AG is suppressed due to zero padding. NSCT is better in comparison with other two methods. In our proposed NSST scheme, all the information of source images is restored properly with less computational complexity.

4 Conclusion

In this study, we have decomposed source images in NSST domain to capture all the directional information successfully. In addition to that, we have proposed PCA-based fusion rule of lower subband to overcome the contrast degradation problem of simple averaging process. However, we have found that the information of skull boundary of MR-T2 images is blurred in the fused images of proposed technique which will get to be improved in future. Furthermore, some other advanced fusion rule may be formulated in future for better integration of source images.

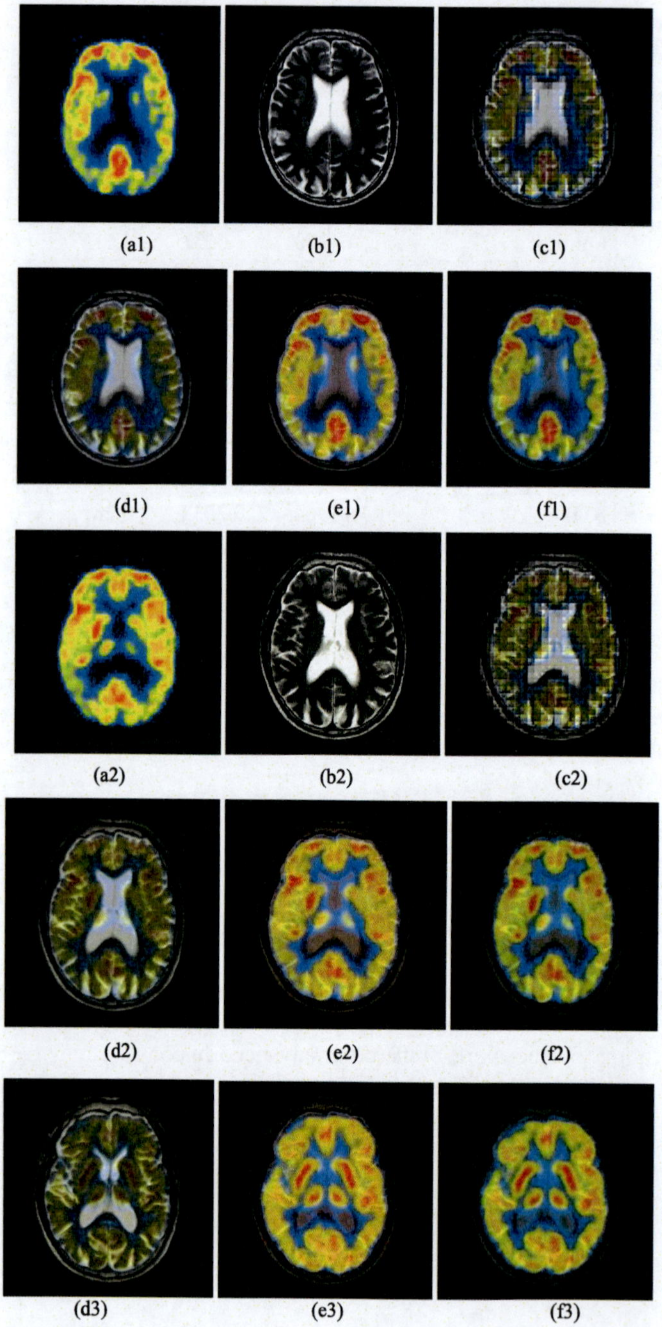

Fig. 3 Fusion of PET-FDG and MR-T2 images: **a1–a3** PET-FDG; **b1–b3** MR-T2; **c1–c3** DWT; **d1–d3** SWT; **e1–e3** NSCT; **f1–f3** NSST

Acknowledgements. This work is supported by the Center of Excellence (CoE), in Systems Biology and Biomedical Engineering, University of Calcutta, funded by the World Bank, TEQIP Phase-III, MHRD India.

References

1. Das, A., Bhattacharya, M.: Study on neurodegeneration at different stages using MR images: computational approach to registration process with optimisation techniques. Comput. Methods Biomech. Biomed. Eng. Imaging Vis. **5**(3), 165–182 (2017)
2. Das, A., Bhattacharya, M.: Effective image fusion method to study Alzheimer's disease using MR, PET images. In: IEEE International Conference on Bioinformatics and Biomedicine (BIBM), pp. 1603–1607 (2015)
3. Singh, R., Khare, A.: Redundant discrete wavelet transform based medical image fusion. Adv. Signal Process. Intell. Recogn. Syst. **26**(1), 505–515 (2014)
4. Singh, R., Srivastava, R., Prakash, O., Khare, A., Srivastava, R.: Multimodal medical image fusion in dual tree complex wavelet transform domain using maximum and average fusion rules. J. Med. Imaging Health Inform. **2**(2), 168–173 (2012)
5. Cunha, A.L.D., Zhou, J., Do, M.N.: The nonsubsampled contourlet transform: theory, design, and applications. IEEE Trans. Image Process. **15**(10), 3089–3101 (2006)
6. Easley, G., Labate, D., Lim, W.: Sparse directional image representations using the discrete shearlet transform. Appl. Comput. Harmon. Anal. **25**, 25–46 (2008)
7. Miao, Q.G., Shi, C., Xu, P.F., Yang, M., Shi, Y.B.: A novel algorithm of image fusion using shearlets. Optics Commun. **284**(6), 1540–1547 (2011)
8. Wang, A., Sun, H., Guan, Y.: The application of wavelet transform to multi-modality medical image fusion. In: IEEE International Conference on Networking, Sensing and Control, pp. 270–274 (2006)
9. Johnson, K.A., Becker, J.A.: The whole brain atlas. Available online at: https://www.med.harvard.edu/aanlib/home.html
10. Gonzalez, R.C., Woods, R.E., Eddins, S.L., Digital Image Processing Using MATLAB. Pearson-Prentice-Hall, Upper Saddle River, New Jersey (2004)

System Stability Performance Analysis on an Artificial Lower Limb

Susmita Das[1(✉)], Dalia Nandi[2], and Biswarup Neogi[3]

[1] Electronics and Instrumentation Engineering, Narula Institute of Technology, Kolkata, West Bengal, India
susmitad2k17@gmail.com
[2] Electronics and Communication Engineering, Indian Institute of Information Technology Kalyani, Kalyani, West Bengal, India
[3] Electronics and Communication Engineering, JIS College of Engineering, Kalyani, West Bengal, India

Abstract. An approach of performance analysis-based system modeling of artificial lower limb has a very significant role in the control action applications. The control system stability checking is performed in this research paper to identify the system efficiency and design requirements. The efforts have been made to initiate different approaches to achieve the best effective model using the mathematical model. The state-space analysis has been incorporated to analyze different control parameter performances such as controllability, observability and stability testing of the artificial limb model. The closed loop representation by the simulation approach using Jury stability and Lyapunov stability analysis has been involved. The control action analysis shows the relevancy and the precision of the proposed mathematical model.

Keywords: System modeling · Control performance · State-space analysis · Jury stability · Lyapunov stability · Artificial limb

1 Introduction

A construction of a mathematical model is a vital requirement as per the applications of artificial limbs. A precise control mechanism is also an indispensable criterion to achieve the effective system performance. The design process can be made simplified with some approximation over the modeling. In this study, an open loop transfer function has been considered for the angular positions of the artificial limb joints [1]. Initially, closed loop model is tuned by proportional–integral–derivative (PID) tuning [2]. The study of control parameters through closed loop model is set up to select the further methodology. The state-space analysis is performed to show the system controllability and observability. The discrete domain stability analysis is attempted by Jury testing in Jury simulator [3] for the system stability checking. Finally, Lyapunov stability [4] method is established. This can be extended to provide a strategy for constructing a stabilized feedback controller. The most important reason behind using a PID controller

© Springer Nature Singapore Pte Ltd. 2021
N. R. Das and S. Sarkar (eds.), *Computers and Devices for Communication*,
Lecture Notes in Networks and Systems 147,
https://doi.org/10.1007/978-981-15-8366-7_8

is due to its simple control structure and satisfactory results. Here, Ziegler–Nichols (Z-N) tuning rules have been considered for tuning as it is one of the premier optimization algorithms and also as it can be easily implemented in MATLAB [4] simulation. The z-transform is the discrete-time counterpart of the Laplace transform and a generalization of the Fourier transform of a sampled signal. The theory of Lyapunov stability is a standard theory for nonlinear systems and one of the most important mathematical tools in the analysis of nonlinear systems [5]. A sampled data system involves both continuous-time and discrete-time signals in its operation. The concepts of controllability and observability [6] were introduced by Kalman and have been employed primarily in the study of optimal control of the artificial limb movements with stability in the scale of performance. Key advantages of the state-space approach are that a time-domain formulation exploited the advances in digital computer technology and the analysis and design methods are well-suited to multiple-input, multiple-output systems. System stability analysis for linear systems and continuing for the nonlinear approach especially for the humanoid limb movements are the motivations of this research work (Fig. 1).

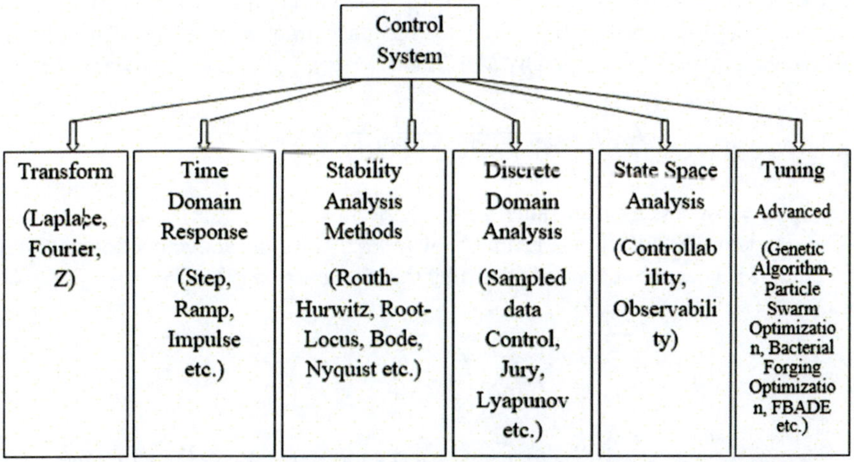

Fig. 1 Flow diagram of different types of control system performance analysis

2 Retuning and Performance Study on Standard Artificial Limb

The requirement of controller is to reduce the error by adjusting control variables. As the constants of proportionality, integration and derivation are very useful metrics in PID controller, therefore precise and finest control is possible by its applications only. Initially, estimated values of the constants can be taken significantly as per application and then they are generally developed by monitoring the response of the system. Though the PID algorithm may not assure optimized control, but it simply relies on the response obtained from the calculated process. Here, an open loop transfer function [1] of a knee joint output followed by its angular position is given in Eq. (1) (Fig. 2).

$$G(s) = \frac{7.094}{s^2 + 5.388s + 7.016} \tag{1}$$

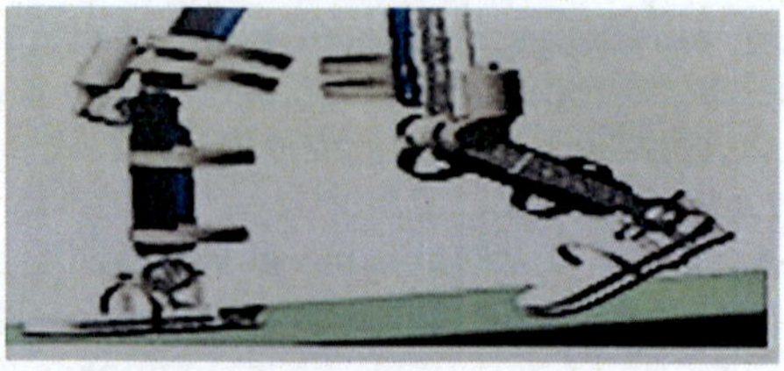

Fig. 2 Schematic diagram of lower extremity design [1]

Final closed loop transfer function = $\frac{G(s)}{1+G(s)H(s)}$ as per control theory where $G(s)$ = system transfer function and $H(s)$ = unity feedback transfer function. Closed loop mathematical model [7] is formed by MATLAB (Matrix Laboratory) shown in Eq. (2).

$$G(s) = \frac{7.094}{s^2 + 5.388s + 14.11} = G_p(s) \tag{2}$$

where $G_p(s)$ = process transfer function.

The standard PID equation is given below in Eq. (3) for the abovementioned closed loop output [8]. Then the generated controller transfer function [9] is presented in Eq. (4) for $G_c(s)$.

$$G_c(s) = k_p\left(1 + \frac{k_i}{s} + k_d \cdot s\right) \tag{3}$$

$$= 4.9256\left(1 + \frac{12.5848}{s} + 0.4694s\right) = 2.3120s^3 + 4.9256s^2 + 61.9876s \tag{4}$$

where $G_c(s)$ = controller transfer function and k_p = Proportional constant = 4.9256, k_i = Integral constant = 12.5848, k_d = Derivative constant = 0.4694.

Now, the tuned output [10, 11] is shown in Eq. (5),

where

$$G(s) = G_c(s) \cdot G_p(s) = \frac{16.4013s^3 + 34.9422s^2 + 439.74s}{s^2 + 5.388s + 14.11} \tag{5}$$

Therefore, the mathematical model presented in Eq. (4) has been chosen for further proceeding due to its stability factors such as rise time and settling time. A dynamic system has to be controllable for further processing with control input. The system is identified as controllable if the states of the system can be configured by altering the system input.

State-space matrices are given below for tuned transfer function mentioned in Eq. (5):

$$A = \begin{bmatrix} -5.3880 & -14.1100 \\ 1.0000 & 0 \end{bmatrix} \quad B = \begin{bmatrix} 1 \\ 0 \end{bmatrix}$$

$$C = \begin{bmatrix} -53.4280 & 208.3177 \end{bmatrix} \quad D = [16.4013]$$

Now, the controllability testing has been performed. Rank of the controllability matrix is 2. The output performance of the system is observed to establish the internal system states. This type of system is called observable system [12]. The observability testing has been performed. Rank of the observability matrix is 2. The rank of both the controllability and observability is 2. So, it is controllable and observable also.

3 Discrete Analysis of Artificial Lower Limb Model Toward Stability

The stability of a linear discrete-time system [13] can be determined by Jury stability criterion. Here the sampling time is 0.8 s as per Eq. (5), which is chosen for the better dynamic characteristics of the system compared to the system having 0.1 s sampling time. The step responses of the systems in discrete domain are shown in Fig. 3a, b.

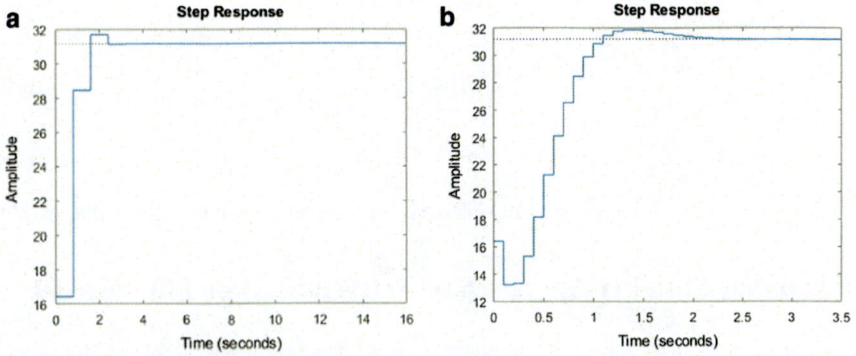

Fig. 3 **a** Step response of system in discrete domain (0.8 s sampling time). **b** Step response of system in discrete domain (0.1 s sampling time)

Discrete-time transfer function of zero order hold is given below in Eq. (6).

$$G(z) = \frac{16.4z^2 + 13.95z + 4.847}{z^2 + 0.1158z + 0.01343} \tag{6}$$

Jury test is an algebraic test. For a polynomial, if $F(1) > 0$ and $(-1)^n F(-1) > 0$, then the system in z domain is stable. Now, the sufficient conditions for stability are obtained by forming a table shown in Table 1 using Jury simulator. The sufficient conditions for stability are given by $|a_0| < a_n$ and through Eqs. (7), (8) and (9). Jury stability output has been demonstrated using Jury simulator [8].

$$F(1) = 0.33808, \tag{7}$$

$$(-1)^n F(-1) = 0.07224 \tag{8}$$

$$4.847 < 16.4 \tag{9}$$

Table 1 Jury stability output analysis through Jury simulator with the generated iteration table

Coefficient of z	Terms of z		
	z^0	z^1	z^2
A	4.847	13.95	16.4
B	16.4	13.95	4.847

The above conditions are satisfied, and through Table 1 Jury stability output has been demonstrated with the generated iteration table using Jury simulator.

Through Eqs. (10) and (11), final Jury stability output has been presented using Jury simulator.

From Table 1, it is shown that

$$|a_0| < |a_2| \tag{10}$$

$$|b_0| > |b_1| \tag{11}$$

Since, these conditions are satisfied. Therefore, the system can be declared as stable.

4 Lyapunov Stability Approach of Artificial Lower Limb Model

As the Jury test is successful to define the stability, the Lyapunov stability analysis is proposed hereafter [14, 15] for ultimate justification. The controlled output [16] is shown in Eq. (12).

$$G(s) = \frac{16.4013s^3 + 34.9422s^2 + 439.74s}{s^2 + 5.388s + 14.11} \tag{12}$$

Since,

$$e(s)G_c(s) \cdot G_p(s) = Y(s) \tag{13}$$

and $r(s) - y(s) = e(s)$ and assuming $r(s) = 0$, it can be stated that $y(s) = -e(s)$.

Where $r(s)$ = reference input, $y(s)$ = system output, $e(s)$ = error value, $G_c(s)$ = controller transfer function and $G_p(s)$ = process transfer function. Now, the mathematical model is shown in Eq. (14) according to Eq. (13).

$$e(s)\left(16.4013s^3 + 34.9422s^2 + 439.74s\right) = -e(s)\left(s^2 + 5.388s + 14.11\right) \tag{14}$$

or

$$e(s)\left(16.4013s^3 + 35.9422s^2 + 445.128s + 14.11\right) = 0 \tag{15}$$

Using inverse Laplace transform, the transfer function is shown in Eq. (16)

$$(16.4013\dddot{e} + 35.9422\ddot{e} + 445.128\dot{e} + 14.11e) = 0 \tag{16}$$

Let

$$x_1 = e,\ x_2 = \dot{e},\ x_3 = \ddot{e},\ x_4 = \dddot{e} \tag{17}$$

$$\dot{x}_1 = x_2,\ \dot{x}_2 = x_3,\ \dot{x}_3 = x_4 \tag{18}$$

$$\dot{x}_3 \cdot a + \dot{x}_2 \cdot b + \dot{x}_1 \cdot c + x_1 \cdot d = 0 \tag{19}$$

Here, $a = 16.4013,\ b = 35.9422,\ c = 445.128,\ d = 14.11$.
Equation (19) can be written as

$$\dot{x}_3 = -\{x_3 \cdot a_1 + x_2 \cdot b_1 + x_1 \cdot c_1\} \tag{20}$$

Here, $a_1 = 2.1914,\ b_1 = 27.1397,\ c_1 = 0.8602$
Now taking scalar positive definite function which is given by,

$$V(x) = \frac{1}{2}S_1 \cdot x_1^2 + \frac{1}{2}S_2 \cdot x_2^2 + \frac{1}{2}S_3 \cdot x_3^2 \tag{21}$$

where $S_1 > 0,\ S_2 > 0$.
Now taking the derivative with respect to time t yields

$$V(x) = S_1 \cdot x_1 \cdot \dot{x}_1 + S_2 \cdot x_2 \cdot \dot{x}_2 + S_2 \cdot x_3 \cdot \dot{x}_3 \tag{22}$$

$$\begin{aligned} &= S_1 \cdot x_1 \cdot x_2 + S_2 \cdot x_2 \cdot x_3 + S_3 \cdot x_3 \cdot x_4 \\ &= S_1 \cdot x_1 \cdot x_2 + x_2 \cdot x_3(S_2 - b_1S_3) - x_3 \cdot (a_1S_3x_3 + c_1S_3x_1) \end{aligned} \tag{23}$$

For the positive definite function V, another positive definite function U is needed such that $\dot{V}(x) = -U(x)$. Now, the coefficients are taken in such a manner that $\dot{V}(x) = -U(x)$. So, it is taken as

$$(S_2 - b_1S_3) = 0 \tag{24}$$

and

$$S_1 = 0 \tag{25}$$

Now, substituting Eqs. (24) and (25) in Eq. (23) and also $\dot{V}(x) = 0$, from Eq. (22) $V(x) \to \infty$ as $x \to \infty$ and $\dot{V}(x) = 0$ is obtained. A way of showing that $\dot{V}(x)$ being negative semi-definite is sufficient for asymptotic stability is to show that x-axis is not a trajectory of the system. For $\dot{x}_1 = x_2 = 0$ and $\dot{x}_2 = x_3 = 0$, it is shown that $x_1 = m$ (constant). The equilibrium state at the origin of the system is asymptotically stable. Therefore, the mentioned system in this work is asymptotically stable [17]. The overall control law design [18] and development of the efficient system for artificial lower limb are the main focus of the work.

5 Conclusions

In this research work, the closed loop system modeling and retuning are performed with standard artificial lower limb model. The state-space model is formed to achieve controllable and observable system with rank determination. Jury stability testing is attempted to establish the discrete domain aspect. Then Lyapunov stability process is done to show the asymptotically stable system. A significant area of discrete domain control system analysis has been incorporated in this work for better efficacy and digital presentation. Utilizing all the possible ways for the overall stability achievement in different domain has been involved in this research work for the improvement in the field of hardware designing. In future, there are some other deterministic approaches which can be involved for control system analysis toward the nonlinear approach for effective system design. This work enlightens the path to choose control parameter range in which a system can be analyzed to show the performance. System modeling and simulation can create a resource place for designing a controlled and tuned system.

Acknowledgements. We are grateful for the help of IIIT Kalyani to complete the research work and also very much thankful for the contribution of N. Aliman et al.

References

1. Aliman, N., Ramli, R., Haris, S.M.: Modeling and co-simulation of actuator control for lower limb exoskeleton. In: 2018 3rd International Conference on Control and Robotics Engineering (ICCRE), Nagoya, pp. 94–98 (2018)
2. Das, S., Chakraborty, A, Ray, J.K., Bhattacharjee, S., Neogi, B.: Study on different tuning approach with incorporation of simulation aspect for Z-N (Ziegler-Nichols) rules. Int. J. Sci. Res. Publ. (IJSRP) **2**(8) (2012). ISSN 2250-3153
3. Neogi, B., Singha, T., Das, S., Ghosh, S., Manna, N., Bhattacharyya, S.: Performance study and analysis towards discrete system introducing jury test simulator. Afr. J. Comput. ICT (AJOCICT) **8**(1) (2015). ISSN: 2006-1781, Index Copernicus
4. Neogi, B., Ghosal, S., Sarkar, S.: Analysis & control towards limb prosthesis for paraplegic & fatigued conditions by introducing Lyapunov & sample data domain aspects. Int. Rev. Autom. Control (IREACO Praise Worthy Prize Publication) **5**(4), 548–552 (2012). ISSN: 1974-6059

5. Aliman, N., Ramli, R., Haris, S.: Design and development of lower limb exoskeletons: a survey. Rob. Auton. Syst. **95**, 102–116 (2017)
6. Dorf, R.C., Bishop, R.H.: Modern Control Systems, 10th edn. Prentice Hall (2008)
7. Das, S., Ghosh, S., Neogi, B., Ashour, A.S., Dey, N.: Control action simulation for industrial boiler performance analysis. World J. Model. Simul. (WJMS), Scopus Indexed **13**(1), 27–36 (2017). ISSN: 1746-7233
8. Nise, N.S.: Control System Engineering, 6th edn. Wiley (2011)
9. Kang, C.G.: Origin of stability analysis: on governors by J.C. Maxwell [historical perspectives]. Control Syst. IEEE **36**, 77–88 (2016). ISSN 1066-033X
10. Hang, C.C., Åström, K.J., Ho, W.K.: Refinements of the Ziegler-Nichols tuning formula. IEEE Control Theory Appl. **138**(2), 111–118 (1991)
11. Cominos, P., Munro, N.: PID controllers: recent tuning methods and design to specification. IEEE Proc. Control Theory Appl. **149**(1), 46–53 (2002)
12. Murray, R.M., Li, Z., Sastry, S.S.: A Mathematical Introduction to Robotic Manipulation. CRC Press (1993)
13. Ghan, J., Steger, R., Kazerooni, H.: Control and system identification for the Berkeley lower extremity exoskeleton (BLEEX). Adv. Robot. **20**, 989–1014 (2006)
14. Khansari-Zadeh, S.M., Billard, A.: Learning control Lyapunov function to ensure stability of dynamical system-based robot reaching motions. Rob. Auton. Syst. **62**(6) (2014), 752–765
15. Panich, S.: Design and simulation of leg-exoskeleton suit for rehabilitation. Glob. J. Med. Res. **12** (2012)
16. Guo, Q., Li, S., Jiang, D.: A lower extremity exoskeleton: human-machine coupled modeling, robust control design, simulation, and overload-carrying experiment. Math. Probl. Eng. 1–15 (2015). https://doi.org/10.1155/2015/905761
17. Velandia, C., Tibaduiza, D., Vejar, M.: Proposal of novel model for a 2 DOF exoskeleton for lower-limb rehabilitation. Robotics **6** (2017)
18. Ajayi, M.O., Djouani, K., Hamam, Y.: Bounded control of an actuated lower-limb exoskeleton. J. Robot. **2017** (2017). Article ID 2423643, 20 pages

Identification of Satellite DNA in Different Species

Rachita Ghoshhajra[1], Sanghamitra Chatterjee[2(✉)], and Soma Barman (Mandal)[3]

[1] MCKV Institute of Engineering, Liluah, Howrah, India
rghhajra@gmail.com
[2] Camellia Institute of Technology, Kolkata, India
sangha3030@gmail.com
[3] Institute of Radio Physics and Electronics, University of Calcutta, Kolkata, India
barmanmandal@gmail.com

Abstract. Satellite DNA is a very small DNA sequence which repeats in tandem (array) inside noncoding DNA. We have identified two different Satellite DNA (SatDNA) structures in Transposon of different species and detected its Frequency of Occurrence (FO) using 'IdenSatDNA' algorithm. We have found that SatDNA structures are present in both AT-rich and GC-rich sequences, and most of the samples we have tested contain at least one SatDNA structure. We have also pointed out that Frequency of Occurrence (FO) of both DNA structures is very high in at least one NCBI reference sequence for every species. This may play a significant role to unfold the hidden riddle of human evolution.

Keywords: Satellite DNA · Tandem repeat · Noncoding DNA · Transposon · Evolution

1 Introduction

The name 'Satellite DNA', popularly known as SatDNA, refers to highly repetitive non-transcript short DNA sequence found in noncoding region of eukaryotic genome.

Work on Satellite DNA was started by Bonnewell et al. [1], Skinner and their team [2, 3] in 1980s, and they published several papers during this period. But research in this field was accelerated after successful completion of Human Genome Project and availability of efficient computational techniques afterwards. In 2014, Rich et al. [4] worked on the classification of Satellite DNA and its impact on genetic disease. In 2015, Garrido-Ramos showed in his review work that Satellite genome contains huge information related to plant genome [5]. At the same time, Araujo et al. [6] identified SatDNA as a part of telomeric region which is a part of noncoding DNA, and Jagannathan and Yamashita [7] showed that it also takes part in chromosome maintenance. The review work of Garrido-Ramos [8] provided information in support that SatDNA is the most evolving topic of recent years. Last year, Šatović et al. [9] studied the characteristics and evolution of SatDNA sequences in bivalve molluscs. Recently, a paper was published by Belyayev et al. [10] on the history of Satellite DNA family.

© Springer Nature Singapore Pte Ltd. 2021
N. R. Das and S. Sarkar (eds.), *Computers and Devices for Communication*,
Lecture Notes in Networks and Systems 147,
https://doi.org/10.1007/978-981-15-8366-7_9

Scientists have found that noncoding part of DNA (NCDNA) has significant contribution in human evolution. Among all classes of NCDNA, Transposon plays the most crucial role. It is available in the literature that Transposon is divided into four subclasses, LINE, SINE, *Alu* and SVA. In our previous work, we have identified and located the structure of LINE, SINE and *Alu* efficiently in noncoding DNA sequences of different primates and non-primates to find their role in evolution [11, 12]. While working on the structure of SVA (where S stands for SINE, V stands for VNTR and A stands for *Alu*), we have noted that VNTR signifies variable number tandem repeats, and in-depth study on tandem repeats leads to Satellite DNA, a typical class of noncoding DNA.

In this paper, we have searched for the presence of SatDNA in Transposon and its contribution in evolution process using 'IdenSatDNA' algorithm. Methodology in Sect. 2 describes brief introduction of SatDNA, its structure and proposed 'IdenSatDNA' algorithm. Results are presented in Sect. 3 after applying the algorithm on sample database of non-vertebrates and vertebrates collected from NCBI site [13]. Based on the findings, conclusions are drawn in Sect. 4.

2 Methodology

Noncoding DNA consists of two types of repeated sequences, one is 'interspersed repeated sequence' (LINE, SINE) and the other one is 'tandem repeated sequence' or 'satellite DNA'. The size of one tandem repeat unit may vary from one base pair (bp) to several hundred bps. SatDNA can be classified into micro/macro satellite based on their size (length of the sequence). According to Rich et al. [4], the structure of microsatellite DNA (sequence length 6–10 bp) is defined as 'GCTGTGG' sequence.

But last year Šatović [9] stated that main building block of satellite DNA is a 'TTAGGG' motif and its variations. 'TTAGGG' is a characteristic sequence in vertebrates. He stated that, at this moment, it is not clear whether and how these sequences present in bivalve mollusc influence the biology of these organisms and other species.

In this paper, we have identified the two SatDNA structures 'GCTGTGG' proposed by Rich et al. [4] and 'TTAGGG' proposed by Šatović [9] from different species using 'IdenSatDNA' algorithm. One of the structures is described in Fig. 1 [13].

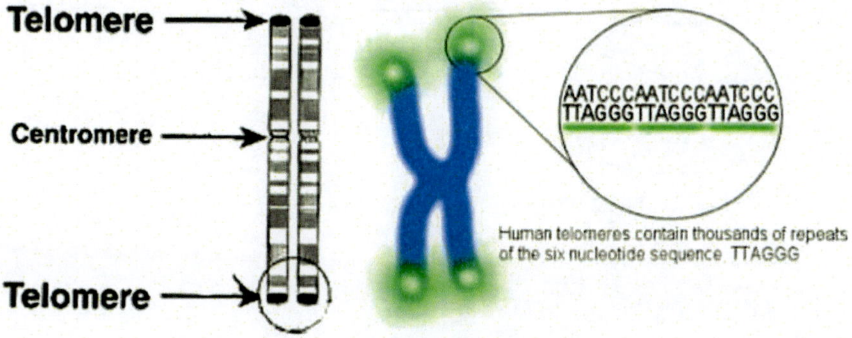

Fig. 1 Structure of satellite DNA

'IdenSatDNA' Algorithm: Algorithm for identification of different structures of Satellite DNA ('IdenSatDNA') is presented in Fig. 2. Objective of the proposed method is as follows: (i) Identification of SatDNA structures in Transposon, (ii) Frequency of Occurrence (FO) of SatDNA and (iii) Whether SatDNA is present only in AT-rich sequence or not. At the preprocessing step (Step 1), unwanted noise is removed from sample dataset. Then, AT% and GC% of the sample data are calculated (Step 2). Afterwards, pattern search with continuous sliding window is employed to find the presence of both SatDNA structures (Step 3) and their Frequency of Occurrence (FO) (Step 4) in each sample data.

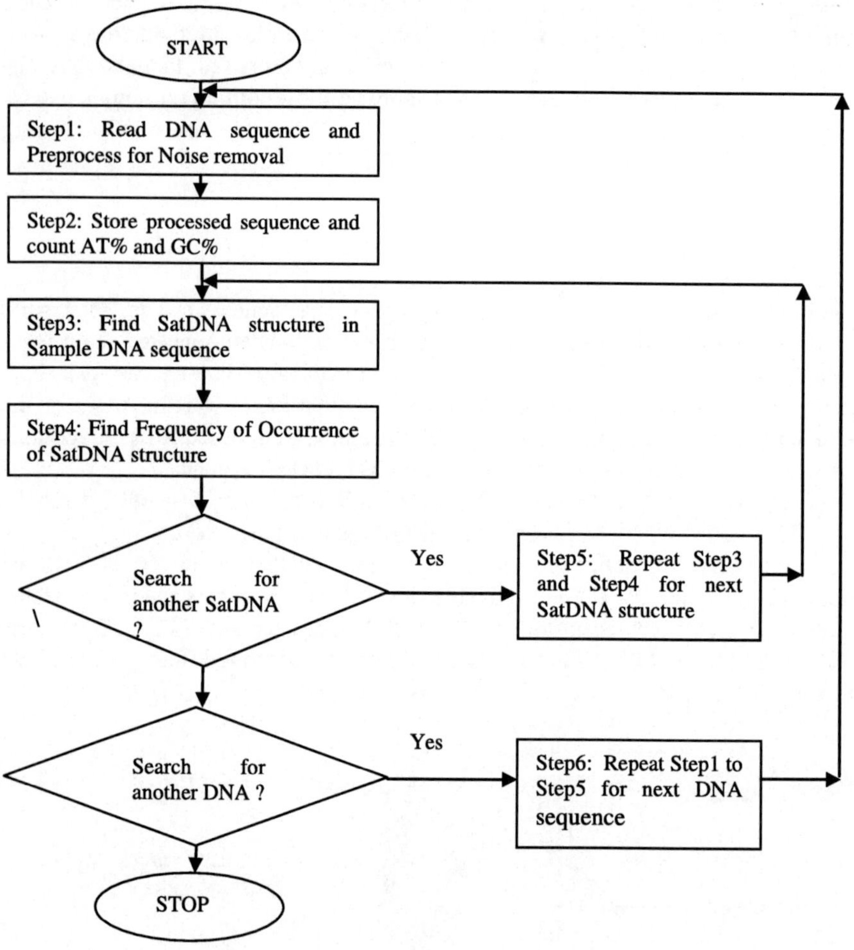

Fig. 2 Flowchart for IdenSatDNA algorithm

3 Results and Discussion

Datasets are collected for mollusc variants like Sea-hare, Mussel, Octopus, Oyster and Squid in the Dataset1. Also, datasets are collected for House Mouse (Mus musculus), Rhesus Monkey (Macaca mulatta), Chimpanzee (Pan troglodytes), Bonobo/Pygmy Chimpanzee (Pan paniscus) and Human (Homo sapiens) from the NCBI Web site https://www.ncbi.nlm.nih.gov. Pattern for SatDNA is searched on different datasets using 'IdenSatDNA' algorithm, and results are tabulated in Tables 1, 2 and 3.

Table 1 Frequency of occurrence (FO) of SatDNA in mollusca dataset

S. No.	NCBI reference sequence	%AT	%CG	FO of 'TTAGGG'	FO of 'GCTGTGG'
1	NW_004798128.1	59.46	40.54	6	1
2	NW_004797396.1	62.01	37.99	9	1
3	NW_004797413.1	59.74	40.26	7	2
4	NW_004797357.1	58.11	41.89	14	11
5	NC_006886.2	61.65	38.35	1	0
6	NC_006353.1	68.97	31.03	1	0
7	NW_011935168.1	65.08	34.92	1	0
8	NC_002507.1	62.82	37.18	0	0

Table 2 Presence and absence of two different SatDNA structures in vertebrates

Species	Total sample taken	No. of AT-rich sample	No. of GC-rich sample	Presence of 'TTAGGG' (%)	Presence of 'GCTGTGG' (%)	Presence of both seq (%)	Absence of both seq (%)
Mouse	9	5	4	66	55	78	22
Monkey	12	7	5	50	66	83	17
Chimpanzee	6	5	1	83	50	83	17
Bonobo	8	7	1	88	62	87	13
Human	8	6	2	75	40	87	13

The presence of SatDNA and its Frequency of Occurrence (FO) in mollusca dataset are shown in Table 1. NCBI reference sequence of mollusca variants like Sea-hare (Sl. No. 1_4), Mussel (Sl. No. 5), Octopus (Sl. No. 6), Oyster (Sl. No. 7) and Squid (Sl. No. 8) are also displayed in Table 1.

The interesting features we observed from result shown in Table 1 are (i) SatDNA structure 'TTAGGG' is present in most of the tested data, whereas 'GCTGTGG' is present only in Sea-hare and (ii) all mollusca samples are AT rich, which concludes that, SatDNA is present in AT-rich sequences.

Table 3 NCBI reference sequence identification for highest frequency of occurrence (FO) of both SatDNA

Species	NCBI ref sequence	Highest FO of 'TTAGGG'	Highest FO of 'GCTGTGG'
Molluscs	NW_004797357.1	14	11
Mouse	NC_000074.6	22	16
Monkey	NC_027903.1	24	7
Chimpanzee	NC_036885.1	21	8
Bonobo	NC_027881.1	45	8
Human	NC_000006.12	8	2

We have tested our algorithm for different non-primate and primate species also to find the presence of SatDNA structures and its Frequency of Occurrence (FO). The result is shown in Table 2.

From the tested results, it is observed that,

1. SatDNA is absent in very few DNA sequence and
2. SatDNA is present in both AT-rich and GC-rich sequence.

Figure 3 represents the presence of two different structures of SatDNA sequence ('TTAGGG' and 'GCTGTGG') in vertebrate species like Mouse, Monkey, Chimpanzee, Bonobo and Human. It can be concluded from the graph that Transposon of higher order species, i.e., primates contains more number of SatDNA ('TTAGGG') than lower order species, i.e., non-primates.

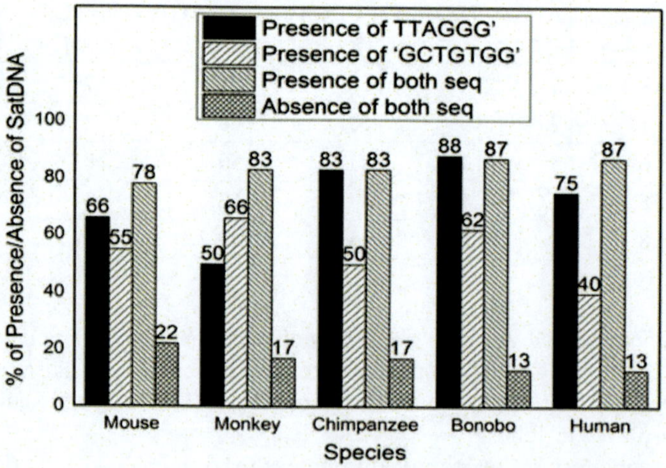

Fig. 3 Comparative study of presence of two SatDNA sequences in different vertebrates

NCBI Reference sequence listed in Table 3, identified by our algorithm, has highest Frequency of Occurrence (FO) for both SatDNA structure. These samples may have significant role in analysis of SatDNA and contribution of SatDNA in evolution or any other field.

4 Conclusions

In this paper, we have identified the SatDNA structures in a number of vertebrates and non-vertebrates using 'IdenSatDNA' algorithm. We have considered from phylum mollusca to human, in order to study the impact of SatDNA on evolution. It is found that SatDNA not necessarily present in only GC-rich sequences but it is also present in AT-rich sequences. In each species, there is only one sample, which has multiple occurrences of SatDNA sequences and is significantly large in number. Also, primate samples contain more number of SatDNA ('TTAGGG') than lower order species, i.e., non-primates. These samples may play significant role in analysis of SatDNA as well as evolution. Though more study is required in this field, this work will definitely ignite some avenues to carry out future research on SatDNA and may help to find some strong link between SatDNA and evolution of species.

References

1. Bonnewell, V., Fowler, R.F., Skinner, D.M.: An inverted repeat borders a fivefold amplification in satellite DNA. Science **221**(4613), 862–865 (1983)
2. Stringfellow, L.A., Fowler, R.F., La Marca, M.E., Skinner, D.M.: Demonstration of remarkable sequence divergence in variants of a complex satellite DNA by molecular cloning. Gene **38**(1–3), 145–152 (1985)
3. Fowler, R.F., Skinner, D.M.: Cryptic satellites rich in inverted repeats comprise 30% of the genome of a hermit crab. J. Biol. Chem. **260**(2), 1296–1303 (1985)
4. Rich, J., Ogryzko, V.V., Pirozhkova, I.V.: Satellite DNA and related diseases. Biopolym. Cell **30**(4), 249–259 (2014). ISSN 0233–7657
5. Garrido-Ramos, M.A.: Satellite DNA in plants: more than just rubbish. Cytogenet. Genome Res. **146**, 153–170 (2015). https://doi.org/10.1159/000437008
6. Araujo, N.P., Lima, L.G., Dias, G.B., Silva Kuhn, G.C., et al.: Identification and characterization of a subtelomeric satellite DNA in Callitrichini monkeys. DNA Res. **24**(4), 377–385 (2017). https://doi.org/10.1093/dnares/dsx010
7. Jagannathan, M., Yamashita, Y.M.: Function of junk: pericentromeric satellite DNA in chromosome maintenance. In: Cold Spring Harbor Symposia on Quantitative Biology, vol. LXXXII (2017). https://doi.org/10.1101/sqb.2017.82.034504
8. Garrido-Ramos, M.A.: Satellite DNA: an evolving topic. Genes **8**, 230 (2017). https://doi.org/10.3390/genes8090230
9. Šatović, E., Zeljko, T.V., Plohl, M.: Characteristics and evolution of satellite DNA sequences in bivalve molluscs. Eur. Zool. J. **85**(1), 94–103 (2018). https://doi.org/10.1080/24750263.2018.1443164
10. Belyayev, A., Josefiová, J., Jandová, M., Kalendar, R., Krak, K., Mandák, B.: Natural history of a satellite DNA family: from the ancestral genome component to species-specific sequences, concerted and non-concerted evolution. Int. J. Mol. Sci. **20**, 1201 (2019). https://doi.org/10.3390/ijms20051201

11. Ghoshhajra, R., Chatterjee, S., Barman (Mandal), S.: Identification of some transposable elements of DNA using "BP Suche" algorithm. In: Das, A., Nayak, J., Naik, B., Pati, S., Pelusi, D. (eds.) Computational Intelligence in Pattern Recognition. Advances in Intelligent Systems and Computing, vol. 999. Springer, Singapore (2020)
12. Ghoshhajra, R., Chatterjee, S., Barman Mandal, S.: Study of evolution by searching Alu pattern from primate transposon. In: Proceedings of International Conference on Communication Devices & Computing, ICCDC2019. Lecture Note in Electrical Engineering, vol. 602. Springer, Singapore (2019)
13. https://www.hammiverse.com/lectures/19/2.html

Modeling and Simulation of p53-Mdm2 Protein Pathway in Normal Cells

Trisha Patra[1], Sanghamitra Chatterjee[2], Soumya Pandit[1], and Soma Barman (Mandal)[1](✉)

[1] Institute of Radio Physics and Electronics, University of Calcutta, 92 APC Road, Kolkata 700009, India
trishapatra92@gmail.com, soumya.pandit.rpe@gmail.com, barmanmandal@gmail.com

[2] Camelia Institute of Technology, Kolkata 700129, India
sangha3030@gmail.com

Abstract. The p53, "Guardian of Genome," is a special type of protein acting as a tumor suppressor and transcription factor for Mdm2. The interaction between p53-Mdm2 protein pathway forms a negative feedback network where p53 activates Mdm2 and Mdm2 deactivates p53 through ubiquitination. The interaction between p53-Mdm2 plays an important role in cancer progression. In this paper, ordinary differential equation (ODE) is derived based on mass action kinetics of the chemical reactions of the protein pathway under normal condition of DNA. A system model of p53-Mdm2 pathway is derived for normal cells using Michcalis-Menten kinetics. The system model describing the dynamics of the pathway is simulated in MATLAB R2014a environment.

Keywords: p53 · Mdm2 · Ubiquitination · Negative feedback · Michaelis–Menten kinetics

1 Introduction

A cell is an assembly of several thousands of interacting proteins, and each of these nanometer-sized structures performs some specific functions that are unique. Tumor protein p53 is a transcription factor which is responsible for cell repair, cell division and cell death. Discovered in 1979, it is a crucial tumor suppressor protein and causes cell arrest and apoptosis during stress. The activities of p53 are controlled by a primary negative regulator, Murine double minute 2 (Mdm2), an ubiquitin ligase.

In the absence of stress, Mdm2 limits the action of p53 by binding with it and forming a complex. As the complex is formed, p53 is degraded by ubiquitination instigated by Mdm2. Since p53 controls cell cycle arrest and apoptosis, ubiquitination is important for its stability and transcriptional activities; otherwise, Mdm2 level will keep on rising, leading to loss in p53-dependent activities [1]. Mdm2 and p53 thus form a negative feedback loop (Fig. 1), in which p53 induces the expression of Mdm2, which in turn promotes the degradation of p53 and quenches cellular p53 activity [2].

© Springer Nature Singapore Pte Ltd. 2021
N. R. Das and S. Sarkar (eds.), *Computers and Devices for Communication*, Lecture Notes in Networks and Systems 147,
https://doi.org/10.1007/978-981-15-8366-7_10

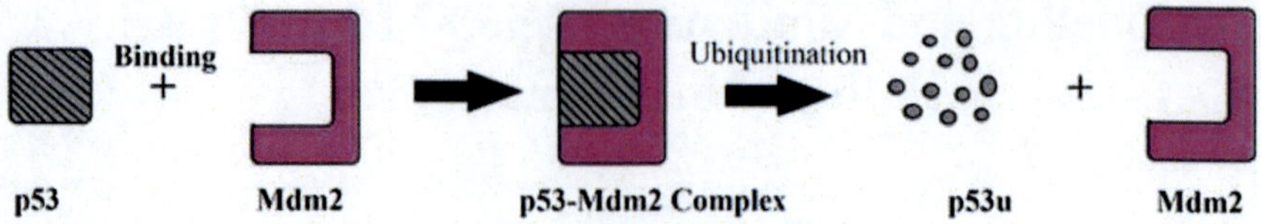

Fig. 1 p53-Mdm2 pathway

In around 50% cases, growth of cancer is discovered due to mutated form of p53. Mdm2 gene amplification is responsible for 17% of tumor growth, since it seizes p53-dependent apoptosis and cell cycle arrest [3]. Therefore, in current scenario, study of p53-Mdm2 interaction is a focal point for research in bioinformatics. Shi et al. presented a dynamic model for the p53-Mdm2 feedback loop without time lag effect of transcription [4]. Hunziker et al. constructed a mathematical model of the negative feedback loop involving p53 and Mdm2 and examines the effect of different stresses that trigger p53 [2]. Teo et al. demonstrated a cytomorphic approach for building analog circuits, derived from ordinary differential equations (ODE) and partial differential equations (PDE) based on living cells and its mechanisms [5]. Eliaš et al. proposed ODE and PDE models of p53 activation and regulation in single cells following DNA damage [6]. Understanding the significance of p53 protein as tumor suppressor gene, scientists have worked with it under DNA-damaged condition very minutely in the past few years. In this paper, authors present p53-Mdm2 protein pathway under normal or damage-free condition. Mathematical models for individual p53 and Mdm2 pathway are derived based on mass action and Michaelis–Menten kinetics of its chemical reactions, and the response of the model is studied in time domain. A system model is thereby derived from the protein pathway and simulated in MATLAB R2014a environment.

The paper consists of five sections. In Sect. 2, mechanism of p53-Mdm2 interaction is presented. Mathematical model and system model are presented in Sect. 3. In Sect. 4, simulated results of the models are presented and discussed subsequently. Finally, in Sect. 5, the paper is concluded along with future scopes.

2 Mechanism of p53-Mdm2 Interaction

For converting a chemical reaction to a mathematical model, the dynamics of the system is considered where state is changing with time [7]. The differential Eq. (1) describes change in concentration of reactant X over time due to its interaction with the other reactants involved in the reaction.

$$\frac{dX}{dt} = \text{Synthesis} - \text{Degradation} - \text{Phosphorylation} + \text{Dephosphorylation etc.} \quad (1)$$

2.1 p53 Production and Degradation

p53 protein is assumed to be produced at a constant rate in the cytoplasm. Also, the abundance of p53 in a cell is determined by degradation rather than its production [8].

Here, a constant rate of p53 synthesis is considered for calculations. The main source of p53 protein degradation is through E3 ligase Mdm2 which results in ubiquitination and hence p53 level decreases. Additionally, other proteins in cells, such as the hydrolase Hausp, are also responsible for p53 degradation [9]. For calculations, degradation rate of p53 in both cases, i.e., Mdm2-dependent and Mdm2-independent are considered.

2.2 Mdm2 Synthesis and Degradation

In the cell, p53 is responsible for Mdm2 production where it acts as a transcription factor and upregulates Mdm2-mRNA in nucleus. Mdm2-mRNA is then transported to cytoplasm and converted into Mdm2 protein. It is now ready to bind with p53 for ubiquitination. Mdm2 is degraded by the action of ARF. Other mechanism, responsible for Mdm2 degradation is stress-induced phosphorylation [10]. Constant rates of production and degradation of Mdm2 are considered here.

2.3 Parameters Assumptions

The mathematical equations are formulated based on Table 1 [4]. Since law of mass action is relevant for free diffusion, the rate of mRNA translation, k4, is highly dependent on the nature of cytoplasm, since sometimes it behaves like a gel and not a liquid. This might alter the mathematical calculations [11]. The medium of cytoplasm, considered here, is in liquid form.

Table 1 Parameters and their values

Parameters	Description	Value
p_0	p53 production rate	1000 nMh^{-1}
m_0	Mdm2-mRNA production rate	0.2 nM h^{-1}
k_1	Mdm2 dependent p53 degradation rate	11 h^{-1}
k_2	Mdm2 independent p53 degradation rate	0.1 h^{-1}
k_3	Rate of Mdm2-mRNAdegradation	0.6 h^{-1}
k_4	Rate of mRNA translation	1.4 h^{-1}
k_5	Rate of Mdm2 degradation	0.2 h−1

3 Mathematical Models

Individual mathematical models for p53, Mdm2-mRNA, Mdm2 and the system transfer function are described in the following sections.

3.1 Mathematical Model for p53 and Mdm2

The variations in concentration of p53 and Mdm2 with respect to time are described by the following differential equations. Under normal condition of DNA, only the synthesis and degradation of reactant are considered in Eq. (1).

$$\frac{dP}{dt} = p_0 - k_1MP - k_2P. \tag{2}$$

$$\frac{dm}{dt} = m_0 - k_3m. \tag{3}$$

$$\frac{dM}{dt} = k_4m - k_5M. \tag{4}$$

where P, m and M denote concentrations of inactive p53, Mdm2-mRNA and Mdm2 protein, respectively.

3.2 Mathematical Model for Chemical Reactions

Considering k_f, k_{-f} and k_r to be forward reaction rate of forming complex from p53 and Mdm2, backward reaction rate of complex and reaction rate for dissociation of complex by ubiquitination, respectively, the association and dissociation reaction for p53 and Mdm2 during normal condition, is described by

$$\text{p53} + \text{Mdm2} \rightleftharpoons \text{Complex} \rightarrow \text{p53}_u + \text{Mdm2}. \tag{5}$$

where p53_u denotes the degraded or ubiquitinated form of p53.

Using mass action kinetics of a chemical reaction, the concentration of complex, with respect to that of p53 and Mdm2 is expressed as

$$\frac{d[\text{Complex}]}{dt} = k_f[\text{p53}][\text{Mdm2}] - k_r[\text{Complex}] - k_{-f}[\text{Complex}]. \tag{6}$$

Following the steady-state assumption and Michaelis–Menten kinetics of enzyme–substrate saturation

$$[\text{Complex}] = [E_t]\frac{K[\text{p53}]}{K[\text{p53}] + 1}. \tag{7}$$

where

$$[E_t] = [\text{Mdm2}] + [\text{Complex}]. \tag{8}$$

and

$$K = \frac{k_f}{k_{-f} + k_r}. \tag{9}$$

3.3 System Model of the Pathway

Figure 2 is the system model of the pathway, represented by Eq. (7). There are two negative feedback loops. First loop, L1 represents slowing down of forward reaction due to fall in reactant concentration and second loop, L2 signifies speeding up of backward reaction as the product concentration rises with time.

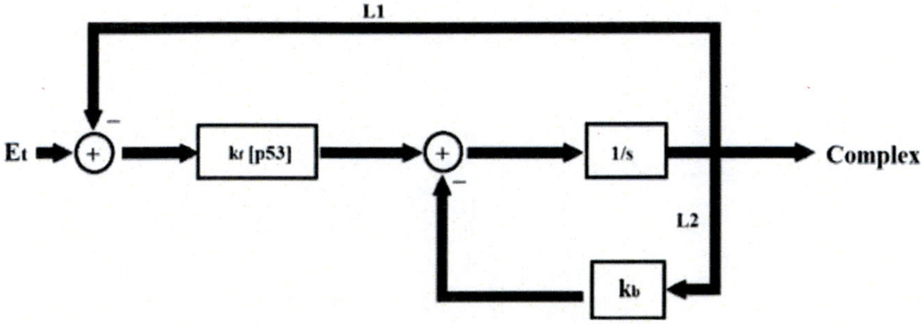

Fig. 2 System model of the pathway

4 Results and Discussion

Simulated results of p53, Mdm2-mRNA and Mdm2 based on Eqs. (2), (3) and (4) are displayed in Fig. 3a–c, respectively. In Fig. 3a, the concentration level of p53 degrades with time due to action of Mdm2. Figure 3b shows concentration of Mdm2-mRNA decreases with time due to action of p53 as a transcription factor for Mdm2 formation. In Fig. 3c, concentration of Mdm2 increases with time and reaches to a constant value since after p53 degradation, and it binds with other p53 proteins, causing further ubiquitination.

Figure 3d depicts the oscillatory nature of p53-Mdm2 pathway obtained by simulating Eq. (7). Due to p53-Mdm2 interaction, the concentration of complex increases initially. But, after ubiquitination, the concentration decreases. The nature of the response obtained is oscillatory in nature due to the two feedback loops in the system [12]. Plots are simulated in MATLAB R2014a environment.

5 Conclusion and Future Work

In this paper, the authors present a mathematical model of p53-Mdm2 pathway under damage-free condition of DNA. Mathematical expressions of p53, Mdm2-mRNA and Mdm2 are simulated, and the responses show the change of concentration with time which validates the results of the research work of [12]. The response of the model in Fig. 3d justifies the biological nature of proteins under normal or damage-free condition of DNA. Under damaged condition of DNA, the mathematical expressions will be different. In future, other proteins, ARF and E2F1, will also be considered along with p53 and Mdm2 in the pathway. Further each block of system model of the pathway will be transformed into equivalent electrical circuits to validate our results.

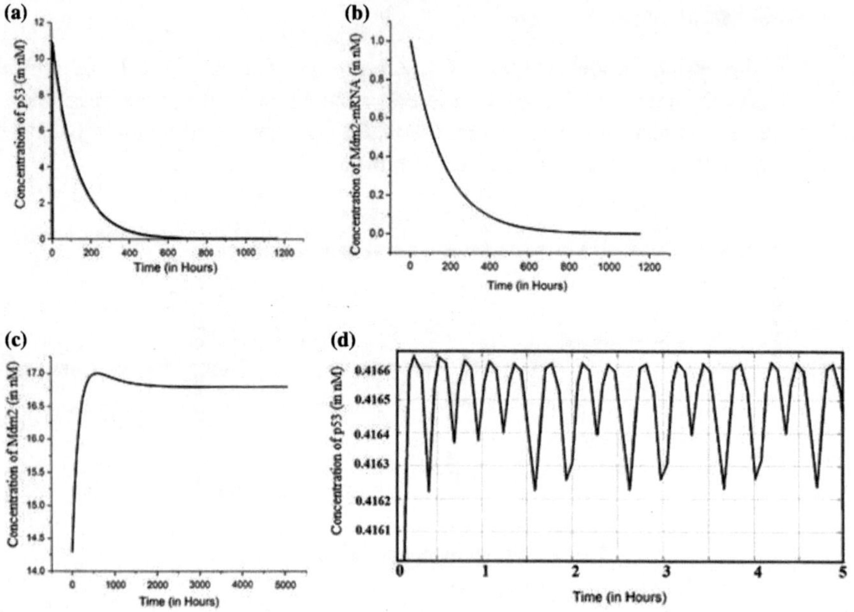

Fig. 3 Variation of concentration of (**a**) p53, (**b**) Mdm2-mRNA, (**c**) Mdm2 with time (in hours) and (**d**) simulation result of system model

Acknowledgements. The authors would like to thank West Bengal Higher Education, Science and Technology and Biotechnology (Science and Technology) funded project "Cytomorphic Circuit Modeling of p53 Protein Pathway for Synthetic Biology Applications" for funding support of the research work.

References

1. Brooks, C.L., Gu, W.: p53 regulation by ubiquitination. FEBS Lett. 2803–2809 (2011)
2. Nag, S., Qin, J., Srivenugopal, K.S., Wang, M., Zhang, R.: The Mdm2-p53 pathway revisited. J. Biomed. Res. 254–271 (2013)
3. Shi, X., Qin, K., Liu, Z., Zheng, Y.: A simplified dynamic model for p53-Mdm2 feedback loop. In: 10th IEEE International Conference on Control and Automation, pp. 264–267 (2013)
4. Hunziker, A., Jensen, M., Krishna, S.: Stress-specific response of p53-Mdm2 feedback loop. BMC Syst. Biol. (2010)
5. Teo, J.J.Y., Woo, S., Sarpeshkar, R.: Synthetic biology: a unifying view and review using analog circuits. IEEE Trans. Biomed. Circuits Syst. 1–22 (2015)
6. Eliaš, J., Dimitro, L., Clairambault, J., Natalini, R.: The Dynamics of p53 in Single Cells: Physiologically Based ODE and Reaction-Diffusion PDE Models. IOP Publishing (2014)
7. Conrad, E., Tyson, J.: Modeling molecular interactions and non-linear ODEs. In: System Modeling in Cellular Biology, pp. 97–124 (2006)
8. Monk, N.: Oscillatory expression of Hes1, p53 and NH-κB driven by transcriptional time delays. Curr. Biol. 1409–1413 (2003)
9. Vogelstein, B., Lane, D., Levine, A,J.: Surfingthep53 network. Nature 307–310 (2000)

10. Stommel, J.M., Wahl, G.M.: Accelerated MDM2 auto-degradation induced by DNA-damage kinases is required for p53 activation. J. EMBO 1547–1556 (2004)
11. https://en.m.wikipedia.org/wiki/Michaelis%E2%80%93Menten_kinetics
12. Azam, M.R, Fazal, S., Ullah, M., Bhatti, A.I.: System-based strategies for p53 recovery. IET Syst. Biol. 101–107 (2017)

An Empirical Study of Incremental Learning in Neural Network with Noisy Training Set

Shovik Ganguly[1], Atrayee Chatterjee[2], Debasmita Bhoumik[3], and Ritajit Majumdar[3](✉)

[1] Lexmark International India Pvt. Ltd, Kolkata, India
gangulyshovik@gmail.com

[2] Asutosh College, University of Calcutta, Kolkata, India
atrayeechatterjeeiii@gmail.com

[3] Advanced Computing and Microelectronics Unit, Indian Statistical Institute, Kolkata, India
debasmita.ria21@gmail.com, majumdar.ritajit@gmail.com

Abstract. The notion of incremental learning is to train an ANN algorithm in stages, as and when newer training data arrives. Incremental learning is becoming widespread in recent times with the advent of deep learning. Noise in the training data reduces the accuracy of the algorithm. In this paper, we make an empirical study of the effect of noise in the training phase. We numerically show that the accuracy of the algorithm is dependent more on the location of the error than the percentage of error. Using perceptron, feedforward neural network and radial basis function neural network, we show that for the same percentage of error, the accuracy of the algorithm significantly varies with the location of error. Furthermore, our results show that the dependence of the accuracy with the location of error is independent of the algorithm. However, the slope of the degradation curve decreases with more sophisticated algorithms.

Keywords: Incremental learning · Noisy dataset · Artificial neural network

1 Introduction

Machine learning (ML) algorithms aim to train a computer to make decisions [1]. These algorithms are used in various fields such as image processing [2, 3], object recognition [4], handwriting recognition [5], natural language processing [6] and even quantum computing [7]. There are two major techniques for decision making used by ML algorithms—(i) supervised learning where the algorithm is initially trained with a set of labeled data [8], and (ii) unsupervised learning where the algorithm looks for patterns and similarities and tries to group similar data in the same cluster without any prior learning phase [9]. Although these two are the major types of learning algorithms, other forms of learning algorithms such as semi-supervised learning [10] and reinforcement learning [11] are also widely studied.

Artificial neural network (ANN) algorithms are supervised algorithms which mimic the working principles of neurons in the human brain. The most basic ANN algorithm

© Springer Nature Singapore Pte Ltd. 2021

N. R. Das and S. Sarkar (eds.), *Computers and Devices for Communication*,
Lecture Notes in Networks and Systems 147,
https://doi.org/10.1007/978-981-15-8366-7_11

is the single-layer perceptron. However, more sophisticated algorithms use multiple layers, complicated functions for decision making, and take the error of the output into consideration in order to update the weights. Although these algorithms assume a single initial training phase, in real world, data often arrives in batches, which requires the training to be performed in multiple stages. The model is first trained with the initial set of training data, and when more training data is available, the model is trained further. Such a model of training is called incremental learning [12].

In this paper, we study incremental learning in ANN where the training dataset may be noisy, i.e., some of the training data has incorrect label. It is obvious that if the training dataset is erroneous, the training will be less effective and the ANN will be less accurate in its prediction. In this paper, we show, however, that not only the number of erroneous data, but also the location of the error in incremental learning plays a vital role in the performance of the algorithm. The accuracy of an algorithm varies even when the percentage of error is the same, but the errors are concentrated in different locations of the training set.

In this paper, we show numerically by using three ANN algorithms (perceptron [13], feedforward neural network (FFN) [14, 15] and radial basis function neural network (RBF) [15]) that (i) for the same percentage of erroneous data, the location of error clusters can significantly alter the performance of the algorithm, (ii) the performance degradation is independent of the number of features per data and (iii) although more sophisticated algorithms are more robust to errors, the degradation in performance due to concentrated error has a similar nature for all algorithms.

The rest of the paper is organized as follows—in Sect. 2, we give a brief description of the three ANN algorithms used. In Sects. 3 and 4, we show the performance of the algorithms for two-step and three-step incremental learning. We conclude in Sect. 5.

2 Brief Review of the ANN Algorithms Used

An m-class classification problem [16], where $C_1, C_2, \ldots, C_m$ are the object classes, is associated with k training samples and n testing samples. Each training sample is a vector (s_i, C_{si}), where C_{si} is the designated class of the sample s_i. After training the algorithm with this set of training data, for each testing sample t_i, such that t_i belongs to class C_j, the algorithm is expected to produce

$$\text{Prob}\big(t_i \in C_j\big) > \text{Prob}(t_i \in C_l) \quad \forall l \neq j. \tag{1}$$

The algorithm is said to have made an error in the prediction if, for some testing sample $t_p \in C_p$, it produces $\text{Prob}(t_p \in C_p) < \text{Prob}(t_p \in C_q)$ for some $q \neq p$. The objective of learning is to minimize the number of errors. In the following part of this subsection, we briefly discuss the working principles of the three ANN algorithms (perceptron, FNN and RBF) used in this paper. Our motivation behind using these three algorithms is to show that although sophistication of the algorithm enhances the robustness to training errors, the performance loss due to the location of error concentration remains invariant under the type of algorithm used.

Perceptron Learning Algorithm. Perceptron is one of the simplest ANN algorithms. In this algorithm, an input is a vector $(x_1, x_2, \ldots, x_n)$, where each x_i is called a feature and associated with each feature is a weight w_i. For a 2-class classification problem, the output

$$y = \begin{cases} 1 \text{ if } \sum_{i=1}^{n} w_i x_i \geq \theta \\ 0 \text{ otherwise} \end{cases} \tag{2}$$

where θ is a threshold value. The weights w_i are initialized randomly and are modified during the training phase to match the class labels of each training sample. This algorithm can be easily modified for multi-class classification.

Feedforward Neural Network Algorithm. In FFN, the perceptrons are arranged in layers, with the first layer taking in inputs and the last layer producing outputs. The middle layers are called hidden layers. Each perceptron in one layer is connected to every perceptron on the next layer, but there is no interconnection among the perceptrons of the same layer. The information is constantly fed forward from one layer to the next. A single perceptron can classify points into two regions which are linearly separable, whereas by varying the number of layers, the number of input, output and hidden nodes, one can classify points in arbitrary dimension into an arbitrary number of groups.

Radial Basis Function Algorithm. A drawback of perceptron is that the activation function is linear and hence fails to classify nonlinearly separated data. The RBF algorithm uses radial basis functions as activation functions. RBF is a real-valued function φ whose value depends only on the distance from the origin $\varphi(x) = \varphi(\|x\|)$. Its characteristic feature is that the response decreases or increases monotonically with distance from a central point. A typical radial function is the Gaussian function. RBF transforms the input signal into a different form, which can then be fed to the ANN to get linear separability. RBF has an input layer, a single hidden layer and an output layer. The sophistication of the activation function of this algorithm usually leads to better classification accuracy than perceptron or FFN.

3 Performance of ANN in Noisy Two-Step Incremental Learning

We have performed our study on the standard benchmark dataset IRIS [17], which contains 120 data samples, where each sample is a vector containing four features. Forty samples have been used to train each ANN algorithm, while the other 80 have been used to test their performance.

Error in training data implies that for a particular training sample $s_i \in C_i$, the class is incorrectly labeled as $C_j, j \neq i$. For this dataset, containing $2n$ training samples (here $n = 20$), each ANN algorithm is trained twice sequentially with the first n and the last n training samples. We have varied the errors in the training set from 0 to 50% by a gap of 10. Three scenarios are considered, where the erroneous data is (i) uniformly distributed in the entire training set, (ii) uniformly distributed in the first half of the training set and (iii) uniformly distributed in the second half of the training set.

In Table 1, we show the accuracy obtained by perceptron, FFN and RBF algorithms, respectively, as the error in the training sample is varied as discussed above.

Table 1 Accuracy of perceptron, FFN and RBF in the presence of clustered noise

Error %	Error distributed uniformly in								
	Perceptron			FFN			RBF		
	Entire dataset	First half	Second half	Entire dataset	First half	Second half	Entire dataset	First half	Second half
0	100	100	100	100	100	100	100	100	100
10	96.45	96.67	100	96.67	96.67	100	97.8	97	100
20	60	60	100	60	60	100	70.65	72	100
30	46.67	45.2	93.33	49	43.67	96.67	48.9	52.45	98.2
40	33	30	92.4	33	30	92.52	40.2	38	96.47
50	30.44	28.55	85.69	30	28.40	88.54	36	36	94.44

4 Performance of ANN in Noisy Three-Step Incremental Learning

We have performed this study on the WINE dataset [17], which contains 150 data samples, where each sample is a vector containing 13 features. The IRIS dataset chosen in the previous two-step experiment does not contain enough samples to effectively divide into three training sets. As such, we chose the WINE dataset for this experiment. For this dataset, the ANN algorithms have been trained using 60 samples and the other 90 have been used to test their performance. The motivation to use these two different datasets (IRIS and WINE) is to show that the performance degradation due to error clusters is independent of the number of features per sample. For the WINE dataset containing $3n$ training samples, each ANN algorithm is trained thrice sequentially with the first n, second n and the last n training samples. We have varied the errors in the training set from 0 to 40% by a gap of 10. Three scenarios are considered in each case, where the erroneous data is (i) uniformly distributed in the entire training set, (ii) uniformly distributed in the first 20 entries, (iii) uniformly distributed in the second 20 entries and (iv) uniformly distributed in the last 20 entries of the training set.

In Fig. 1, we show the graph of the accuracy of the three algorithms for incremental training. The first row shows the performance for two-step training, and the second row shows the same for three-step training. The graphs readily show that the degradation in performance is heavily dependent on the location of error. The performance of all the algorithms, when errors are distributed uniformly or are clustered in the first training set, is almost the same. However, as the errors move to later training sets, the performance of the algorithm increases significantly. Moreover, although the performance of RBF is better than FFN, which in its turn is better than perceptron, the nature of degradation remains similar for all the algorithms, irrespective of its sophistication.

For the three-step learning, we have also studied the accuracy of the said algorithms when the error is uniformly distributed in two of the three steps. The graph of the accuracy in this scenario is shown in Fig. 2. The performance of the algorithms due to error in two of the three halves also has a similar nature.

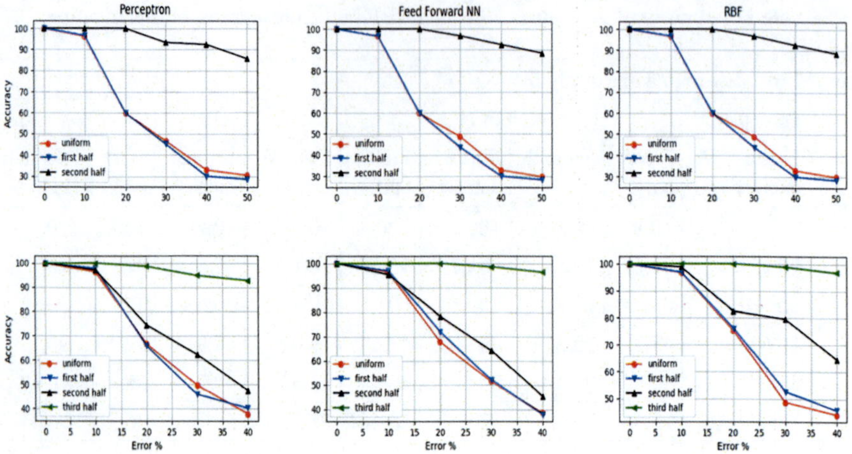

Fig. 1 Accuracy of perceptron, FFN and RBF for two- and three-step incremental learning

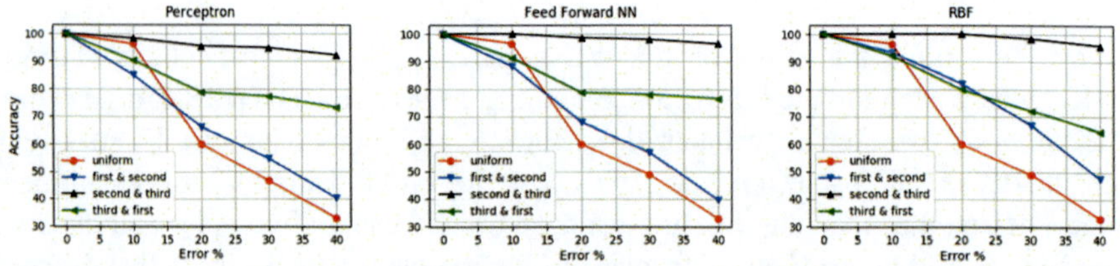

Fig. 2 Accuracy of ANN algorithms when errors are uniformly distributed in two of the three steps

5 Conclusion

In this paper, we have numerically studied the accuracy of perceptron, FFN and RBF for two-step and three-step incremental learning in the presence of noisy dataset. We show that the accuracy of the algorithms depends not only on the percentage of error on the training set, but also on the location of the error concentration. In fact, for the same percentage of error, the location plays a significant role in the accuracy of the algorithms. Moreover, we also show that the nature of degradation due to concentrated error is invariant of the number of features in the data. The accuracy obtained from the most basic ANN (perceptron) and more sophisticated ANNs (FFN, RBF) shows that although sophistication makes the algorithm more robust to errors, the nature of the performance degradation due to location of error is similar for all the algorithms. Therefore, the concentration of error is a more acute threat to incremental learning with noisy training data than the percentage of error.

References

1. Nasrabadi, N.M.: Pattern recognition and machine learning. J. Electron. Imaging **16**(4), 049901 (2007)

2. Rosten, E., Drummond, T.: Machine learning for high-speed corner detection. In: European Conference on Computer Vision, pp. 430–443. Springer (2006)
3. Dietterich, T.G.: Ensemble methods in machine learning. In: International Workshop on Multiple Classifier Systems, pp. 1–15. Springer (2000)
4. Duygulu, P., Barnard, K., Freitas, J., Forsyth, D.: Object recognition as machine translation: learning a lexicon for a fixed image vocabulary. In: European Conference on Computer Vision, pp. 97–112. Springer (2002)
5. Lei, Xu., Krzyzak, A., Suen, C.Y.: Methods of combining multiple classifiers and their applications to handwriting recognition. IEEE Trans. Syst. Man Cybern. **22**(3), 418–435 (1992)
6. Berger, A.L., Pietra, V.J.D., Pietra, S.A.D.: A maximum entropy approach to natural language processing. Comput. Linguist. **22**(1), 39–71 (1996)
7. Rebentrost, P., Mohseni, M., Llyod, S.: Quantum support vector machine for big data classification. Phys. Rev. Lett. **113**(13), 130503 (2014)
8. Caruana, R., Niculescu-Mizil, A.: An empirical comparison of supervised learning algorithms. In: Proceedings of the 23rd International Conference on Machine Learning, pp. 161–168. ACM (2006)
9. Barlo, H.B.: Unsupervised learning. Neural Comput. **1**(3), 295–311 (1989)
10. Chapelle, O., Scholkopf, B., Zien, A.: Semi-supervised learning. IEEE Trans. Neural Netw. **20**(3), 542–542 (2009)
11. Kaelbling, L.P., Littman, M.L., Moore, A.W.: Reinforcement learning: a survey. J. Artif. Intell. Res. **4**, 237–285 (1996)
12. Pan, S.J., Yang, Q.: A survey on transfer learning. IEEE Trans. Knowl. Data Eng. **22**(10), 1345–1359 (2009)
13. Stephen, I.: Perceptron-based learning algorithms. IEEE Trans. Neural Netw. **50**(2), 179 (1990)
14. Gurney, K.: An Introduction to Neural Networks. CRC Press (2014)
15. Haykin, S.: Neural Networks: A Comprehensive Foundation. Prentice Hall PTR (1994)
16. Shalev-Shwartz, S., Srebro, N.: Understanding Machine Learning. Cambridge University Press (2014)
17. UCI repository of machine learning databases. https://archive.ics.uci.edu/ml/datasets.php

Check for updates

Resistivity, Dielectric, Activation and Optical Behaviour of $Y_{1-x}Nd_xCrO_3$ Nanoparticles

R. Sinha and Sandip Haldar(✉)

Department of Physics, Asansol Engineering College, Asansol, Kanyapur, West Bengal 713305, India
sand_ju@yahoo.com

Abstract. In this article, the conductivity, dielectric, activation and optical behaviour of $Y_{1-x}Nd_xCrO_3$ nanoparticles have been explained. The characterization is followed by XRD and TEM and measurements. The resistivity behaviour shows that the samples are semiconducting in nature and the resistivity decreases with the increase of doping concentration. From the resistivity analysis, it has been observed that the value of the activation energy decreases with the increase of the doping concentration. The band gap of the samples decreases with increasing doping concentration. The frequency variation of dielectric permittivity for different temperatures is measured for the samples. The variation of dielectric loss with temperature for different frequencies is also estimated.

Keywords: Nanomaterial · X-ray diffraction · Electrical properties

1 Introduction

In recent times, nanostructured materials are renowned for their potential application in various electronic devices. Multiferroic materials have simultaneous existence of ferromagnetic, ferroelectric and/or ferroelastic orderings. Nowadays, the coexistence of ferroelectricity and ferromagnetism in the perovskites compound is the focus of the study because it creates rich functionality in a single device. Earlier studies reveal the magnetic and electronic properties of the perovskites $RCrO_3$ type compound (where R = Yttrium or any rare earth element) at low temperatures. The later investigation was carried out to find out the electrical and chemical properties as well as structural stabilities for high temperature. These investigations had potential applications in high-temperature electrodes, thermistors and different thermoelectric materials. Previously, Serrao et al. [1] investigated the multiferroism in $YCrO_3$ compound.

Considering the importance of yttrium Ccromite in this work, we have prepared nanostructured Nd-doped $YCrO_3$ compound to improve its features for the potential applications in new electronic devices. The dielectric and conductivity study of the material is discussed in this work, which is very important for device application.

© Springer Nature Singapore Pte Ltd. 2021
N. R. Das and S. Sarkar (eds.), *Computers and Devices for Communication*,
Lecture Notes in Networks and Systems 147,
https://doi.org/10.1007/978-981-15-8366-7_12

2 Synthesis of $Y_{1-x}Nd_xCrO_3$

$Y_{1-x}Nd_xCrO_3$ nanoparticles are prepared following sol–gel synthesis method using weighted amount of yttrium nitrate hexahydrate [$Y(NO_3)_3$ x $6H_2O$], chromium nitrate nonahydrate [$Cr(NO_3)_3$ x $9H_2O$], neodymium nitrate hexahydrate [$Nd(NO_3)_3$ x $6H_2O$], PVA solution (12.5 g/litre) and 50 ml distilled water. Considering the content of neodymium here, 0% Nd and 10% Nd are written for $x = 0$ and $x = 0.10$. A homogeneous mixture is prepared with the above materials by continuous stirring. Then the solution is evaporated to get dry precursor powder. After grinding properly, the dry powder is calcined in air at 800 °C.

3 Results and Discussions

3.1 X-Ray Diffraction (XRD) and TEM (Transmission Electron Microscope) Analysis

The powder X-ray diffraction (XRD) patterns prove the well crystalline nature of the samples. Figure 1 shows the XRD patterns of 0% Nd and 10% Nd-doped samples.

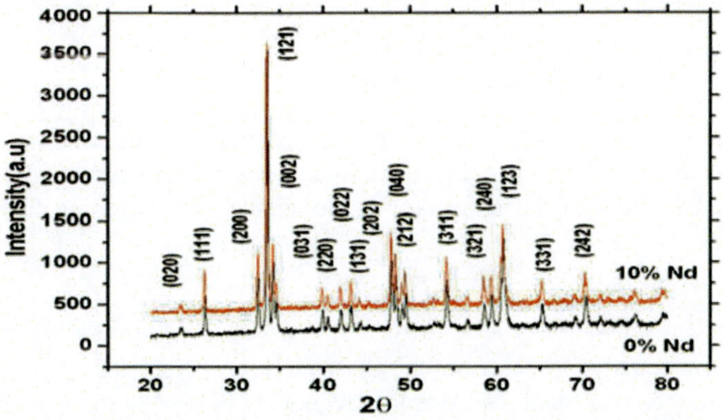

Fig. 1 XRD patterns of 0% Nd and 10% Nd-doped samples

The peaks show the (*hkl*) planes, and it matches with "ICDD PDF No. 34-0365." The calculated average crystallite size is 30 nm following the Debye Scherer formula [2, 3] given as follows:

$$D = 0.9\lambda/\beta\cos\theta \quad (1)$$

In the above equation, D is the average crystallite size, λ is the wavelength of X-ray, and β is the full width at half maximum of diffraction peak, respectively.

A typical TEM image of 10% Nd-doped sample is shown in Fig. 2. The average particle size is calculated as 35 nm.

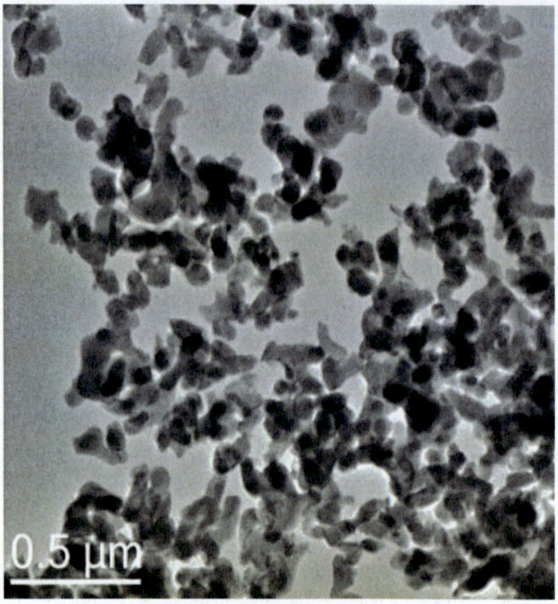

Fig. 2 Typical TEM image of 10% Nd-doped sample

3.2 UV–Vis Absorption Spectra and Optical Band Gap

The data for the UV–Vis absorption spectroscopy are taken within the range of 250–800 nm. The position of the absorption peak for all the samples lies around 305 nm and is given in Fig. 3. The optical band gap for all the samples is found using the relation [4, 5]

$$(\alpha h\nu)^u = H\left(h\nu - E_g\right) \tag{2}$$

In the above equation, α describes the absorption coefficient and h is Planck's constant. H represents a constant, and ν is the frequency. Following the equation, u may have values such as 2, 2/3, 0.5 and 3, and it is dependent on the mode of transition. The direct allowed transition has $u = 2$, and the direct forbidden transition has 2/3 u value.

The u value is 0.5 for direct forbidden transition, and it is 3 for indirect forbidden transition [6]. Here the mobility of electrons influences the constant H. The optical band gap has been measured of the different samples plotting $(\alpha h\nu)^u$ with $h\nu$ with different values of u as described above. The best fit of the optical absorption data is found for $u =$ 2. The band gap of 10% Nd-doped sample is shown in Fig. 4, and it supports the allowed direct band transition for our sample. It is found that the band gap decreases from 2.96 to 2.66 eV for 0% Nd to 10% Nd-doped samples with the increasing doping concentration of neodymium, which can be for the creation of mid-gap states. The values of the band gap are shown in Table 1.

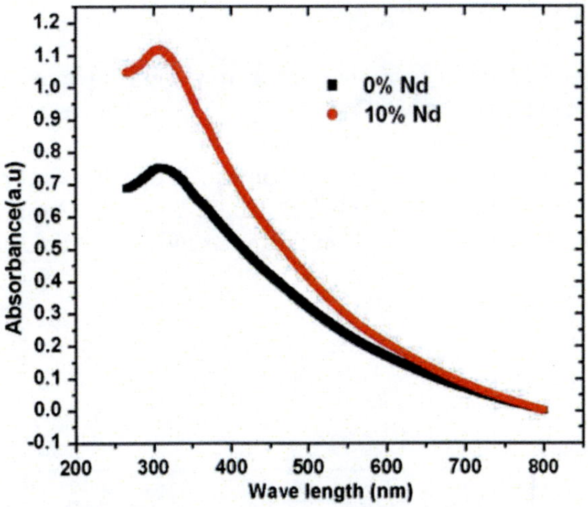

Fig. 3 UV–Vis absorption spectra of 0% Nd and 10% Nd-doped samples

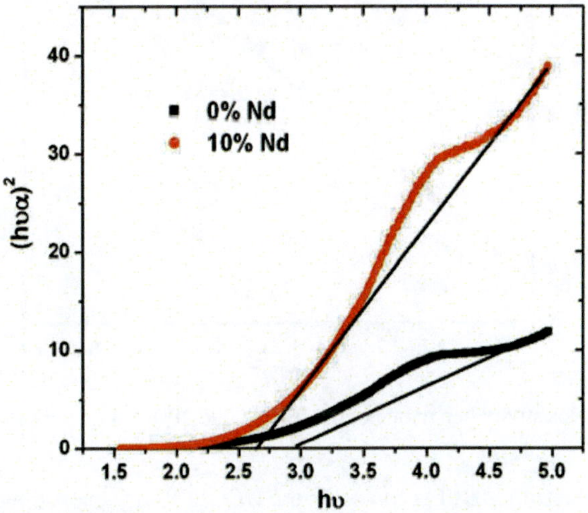

Fig. 4 Optical band gap of 0% Nd and 10% Nd-doped sample

Table 1 Band gap and activation energy of 0% Nd and 10% Nd-doped $YCrO_3$ samples

Sample	Band gap (eV)	Activation energy (eV)
0% Nd	2.96	0.43
10% Nd	2.66	0.36

3.3 Resistivity and Activation Behaviour

The DC resistivity of the samples has been studied within the temperature range 298–523 K. The temperature variation of DC resistivity is depicted in Fig. 5 where ln $\rho_{dc}(T)$ is plotted against 1000/*T*. It is clear from Fig. 5 that the resistivity decreases by a significant amount with the increase of the neodymium content. The temperature variation of DC resistivity indicates the semiconducting behaviour of the samples. From the slope of the graph in Fig. 5, DC activation energy is measured using the Arrhenius relation written as follows:

$$\rho_{dc}(T) = \rho_0 \exp\left(\frac{E_a}{kT}\right) \tag{3}$$

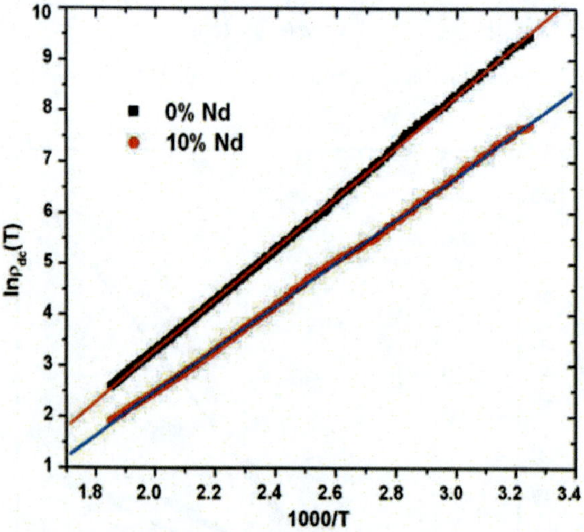

Fig. 5 Resistivity behaviour of the 0% Nd and 10% Nd-doped samples

In the above equation, ρ_{dc}(T) represents the DC resistivity, ρ_0 is the pre-exponential factor, and E_a and k indicate the activation energy and Boltzmann constant, respectively. It has been observed that the activation energy decreases for 10% Nd-doped sample with respect to 0% Nd-doped sample, and the values are shown in Table 1. It varies from 0.43 to 0.36 eV for 0% Nd-doped sample to 10% Nd-doped sample.

3.4 Dielectric Property Analysis

The real part of dielectric permittivity $\varepsilon'(f)$ and the imaginary part of dielectric permittivity $\varepsilon''(f)$ follow the relation

$$\varepsilon'(f) = \frac{Cd}{\varepsilon_0 A} \tag{4}$$

$$\varepsilon''(f) = \varepsilon'(f)D \quad (5)$$

In the above equations, C and D represent the capacitance and dissipation factor, respectively. ε_0 represents the free space permittivity, A denotes the area of the electrode, and d is the thickness of the sample. The frequency variation of $\varepsilon'(f)$ and $\varepsilon''(f)$ for different temperatures for 10% Nd-doped sample is shown in Fig. 6a, b. It is evident from Fig. 6 that the real and imaginary part of dielectric permittivity fall sharply with the increase of frequency [7, 8] but have weak variation at higher frequency. Because at lower-frequency electronic, ionic, interfacial and orientational polarization [9–11] give contribution to the dielectric permittivity. The dipoles cannot follow the applied electric field for rapid frequency variation, and the orientational polarization is stopped at high frequency. As a result, the dipole oscillation lags behind the applied field, which causes a weak variation of $\varepsilon'(f)$ and $\varepsilon''(f)$ at higher frequency. Figure 7 shows the temperature variation of dielectric loss of 10% Nd-doped sample for different frequencies, and it can be concluded that the dielectric loss of the sample increases with temperature for a particular frequency. For a fixed temperature, dielectric loss increases when the frequency decreases.

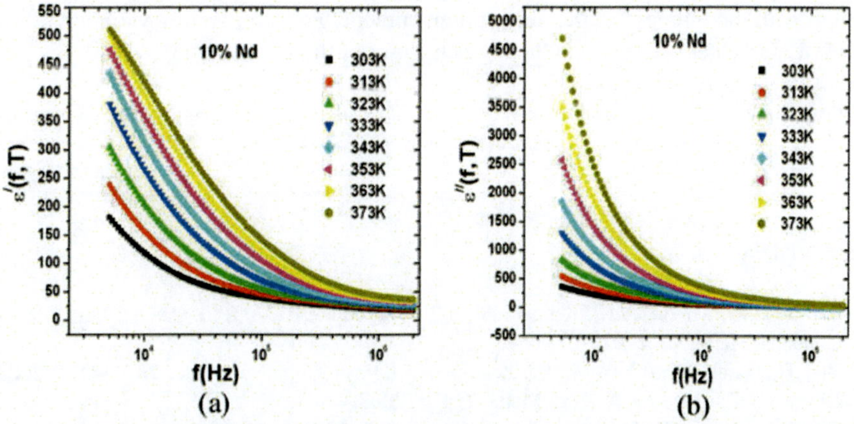

Fig. 6 **a** Frequency variation of the real part of dielectric permittivity for different temperatures for 10% Nd-doped sample. **b** Frequency variation of the imaginary part of dielectric permittivity for different temperatures for 10% Nd-doped sample

4 Conclusions

Neodymium-doped yttrium chromite nanoparticles are synthesized by the sol–gel method. X-ray diffraction patterns and the TEM analysis confirm the nanocrystalline nature of the samples. The optical band gap is measured using UV–Vis absorption spectroscopy data, and it is observed that the optical band gap decreases when the doping concentration of neodymium increases. At high frequency, weak variation of dielectric

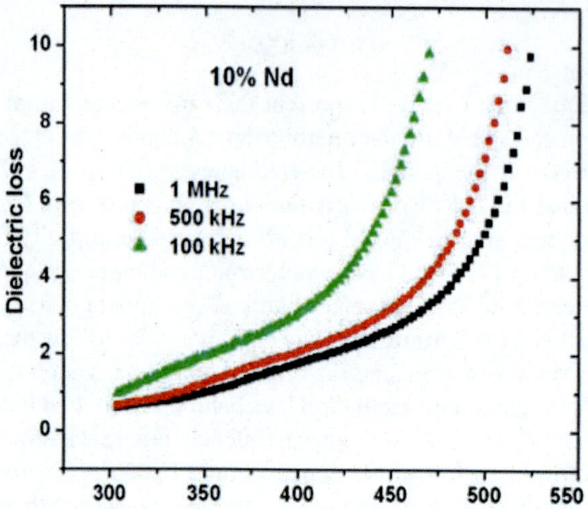

Fig. 7 Temperature variation of dielectric loss of the 0% Nd and 10% Nd-doped samples

permittivity has been observed. The resistivity of the sample decreases by a significant amount with the increase of the neodymium content. The temperature variation of DC resistivity indicates the semiconducting behaviour of the samples.

Acknowledgements. The authors are very much thankful to Asansol Engineering College and NIT, Durgapur, for all kinds of support during the work.

References

1. Serrao, C.R., Kundu, A.K., Krupanidhi, S.B., Waghmare, U.V., Rao, C.N.R.: Phys. Rev. B Condens. Matter **72**, 220101 (2005)
2. Bea, H., Gajek, M., Bibes, M., Barthelemy, A.: J. Phys. Condens. Matter **20**, 434231 (2008)
3. Bibes, M., Barthélémy, A.: Nat. Mater. **7**, 425 (2008)
4. Gupta, K., Jana, P.C., Meikap, A.K.: Solid State Sci. **14**, 324 (2012)
5. Sinha, R., Basu, S., Meikap, A.K.: Phys. E **69**, 47 (2015)
6. Shkir, M., Abbas, H., Siddhartha, Khan, Z.R.: J. Phys. Chem. Solids **73**, 1309 (2012)
7. Cross, L.E.: Relaxor ferroelectrics. Ferroelectrics **76**, 241–267 (1987)
8. Smolenskii, G.A., Isupov, V.A., Agranovskaya, A.I., Popov, S.N.: Sov. Phys. Solid State **2**(11), 2584 (1961)
9. Zhao, Y., Liu, X.Q., Wu, J.W., Wu, S.Y., Chen, X.M.: J. Alloys Compd. **729**, 57 (2017)
10. Correa, M., Kumar, A., Katiyar, R.S.: Appl. Phys. Lett. **91**, 082905 (2007)
11. Li, K., Zhu, X.L., Liu, X.Q., Chen, X.M.: J. Appl. Phys. **114**, 044106 (2013)

An Approach to Geometric Modeling Using Genetic Programming

Snigdhajyoti Ghosh(✉), Damodar Goswami, and Chira Ranjan Datta

Department of Electronics and Communication Engineering, Netaji Subhash Engineering College, Kolkata 700152, India

snigdhajyoti.ghosh@gmail.com, goswami.damodar@gmail.com, crdatta@gmail.com

Abstract. In this work, we 'derived' the famous Pythagorean theorem from the measurements of the sides of right-angled triangles with machine learning. In classical Euclidean geometry, this result is proved with rigorous geometrical argument, but we have followed a data-driven approach and got the same result without entering a single step in the domain of geometry. We used symbolic regression with genetic programming to reach the model. As far as our knowledge goes, this result is a novel one and may open up a new avenue of applying machine learning tool in geometry. We have used Python programming language 3.7 and libraries such as DEAP (v1.2) and pygraphviz. The whole project can be found on https://github.com/snigdhasjg/Pythagorean-Triplate.git.

Keywords: Genetic programming · DEAP · Genetic algorithm · Pythagorean triple

1 Introduction

Genetic programming is a machine learning tool that is driven by the evolutionary principle. Based on that principle, it tries to find patterns in the data automatically without human intervention.

Symbolic regression is a type of regression analysis that searches the space of mathematical expressions to find the model that best fits a given dataset, both in terms of accuracy and simplicity.

First, we started experimenting with 'GPLAB', a MATLAB-based software developed by Sara Silva. But eventually, we moved to Distributed Evolutionary Algorithms in Python abbreviated as 'DEAP'.

Our motive is to examine whether we can 'prove' the celebrated Pythagoras theorem without having the domain knowledges such as the properties of triangles, axioms of geometry and way of geometrical inferencing , just from the numerical measurements of the sides of triangles applying a machine learning algorithm. We have used genetic programming for this purpose.

© Springer Nature Singapore Pte Ltd. 2021
N. R. Das and S. Sarkar (eds.), *Computers and Devices for Communication*, Lecture Notes in Networks and Systems 147,
https://doi.org/10.1007/978-981-15-8366-7_13

1.1 Genetic Algorithm and Related Works

Genetic algorithm (GA) is a bio-inspired optimization technique, first introduced by Holland [1, 3] and widely used in engineering design [2]. The algorithm first creates a random population, evaluates its fitness for selection and uses two genetic operators 'crossover' and 'mutation' on those selected individuals to generate a new population.

1.2 Genetic Programming and Related Works

Genetic programming is a machine learning tool that is driven by the evolutionary principle. Based on that principle, it tries to find patterns in the data automatically without human intervention.

GP has been successfully used as an automatic programming tool, a machine learning tool and an automatic problem-solving engine. John R. Koza mentions 76 instances where genetic programming has been able to produce results that are competitive with human-produced results [4].

1.3 Symbolic Regression and Related Works

Symbolic regression is a type of regression analysis that searches the space of mathematical expressions to find the model that best fits a given dataset without any prior assumption on the model structure. Genetic programming is widely used for this purpose [5, 6].

2 Methodology

We have a set of right-angled triangles and length of each sides of those triangles. So, in numerical terms, we have a set of triplets and we have to extract the relationship among the three sides represented by the triplets.

2.1 Configuration of DEAP Software

2.1.1 Input Preparation

We start with 200 triplets of (a, b, c) where (a, b, c) are the three sides of a right-angled triangle with the hypotenuse c. The Pythagorean theorem tells us that the relation between these three variables is $c^2 = a^2 + b^2$. We try to see whether the same relation can be found out with genetic programming only. For this purpose, we tried to establish the equation as $c^2 - \left(a^2 + b^2\right) = 0$, so technically, we are searching for a function of (a, b, c) where the function returns 0. But the evaluation process does not know much about what function should it return rather than a function which has fitness score near a threshold (in this case, it is 0). So, it was returning a function like $f(a, b, c) = a - a$ or $f(a, b, c) = b - b$, etc. This type of function is not acceptable because with any value of a, b or c, it always returns 0.

Then we moved on to $c = f(a, b)$. Here, the inputs are 'a' and 'b', and for every unique set of (a, b), we are getting a different 'c'. The fitness criteria are to match $f(a, b)$ with respective 'c'.

2.1.2 Creating Primitive Set

Primitive set refers to a set of operators used to construct the output function. So, a set of basic math operators (addition, subtraction, multiplication, division, etc.) can be used as the primitive set. The leaf or terminal node consists of random numbers and input variables.

We have used strongly typed GP where every primitive and terminal are assigned a specific type. The output type of a primitive must match the input type of another one for them to be connected. For example, if a primitive returns a Boolean, it is guaranteed that this value will not be multiplied with a float if the multiplication operator operates only on floats.

Choosing primitive set is one of the most crucial aspects of a GP. In this problem, we use a classical set of primitives, which are basic arithmetic functions (i.e., addition, subtraction, multiplication, division and power). We have created our own division for overcoming 'ZeroDivisionError' and power for taking rational exponents and added a 'Terminal' of multiple value, i.e., called 'EphemeralConstant'.

2.1.3 Preparation of Toolbox and Statistics

As any evolutionary program, symbolic regression needs (at least) two object types, an individual containing the genotype and a fitness. Genotype in simple word is each node of a tree where the tree itself is an individual. An individual should have set of rules and set of parameters. Next, we added 'Toolbox' for evolution that contains the evolutionary operators (i.e., selection, crossover and mutation). With 'Toolbox', we can register how a selection, crossover and mutation will happen and can decorate as needed. Fitness function evaluates how close a given solution is to the optimum solution of the desired problem. It determines how fit a solution is. Each problem has its own fitness function. To evaluate fitness, we used mean square error (MSE). For statistical calculations, we used the module 'NumPy'.

2.1.4 Launching the Evolution

At this point, DEAP had all the information needed to begin the evolutionary process, but nothing has been initialized. We started the evolution by creating the population and then calling a complete algorithm.

2.2 Extension of DEAP Software

We needed to modify the software to modulate the stopping criteria.

Sometimes, it becomes difficult to achieve global optima with simple GP algorithm. A basic multi-start procedure can help GP improve the probability of jumping out of the local optima and finding the global optimal solution [7]. This procedure starts the algorithm multiple times and then picks the best solution among those found over all runs [8–10].

2.3 Experimentation

We started with 50 data points, but the success rate was very low. It was taking too much time to reach the optima, and most of the time, it converged to local optima instead of finding the global one. The probability of jumping out from local optima was low because the number of input points was not enough in number.

So, then, we moved on to 500 points. In this case, the computation time for each generation became very high. After many runs, we figured out that for number of points around 200–300, the algorithm was not taking too much time for computation and able to jump out from local optima as it was able to see more vector space. We took minimal approach to search for solution. We decorate the mate and mutate method to limit the size (size can be expressed as the number of nodes in the tree representation) of generated individuals.

This is done to avoid an important drawback of genetic programming: bloat [11, 12]. Koza in his book on genetic programming suggests to use a max depth of 17 [13]. So, we started from tree depth 3 as this is the minimal tree needed for this problem. Then, we gradually move on to 4 and 5. Then, we stick to 5 because larger tree depth leads to more complex tree. Also, this has the dual benefits of providing the simplest/smallest solutions and preventing GP bloat, thus shortening run times. GP search should be limited to program lengths that are within the limit and that can achieve optimum fitness [14]. In Fig. 1, the maximum tree depth is 17, and it runs for 10,000 generation and finds a solution which has fitness value 0.0117 near our optimum fitness, i.e., less than 0.01, but the tree became very complex. The average tree size over 10,000 generation is 27.997. On the other hand, in Fig. 2, the maximum tree depth is 5, and it runs 811 generation and finds the solution. The average tree size over 811 generation is 10.511. This height limit leads to less complex tree which allows to get simples models.

We fixed a resetting point after 20 generation. As we observed the maximum time, the fitness stuck around 10th generation (In Fig. 3, the model converges after 10th gen).

So, every after 20th generation, the population gets reset. With that, we got success in finding solution.

In Fig. 4, we clearly see the effects of selection, crossover and mutation in the new population. We have used tournament-based selection (tournament size 5) with 80% crossover probability where crossover happens via exchanging subtree with the point as root between each individual. Uniform mutation happens at rate of 20%. As a result, almost 85% of the population changes in the next generation.

3 Results

With this approach, we have achieved 8 successful runs from 10 runs. And got results like $c = \sqrt{A^2 + B^2}$, $c = \sqrt{A^2 + B^2 - B^{-7}/5}$, $c = \sqrt{A^2 + B^2 + \frac{B-B}{A}}$, etc. All of these equations closely represent the Pythagorean theorem. The program took around 1 h and 32 min for 10 runs.

Detailed output mentioned in Sect. 4.1 also can be found in GitHub repository.

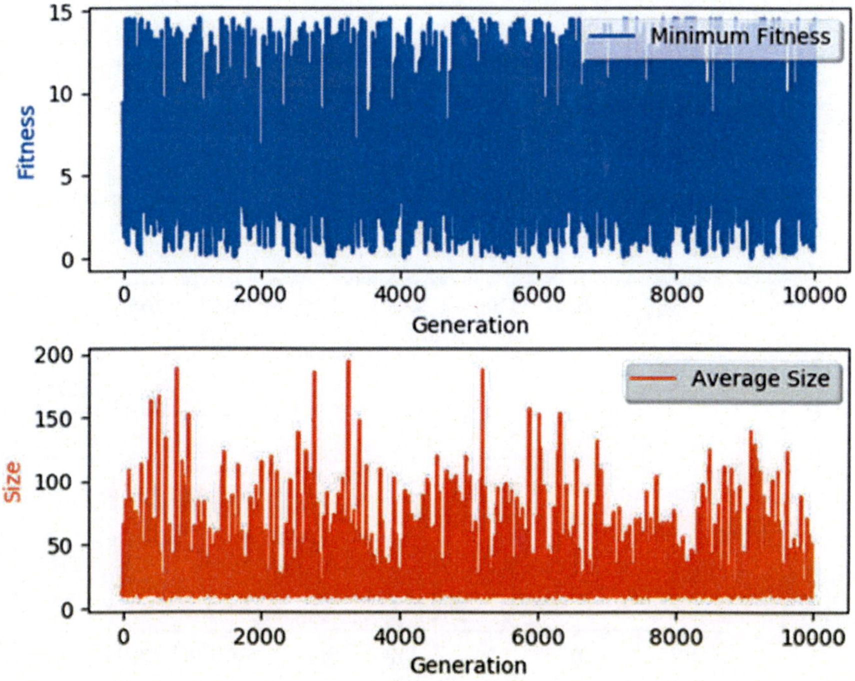
15
10
5
0
Fitness
Minimum Fitness
0
2000
4000
6000
8000
10000
Generation
200
150
100
50
0
Size
Average Size
0
2000
4000
6000
8000
10000
Generation

Fig. 1 .

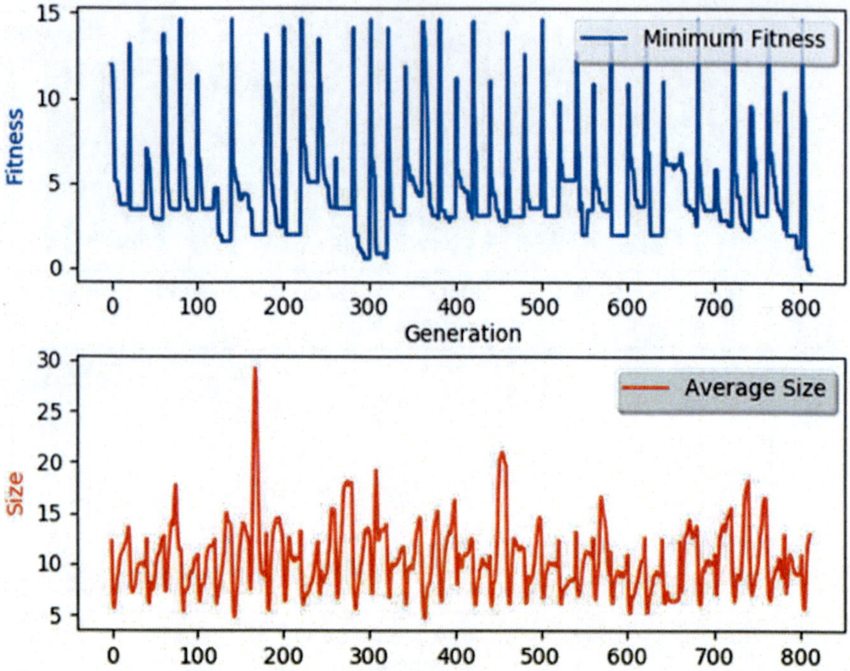

Fig. 2 .

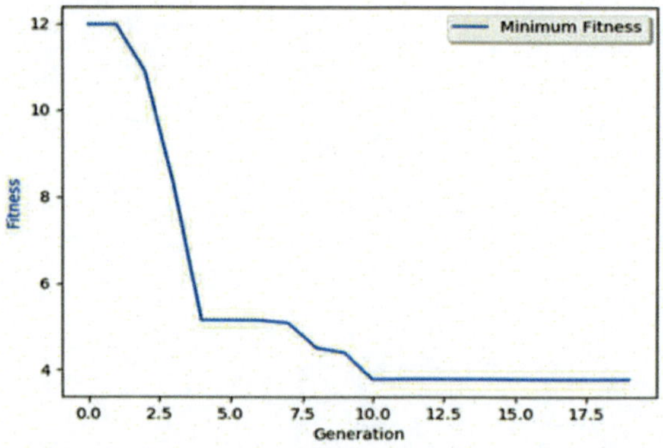

Fig. 3 Generation versus fitness

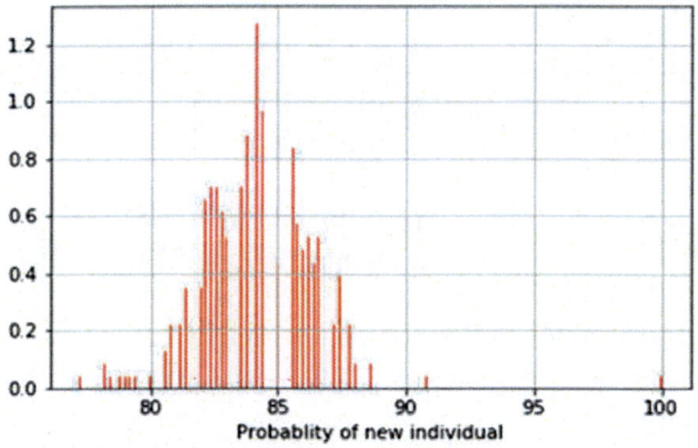

Fig. 4 Effects of selection crossover and mutation

Appendix

The algorithm gives output as string representation of tree like $power(add(mul(A, A), mul(B, B)), 1, 2)$ that can be expressed as $\sqrt{A^2 + B^2}$. In Fig. 5, the same equation is represented as tree.

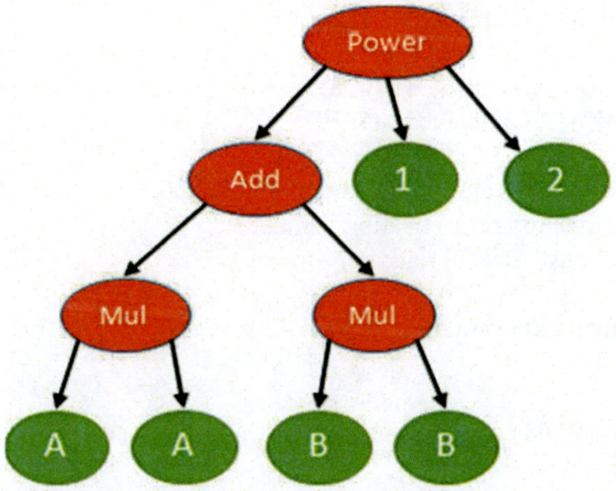

Fig. 5 A typical tree structure of Pythagorean equation

Not every time this algorithm gives the same result. Some non-trivial and apparently complicated trees which finally leads to the Pythagorean theorem are:

- power(add(mul(A,A),mul(B,B)),1,2)
- power(add(mul(B,B),mul(A,A)),2,4)
- power(add(mul(B,B),mul(A,A)),3,6)
- power(add(mul(B,B),mul(A,A)),4,8)
- power(add(mul(A,A),mul(B,B)),5,10)

- power(add(mul(A,A),power(B,8,4)),1,2)
- power(add(power(A,8,4),mul(B,B)),2,4)
- power(add(power(A,4,2),mul(B,B)),5,10)
- power(add(sub(mul(A,A),sub(A,A)),mul(B,B)),1,2)
- power(add(sub(A,A),add(mul(A,A),mul(B,B))),1,2)
- power(add(mul(power(B,4,2),safe_div(A,A)),add(mul(A,A),sub(B,B))),4,8)
- power(sub(add(mul(A,A),sub(A,B)),sub(sub(A,B),mul(B,B))),5,10)
- power(add(mul(B,B),mul(A,power(power(A,4,3),3,4))),3,6)
- power(sub(mul(add(B,A),A),sub(mul(B,A),mul(B,B))),2,4)
- add(power(add(add(mul(B,B),sub(A,A)),mul(A,A)),2,4),safe_div(A,add(mul(B,B), mul(power(B,10,10),A))))
- power(add(add(mul(A,A),mul(B,B)),safe_div(add(B,A),power(B,5,1))),4,8)

Run Results

We have achieved 8 successful runs out of 10 test runs. The results are as follows.

1. power(add(mul(A, A), mul(B, B)), 1, 2)
2. power(add(mul(B, B), mul(A, A)), 3, 6)
3. power(sub(mul(B, B), sub(safe_div(power(B, 3, 5), mul(B, B)), mul(A, A))), 4, 8)
4. power(sub(add(mul(A, A), mul(B, B)), safe_div(sub(B, B), A)), 2, 4)
5. power(add(mul(A, A), add(sub(B, B), mul(B, B))), 4, 8)
6. power(mul(add(mul(A, A), power(B, 6, 3)), add(mul(A, A), power(B, 6, 3))), 2, 8)
7. power(add(mul(A, A), mul(B, B)), 4, 8)
8. power(add(mul(A, A), mul(B, B)), 5, 10)

All these generated trees are having fitness value near 0, i.e., 0.0407. This small error is due to choosing input points as floating number.

And the other 2 unsuccessful runs are:

1. add(B, safe_div(add(power(A, 5, 6), mul(A, A)), add(safe_div(mul(A, A), add(A, B)), add(B, B))))

   which has fitness value of 3.795.

2. add(B, safe_div(safe_div(add(mul(B, A), B), safe_div(B, A)), add(add(A, B), safe_div(mul(B, B), add(B, A)))))

   which has fitness value of 3.816.

References

1. Holland, J.H.: Adaptation in Natural and Artificial Systems.: University of Michigan Press, Ann Arbor (1975)
2. Gen, M., Cheng, R.: Genetic Algorithms and Engineering Design. Wiley, New York (1997)
3. Holland's schema theorem

4. Human-competitive results produced by genetic programming by John R Koza
5. Smits, G., Kotanchek, M. : Pareto-front exploitation in symbolic regression. In: Genetic Programming Theory and Practice II. Springer, Ann Arbor, pp 283–299 (2004)
6. Martí, R., Resende, M.G.C., Ribeiro, C.C.: Multi-start methods for combinatorial optimization. Eur. J. Oper. Res. **226**(1), 1–8 (2013)
7. Dao, S.D., Abhary, K., Marian, R.: An Adaptive Restarting Genetic Algorithm for Global Optimization
8. Vladislavleva, E.J., Smits, G.F., Hertog, D.: Order of nonlinearity as a complexity measure for models generated by symbolic regression via pareto genetic programming. In: IEEE Trans. Evolut. Comput. **13**(2), 334 (2009)
9. Kessaci, Y., et al.: Parallel evolutionary algorithms for energy aware scheduling. In: Bouvry, P., González-Vélez, H., Kołodziej, J. (eds.) Intelligent Decision Systems in Large-Scale Distributed Environments, pp. 75–100. Springer, Berlin (2011)
10. Dao, S.D., Abhary, K., Marian, R.: Optimisation of partner selection and collaborative transportation scheduling in virtual enterprises using GA. Expert Syst. Appl. **41**(15), 6701–6717 (2014)
11. Purohit, A., Choudhari, N.S., Tiwari, A.: Code Bloat Problem in Genetic Programming
12. Trujillo, L., Naredo, E., Martínez, Y.: Preliminary Study of Bloat in Genetic Programming with Behaviour-Based Search
13. Koza, J.R.: Genetic Programming: On the Programming of Computers by Means of Natural Selection (Chap. 6), p 114
14. Dignum, S., Poli, R.: Operator Equalisation and Bloat Free GP

Check for updates

Detecting Different Emotional States of Human Brain Using Bio-potential Signals

Prithwijit Mukherjee(✉) and Anisha Halder Roy

Institute of Radio Physics and Electronics, University of Calcutta, Kolkata, India
mukherjeprithwijit@gmail.com, halder.anisha@gmail.com

Abstract. The essence of this paper is to design a mechanism for detecting emotional state of person using different bio-potential signals like electroencephalogram (EEG) signals of frontal lobe, pulse rate and S_pO_2. We record EEG signals of Fp1, Fp2, F3 and F4 electrodes, pulse rate and S_pO_2 of thirty subjects and extract twenty-two features from the recorded bio-potential signals. We design a *k*-nearest neighbor (*K*NN) classifier model for predicting the emotional state of a person. The designed *K*NN classifier model is trained with the extracted feature values of the thirty subjects. We again record the same bio-potential signals of ten new subjects and extract features. These extracted feature values are used for validating the performance of the trained *K*NN classifier model. The obtained overall efficiency of our designed emotion detection mechanism is 95.4%.

Keywords: Emotion detection · EEG · S_pO_2 · Pulse rate · *K*-nearest neighbor (*K*NN)

1 Introduction

Brain–computer interface (BCI) is one of the major applications of artificial intelligence [1–5]. At present, BCI is the most explored research field by the researchers. External hardware devices can be supervised by using brain signals of humans with the help of BCI techniques [1, 4, 5]. BCI can be used for understanding the emotional states of a human being using computers. In this paper, we aim to find the emotional state of a person using electroencephalogram (EEG) signal, pulse rate and peripheral capillary oxygen saturation (S_pO_2). We classify the emotional state of a person in five categories, i.e., happy, sad, relaxed, angry and scared. We use video stimuli for eliciting different emotions, i.e., happy, sad, relaxed, anger and scared in the mind of the subjects. We record EEG signals of Fp1, Fp2, F3, F4 electrodes, pulse rate and S_pO_2 values of thirty subjects while they are subjected to stimuli. After that, the features are extracted and processed from the recorded EEG signals, pulse rate and S_pO_2. A *k*-nearest neighbor classifier model is designed and trained with the extracted feature values. The trained *K*NN classifier model is capable of predicting the emotional state of a person.

We again process the same features extracted from the recorded EEG signals of Fp1, Fp2, F3, F4 electrodes, pulse rate and S_pO_2 values of ten new subjects while they are

© Springer Nature Singapore Pte Ltd. 2021
N. R. Das and S. Sarkar (eds.), *Computers and Devices for Communication*,
Lecture Notes in Networks and Systems 147,
https://doi.org/10.1007/978-981-15-8366-7_14

watching the same video stimuli. These extracted feature values are used for validating the performance of the trained *K*NN classifier model.

The paper is ordered as follows. Section 2 describes our proposed work in detail. Section 3 explains *k*-nearest neighbor (*K*NN) classifier. Section 4 illustrates the experimental details. Section 5 illustrates the results and discussion. Section 6 concludes the paper.

2 Proposed Work

In this paper, we wish to design a mechanism for finding the present emotional state of a person using electroencephalogram (EEG) signal, pulse rate and S_pO_2 values. At first, we prepare video stimuli for eliciting different emotions such as happy, sad, relaxed, anger and scared in the mind of the subjects. In our experiment, two different video clips are used for eliciting each of the emotional conditions (i.e., happy, sad, angry, relaxed and scared) in the mind of the subjects. We select total thirty subjects in the age group of 18 to 35 years. We record the electroencephalogram (EEG) signals of Fp1, Fp2, F3, F4 electrodes, pulse rate and S_pO_2 values of the subjects while they are subjected to the stimuli. After that, we process the recorded EEG signals, pulse rate and S_pO_2 values. We extract twelve features such as power spectral density (PSD) of delta wave, PSD of theta wave, PSD of alpha wave, PSD of beta wave, mean, standard deviation, kurtosis, Hjorth activity, skewness, maximum amplitude and minimum amplitude and root mean square from the processed EEG signals.

We extract five features such as maximum value, minimum value, standard deviation, kurtosis and skewness from the recorded values of pulse rate. We extract five features, i.e., maximum value, minimum value, standard deviation, kurtosis and skewness from the recorded S_pO_2 values.

We design a *K*NN classifier model for detection of the present emotional state of a person. The designed *K*NN model is trained using the extracted feature values of the thirty subjects. Figure 1 illustrates the training process of the designed *K*NN classifier model. Figure 2 shows the recording of EEG signals, pulse rate and S_pO_2 of a subject.

After that, we select ten new healthy subjects for validating the performance of the trained *K*NN classifier model. We record the EEG signals of Fp1, Fp2, F3, F4 electrodes, pulse rate and S_pO_2 values of the subjects. Similarly, we process and extract features from the recorded bio-potential signals of the subjects. The extracted feature values of these ten subjects are used for validating the performance of the trained *K*NN classifier model.

3 *K*-Nearest Neighbor

K-nearest neighbor (*K*NN) algorithm is a supervised machine learning algorithm. We can determine the class of an unknown data point using *K*NN algorithm. Initially, some labeled data is given. At first, Euclidean distance of the unknown data point from the known labeled data points is computed. After that, we select *K* number of nearest neighbors. Nearest neighbors are selected depending on the computed Euclidean distance. The class of the unknown point is the class that contains most number of neighbors.

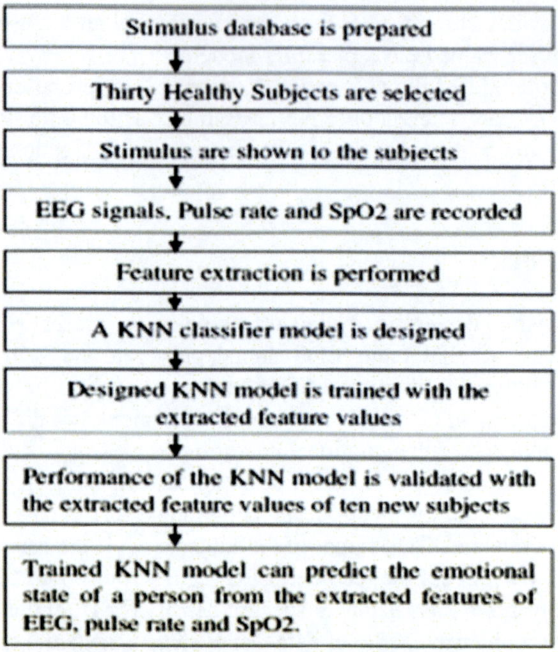

Fig. 1 Training process of our *K*NN model

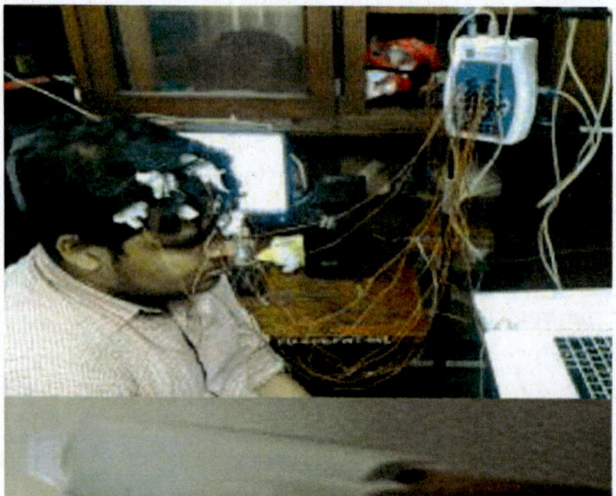

Fig. 2 Recording of EEG, pulse rate and S_pO_2

4 Experimental Details

In our experiment, we first select thirty healthy subjects in between 18 and 30 years of age group. Then, we prepare a database of ten movie clips and use them as stimulus in our experiment. These movie clips rouse five different emotions such as happy, sad, relaxed, anger and scared in the mind of the subjects. Two movie clips are used for rousing one emotion in the subject's mind. Each of the prepared movie clips is shown to the subjects on different days. We record the EEG signals of Fp1, Fp2, F3 and F4 electrodes, pulse rate and S_pO_2 values of the subjects while they watch the prepared video stimuli. A 32-channel EEG machine is used for recording the EEG signals of the subjects. Impendence level of the EEG electrodes is kept below 10 K while recording the EEG signal. We record pulse rate and S_pO_2 value of the subjects using RMS Relax 701. So, our experimental data contains EEG data, pulse rate and S_pO_2 values of the subjects while they are happy, sad, relaxed, angry and scared.

After that, we process the recorded EEG signals of Fp1, Fp2, F3 and F4 electrodes, pulse rate and S_pO_2 values of the subjects. Totally, twenty-two features are extracted from the processed EEG signals, pulse rate and S_pO_2 values. We extract four frequency domain features like power spectral density (PSD) of delta wave, PSD of theta wave, PSD of alpha wave, PSD of beta wave and eight time domain features such as mean, standard deviation, kurtosis, Hjorth activity, skewness, maximum amplitude and minimum amplitude and root mean square from the processed EEG signals of Fp1, Fp2, F3 and F4 electrodes. We extract five time domain features from the recorded pulse rate and S_pO_2 values. The extracted five time domain features of the pulse rate are maximum value, minimum value, standard deviation, kurtosis and skewness. The extracted five time domain features from the recorded S_pO_2 values are maximum value, minimum value, standard deviation, kurtosis and skewness. Tables 1, 2 and 3 show some extracted values of EEG signals, pulse rate and S_pO_2 (of thirty subjects), respectively.

A *k*-nearest neighbor classifier model is designed in MATLAB 18b for prediction of emotional state of a person. The designed *K*NN model is trained with the extracted EEG features, pulse rate features and S_pO_2 features of the thirty subjects. Designed *K*NN classifier model is trained with 15,360 labeled data. Each labeled data consists of twenty-two feature values. These 15,360 labeled data, which are used for training purpose, consist of 5120 labeled data for each of the emotional states. Trained *K*NN classifier model is capable of predicting the present emotional condition of a person.

In the next step, ten new healthy subjects are selected for validating the performance of the trained *K*NN model. Experimental setup is shown in Fig. 3. We record the EEG data of Fp1, Fp2, F3, F4 electrodes, pulse rate and S_pO_2 while they are subjected to video stimuli. Then, we process the recorded EEG signals, pulse rate and S_pO_2 values of the ten new subjects. Using similar process, we extract twelve features from the processed EEG signals, five features from the recorded pulse rate and five features from the recorded S_pO_2 of the subjects. These extracted twenty-two feature values of the ten new subjects are used for computing the efficiency of the trained *K*NN classifier model.

Table 1 Some extracted features of EEG signal

PSD (δ)	PSD (θ)	PSD (α)	PSD (β)	Mean	SD	Kurt	Skew	Min	Max	RMS	Hjorth activity	Emotion
316.331	82.795	14.53	3.12	1.53	53.77	1.27	0.791	−20.5	167	32.61	1120.4	Angry
964.04	111.04	21.43	4.11	62.53	69.52	−0.59	0.93	−32.5	174	186.2	2144.8	Scared
875.5	106.11	25.42	9.06	963	192	8.13	2.59	0.0625	1105	126.2	1781.1	Happy
869.47	181.07	24.43	4.11	62.53	59.36	−0.59	0.93	−32.5	180.5	257.4	2564.8	Relaxed
977.5	174.11	27.42	9.06	963	784	6.1	2.51	0.0475	1711	182.6	1647.1	Sad

Table 2 Some extracted features of pulse rate

SD	Kurt	Skew	Min	Max	Emotion
19.6	−2.04	0.103	105	135	Angry
7.56	0.60	0.746	92	124	Scared
3.48	−1.5	0.077	71	89	Happy
4.51	−1.43	0.062	64	76	Relaxed
5.09	−0.552	0.506	55	72	Sad

Table 3 Some extracted features of S_pO_2

SD	Kurt	Skew	Max	Min	Emotion
0.487	−8.42	−9.48	99	98	Angry
0.534	−2.87	0.275	97	96	Scared
0.536	−2.4	−2.81	96	95	Happy
0.542	−2.57	−2.73	100	97	Relaxed
50.63	−2.7	0.288	96	94	Sad

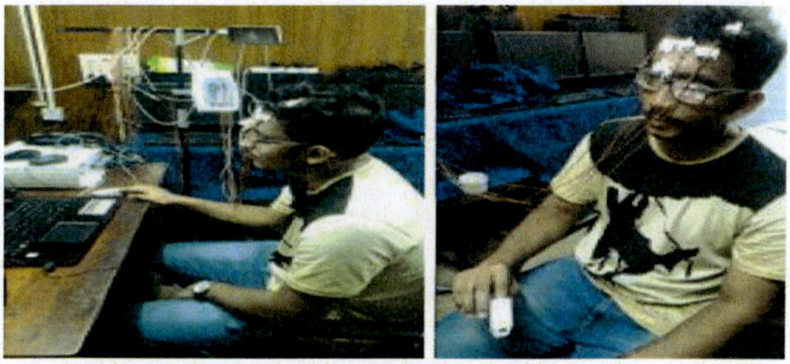

Fig. 3 Experimental setup

5 Results and Discussion

We have conducted our experiment in a tranquil environment. A trained k-nearest neighbor classifier model is used for predicting the emotional state of a person. The designed KNN classifier has considered eleven nearest data points for classifying the emotional state of a person. Extracted EEG, pulse rate and S_pO_2 features of the new ten subjects are used for validating the performance of the trained KNN classifier model. Totally, 5120 labeled data are used for validation purpose. These 5120 labeled data, which are used for validation purpose, consists of 1024 labeled data for each of the emotional states (i.e., happy, sad, scared, relaxed and angry).

The trained *K*NN model classifies 1024 samples of 'angry' emotion with 99% efficiency, 1024 samples of 'scared' emotion with 100% efficiency, 1024 samples of 'happy' emotion with 78.8% efficiency, 1024 samples of 'relaxed' emotional state with 98.4% efficiency and 1024 samples of 'sad' emotion with 99.9% efficiency. The overall obtained efficiency of our trained stacked autoencoder network model is 95.6%. Figure 4 shows the prediction of emotional state of an unknown subject using our trained *K*NN model. Table 4 illustrates the confusion matrix of the validation result of the trained *K*NN model.

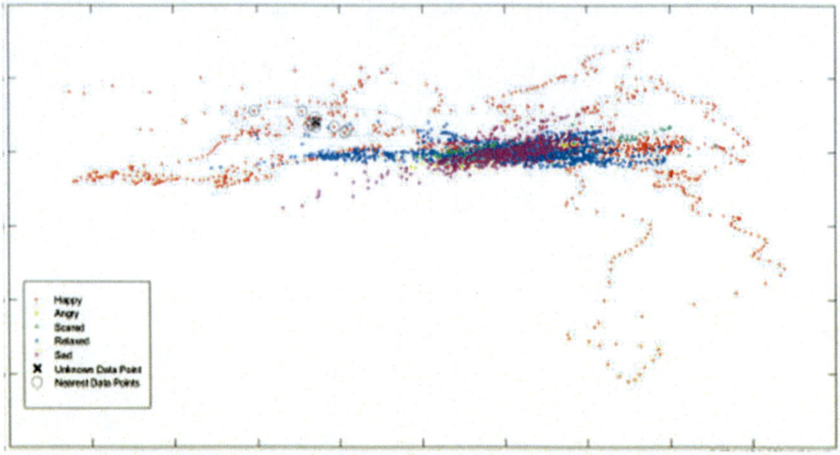

Fig. 4 Prediction of emotional state using the trained *K*NN model

Table 4 Confusion matrix

Emotion	Number of validation data	Number of correctly predicted data	Efficiency (%)
Angry	1024	1020	99.6
Scared	1024	1024	100
Happy	1024	807	78.8
Relaxed	1024	1008	98.4
Sad	1024	1023	99.9
Overall efficiency: 95.4%			

6 Conclusion

In this paper, we have designed a mechanism for detecting emotional state of a person. Our designed mechanism can classify human emotions in five categories, i.e., happy, scared, angry, relaxed and sad. We have used a trained *k*-nearest neighbor classifier model for detection of emotional state of a person. Our designed emotion detection mechanism can classify emotions such as 'angry,' 'scared,' 'happy,' 'relaxed' and 'sad' with an accuracy of 99.6%, 100%, 78.8%, 98.4% and 99.9%, respectively. The obtained overall efficiency of our designed emotion detection mechanism is 95.4%. In the future, we can add other bio-potential signals such as electromyography (EMG) signals and galvanic skin response (GSR) with our designed emotion detection mechanism for better performance.

References

1. Alhagry, S., Fahmy, A.A., El-Khoribi, R.A.: Emotion recognition based on EEG using LSTM recurrent neural network. IJACSA 8:355–358 (2017)
2. Kehri, V., Ingle, R., Patil, S., Awale, R. N.: Analysis of Facial EMG Signal for Emotion Recognition Using Wavelet Packet Transform and SVM, pp. 355–358 (2019)
3. Jerrittaa, S., Murugappana, M., Wana, K., Yaacoba, S.: Emotion recognition from facial EMG signals using higher order statistics and principal component analysis. J. Chin. Inst. Eng. **37**(3) (2013)
4. Ren, T., Ruan, H., Tao, W., Tao, D.: Video-based emotion recognition using multi-dichotomy RNN-DNN. ACII Asia (2018)
5. Jirayucharoensak, S., Pan-Ngum, S., Israsena, P.: EEG-based emotion recognition using deep learning network with principal component based covariate shift adaptation. Sci. World J. (2014)

A Comparative Study Between True Color and Grayscale Radar Imageries of Thundercloud

Sonia Bhattacharya[1](✉) and Himadri Bhattacharyya Chakrabarty[2,3]

[1] Department of Radio Physics and Electronics, University of Calcutta, Kolkata, India
sonia.rpe2020@gmail.com
[2] Jibantala Rokeya Mahavvidyalaya, Jibantala, India
himi0201@gmail.com
[3] Department of Computer Science, Surendranath Collage (On Lien), University of Calcutta, Kolkata, India

Abstract. Severe thunderstorm is the convective weather feature, originated from cumulonimbus cloud. It has devastating effect in the society. Accurate prediction with enough lead time is needed to make the people alert from such destructive event. In this study, analysis of grayscale cloud imageries and histograms has been used to nowcast severe thunderstorm.

Keywords: Severe thunderstorm · Convective cloud · Image processing · Histogram

1 Introduction

Thunderstorm is one of the most devastating types of mesoscale, convective weather phenomenon, generated from the cumulonimbus cloud. Severe squall storms can cause destructions of various dimensions, like uprooting of trees, electric polls, electro-cuttings, damages of weak structures and crops, blockage of roads and railway traffic, and water logging, and affect aviation system mostly [1]. Additional hazards that can occur include large hail, flash flooding associated with heavy rainfall, and destructions occurred by high wind like tornadoes and/or non-rotational (straight-line) wind [2, 3]. Correct prediction of thunderstorm is very much needed to inform the people with sufficient lead time because of the adverse socioeconomic impact associated with thunderstorms [3]. It happens in many subtropical places of the world [4]. In India, occurrences of severe thunderstorms are the common feature during pre-monsoon season consisting of generally March–April–May. Analysis of convective cloud imageries can give correct prediction of such severe weather event [5]. Cloud imageries are able to simulate the life cycle of each convective clouds [6]. Some papers [7, 8] concerned the propagation and regeneration of cells within a multi-cell storm.

Every thunderstorm cloud has a core region, a spreading anvil top, and an inflow–outflow region. The core is that part of the cloud where sustained strong up draughts of

© Springer Nature Singapore Pte Ltd. 2021
N. R. Das and S. Sarkar (eds.), *Computers and Devices for Communication*,
Lecture Notes in Networks and Systems 147,
https://doi.org/10.1007/978-981-15-8366-7_15

relatively warm and moist air condense to produce rain, hail and/or snow and associated downdraughts [9]. Underneath the core, we see a rain curtain, while above it, the tallest part of the thunderstorm can be found. The dark flat cloud base that extends away from the core (usually to the west or north) is called the flanking line or rain-free base, along which air fueling up draughts into the thunderstorm rises in successive cumulus towers [10]. A large number of research works have been done to analyze clouds based on satellite images [11]. Since the number of droplets of cloudy liquid and ice crystals increases in the cloud, reflectivity value increases, and this has a direct relation with the rain [12]. Color is an important feature for image representation [13]. Color moments, color histogram and coherent color vector are the important components of the extraction of color characteristics [13]. Any satellite picture of convective cloud fields has a great variety of cloud patterns and cloud sizes, [14]. The size distribution of convective clouds might give hints of the precipitation observed at ground stations [14].

In our previous study, it has been obtained density of the pixel which changes with formation of cloud. In our current study, the above-mentioned cloud imageries have been converted into gray scale and a comparative study is done.

2 Data

Here, in this study, ten images of a thundercloud having different stages from time to time are considered for analysis. All these images are taken over Kolkata (22.3° N/88.3° E) on April 10, 2005, from 07:12:04 to 14:12:04. The cloud imageries were obtained from the observations through Doppler Radar of Regional Meteorological Center, India Meteorological Dept., Kolkata.

3 Methodology

3.1 Image Processing

Here, in this study, ten GIF images have been taken into consideration for severe thunderstorm prediction. All these imageries are TRUE color image. These imageries have been converted into gray scale using ind2gray() function. The pixel values of two consecutive images have been obtained using impixel function which gives two to three columnar matrices. A comparison has been done between these two matrices of two consecutive imageries which revealed that there is a noticeable increase between their number of pixel values.

3.2 Observations on Cloud Imageries

As the first three images showed that formation of convective cloud, in which only three types of colors are observed, these are White, Black, and Gray. The above imageries revealed that with the formation of cloud, intensity of Black and Gray color increased. Our previous study [5] on the same color imageries showed that intensity of Red color increased from image to image (Figs. 1, 2, 3, 4, 5 and 6).

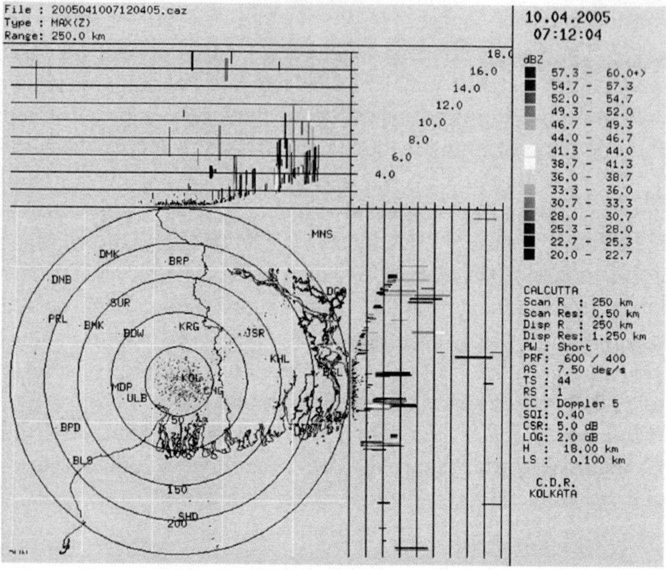

Fig. 1 First image of cloud formation

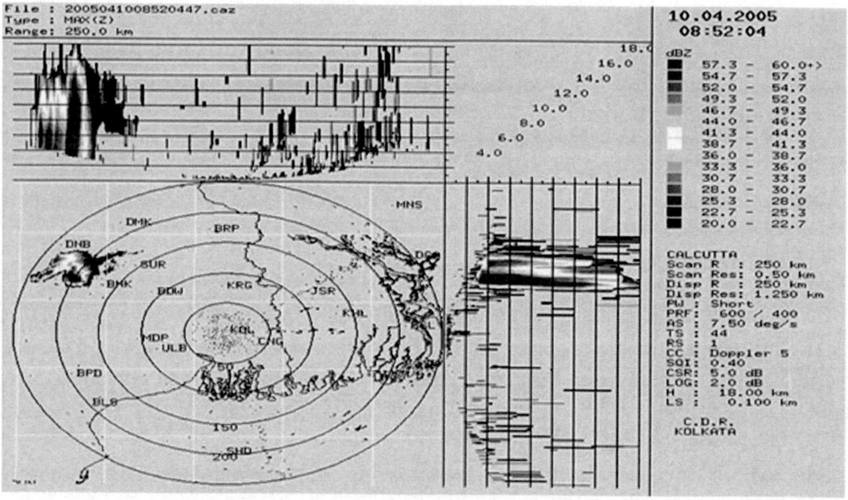

Fig. 2 Second image of cloud formation

In the fourth, fifth, and sixth images, it can be observed that density of Black color pixels has been increased noticeably and the cloud has been moved more toward Kolkata. It can be observed that from seventh image, the density of Black colored pixels was decreasing and Gray colored pixels increasing as it moves over Kolkata. Since Black colored pixel denotes water content of the cloud, reduction of Black color pixel density

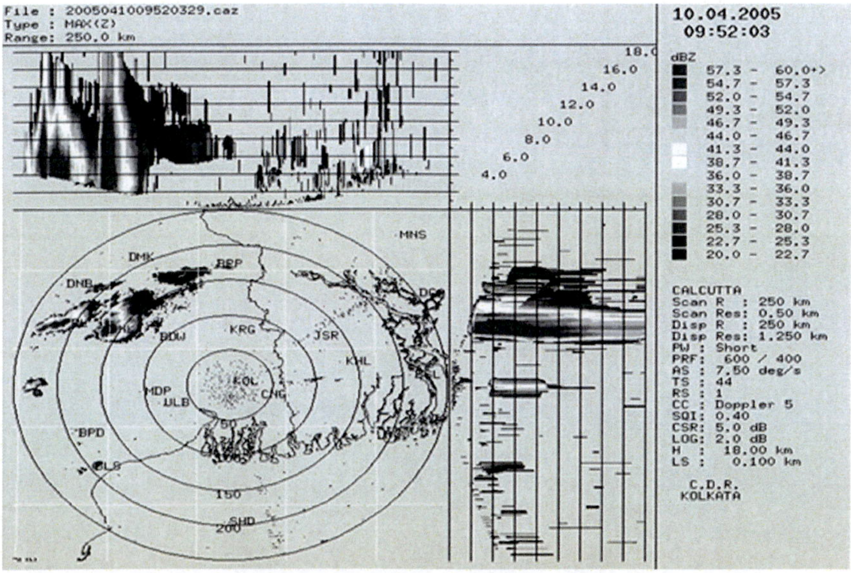

Fig. 3 Third image of cloud formation

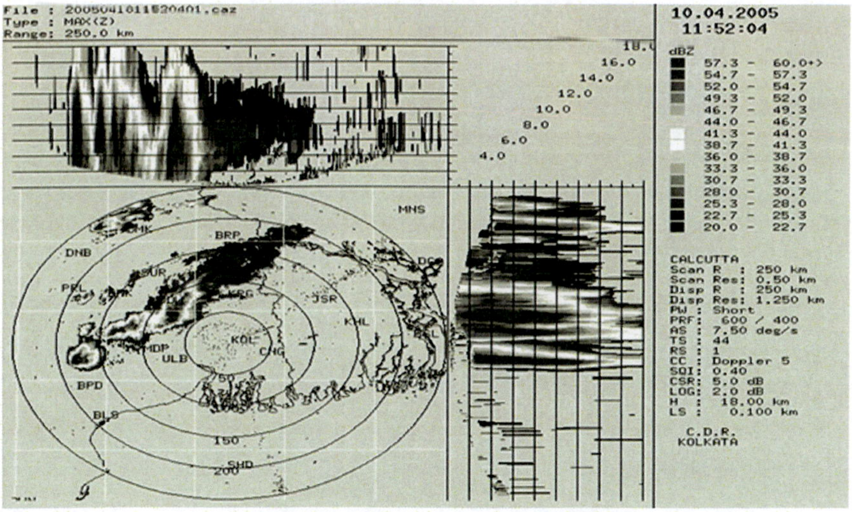

Fig. 4 Fourth image of cloud formation

and increase in Gray colored pixel intensity indicate toward the precipitation (Figs. 7 and 8).

It can be clearly observed that in ninth and tenth image, Black colored pixels were noticeably reduced leaving Gray colored pixels. This signifies that precipitation has been done (Figs. 9 and 10).

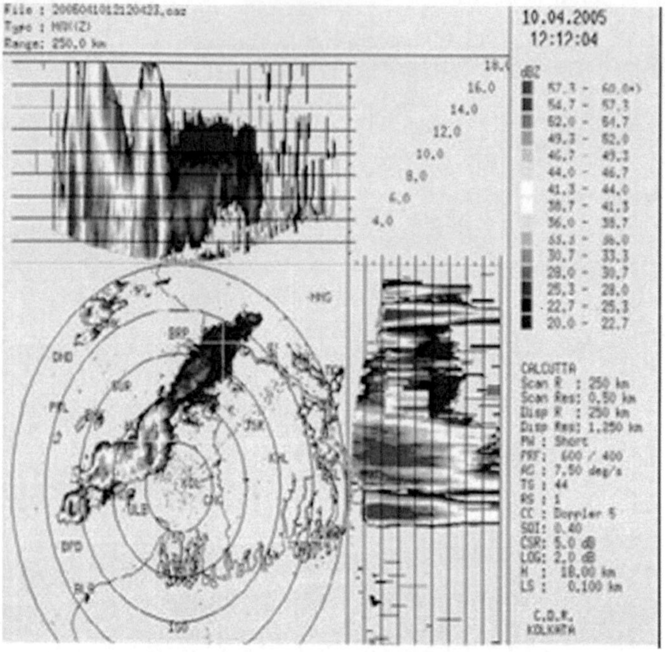

Fig. 5 Fifth image of cloud formation

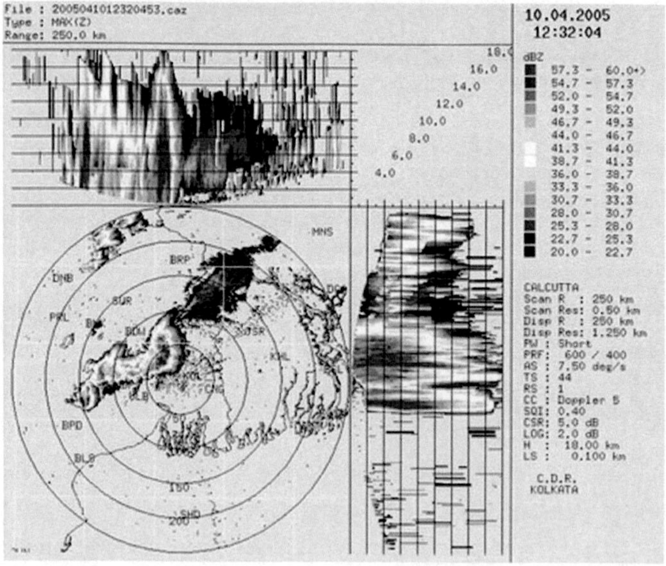

Fig. 6 Sixth image of cloud formation

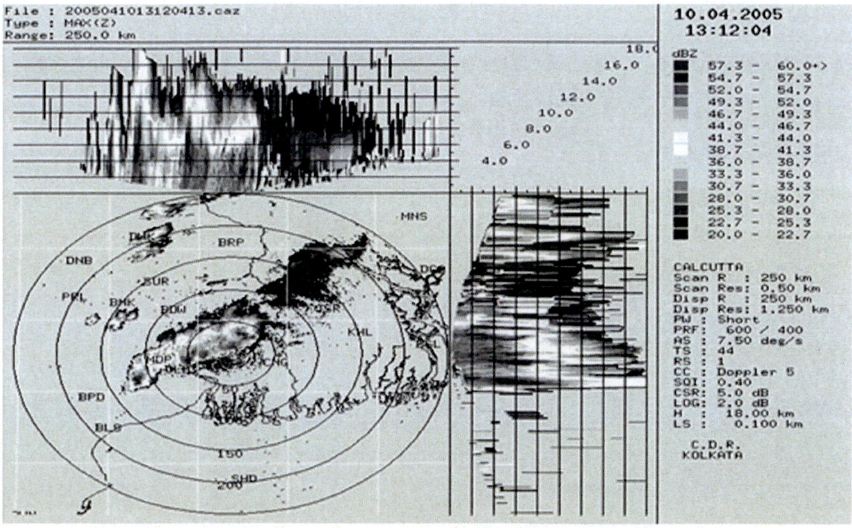

Fig. 7 Seventh image of cloud formation

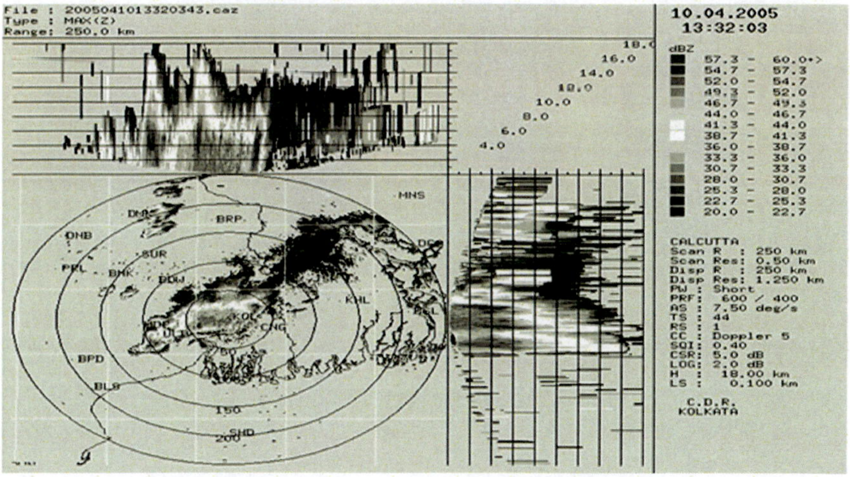

Fig. 8 Eighth image of cloud formation

4 Results

In our previous study [5] using impixel function, RGB value of each pixel has been obtained and a comparison between the RGB values of the pixels of two consecutive images yields the following result. In our present study, comparison of each pixel values between two consecutive grayscale images yields the following result.

It can be observed from the above table that pixel value decreases after precipitation for true color images. But, in case of grayscale image, pixel value increases even

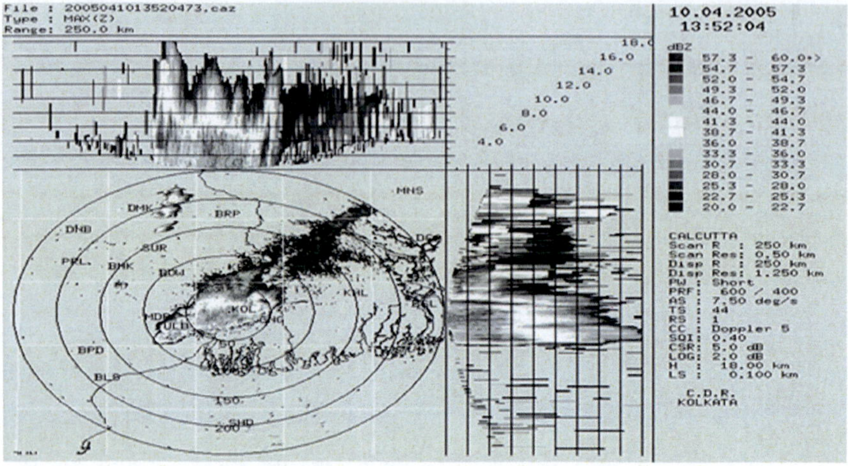

Fig. 9 Ninth image of cloud formation

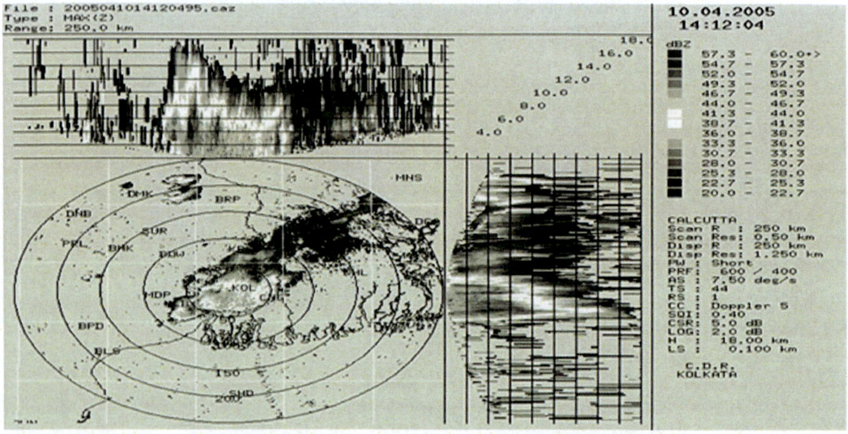

Fig. 10 Tenth image of cloud formation

after precipitation. It can be concluded that true color imageries give better result than grayscale images.

4.1 Analysis of Histograms

See Figs. 11, 12, 13, 14, 15, 16, 17, 18, 19 and 20

5 Discussions

The above study reveals a comparison between true color image and grayscale image. It can be concluded that with the analysis of such true color and grayscale imageries, we

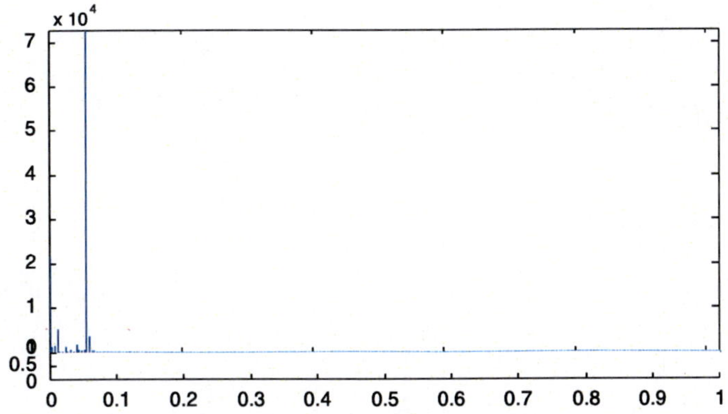

Fig. 11 Histogram of the first image

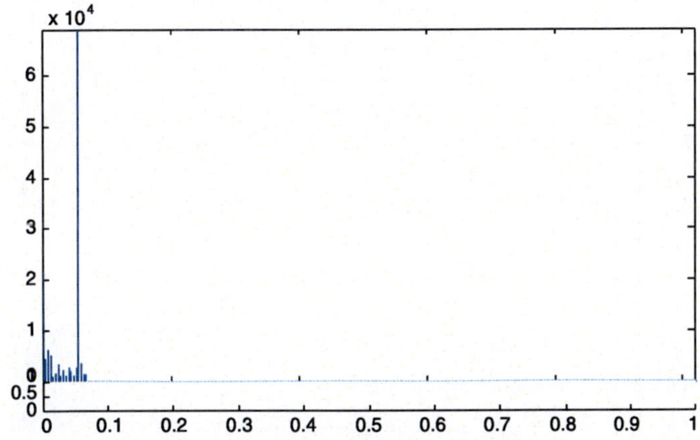

Fig. 12 Histogram of the second image

can alert people from the disaster caused by severe thunderstorm. Such forecast has a lead time of 6–8 h which is sufficient to save society from severity of damage.

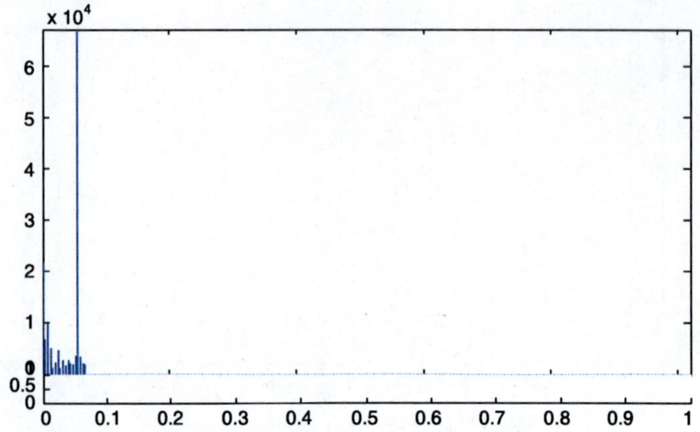

Fig. 13 Histogram of the third image

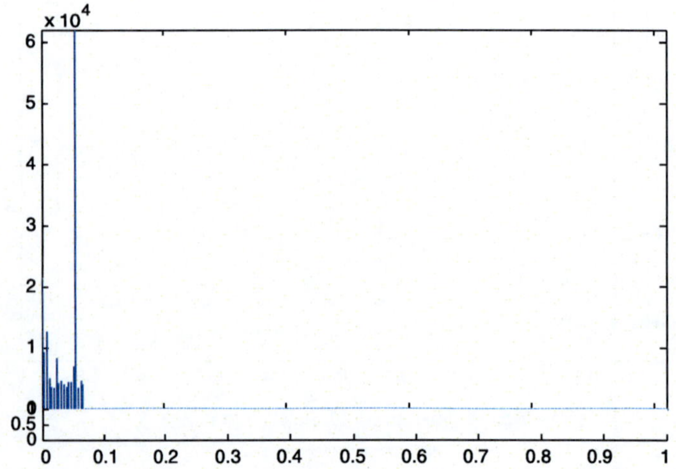

Fig. 14 Histogram of the fourth image

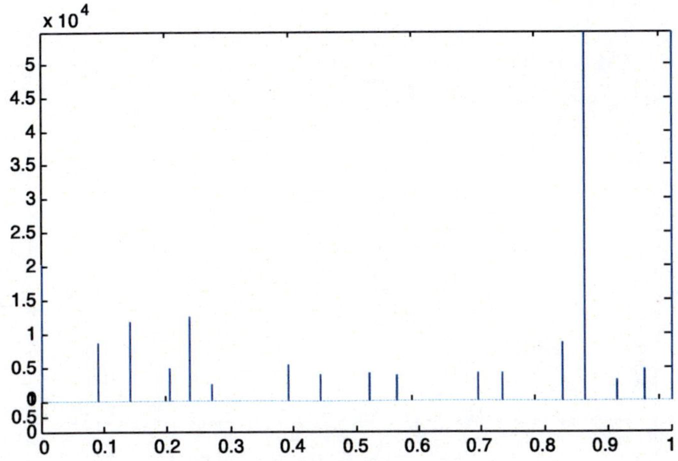

Fig. 15 Histogram of the fifth image

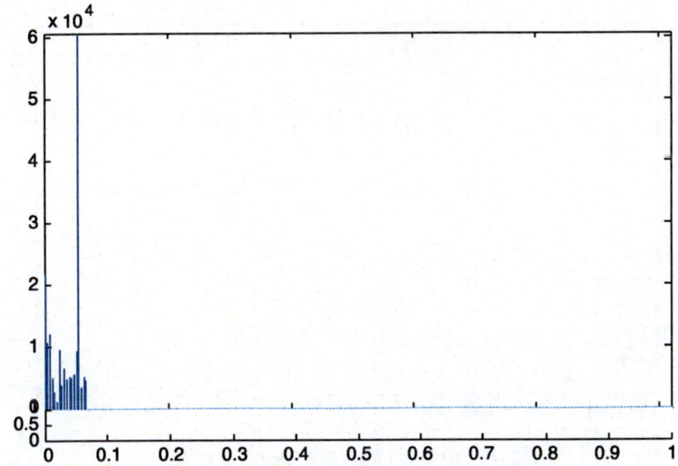

Fig. 16 Histogram of the sixth image

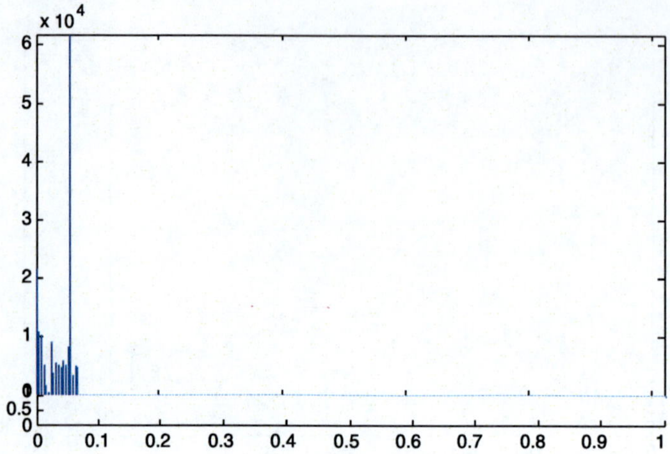

Fig. 17 Histogram of the seventh image

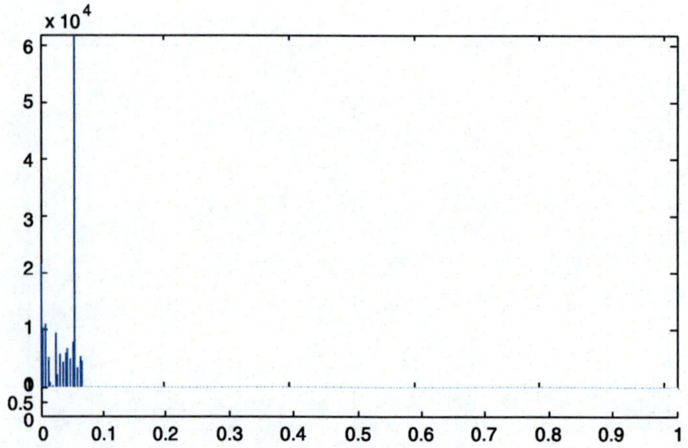

Fig. 18 Histogram of the eighth image

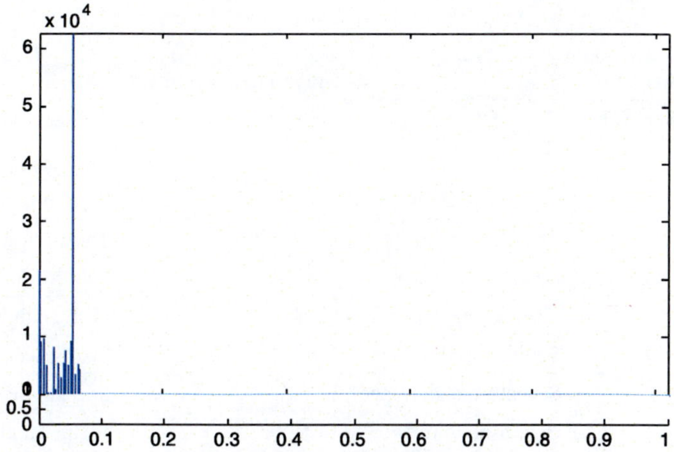

Fig. 19 Histogram of the ninth image

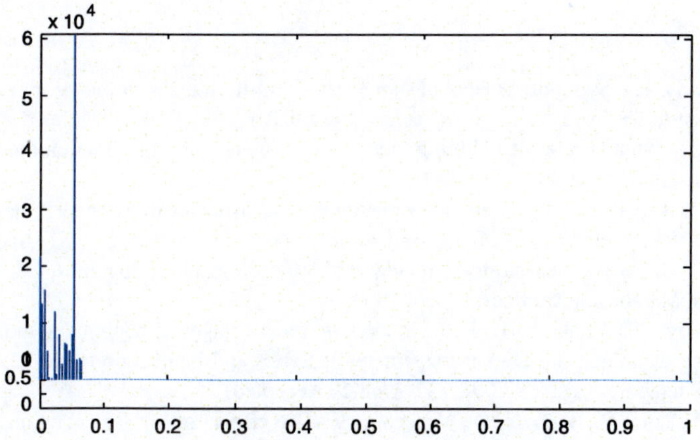

Fig. 20 Histogram of the tenth image

Table 1 Compared pixel value between two consecutive true color and grayscale images

True color imageries		Grayscale imageries	
Increase in pixel value	Consecutive images	Increase in pixel value	Consecutive images
218	First-second image	4311	First-second image
240	Second-third image	4546	Second-third image
829	Third-fourth image	4886	Third-fourth image
1336	Fourth-fifth image	5585	Fourth-fifth image
2126	Fifth-sixth image	6623	Fifth-sixth image
3472	Sixth-seventh image	8762	Sixth-seventh image
2539	Seventh-eighth image	11,607	Seventh-eighth image
1837	Eighth-ninth image	14,577	Eighth-ninth image
1845	Ninth-tenth image	16,870	Ninth-tenth image

References

1. Chakrabarty, H.: Synoptic aspects of Nor'wester and its impact to the people in Kolkata, North-East India. Int. J. Sci. Soc. **1**(4), 135–148 (2010)
2. Holle, R.L.: Annual rates of lightning fatalities by country. In: 20th International Lightning Detection Conference, Tucson , 21–23 Ap 2008
3. Collins, W.G., Tissot, P.: Thunderstorm predictions using artificial neural networks. In: Artificial Neural Networks—Models and Applications. https://doi.org/10.5772/63542
4. Ludlam, F.H.: Severe local storms: a review. In: Meteorological Monographs, vol. 5, pp. 1–30. American Meteorological Society Sept 1963
5. Bhattacharya, S., Chakrabarty, H.B.: Studies on radar imageries of thundercloud by image processing technique. In: Data management, analytics and innovation (Chap. 25). Springer Nature, Singapore. ISBN: 978-981-32-9948-1 (Accepted)
6. Curic, M., Janc, D., Vujovic, D., Vuckovic, V.: The effects of a river valley on an isolated cumulonimbus cloud development. Atmos. Res. **66**, 123–139 (2003)
7. Lin, Y.L., Joyce, L.E.: A further study of mechanisms of cell regeneration, development and propagation within a two-dimensional multicell storm. J. Atmos. Sci. **58**, 2957–2988 (2001)
8. Lin, Y.L., Deal, R.L., Kulie, M.S.: Mechanisms of cell regeneration, propagation, and development within two-dimensional multicell storms. J. Atmos. Sci. **55**, 1867–1886 (1998)
9. Rotunno, R., Klemp, J.B.: The influence of the shear-induced pressure gradient on thunderstorm motion. Mon. Weather Rev. **110**, 136–151 (1982)
10. Tajbakhsh, S., Ghafarian, P., Sahraian, F.: Instability indices and forecasting thunderstorms: the case of 30 April 2009. Nat Hazards Earth Syst Sci **12**, 403–413 (2012)
11. Anil Kumar, P., Anuradha, B., Arunachalam, M.S.: Extraction of time series convective cloud profile from Doppler weather radar MAX (Z) product using a novel image processing technique. Int. J. Adv. Eng. Res. Dev. **4**(7), e-ISSN (O): 2348-4470
12. Gil, J.Y., Kimmel, R.: Efficient dilation, erosion, opening, and closing algorithms. IEEE Trans. Pattern Anal. Mach. Intell. **24**, No. 12 (2002)

13. Skamarock, W.C., Klemp, J.B., Dudhia, J., Gill, D.O., Barker, D.M., Wang, W., Powers, J.G.: A description of the advanced research WRF, Version 2. NCAR Technical Note NCAR/TN-4681STR (88 pp) (2008)
14. Gryschka, M., Witha, B., Etling, D.: Scale analysis of convective clouds. In: Borntraeger, G. (ed.) Meteorologische Zeitschrift, vol. 17(6), pp. 785–791, Dec 2008

A Compact Multiband Antenna for Mobile Handset Application

Juin Acharjee[1](✉), Mihir Kumbhakar[2], Kaushik Mandal[3], and Sujit Kumar Mandal[4]

[1] Department of ECE, ST. Thomas' College of Engineering and Technology, Kolkata 700023, India
juin.ece@gmail.com
[2] Department of Telecom, Kolkata, India
[3] Radio Physics and Electronics, University of Calcutta, Kolkata 700009, India
[4] Department of Electronics and Communication Engineering, NIT Durgapur, Durgapur 713209, India

Abstract. This paper presents the design of a triple-band antenna for mobile handset covering frequency bands of GSM 900, GSM 1800, and ISM 2450. The proposed antenna includes two open rings attached with the transmission line and a slotted partial ground plane for the three operating bands. This antenna is printed on a less expensive FR4 substrate, occupying very small volume of $26 \times 27 \times 1.6\ \text{mm}^3$. The ground plane of the proposed structure is extended to meet the dimension of a standard smart mobile phone to verify its applicability as a mobile handset antenna. The results of the proposed triple-band antenna and extended ground plane antenna for the mobile handset are in good agreement. The details of design consideration, working principle and the parametric analysis are discussed.

1 Introduction

With the technological advancement and increasing user demands, it is essential to incorporate more and more advanced features on the mobile handset by maintaining a very compact size with lighter in weight. One of the attractive features is to include more RF bands in a single antenna. To mitigate this issue, it is essential to design an antenna that can be able to cover all the useful operating bands. Some conventional multiband antenna in the form of monopole [1–3] provides very narrow bandwidth. Different techniques are already reported to realize multiband antennas with wide operating bands. Some of these are the use of parasitic elements [4], the combination of G-shaped radiator and slotted ground plane [5], embedding E-shaped slot on the ground plane [6], and employing four pad elements [7]. A low SAR multiband antenna is also proposed in [8] by adding an open ring slot with the rectangular monopole. Design of antenna for mobile handset application covering the 2G/3G bands and also considering other aspects of mobile phones is a very challenging task to the researchers. In this context, a combination of the resonating and tuning branch [9], combination of loop antenna and monopole antenna

© Springer Nature Singapore Pte Ltd. 2021
N. R. Das and S. Sarkar (eds.), *Computers and Devices for Communication*, Lecture Notes in Networks and Systems 147,
https://doi.org/10.1007/978-981-15-8366-7_16

[10], use of different alphabetic-shaped elements [11] have been reported to design antennas for the mobile phone.

In this paper, a triple-band printed antenna for mobile handset application is reported. This structure operates over the frequency bands 0.85–1.0 GHz, 1.8–1.91 GHz, and 2.37–2.8 GHz, which are applicable for GSM900, GSM1800, and ISM2450 bands, respectively. The −10 dB bandwidth of the operating bands is 150 MHz, 110 MHz, and 430 MHz, respectively.

2 Antenna Design and Working Principle

The proposed planar antenna is printed on a FR4 substrate with a dimension of 26 × 27 × 1.6 mm^3, dielectric constant (ϵ_r) of 4.4 and loss tangent of 0.02. Figure 1 shows the geometry of the proposed multiband antenna. The top side of the substrate is composed of two open rings and one microstrip line. The arc with larger electrical length (Arc_{L1}) acts as a radiator for the GSM 900 band and the length of this arc is set to 0.16λ, where λ is the wavelength corresponding to the resonating frequency 930 MHz. Consequently, the smaller arc (Arc_{L2}) acts as a radiator for the GSM1800 band with the electrical length of ~0.11λ where λ is the wavelength corresponding to the resonating frequency 1.86 GHz. The optimized length of the outer and inner arc is 52.7 and 22 mm, respectively. The optimized values of all the design parameters are given in Table 1. In the design procedure, angle θ_1, θ_2, arc length Arc_{L1} and Arc_{L2} can be calculated by the following equations.

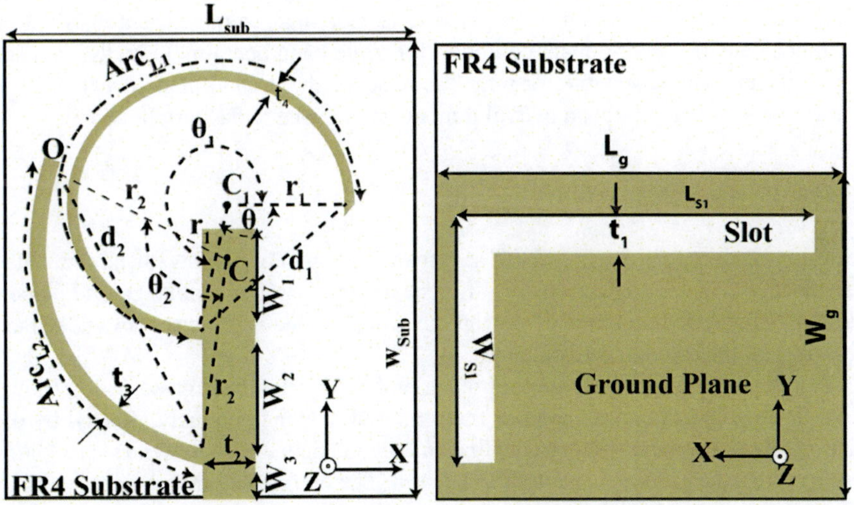

Fig. 1 Schematic diagram of the proposed multiband antenna. **a** Top view, **b** bottom view

$$\theta = \cos^{-1}\frac{2r_1^2 - d_1^2}{2r_1^2} \tag{1}$$

Table 1 Dimensions of the proposed multiband antenna (all values are in mm)

L_{sub}	W_{Sub}	L_g	W_g	L_{S1}	W_{S1}	W_1
26	27	26	17.7	24	11	7.5
W_2	W_3	Arc_{L1}	Arc_{L2}	t_1	r_1	r_2
4.2	1	52.7	22	2.4	10.3	12.9
t_2	t_3	t_4	d_1	d_2	θ_1	θ_2
2.86	0.8	0.3	12.19	20	287.5°	101.53°

$$\theta_1 = 360^{\circ} - \theta \tag{2}$$

$$\theta_2 = \cos^{-1}\frac{2r_2^2 - d_2^2}{2r_2^2} \tag{3}$$

$$Arc_{L1} = \frac{2\pi r_1 \theta_1}{360} \tag{4}$$

$$Arc_{L2} = \frac{2\pi r_2 \theta_2}{360} \tag{5}$$

where Arc_{L1}, Arc_{L2}, r_1, r_2, θ_1, and θ_2 are indicated in Fig. 1.

After calculating the arc lengths and the corresponding angles, two open-ended rings of calculated length are attached to the microstrip line. An inverted L-shaped slot is etched out from the partial ground plane. This slot changes the current distribution path to generate the new resonant condition for triple-band operation. Finally, to make this multiband antenna suitable for a mobile handset, the ground plane is extended in the negative x-direction with an overall dimension of a standard smartphone.

3 Results and Discussion

The design aim is to realize a multiband antenna where operating bands can be controlled independently. Parametric analysis of different sensible parameters like ground plane's width, arc length, and presence of slot on the ground plane is important to understand their effect on the antenna performances.

The effect of changing the ground plane width (W_g) is investigated and is shown in Fig. 2. The third operating band of the proposed antenna is mostly affected by the variation of ground plane width. In the first and second operating bands, no such change is observed for the ground plane width variation. The optimized ground plane width of 17.7 mm is considered for the proposed antenna.

Inverted L-shaped slot on the ground plane also plays an important role for the triple-band operation of the proposed antenna. In absence of the slot, the proposed antenna confirms only the GSM900 and GSM 1800 bands. Incorporating the inverted L-shaped slot on the partial ground plane, the third band impedance matching is improved and this is summarized in Fig. 3.

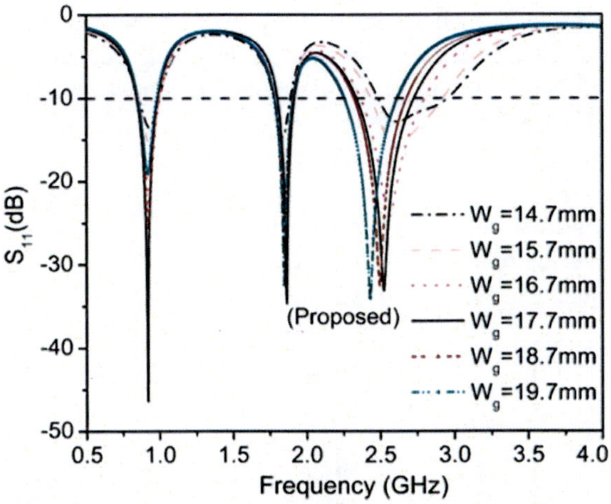

Fig. 2 Effect of ground plane width variation on simulated reflection coefficient (S_{11})

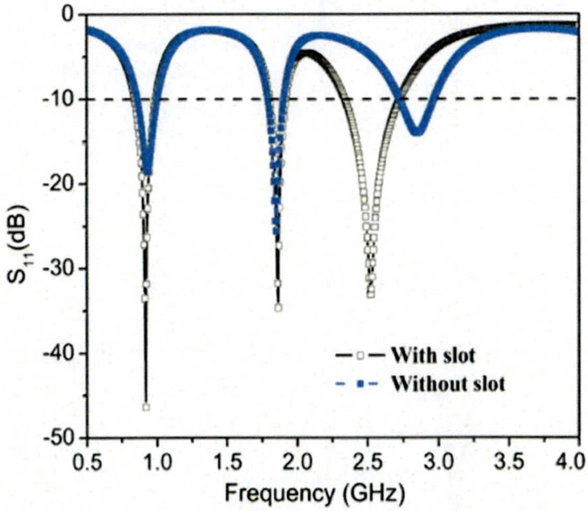

Fig. 3 S_{11} with and without inverted L-shaped slot

The effects of Arc_{L1} and Arc_{L2} variation on the GSM 900 and GSM 1800 are shown in Figs. 4 and 5, respectively. It confirms that the longer open ring is responsible for GSM 900 and the shorter open ring is responsible for the GSM 1800 band. The surface current distribution on the antenna elements is shown in Fig. 6. At 0.92 GHz, the surface current is concentrated across the larger arc, and at 1.86 GHz, it is concentrating across the shorter arc, as shown in Fig. 6a, b, respectively. The ground slot is used to match the impedance of the third operating band at 2.56 GHz. Hence, at this frequency, the current is concentrated around the slot, as shown in Fig. 6c. Thus, the more current concentration across the different parts of the structure as observed in study of current distribution at respective frequencies justify the proper resonating nature of the designed structure that leads to radiate at the corresponding desired frequencies. The performance of the proposed multiband antenna for mobile applications has been verified by extending the ground plane. Figure 7 shows that the first band is not affected, but for the second (1.65–1.89 GHz) and third (2.28–3.11 GHz) bands, the bandwidth has been increased by 130 and 440 MHz, respectively. The simulated and measured reflection coefficients of the proposed multiband antenna are shown in Fig. 8. It shows that the proposed antenna covers the three bands 0.85–1.0 GHz, 1.8–1.91 GHz, and 2.37–2.8 GHz with the resonating frequency of 920 MHz, 1.86 GHz, and 2.52 GHz, respectively.

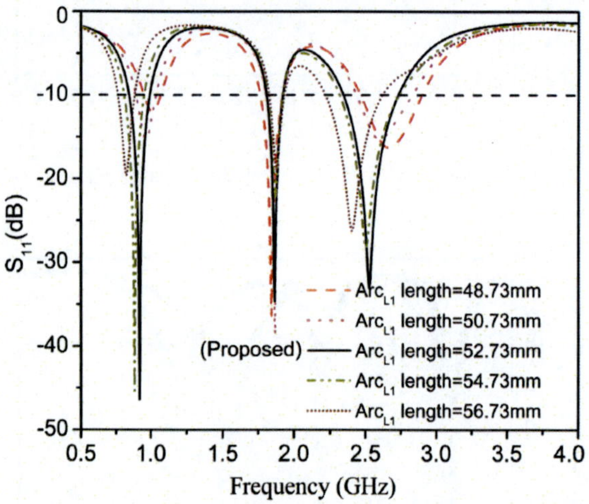

Fig. 4 Simulated S_{11} for different values of Arc_{L1}

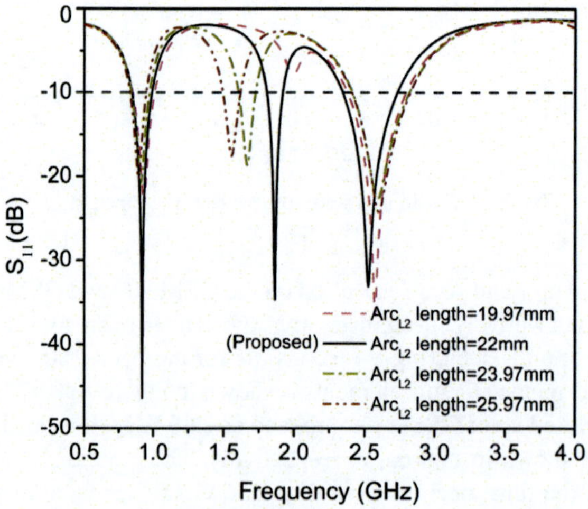

Fig. 5 Simulated S_{11} for different values of Arc_{L2}

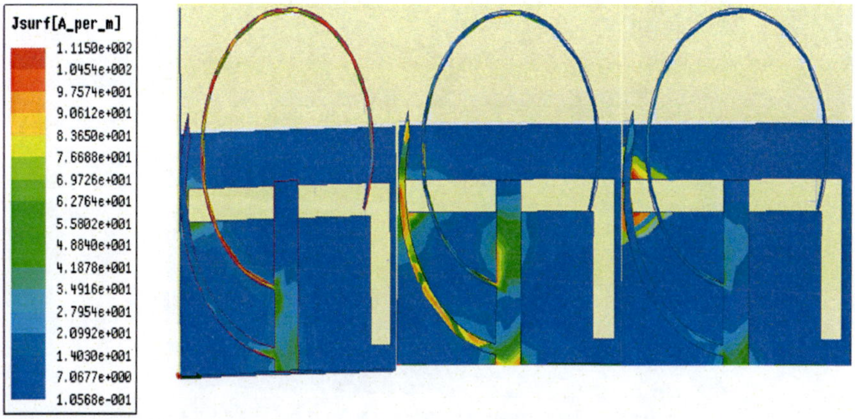

Fig. 6 Simulated surface current distribution at **a** 0.92 GHz, **b** 1.86 GHz, and **c** 2.56 GHz

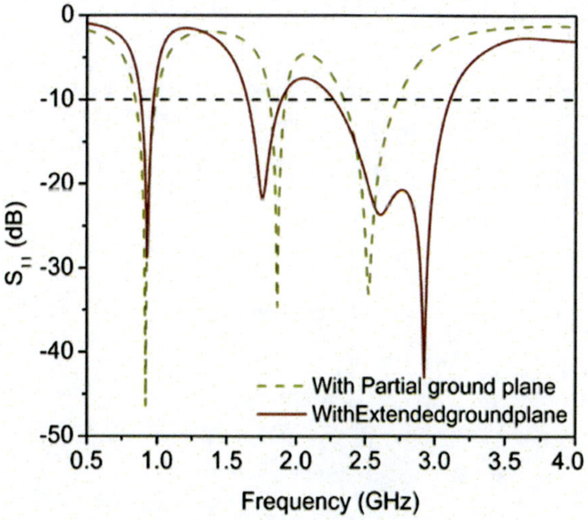

Fig. 7 Effect of extended ground plane on simulated S_{11}

Figure 9 shows the simulated radiation patterns at 960, 1.86, and 2.5 GHz for the two principal planes. The antenna with such nearly omnidirectional patterns is useful for the mobile phone application. The mobile devices are generally operated in a multipath environment where the signals are received from different directions. Therefore, the proposed antenna is a suitable candidate for mobile applications. The performances of the proposed structure are compared with the other reported structures of its kind in Table 2. It reveals that the proposed design is simple, compact in size, and able to cover the three intended essential frequency bands for mobile communication.

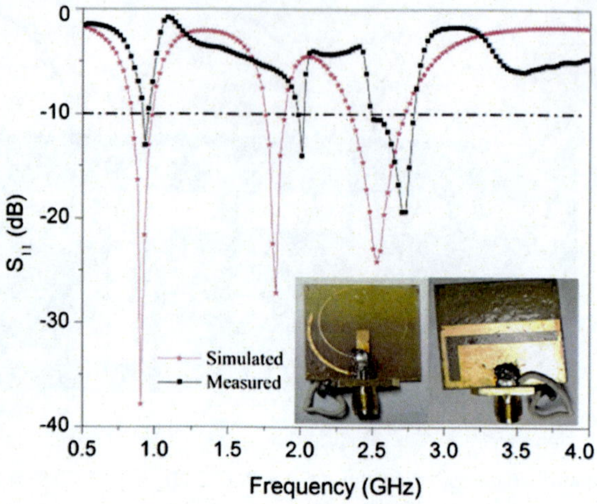

Fig. 8 Comparison between simulated and measured S_{11} of the proposed antenna

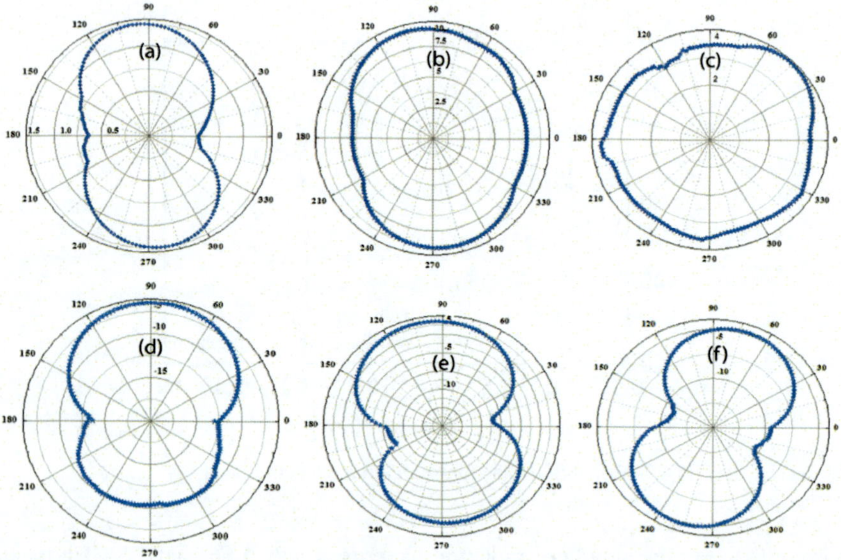

Fig. 9 Simulated radiation patterns at E-plane. **a** 0.96 GHz, **b** 1.86 GHz, **c** 2.52 GHz and at H-plane, **d** 0.96 GHz, **e** 1.86 GHz, and **f** 2.52 GHz

4 Conclusion

A simple and compact multiband antenna suitable for a mobile phone application has been presented. Two open-ended arc-shaped radiators in the microstrip line fed proposed structure are crucial to controlling the operating bands for GSM900 and GSM 1800. A slotted partial ground plane is used in the proposed structure to generate the third application band ISM 2450. Performance of the proposed structure is also verified with

Table 2 Comparison of the proposed structure with other reported structures

Ref.	Methodology used	Operating frequency bands (MHz)	Overall size (mm^2)
[9]	Tuning branch	(891–961), (1705–2180), (2341–2980)	38.5 × 15
[10]	Combination of loop and monopole antenna	(877–966), (1210–3632), (4688–7269)	80 × 35
[11]	C-shaped and inverted L-shaped elements	(745–800), (1700–3000), (3400–3840)	20 × 70
Pro	Combining two open rings and grounded slot	(850–1000), (1800–1910), (2370–2800)	26 × 27

extended ground plane to make it suitable for mobile handset. Overall dimension and nearly omnidirectional coverage of the proposed structure make it a suitable candidate for the mobile handset antenna.

References

1. Wong, K.L., Lin, Y.C., Tseng, T.C.: Thin internal GSM/DCS patch antenna for a portable mobile terminal. IEEE Trans. Antennas Propag. 54, 238–242 (2006)
2. Wong, K.-L., Lin, Y.C., Chen, B.: Internal patch antenna with a thin air-layer substrate for GSM/DCS operation in a PDA phone. IEEE Trans. Antennas Propag. **55**, 1165–1172 (2007)
3. Guo, Y.-X., Chia, M.Y.W., Chen, Z.N.: Miniature built-in multiband antennas for mobile handsets. IEEE Trans. Antennas Propag. **52**, 1936–1944 (2004)
4. Lin, K.C., Lin, C.-H., Lin, Y.C.: Simple printed multiband antenna with novel parasitic-element design for multistandard mobile phone applications. In: IEEE Trans. Antennas Propag. **61**, 488–491 (2013)
5. Dadgarpour, A., Abbosh, A., Jolani, F.: Planar multiband antenna for compactmobile transceivers. IEEE Antennas Wireless Propag. Lett. **10**, 651–654 (2011)
6. Abutarboush, H.F., Nasif, H., Nilavalan, R., Cheung, S.W., Dadgarpour, A., Abbosh, A., Jolani, F.: Multiband and wideband monopole antenna for GSM900 and other wireless applications. IEEE Antennas Wireless Propag. Lett. **11**, 539–542 (2012)
7. Anguera, J., Andújar, A., García, C.: Multiband and small coplanar antenna system for wireless handheld devices. IEEE Trans. Antennas Propag. **61**, 3782–3789 (2013)
8. Abdullah, H.H., Sultan, K.S.: Multiband compact low sar mobile hand held antenna. Prog. Electromagn. Res. Lett. **49**, 65–71 (2014)
9. Jing, X., Du, Z., Gong, K.: A compact multiband planar antenna for mobile handsets. IEEE Antennas Wireless Propag. Let. **5**, 343–345 (2006)
10. Hsieh, H.W., Lee, Y.C., Tiong, K.K., Sun, J.S.: Design of a multiband antenna for mobile handset operations. IEEE Antennas Wireless Propag. Lett. **8**, 200–201 (2009)
11. Ahmed, F., Chowdhury, M.H.M., Hasan, N.: A compact multiband antenna for 4G/LTE and WLAN mobile phone applications. In: IEEE 3rd International Conference on Electrical Engineering and Information Communication Technology (ICEEICT), pp. 1–4 (2016)

Communication and Space Science (CSS)

Reduced Subcarrier Index Modulation Scheme in OFDM System for Next-Gen Wireless Networks

Ipsita Sengupta[1](✉) and Shounak Dasgupta[2]

[1] Electronics Department, Government College of Engineering and Leather Technology, Block-LB 11, Sector-III, Salt Lake, Kolkata, West Bengal 7000106, India
ipsen1980@yahoo.co.in

[2] ECE Department, Heritage Institute of Technology, Kolkata, West Bengal 700107, India
shounak121@gmail.com

Abstract. Orthogonal frequency division multiplexing (OFDM) has proved itself as a proficient multicarrier transmission technique to be successfully utilized in 5G and next-generation wireless communication. However, conventional OFDM suffers from the limitation of high peak-to-average-power ratio (PAPR). Subcarrier index modulation (SIM) has already been established as one of the efficient techniques to mitigate this high PAPR problem. In this work, we have proposed a novel modulation algorithm named as reduced subcarrier index modulation-OFDM (RIM-OFDM), which offers further considerable reduction in PAPR value and improved bandwidth utilization. In this paper, this newly proposed scheme has been mathematically established and block diagram for its realistic implementation has been described in detail. PAPR performance along with improvement in spectral efficiency of this technique, over SIM, has been investigated. Bit error rate (BER) performance is also studied and compared with that of classical OFDM. The results reveal that RIM-OFDM has come up as an optimized scheme, in terms of PAPR mitigation, bandwidth efficiency enhancement and low BER, for future wireless communication.

Keywords: Orthogonal frequency division multiplexing (OFDM) · Subcarrier index modulation (SIM) · Reduced subcarrier index modulation (RIM) · Peak-to-average-power ratio (PAPR) · Bandwidth efficiency (BWE)

1 Introduction

Orthogonal frequency division multiplexing (OFDM) is used extensively in broadband wired and wireless communication systems as an effective solution to inter symbol interference (ISI), caused in a dispersive channel with multipath effect [1–3]. In OFDM, data is transmitted in parallel, on a number of different orthogonal frequencies, and as a result, the symbol period is much longer than for a serial system with the same total data rate. Because the symbol period is longer, ISI affects minimum number of bits and equalization becomes simplified. In OFDM, the spectra of individual subcarriers overlap,

© Springer Nature Singapore Pte Ltd. 2021
N. R. Das and S. Sarkar (eds.), *Computers and Devices for Communication*,
Lecture Notes in Networks and Systems 147,
https://doi.org/10.1007/978-981-15-8366-7_17

but because of the orthogonality property, the subcarriers can be demodulated without interference, in a linear channel. Due to subcarrier overlapping, in addition, OFDM can increase the bandwidth efficiency (BWE) for a given data rate. For these advantages, OFDM has proved itself as a popular multicarrier transmission technique in different wireless communications, such as mobile worldwide interoperability microwave access for next-generation wireless communication systems (WiMAX) and the long term evolution (LTE). Consequently, OFDM is also recognized as an integral part of IEEE 802.16 standards. On the other hand, OFDM system has high peak-to-average power ratio (PAPR) which leads to distortion on signal if they fall in the nonlinear region of power amplifier [4]. In other words, high PAPR requires a large dynamic range for the power amplifier which increases the size of device and power consumption [5].

The early approaches to mitigate the PAPR problem was mainly focused about the techniques involving coding [6] and nonlinear distortion (clipping) [7]. Coding techniques aim to impose preconceived coding to the input vectors so that OFDM symbols which have high PAPR are avoided. These techniques are computationally rigorous to be useful in most fundamental applications [4]. On the other hand, clipping is done on either the analog signal, or an up sampled version of the digital signal with an oversampling factor of at least two [4]. However, this scheme is also associated to degradation of bit error rate (BER) performance.

More recently, a novel transmission technique, known as, subcarrier index modulation (SIM), integrated with the orthogonal frequency division multiplexing (OFDM) systems, is proposed [8, 9]. This approach reduces PAPR value with improved error performance. The concept of SIM is based on spatial modulation (SM) technique, which has emerged as a potential candidate for transmission in multiple-input multiple-output (MIMO) system [10–12]. In the SM technique, that uses multiple antennas, information bits are conveyed both by the amplitude/phase modulation techniques and by the selection of antenna indices [11, 12]. It improves overall spectral efficiency by the base-two logarithm of the number of transmit antennas [12]. Similarly, in subcarrier index modulation with OFDM (SIM-OFDM), the information data is divided and mapped into two information carrying units: (1) M-QAM modulated bits, as in classical OFDM and (2) the indices bits. The indices bits select which subcarriers are active. Thus, SIM-OFDM transmits only M-QAM modulated bits on the selected active subcarriers [9].The inactive subcarriers are set to zero value. Since the PAPR performance of OFDM systems is related to the number of the sub-carriers, SIM-OFDM can improve the PAPR performance due to only partial subcarriers activation. However, SIM-OFDM system which utilizes SIM method to select active subcarriers, offers flexible BWE. However, a number of subcarriers, engaged in each subblock, remain unutilized as inactive subcarriers.

In this communication, lesser number of subcarriers is allocated to each subblock in comparison to the algorithm proposed in SIM-OFDM [9]. The look up table, for subcarrier assignment, has been modified. This assignment offers better BWE with low PAPR value. The algorithm also reduces the decoding complexity at the receiver and leads to a reduced subcarrier index modulation with OFDM (RIM-OFDM). The improved spectral efficiency and lower bit error rate (BER) indicate improvement in performance of RIM-OFDM over classical OFDM. The RIM-OFDM may be considered as an efficient technique, with optimum PAPR and BWE, to be used in multicarrier wireless communication systems.

2 Theory and Modeling

2.1 OFDM System Description

OFDM modulation technique uses several orthogonal frequency components to transmit information bearing data or symbols [5]. In this approach, incoming bit stream is divided into number of segments. In each segment, N symbols are generated employing M-ASK or M-PSK or M-QAM modulation and are transmitted using N different subcarriers. The block diagram for OFDM transmitter and receiver is shown in Fig. 1. Here, IFFT block in the transmitter and FFT block in the receiver are the main components.

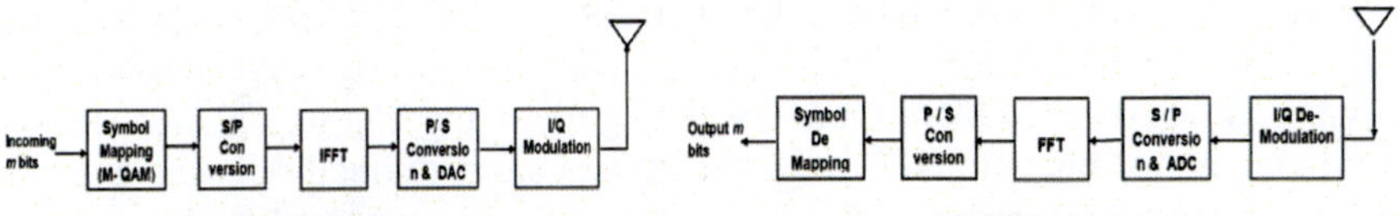

Fig. 1 Classic OFDM transmitter and receiver block diagram

The input symbol sequence to the IFFT block is the complex vector $X = [X_0 X_1 X_2 \ldots X_{N-1}]^{\mathrm{T}}$, here N is the size of the IFFT and also the number of subcarriers used. X_k represents M-QAM modulated symbol to be carried on kth subcarrier. The output of the IFFT is another complex vector $x = [x_0 x_1 x_2 \ldots\ x_{N-1}]^{\mathrm{T}}$; here, x is time domain representation of each OFDM symbol called one OFDM frame. Each element of x is represented by the following equation

$$x_m = \frac{1}{\sqrt{N}} \sum_{k=0}^{N-1} X_k \exp\left(\frac{j2\pi km}{N}\right) \quad \text{for } 0 \le m \le N-1 \tag{1}$$

Forward FFT at the receiver corresponding to Eq. 1 is

$$X_k = \frac{1}{\sqrt{N}} \sum_{m=0}^{N-1} x_m \exp\left(\frac{-j2\pi km}{N}\right) \quad \text{for } 0 \le k \le N-1 \tag{2}$$

However, the most significant limitation encountered in OFDM is its high PAPR due to its large number of subcarriers. This results in clipping of OFDM modulated signal, when passed through any nonlinear system, such as power amplifier at the transmitter. Thus, distortion of the received signal and consequently severe degradation in BER performance occurs at the receiver.

PAPR of OFDM signal at the transmitter is defined as [13]

$$\mathrm{PAPR} = \frac{\max\limits_{0 \le k \le N-1} |x_m|^2}{Av\left[|x_m|^2\right]} \tag{3}$$

One possible solution is to restrict the operation of the transmitter amplifier to its linear region. Unfortunately, this solution is not power efficient and power efficiency is one of the most essential requirements in wireless communication. It is therefore highly important to aim at a power efficient solution to high PAPR problem. To achieve the same objective, several techniques have already been suggested [5–9]. SIM-OFDM [8, 9] is one of those potent methods to offer reduced PAPR of OFDM output by reducing the number of used subcarriers for transmission of modulated data or symbols; thereby, reducing also the size of the IFFT block used at the OFDM transmitter.

2.2 Description of SIM-OFDM

In SIM-OFDM scheme, the incoming bits are segmented into multiple subblocks [9]. The bits in each subblock is further divided into two groups, one for M-QAM signal constellation mapping (as usual in OFDM) and the other for index selection of active subcarriers used in the subsequent subblock. M-QAM modulated symbols are then transmitted by using only active subcarriers in each subblock. The positions of active subcarriers in each subblock are decided by the index selecting bits according to a predefined look up table (Table 1). Although the index selecting bits are not directly transmitted, their information remains preserved in the position of active carriers used for transmission of M-QAM modulated symbols. So there will be no difficulty in detecting them at the receiver from the positions of M-QAM symbols (or active subcarriers) in the subsequent OFDM subblock. Thus, in this technique, less number of subcarriers are kept active for transmission of M-QAM modulated symbols than that of OFDM.

Table 1 Look up table for SIM-OFDM Tx and Rx

$n = 4$ $k = 2, M = 8$
Input bit sequence: $p_1\ p_2\ S_{y1}\ S_{y2}$;
$p_x = p_1\ p_2$ = Index selecting bits
$p_y = S_{y1}\ S_{y2}$ = M-QAM modulated symbols

Index bits p_x		Index combination I_γ	M-QAM symbol Subblock S_γ
p_1	p_2		
0	0	[1 2]	$[S_{y1}\ S_{y2}\ 0\ 0]$
0	1	[2 3]	$[0\ S_{y1}\ S_{y2}\ 0]$
1	0	[3 4]	$[0\ 0\ S_{y1}\ S_{y2}]$
1	1	[1 4]	$[S_{y1}\ 0\ 0\ S_{y2}]$

The block diagram of SIM-OFDM scheme is shown in Fig. 2 [9]. Total m number of incoming bits is split into g number of subblocks, each containing p bits. Those p bits are further divided into two subgroups. First p_x number of bits go to index selector, whose function is to decide active subcarrier indices in each subblock. Next, p_y number of bits out of p bits go to M-QAM mapper, which generates M-QAM constellation symbols to be transmitted by using k number of active subcarriers out of n available subcarriers within individual subblock. Thus, total number of available subcarriers in the system is $N = n \times g$ and total number of active subcarriers used for transmission is $(k \times g)$. The indices of active subcarriers in each subblock are determined from the look up table given in Table 1 [9].

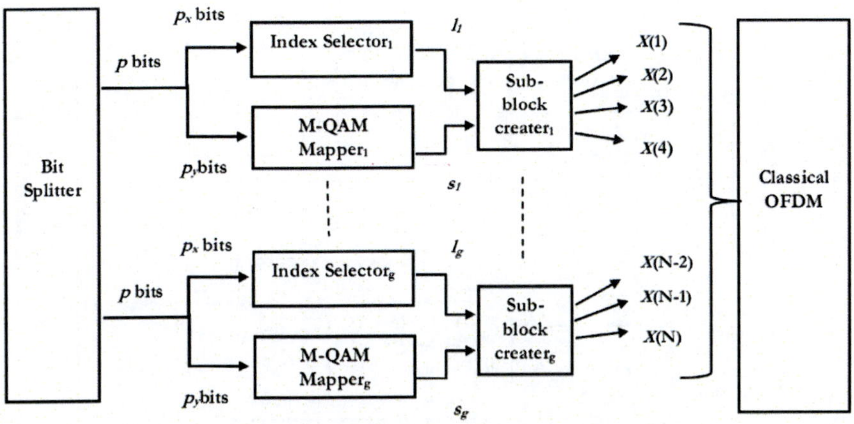

Fig. 2 SIM-OFDM transmitter block diagram

For example, with 10 011 010 incoming bit pattern, index combination of [3 4] and [0 0 $S_{\gamma 1} S_{\gamma 2}$] OFDM subblock are generated. Here $S_{\gamma 1}$ and $S_{\gamma 2}$ are M-QAM mapped symbols corresponding to 011 and 010 bits. This is illustrated in Fig. 3. In the receiver, after undergoing through FFT operation, [0 0 * *] is received as the detected symbol sequence, where * denotes the M-QAM symbols with active subcarriers. From the position of those symbols, the index selecting bits are easily detected with the aid of the look up Table 1. P_y bits are also retrieved by performing M-QAM demodulation of the received symbols.

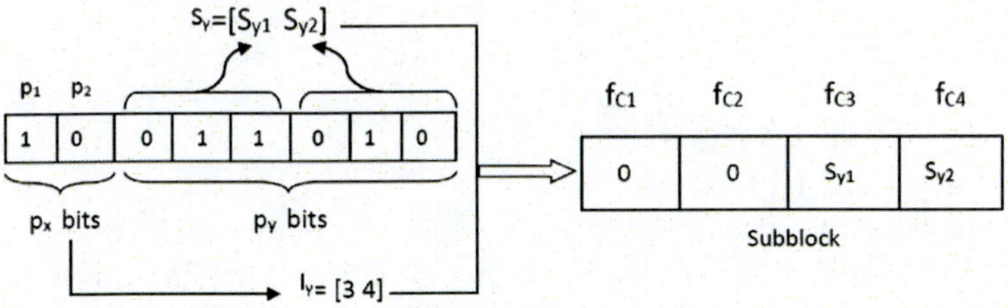

Fig. 3 Generation of SIM-OFDM subblock

2.3 Reduced Subcarrier Index Modulation-OFDM (RIM-OFDM)

2.3.1 Modulation of RIM-OFDM

SIM-OFDM method uses only combinations of active subcarriers in a subblock. It does not consider their orientation. In this paper, we have proposed a new modulation scheme, reduced subcarrier index modulation-OFDM (RIM-OFDM), where a specific orientation of active subcarriers is assigned to each unique combination of index selecting bits within a subblock. The transmitter block diagram is depicted in Fig. 4.

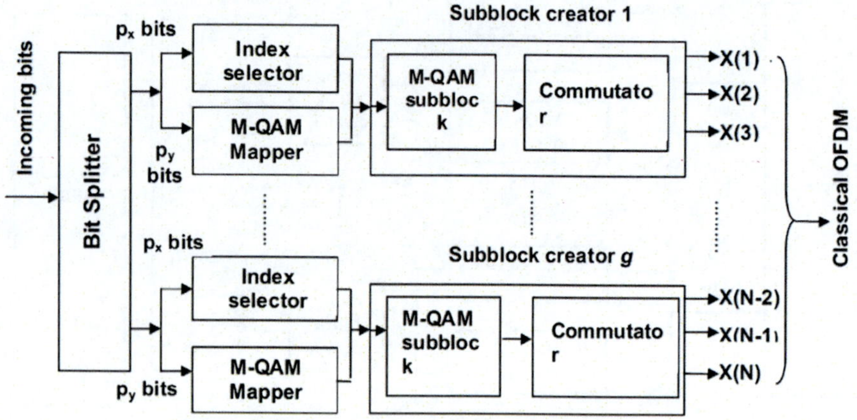

Fig. 4 RIM-OFDM transmitter block diagram

For M-QAM modulation,

$$p_y = k \times \log_2 M \tag{4}$$

p_x is determined by possible combinations of k active subcarriers out of n subcarriers. So,

$$2^{p_x} \leq {}^nC_k \tag{5}$$

The index selector forms a combination I_γ made of k active subcarrier indices out of n available indices in each subblock.

$$I_\gamma = [I_{\gamma 1} I_{\gamma 2} \ldots I_{\gamma k}],$$

where $1 \leq \gamma \leq g$ and $1 \leq k \leq n$.

Total number of bits in any subblock is $p = p_x + p_y$. For the look up table given in Table 1 with $n = 4$, $k = 2$ and $M = 8$, number of possible combinations for k active subcarriers is ${}^nC_k = 6$. Thus, closest possible number, i.e., four combinations are used to satisfy the above equation for detecting the number of index selecting bits, $p_x = 2$. The QAM mapper generates a vector $S_\gamma = [S_{\gamma 1} S_{\gamma 2} \ldots S_{\gamma k}]$, where $1 \leq \gamma \leq g$ and $1 \leq k \leq n$.

The modified look up table used to implement RIM-OFDM modulation is given in Table 2. For the same input bit sequence as used in Table 1, four unique orientations of subcarriers have been used for four different possible combinations of input index bits p_1 and p_2. Unlike SIM-OFDM, the permutations of f_1, f_2 and f_2, f_3 are used. The forward orientation ($f_1 f_2$ and $f_2 f_3$) are employed for appearance of same bits in $p_1 p_2$

Table 2 Look up table used in RIM-OFDM Tx and Rx

$n = 3, k = 2, M = 8$, Input bit sequence: $p_1\ p_2\ S_{y1}\ S_{y2}$;
$p_x = p_1\ p_2$ = Index selecting bits
$p_y = S_{y1}\ S_{y2}$ = M-QAM modulated symbols

Index bits p_x		p_1 XOR p_2	Commutator status	Index combination	Subcarrier allocation	M-QAM symbol subblock
p_1	p_2					
0	0	0	Off	[1 2]	$[f_1\ f_2\ 0]$	$[S_{y1}\ S_{y2}\ 0]$
0	1	1	On	[2 1]	$[f_2\ f_1\ 0]$	$[S_{y1}\ S_{y2}\ 0]$
1	1	0	Off	[2 3]	$[0\ f_2\ f_3]$	$[0\ S_{y1}\ S_{y2}]$
1	0	1	On	[3 2]	$[0\ f_3\ f_2]$	$[0\ S_{y1}\ S_{y2}]$

and reverse orientation ($f_2 f_1$ and $f_3 f_2$) are chosen for occurrence of two different bits in $p_1 p_2$ pattern. These two orientations of subcarrier frequency combinations are ensured by applying a commutator switch controlled by XOR product of p_1 and p_2 bits.

Spectral efficiency (SE) for RIM-OFDM [14, 15] is

$$\mathrm{SE} = \frac{\left[k \log_2 M + \left\lfloor \log_2 ({}^{n}c_k) \right\rfloor\right] \times g}{N} \tag{6}$$

whereas spectral efficiency for Classic–OFDM is $\frac{(n \log_2 M)g}{N}$.

2.4 Demodulation of RIM-OFDM

Demodulation of RIM-OFDM modulated signal is not same as in SIM-OFDM scheme. The modification in block diagram of the classic OFDM required to demodulate RIM-OFDM modulated signal is shown in Fig. 5. When a RIM-OFDM modulated signal arrives at the receiver, the position of active subcarriers can easily be detected after passing through the FFT operation. M-QAM demodulator then executes recovery of transmitted bit pattern designated as p_y. But extraction of index selecting bits p_x will not be similar to that of SIM-OFDM. Identification of proper combination of subcarriers along with their orientation in a subblock is necessary for detection of p_x bits. For example, when the received subblock is $[0\ r_1 r_2\ 0]$ at the output of the FFT block, the position of the symbols r_1 and r_2 confirm the index combination as [2 3] as per look up Table 2, but their orientation [2 3] or [3 2], still remains undecided. Additional circuitry in the classic OFDM receiver block along with another look up table (Table 3) must be employed in order to detect their orientation. The symbols modulated with active subcarriers (in our example r_1, r_2) are sequentially applied to three separate correlators operated at frequencies f_1, f_2 and f_3 respectively, followed by three separate comparators. Occurrence of high output from the subsequent comparator indicates matching of incoming symbol's subcarrier to the respective correlator frequency.

3 Results and Discussion

In this communication, a classic OFDM system with $N = 16$ is primarily considered. As the concept of active subcarriers is absent in classical OFDM, $k = N$ is taken for

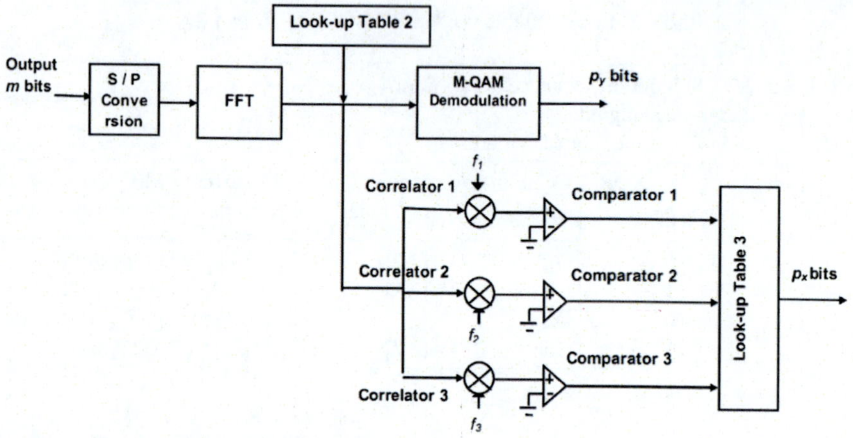

Fig. 5 Modified receiver block diagram for detection of RIM-OFDM

Table 3 Look up table at Rx for identification of index selecting bits in RIM-OFDM modulated signal

Index selecting bits p_x		Transmitted M-QAM symbol sequence with active subcarriers	Check $r = r_1$			If $f_2 = 1, p_2 = 1$; Check $r = r_2$		
			Correlator frequency	p_1	p_2	Correlator frequency	p_1	p_2
p_1	p_2	$r_1\, r_2$	$f_1 = 1$	0	0	$f_1 = 1$	0	1
0	0	$f_1 f_2$	$f_3 = 1$	1	0	$f_3 = 1$	1	1
0	1	$f_2 f_1$	$f_2 = 1$	X	1			
1	1	$f_2 f_3$						
1	0	$f_3 f_2$						

simulation in this case. However, for SIM-OFDM, the bit stream is divided into four subblocks ($g = 4$) with 4 orthogonal carriers ($n = 4$) assigned to each subblock. In this case, two subcarriers are kept active at a time, i.e., $k = 2$. In case of RIM-OFDM, although same number of active subcarriers ($k = 2$) is maintained, the number of orthogonal subcarriers assigned to each subblock, is reduced to three for the same input bit stream. Thus, our proposed scheme leads to reduction in number of total allocated subcarriers by g than that of SIM-OFDM. This feature of RIM-OFDM helps to minimize its PAPR value. Figure 6 shows plot of one OFDM symbol frame in time domain for classic OFDM and RIM-OFDM. It is quite evident from the figure that, our proposed RIM-OFDM scheme presents noticeable shrinkage in peak-to-average power level. Three OFDM systems frequently mentioned in this work, with M-QAM modulation, are simulated in MATLAB with 8-QAM, 16-QAM, and 32-QAM modulation and corresponding PAPR values are presented in Table 4.

The result from Table 4 depicts that the PAPR value is maximum in case of classic OFDM and minimum in our proposed scheme, RIM-OFDM. PAPR is mainly influenced

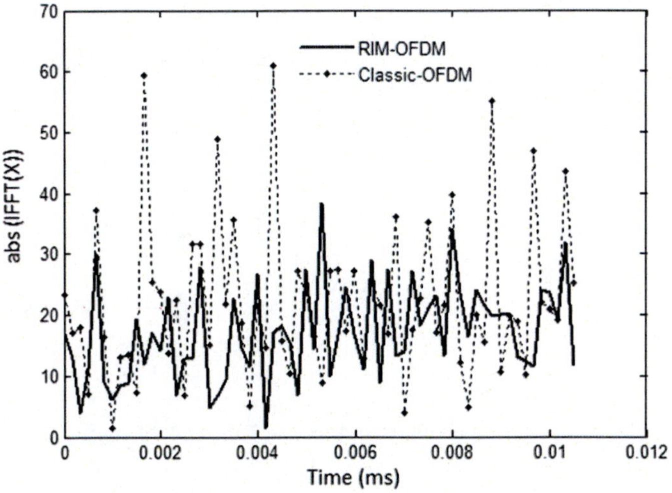

Fig. 6 Time domain plot of one OFDM symbol frame

Table 4 PAPR comparison table

	$\text{PAPR} = 10 \times \log_2\left(\frac{\max(xout)}{Av(xout)}\right)\text{dB}$		
	Classic OFDM $g = 4, n = 4, k = N, N = n \times g = 16$	SIM-OFDM $g = 4, n = 4, k = 2, N = n \times g = 16$	RIM-OFDM $g = 4, n = 3, k = 2, N = n \times g = 12$
8-QAM	4.97	3.82	2.58
16-QAM	6.373	4.384	3.78
32-QAM	5.31	4.01	3.28

by the number of information carrying subcarriers. As proposed in RIM-OFDM, the number of active subcarriers is less than the available subcarriers. Therefore, number of carriers is less in case of RIM-OFDM than classical OFDM and consequently, PAPR value is reduced in RIM-OFDM. However, although the number of active subcarrier per subblock in RIM-OFDM is same to that of SIM-OFDM, the total number of used subcarriers is getting decreased in RIM-OFDM due to the arrangement proposed in this literature. Hence, PAPR value is improved in RIM-OFDM from that of SIM-OFDM. Input and output constellation diagrams for RIM-OFDM with 16-QAM modulation are depicted in Fig. 7.

Plot of spectral efficiency against variation of k/n with n as parameter is shown in Fig. 8. It is evident from these plots that for higher value of n, RIM-OFDM offers improved spectral efficiency than classic-OFDM (at $k/n = 1$).

The optimum choice of number of active carriers (k) can also be made from the same graph for high values of n. As less number of subcarriers is required for transmission of the same number of input bits, RIM-OFDM exhibits an improvement also in spectral

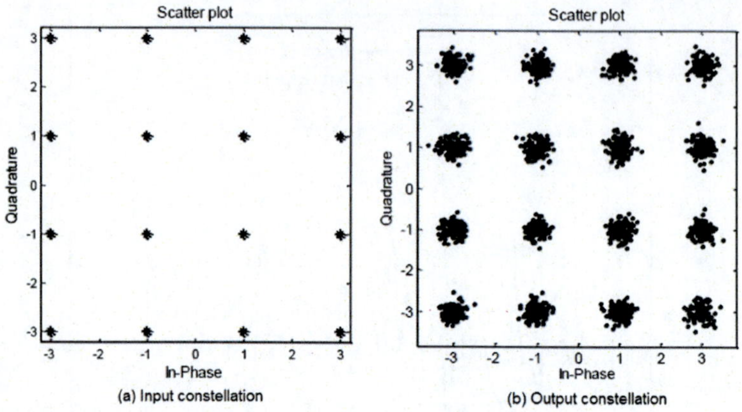

Fig. 7 **a** Input M-QAM constellation diagram; **b** output RIM-OFDM constellation diagram for $M = 16, N = 64, k = 2$

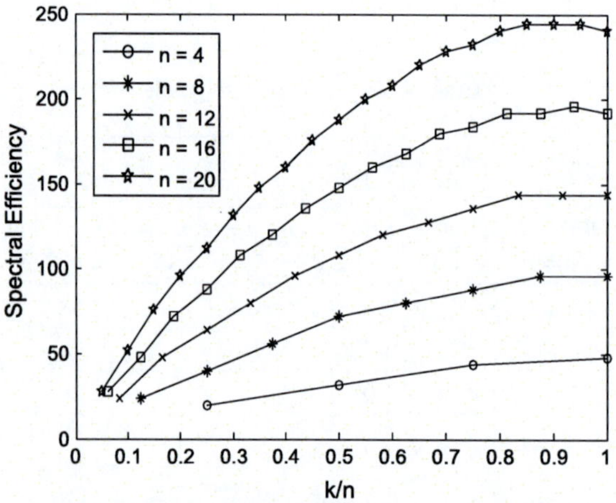

Fig. 8 Plot of spectral efficiency for RIM-OFDM

efficiency in comparison to SIM-OFDM. Mathematically, one can say (refer to Eq. 6)

$$\text{SE}_{\text{improvement}} = \frac{(\text{SE}_{\text{RIM - OFDM}} - \text{SE}_{\text{SIM - OFDM}})}{\text{SE}_{\text{SIM - OFDM}}} \times 100\%$$

Numerically, 33.33% improvement in SE is observed in case of RIM-OFDM with the parameters used and already specified in look up Tables 1 and 2.

BER versus SNR plot (for $N = 64$ and $M = 16$) as shown in Fig. 9 indicates improvement in received signal performance of RIM-OFDM in comparison to classic OFDM.

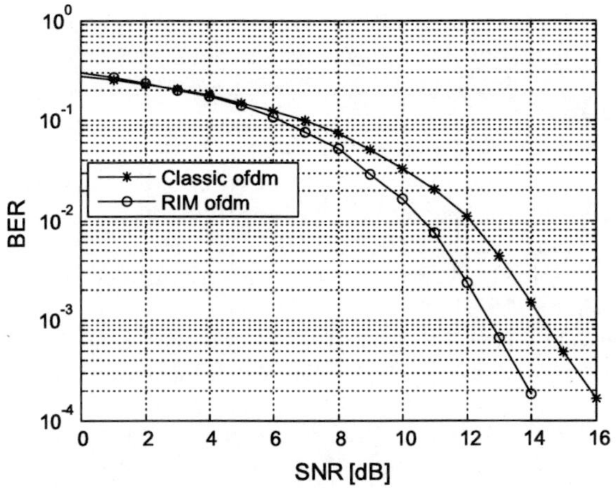

Fig. 9 BER versus SNR plot for classic and RIM-OFDM

4 Conclusion

A new RIM-OFDM modulation technique has been proposed and established mathematically in this work. Implementation of this newly proposed scheme is described in detail through block diagrams of transmitter and receiver along with look up tables. It is explained mathematically that, when compared to SIM-OFDM, a reduction in total allocated subcarriers by total number of subblocks used in the entire system is possible. Study of PAPR performance exhibits significant reduction in PAPR of RIM-OFDM. This feature, in addition with improvement in Bandwidth efficiency, has proved RIM-OFDM as an optimized modulation scheme.

References

1. Van Nee, R., Prasad, R.: OFDM for Wireless Multimedia Communications. Artech House, Boston (2000)
2. Stott, J.H., The how and why of COFDM. EBU Tech. Rev. 43–50 (1998)
3. Wu, Y., William, Z.Y.: Orthogonal frequency division multiplexing: A multi-carrier modulation scheme. IEEE Trans. Consum. Electron. **41**, 392–399 (1995)
4. Nee, R.V., Wild, A.: Reducing the peak-to-average power ratio of OFDM. In: VTC '98. 48th IEEE Vehicular Technology Conference. pp: 2072–2076 (1998)
5. Armstrong, J.: OFDM for optical communications. J. Lightwave Technol. **27**(3), 189–204 (2009)
6. Baig, I., Jeoti, V.: A ZCMT precoding based multicarrier OFDM system to minimize the high PAPR. Wireless Pers. Commun. **68**(3), 1135–1145 (2013)
7. O'Neill, R., Lopes, L.B.: Envelope variations and spectral splatter in clipped multicarrier signals. In: Personal, Indoor and Mobile Radio Communications, PIMRC'95. Wireless: Merging onto the Information Superhighway, Sixth IEEE International Symposium, vol. 1, pp. 71–75, Sept 1995
8. Abu-alhiga, R., Haas, H.: Subcarrier-index modulation OFDM. In: Proceedings of IEEE International Symposium on Personal, Indoor and Mobile Radio Communications, Tokyo, Japan, pp. 177–181, Sep 2009

9. Basar, E., Aygölü, U., Panayırcı, E., Poor, H.V.: Orthogonal frequency division multiplexing with index modulation. IEEE Trans. Signal Process. **61**(22), 5536–5549 (2013)
10. Mesleh, R.Y., Haas, H., Sinanovic, S.: Spatial modulation. IEEE Trans. Veh. Technol. **57**(4), 2228–2241 (2008)
11. Jeganathan, J., Ghrayeb, A., Szczecinski, L., Ceron, A.: Space shift keying modulation for MIMO channels. IEEE Trans. Wireless Commun. **8**(7), 3692–3703 (2009)
12. Fu, J., Hou, C., Xiang, W., Yan, L., Hou, Y.: Generalised spatial modulation with multiple active transmit antennas. In: IEEE GLOBECOM Workshops, pp. 839–844, Dec 2010
13. Tasadduq, I.A., Rao, R.K.: PAPR reduction of OFDM signals using multiamplitude CPM. Electron. Lett. **38**(16), 915–917 (2008)
14. Zhang, H., Ruyet, D.L., Terre, M.: Spectral efficiency analysis in OFDM and OFDM/OQAM based cognitive radio networks. In: VTC-2009. 69th IEEE vehicular technology conference, May 2009
15. Fan, R., Yu, Y.J., Guan, Y.L.: Generalization of orthogonal frequency division multiplexing with index modulation. In: IEEE Trans. Wireless Commun. **14**(10), 5350–5359, Oct 2015

Realization of a 5G Communication System with Rain Fading Mitigation Through Uplink Power Control

Susovan Mondal[1](✉), Dalia Nandi[2], Rabindranath Bera[3], and Subhankar Shome[4]

[1] ECE Department, Surendra Institute of Engineering and Management, Siliguri, West Bengal 734009, India
susovan.1985@gmail.com
[2] ECE Department, Indian Institute of Information Technology Kalyani, Kalyani, West Bengal 741235, India
[3] ECE Department, Sikkim Manipal Institute of Technology, Sikkim Manipal University, Gangtok, Sikkim 737136, India
[4] ECE Department, St. Mary's Technical Campus, Kolkata, West Bengal 700120, India

Abstract. With the objective of improving the reliability of an operational 28 GHz full duplex line of sight (LOS) link even during rain, authors have initiated the link experiment with uplink power control procedure between a 28 GHz transmitter and receiver. The degradation of the said link is monitored at the receiver by measuring its power level which in turn is used to control the transmit power. One threshold level of fade margin has been set at the receiver, below which the control of transmit power is initiated to establish the link. Thus, in reality, the above-experimental facts are ensuring the establishment of a 28 GHz link and are thus useful for the future 5G millimeter-wave communication system which will face strong rain attenuation particularly in a tropical country like India.

Keywords: Millimeter wave · FBMC · Phydyas filter · OQAM

1 Introduction

With the tremendous growth in mobile traffic demand, spectrum shortage becomes a burning problem for telecom operators. On the other hand, huge bandwidth is available in the millimeter (mm) wave band from 30 to 300 GHz. So, for future 5th generation (5G) mobile communications, millimeter waves have been proposed [1, 2]. The spectrum between 24.25 and 86 GHz will be considered for IMT under WRC-19. In addition, the 28 GHz band will be used for millimeter-wave 5G in some countries, such as the USA, South Korea, Japan, India, and Canada [3]. But high propagation loss due to different atmospheric effects, line of sight (LOS) requirement and sensitivity to blockage make millimeter wave very vulnerable for long-distance communication [4, 5]. In order to fulfill the new requirement of 5G, like machine-to-machine communication (MTC) and Internet of things (IoT), new multicarrier modulation methods have been proposed in

© Springer Nature Singapore Pte Ltd. 2021
N. R. Das and S. Sarkar (eds.), *Computers and Devices for Communication*,
Lecture Notes in Networks and Systems 147,
https://doi.org/10.1007/978-981-15-8366-7_18

place of orthogonal frequency-division multiplexing (OFDM). In this paper, we have chosen filter bank multicarrier (FBMC) modulation for its good spectral efficiency than OFDM. Here, we also demonstrated a method to mitigate rain attenuation of millimeter wave of 28 GHz band by controlling its transmitting power.

2 FBMC

In FBMC, a filter is applied per subcarrier and each subcarrier is filtered on its own. It uses a long frequency selective filter which reduces out of band radiation. So, the spectral efficiency of FBMC is better than other multicarrier systems like OFDM [6, 7].

2.1 Prototype Phydyas Filter

FBMC addresses the spectral leakage problem of OFDM with the help of filtering the signal on a subcarrier basis using a long prototype filter $h(n)$. The length of the prototype filter is KN. This prototype filter is K times longer than a rectangular OFDM symbol and N is the number of subcarriers; K is also called the overlapping factor since each FBMC symbol overlaps with K neighboring time-domain symbols. In FBMC, different pulse shaping filters like root raised cosine, Gaussian and Phydyas can be used. In this paper, we implemented a Phydyas filter as it gives low out of band radiation than others. The frequency response of the Phydyas filter is given by Eq. (1)

$$H(f) = \sum_{k=-(K-1)}^{K-1} H_k \frac{\sin\left(\pi\left(f - \frac{k}{\mathrm{MK}}\right)\mathrm{MK}\right)}{\mathrm{MK}\sin\left(\pi\left(f - \frac{k}{\mathrm{MK}}\right)\right)} \tag{1}$$

The impulse response $h(t)$ of the filter is given by the inverse Fourier transform of the $H(f)$, which is

$$h(t) = 1 + 2\sum_{k=1}^{K-1} H_k \cos\left(2\pi\frac{kt}{KT}\right) \tag{2}$$

2.2 OQAM Modulation

To achieve full capacity, offset quadrature amplitude modulation (OQAM) technique is employed. The real and imaginary parts of a complex data symbol are not transmitted simultaneously, as the imaginary part is delayed by half the symbol duration. Transmitted FBMC signal is created using the following steps [6]. For i^{th} subcarrier, OQAM symbols are given by,

For even no. of subcarrier: $s_i(m) = \mathrm{Re}[d_i(l)]$ and $s_i(m+1) = \mathrm{Im}[d_i(l)]$.

For odd no. of subcarrier: $s_i(m) = \mathrm{Re}[d_i(l)]$ and $s_i(m+1) = \mathrm{Im}[d_i(l)]$.

Here, d_i is QAM data symbols and $m = 2l$.

The shifted version of the prototype filter of ith subcarrier is

$$f_i(n) = h(n)e^{j2\pi\left(\frac{in}{N}\right)} \tag{3}$$

The transmitted FBMC signal is given by,

$$y(n) = \sum_{i=o}^{N-1} \sum_{m=-\infty}^{m=+\infty} s_i(m) f_i\left(n - \frac{mN_s}{2}\right) \tag{4}$$

3 Establishment of a 28 GHz Full Duplex LOS Link

In the laboratory, we have established a 28 GHz full duplex LOS link [8], as shown in Fig. 1a. Distance between the transmitter (Tx) and receiver (Rx) is 100 m. For transmission and reception of mm-wave signals, we used two 28 GHz transceivers on both sides. The system parameter specifications are given in Table 1.

4 Control of the Uplink Power to Mitigate the Rain-Induced Large Scale Signal Fading

One large plastic bucket [1.5 ft × 1.5 ft × 3 ft] is placed in front of the LOS link to emulate the rain fading by pouring water into it through a water tap. It is graduated with inch scale from 0 to 24 inch to indicate the height of the water as shown in Fig. 1b. Large scale rain fading is thus emulated at the laboratory and is highly useful in defining the problem of rain-induced large scale signal fading.

4.1 Use of Pilot in the Restoration of LOS Performance After Mitigation of Rain Fading

The above success story motivates the authors strongly and in turn has initiated to automate the above process with the insertion and transmission of a 1 kHz sinusoidal pilot signal. At the receiver, we have set a threshold level of received power. Due to rain, if received signal power degrades and goes below that threshold level, then a signal of 1 kHz will be transmitted from the receiver. At the transmitter end, the 1 kHz signal has been received through a matched filter to clean the pilot from contamination of noise and interference. Different amplitudes of the pilot signal linearly control the transmit power. From the knowledge of received signal amplitude, the transmitter controls its transmitting power and restore the link during rain.

The experiment has been conducted using the above setup and link outage performance has been tested during filling up of water column within the bucket. The received power level has been recorded under different water heights and tabulated as shown in Table 2.

Thus, a very interesting rain mitigation experiment is completed at the laboratory by manually controlling the uplink transmitting power using a laptop remotely.

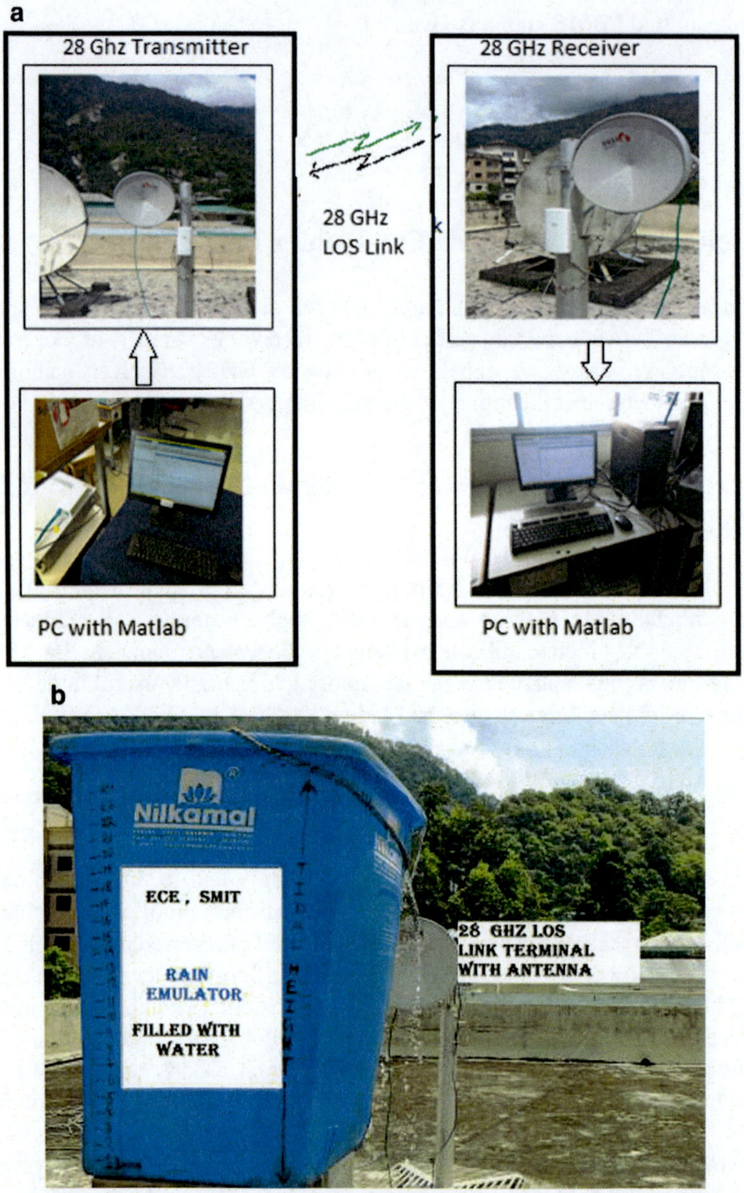

Fig. 1 **a** 28 GHz full duplex 5G LOS link at EC Department, SMIT, Sikkim, **b** rain emulator placed along the line of sight path between two terminals

5 Result

Here, we simulated FBMC transmitter in MATLAB. At the FBMC transmitter side, Phydyas filter with the following filter coefficient: $H1 = 0.971960$; $H2 = \sqrt{2}/2$; $H3 = 0.235147$ are used [9]. To transmit the signal through 28 GHz LOS link, we used the

Table 1 System parameters

Parameter	28 GHz transmitter	28 GHz receiver
Tx frequency	28,780.5 MHz	27,772.6 MHz
Rx frequency	27,772.6 MHz	28,780.5 MHz
Tx power	0 dBm (typical)	0 dBm (typical)
Tx power control range	0–17 dBm	0–17 dBm
System bandwidth	60 MHz	60 MHz
Rx level	−30 to −80 dBm	−30 to −80 dBm
MSE level (mean square error)	−1 to −40 dBm	−1 to −40 dBm
Modem	Up to 1024 QAM	Up to 1024 QAM
Ethernet port	Available	Available
Duplexing mode	FDD	FDD
Antenna	0.3 m dish	0.3 m dish
Remote monitoring and control	Using PC	Using PC

Table 2 Tx and Rx power before and after power control

Water column height (inch)	Tx power (dBm)	Initial Rx power (dBm)	Signal fading (dB)	Final Rx power after power control (dBm)	Remarks
0	0	−44	0	−44	No rain
10	0	−44	0	−44	No rain
14	8	−52	8	−44	Fully restored
15.5	10	−54	10	−44	Fully restored
20	17	−70	26	NA	Link failed

experimental setup of Fig. 1a. The experiment was carried out in two modes. In mode-1, the FBMC signal is transmitted without a rain emulator and from Fig. 2a, we can conclude that in no rain situation signal has been faithfully received. But in mode-2, the rain emulator is filled with water. From Table 2, we can say that when the water column height is below 15 inches, by uplink power control mechanism stated in sect. 4, we can restore the link. Transmitted and received signals in mode-2 are shown in Fig. 2b.

6 Summary and Conclusion

In the above experiment, the authors have established a 28 GHz LOS link with one 5G base station on one side and one 5G UE terminal on the other side. Several steps

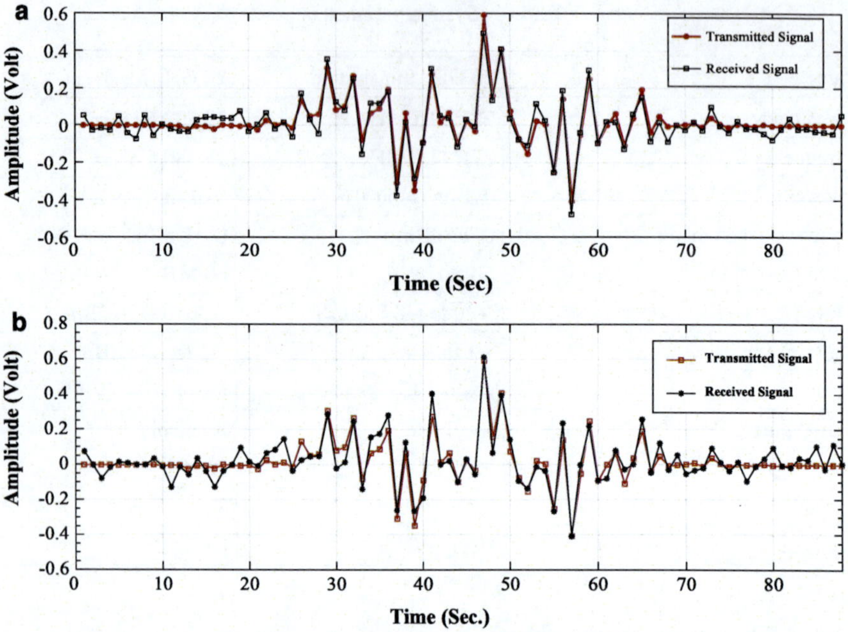

Fig. 2 **a** Transmitted and received signal without rain emulator (no rain condition), **b** transmitted and received signal with rain emulator (water level up to 15 inch)

have been followed during development. In step-1, a basic LOS link has been realized and FBMC-based 5G waveform is tested over the same link. The problems of rain-induced large scale fading were experimented in step-2 by inserting one rain emulator made of a plastic bucket and pouring water into it. The remedial measure was also experimented in step-3 by controlling the transmitter power. Further, the same is automated using a 1 kHz pilot and associated equipment and circuits. In this way, the reliability of the said link was improved.

References

1. Elkashlan, M., Duong, T.Q., Chen, H.H.: Millimeter-wave communications for 5G: fundamentals: Part I [Guest Editorial]. IEEE Commun. Mag. **52**(9), 52–54 (2014)
2. Elkashlan, M., Duong, T.Q., Chen, H.H.: Millimeter-wave communications for 5G–part 2: applications. IEEE Commun. Mag. **53**(1), 166–167 (2015)
3. Agenda and Reference (Resolutions and Recommendations), ITU, Aug 2017
4. Han, C., Duan, S.: The study on characteristics of rain attenuation along 28 GHz and 38 GHz line-of-sight millimeter-wave links. URSI AP-RASC 2019, New Delhi, India, 09–15 March 2019
5. Lam, H.Y., et al.: Impact of rain attenuation on 5G millimeter wave communication systems in equatorial Malaysia investigated through disdrometer data 2017. In: 11th European Conference on Antennas and Propagation (EUCAP)

6. Van Eeckhaute, M., Bourdoux, A.: Philippe De Doncker and François performance of emerging multi-carrier waveforms for 5g asynchronous communications. EURASIP J. Wireless Commun. Netw. (2017). https://doi.org/10.1186/s13638-017-0812-8
7. Bellanger, M.G.: IEEE International Conference on Acoustics, Speech, and Signal Processing. Specification and design of a prototype filter for filter bank based multicarrier transmission, vol. 4, pp. 2417–2420 (2001)
8. Shome, S., et al.: LTE signal relay system using 28 GHz millimeter wave micro cell design for smart home. URSI AP-RASC 2019, New Delhi, India, 09–15 Mar 2019
9. Louveaux, J., Baltar, L., Waldhauser, D., Renfors, M., Tanda, M., Bader, C., Kofidis, E.: PHYDYAS D 3.1. Technical report, PHYDYAS (2008). www.ictphydyas.org/deliverables/PHYDYAS

Check for updates

An ANN Approach in Predicting Solar and Geophysical Indices from Ionospheric TEC Over Indore

Sumanjit Chakraborty(✉) and Abhirup Datta

Discipline of Astronomy, Astrophysics and Space Engineering, Indian Institute of Technolgy Indore, Simrol, Indore 453552, India
{phd1601121006,abhirup.datta}@iiti.ac.in

Abstract. In this paper, preliminary results from the artificial neural network (ANN)-based model developed at IIT Indore have been presented. One year hourly total electron content (TEC) database has been created from the International Reference Ionosphere (IRI)—2016 model. For the first time, a reverse problem has been addressed, wherein the training has been performed for predicting the three indices: 13-month running sunspot number, ionospheric index and daily solar radio flux also called targets to the network when hourly TEC values are the inputs. The root mean square errors (RMSEs) of these targets have been compared and minimized after several training of the dataset using different sets of combinations. Unknown data fed to the network yielded 0.99%, 3.12% and 0.90% errors for Rz12, IG12 and F10.7 radio flux, respectively, thus signifying ~97% prediction accuracy of the model.

Keyword: Solar indices · Geophysical indices · Ionospheric TEC · IRI · Machine learning · ANN

1 Introduction

The Earth's ionosphere is formed as a consequence of ionization by solar radiation. It extends from about 60–1000 km above the Earth's surface. Since nearly 70% of the global ionization is concentrated in and around $\pm 15°$ magnetic latitude crests due to the equatorial ionization anomaly (EIA) [1], it becomes essential to characterize dynamism of the variable ionosphere over this region where sharp latitudinal gradient exists. The total electron content (TEC), which is the columnar number density of electrons expressed in TEC units (1 TEC unit = 10^{16} electrons/m^2), is a fundamental observable parameter [6] which plays an important role in characterization of the ionosphere. The location chosen for the analysis, Indore (22.52° N, 75.92° E geographic) falls near the northern crest of EIA and as a result is a suitable location to study the variability of the ionosphere. Complexity of the spatial and temporal variations of the ionosphere makes it difficult to characterize or model the ionosphere and accurately forecast its impact on the global navigation satellite system (GNSS) signals. Therefore, a requirement arises

© Springer Nature Singapore Pte Ltd. 2021
N. R. Das and S. Sarkar (eds.), *Computers and Devices for Communication*, Lecture Notes in Networks and Systems 147,
https://doi.org/10.1007/978-981-15-8366-7_19

for the development of an artificial neural network (ANN)-based model that would be able to predict ionospheric behavior over the regions where physical data is unavailable.

ANNs, inspired from the neural networks which constitute animal brains, have collection of connected nodes known as artificial neurons. Each connection similar to the synapses in a biological brain can transmit signal from one artificial neuron to another [8]. These connections between neurons are called edges. Artificial neurons and these edges have a weight that self adjust as learning proceeds. This weightage may increase or decrease the strength of the signal at a connection. The neurons have a threshold above which the signal is sent and are aggregated into different layers which perform various transformations on their inputs. The signals travel from the input (first layer), traverses multiple layers (hidden layers) and arrives to the output (last layer) [4]. The ANN discussed in this paper has been developed by preparing dataset from the empirical International Reference Ionosphere (IRI) model. The sources of data to this model are the dense global network of ionosondes, incoherent scatter radars, and Alouette topside sounders in situ instruments on board satellites. Inputs to this model are the date, latitude, longitude and topside electron boundary while the outputs are electron temperature and density, ion temperature and composition and the TEC from 50 to 2000 km altitude range [7].

Studies have been made by several researchers [2, 9, 10] to predict TEC model of the ionosphere by auto-regressive method. Studies have also been made [3, 5] in development of ANN-based TEC model where the solar and geophysical indices are fed as inputs to obtain predicted TEC as output, but for the first time, to the best of our knowledge, the TEC data has been fed as network inputs to obtain the solar and geophysical indices. This work thus addresses a reverse problem, wherein by having the knowledge of TEC variation, one could be able to infer about the indices that would be vital in understanding space weather and ensure flawless service to GNSS users.

2 Methodology

A feed forward network has been used where the signal gets propagated from the input layer to the hidden layer and then to the output layer. The present model is generated by using a single hidden layer of 25 neurons. The model inputs are the hourly vertical TEC values over Indore obtained from the IRI-2016 Web model for the entire year of 2017 which had been in the declining phase of solar cycle 24. The targets set to this model are the 13-month running mean of sunspot number (Rz12), the ionospheric index (IG12) and the F10.7 daily radio flux (sfu). Connections between the nodes are such that they represent the feeding of the output from one node to the other, multiplied by a weight. The weights given to the hidden layer are appropriately modified to obtain relatively less prediction error between the targets and predicted indices. The optimized architecture for the network is obtained by trial and error while the biases and weights are adjusted according to the Levenberg–Marquardt algorithm. The architecture of an ANN with an input layer, a single hidden layer and an output layer is depicted in Fig. 1.

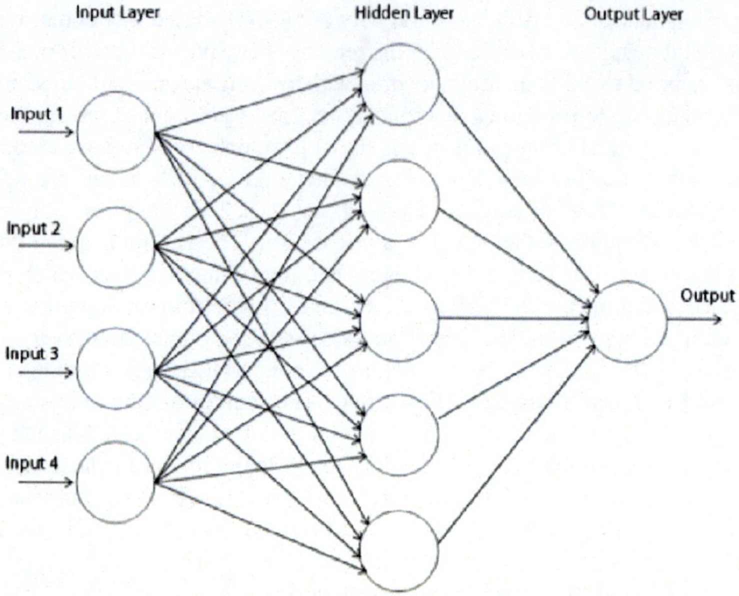

Fig. 1 Typical ANN structure showing the input, hidden and output layers (https://www.analyticsindiamag.com/artificial-neural-networks-101/)

3 Results

For training the neural network, 300 days were randomly selected from a one year database. The splitting of data for training, validation and testing was randomly selected as 70%, 15% and 15%, respectively. The activation function found to be suitable was tan sigmoid given by:

$$\tan \operatorname{sig}(x) = \frac{1 - e^{-2x}}{1 + e^{-2x}} \tag{1}$$

This activation function is used to introduce nonlinearity to the network. This helps the network to understand the complexity and give accurate results. The error function at the end of one feed forward process to check training performance was the mean squared error (MSE) given by the mean of the squared of the error defined as the difference between the predictions and the targets. The idea is to minimize this error function by assigning suitable weights and biases at every step. The network was then trained several times by changing the number of neurons until the cost function was minimum that would perform well when subjected to unknown data. The remaining 65 days that is unknown to the trained network was used in order to check the model performance. Figure 2 shows the scatter plot of predictions and target with the 1:1 red line signifying an accuracy of 100%. The normalized RMSE values obtained are 0.0099, 0.0312 and 0.0090 translating to percentage errors of 0.99%, 3.12% and 0.90% for the indices Rz12, IG12 and F10.7 radio flux, respectively. These RMSE values are computed by using:

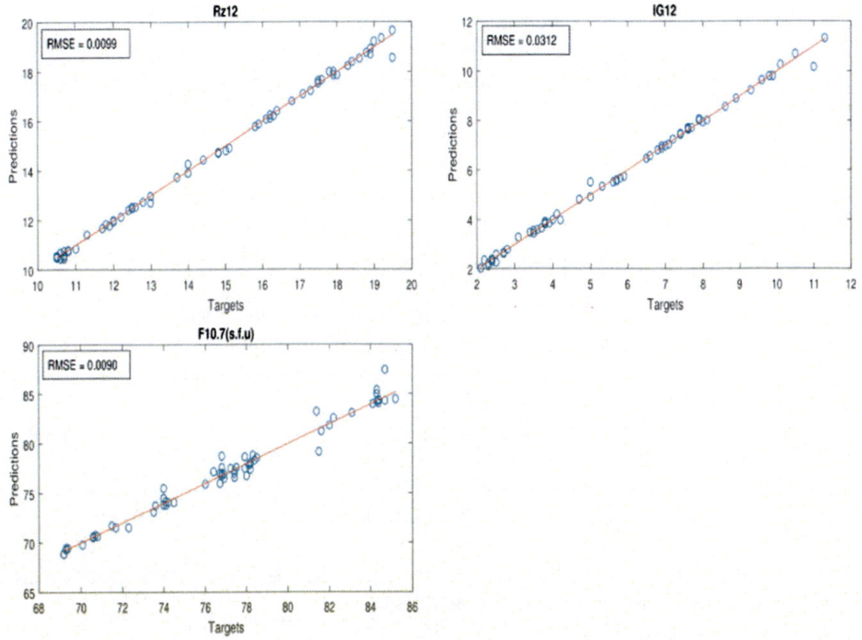

Fig. 2. Scatter plot showing the true values (targets) and the outputs (predictions) values from the neural network with one hidden layer of 25 neurons

$$\sqrt{\frac{1}{N}\sum\left(\frac{\text{targets} - \text{predictions}}{\text{targets}}\right)^2} \tag{2}$$

4 Conclusions

The paper presents initial results from the ANN model developed over Indore, which is near to the anomaly crest, using the IRI model derived database to predict solar and geophysical indices. A single hidden layer has been used and a number of neurons have been varied to obtain the optimized results of this model. The normalized RMSE gave about 0.99, 0.90 and 3.12% errors in Rz12, F10.7 and IG12 index, respectively. Thus, ~97% accuracy has been achieved when unknown data was fed to the trained network. A reverse problem approach is addressed, wherein with the knowledge of TEC, prediction of various indices can be obtained even if real-time indices are not available. Path forward to this work would be to use this approach in training the network with real data and compare with the presented work for validation. This model could help in understanding the variable ionosphere where real data is unavailable.

References

1. Appleton, E.V.: Two anomalies in the ionosphere. Nature 157 (1946)

2. Dick, M., Levy, M.F., Cander, L.R., Kutiev, I., Muhtarov, P.: Short term ionospheric forecasting over Europe. IEE National Conference on Antennas and Propagation. IET (1999)+
3. Habarulema, J.B., McKinnell, L.A., Cilliers, P.J., Opperman, B.D.: Application of neural networks to South African GPS TEC modelling. Adv. Space Res. **43**(11), 1711–1720 (2009). https://doi.org/10.1016/j.asr.2008.08.020
4. Hassoun, M.H.: Fundamentals of artificial neural networks. The MIT Press, Cambridge, MA (1995)
5. Jakowski, N., Hoque, M.M., Mayer, C.: A new global TEC model for estimating transionospheric radio wave propagation errors. J. Geodesy **85**(12), 965–974 (2011). https://doi.org/10.1007/s00190-011-0455-1
6. Maruyama, T.: Solar proxies pertaining to empirical ionospheric total electron content models. J. Geophys. Res. **115** (2010). https://doi.org/10.1029/2009JA014890
7. Rawer, K., Bilitza, D.: Electron density profile description in the international reference ionosphere. J. Atmos. Terr. Phys. **51**, 9–10 (1989)
8. Sarle, W.: Neural networks and statistical models. In: Proceedings of the Nineteenth Annual SAS Users Group International Conference (1994)
9. Stankov, S., Jakowski, N.: Indexing the local ionospheric response to magnetic activity by using total electron content measurements. Acta Geodaetica et Geophysica Hungarica 41 (2006)
10. Tsai, L., Macalalad, E., Liu, C.: Taiwan ionospheric model (TWIM) prediction based on time series autoregressive analysis. Radio Sci. 49 (2014)

Cloud and Rain Attenuation Statistics from Radiosonde and Satellite Observations Over a Tropical Location

Niket Kumar[1], Arijit De[1(✉)], and Animesh Maitra[2]

[1] Netaji Subhash Engineering College, Garia, Kolkata, India
arijitde33@gmail.com

[2] Institute of Radio Physics & Electronics, University of Calcutta, Kolkata, India

Abstract. The objective of this study is to investigate the cloud and rain attenuation at frequencies above 30 GHz, using radiosonde observations and TRMM satellite measurements, respectively, at a tropical location, Kolkata. The location experiences high value of liquid water content (LWC), the main reason for cloud attenuation, during Indian summer monsoon and pre-monsoon season. The rain height variation from TRMM satellite data has also been observed. Rain height varies from 3992 m to 5191 for the years 2004–2010 showing significant interannual variability. The exceedance curve shows higher occurrence in pre-monsoon than in monsoon season for rain attenuation values greater than 10 dB.

Keywords: Rain attenuation · Cloud attenuation · Radiosonde · TRMM satellite

1 Introduction

The communication systems at higher frequency bands are vulnerable to propagation effects which are very much significant at Ka and higher frequency bands. In tropical location, earth-space signal propagation is much affected due to the various atmospheric parameters, like rain and cloud. Rain imposes the most significant detriment on satellite communication links above 10 GHz. The amount of rain-induced fades depends on signal frequency, rain rate, shape, size and relative orientations of the rain drops. Rain rate exceedance for 0.01% of time of the year is required to calculate rain-induced attenuation. Thus, an accurate estimation of yearly variations of rainfall is needed. However, the related studies are mostly from different temperate locations [1–10].

Cloud is the next significant propagation impairment after rain. The effect of cloud should also be considered for higher frequencies and a greater period of time [11]. The liquid water content (LWC) of clouds is the physical cause of cloud attenuation.

Cloud attenuation is an important parameter to be considered for optimal trade-off designing of low fade margin satellite communication links [11]. In this study, cloud and rain attenuation at Ka band has been presented from a tropical location, Kolkata using radiosonde and TRMM data.

© Springer Nature Singapore Pte Ltd. 2021
N. R. Das and S. Sarkar (eds.), *Computers and Devices for Communication*,
Lecture Notes in Networks and Systems 147,
https://doi.org/10.1007/978-981-15-8366-7_20

2 Data

Radiosonde observations are made twice a day, at around 00 and 12 GMT (1830 and 0630 IST). The data of temperature, pressure and dew point temperature have been measured at different heights. The height resolutions are varying from a few tens of meters to a few hundreds of meters up to a height of 15 km. The type of radiosonde is IMD—MK IV.

The temperature sensor is a carbon rod thermistor. It measures temperature in the range—90 to 60 °C with a resolution of 0.1 °C. The pressure is measured with an aneroid barometer which can measure pressure with a resolution of 1 mb. The dew point temperature is obtained from the relative humidity measurement with a carbon hygristor having a resolution of 2% RH. Radiosonde measurements obtained by the Indian Meteorological Department at Kolkata, India (22°34′ N, 88°29′ E), a tropical location, during the period January to December 2010, have been used in the present paper.

TRMM-3B42RT 3-hourly data are analyzed from pre-monsoon to monsoon. The rain height has been calculated from TRMM 3A25 data. The radiosonde and TRMM data for the year 2010 have been considered for the present study.

3 Cloud Attenuation Statistics

ITU-R model: According to ITU-R model, the total attenuation due to cloud is:

$$A_c = \frac{L * K_l}{\sin\theta} \tag{1}$$

L is the total columnar content of liquid water (kg/m^2), K_l depends on frequency and dielectric spectra. Figure 1 shows the profile of liquid water density (LWP) obtained from the Salonen model using radiosonde data on August 26, 2010. In the figure, the cloud layer has been shown to have formed in the height range for which the relative humidity exceeds the critical humidity. The total liquid water content (LWC) has been obtained by integrating the liquid water density profile at every height. The above technique has been used to calculate the number of cloudy days for the year 2010. It has been observed in Fig. 2a that almost all the days are cloudy days for the month August. It is quite obvious that number of cloudy days are more in monsoon season than pre-monsoon. However, number of cloudy days in pre-monsoon season (26 days) are quite significant from communication perspective. The height profile of LWC within cloud has been observed in Fig. 2b. The number of points of LWC is more in lower height than medium or higher height in monsoon. This indicates the significant contribution of low level clouds in cloud attenuation.

3.1 Exceedance Statistics

Figure 3 shows the cloud attenuation exceedance for different seasons over Kolkata. To estimate system reliability from these curves, it is necessary to know the percentage of time that cloud attenuation levels are exceeded.

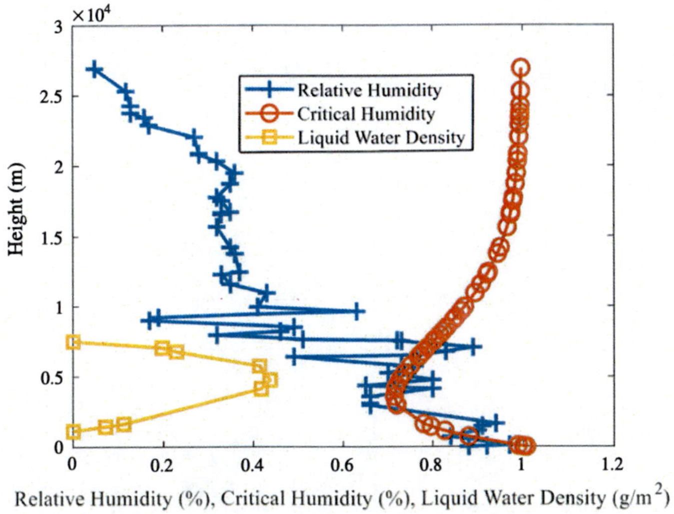

Fig. 1 Technique to calculate liquid water density (g/m^3)

4 Rain Attenuation Statistics

4.1 Rain Height Variation

Tropical Rain Measuring Mission-Precipitation Radar (TRMM-PR) satellite are used to determine the distribution of rain height based on 0 °C isotherm height over Kolkata, India. The results show bright-band height varies between 3992 and 5191 m above mean sea level over the years of observation. The average value of rain height has been calculated for different months. The rain height is more in the month of June (5000 m) (Fig. 4).

4.2 Rain Attenuation Exceedance

Rain attenuation is the prime factor causing degradations in the radio signal above 10 GHz frequency. In order to be able to mitigate the incurred fade, it is necessary to know the percentage of time exceedance of rain attenuation. Figure 5 shows the worst month statistics of rain attenuation during the years 2010. The difference of the rain attenuation values between the monsoon and the pre-monsoon is significant. The pre-monsoon rain occurs due to local intense convective processes. The transport of water vapor from the Indian Ocean from the south-west direction is the main reason for monsoon rain. The monsoon rain is a mixture of convective and stratiform precipitation. The rain characteristics during the two seasons play a major role in causing the variations of the rainfall parameters. The higher exceedance value at higher attenuation (>10 dB) in pre-monsoon season indicates the abundance of extreme events.

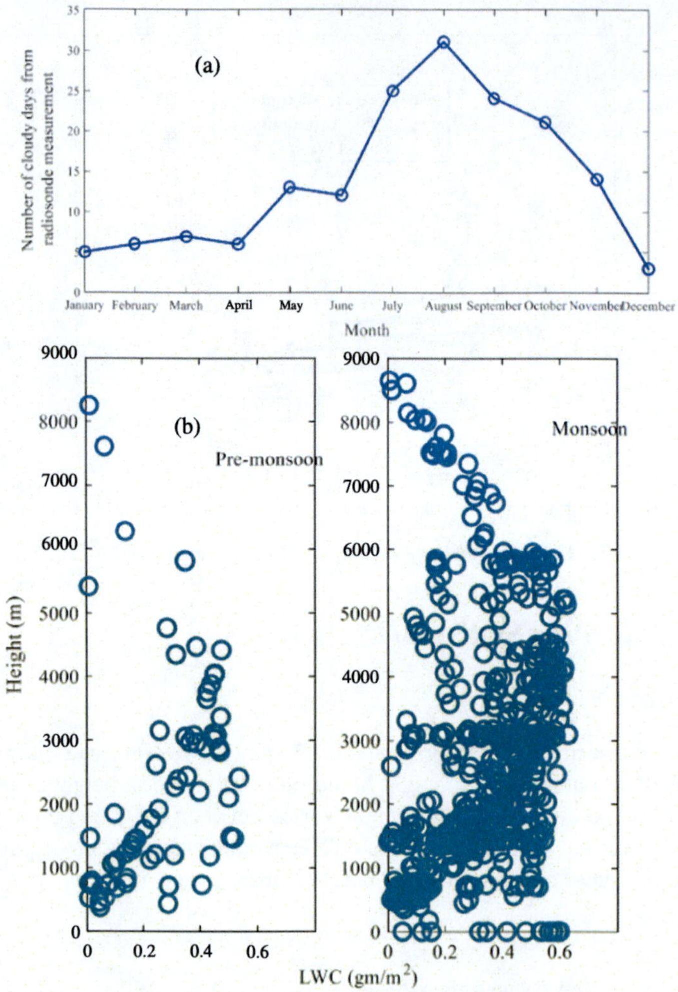

Fig. 2 Cloud attenuation exceedance probability

5 Conclusion

In this paper, cloud and rain attenuation has been studied using radiosonde and satellite data, respectively. The exceedance values of cloud attenuation are quite significant at higher frequency (30 GHz). The contribution of lower level clouds (<3000 m) is more significant in cloud attenuation than upper level cloud. The rain height has been calculated from TRMM satellite data and it lies between 3992 m and 5191 m indicating strong interannual as well as seasonal variation of rain characteristic over this location. Pre-monsoon season shows higher exceedance value in higher attenuation (>10 dB) which indicates more number of extreme events. Long-term analysis of rain and cloud attenuation is required to know the reliable estimation of fade margin at Ka band frequency over this location, Kolkata.

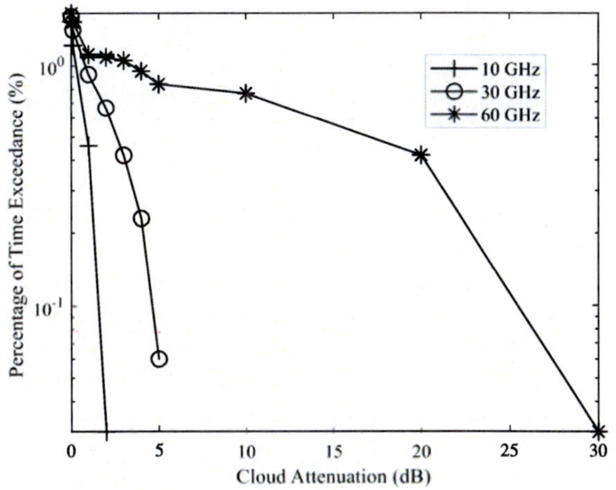

Fig. 3 Cloud attenuation exceedance probability

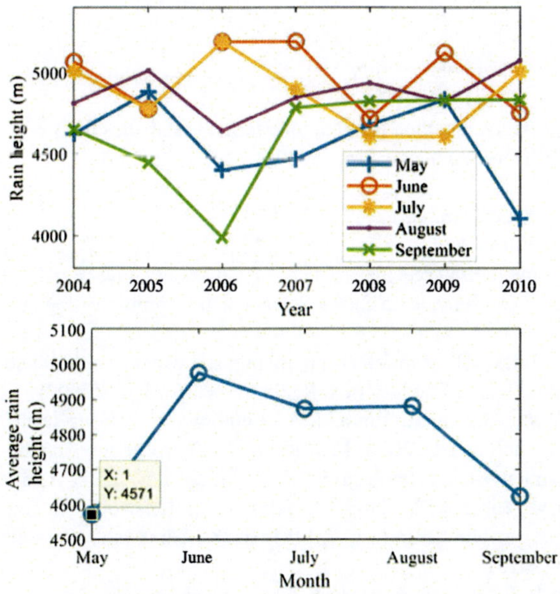

Fig. 4 Rain height from TRMM data

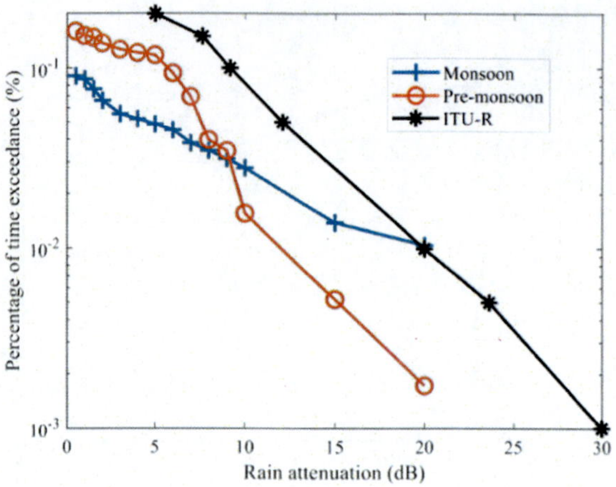

Fig. 5 Rain attenuation exceedance at 30 GHz

References

1. Rice, P., Nettie, H.: Cumulative time statistics of surface-point rainfall rates. IEEE Trans. Commun. **21**(10):1131–1136 (1973)
2. Lee, W.: An approximate method for obtaining rain rate statistics for use in signal attenuation estimating. IEEE Trans. Antennas Propag. **27**(3), 407–413 (1979)
3. Dutton, E., Dougherty, H.: Year-to-year variability of rainfall for microwave applications in the USA. IEEE Trans. Commun. **27**(5), 829–832 (1979)
4. Omotosho, T.V., Willoughby, A.A., Akinyemi, M.L., Mandeep, J.S., Abdullah, M.: One year results of one minute rainfall rate measurement at Covenant University Southwest Nigeria. In: IEEE International Conference on Space Science and Communication (IconSpace), pp. 98–101 (2013)
5. Owolawi, A.P., Afullo, T.J.: Rainfall rate modeling and worst month statistics for millimetric line-of-sight radio links in South Africa. Radio Sci. **42**(6), 1–11 (2007)
6. García Rubia, J.M., García, P., Riera, J.M., Benarroch, A.: Long term measurements of slant-path propagation at 20 GHz in Madrid. In: 2010 Proceedings of the Fourth European Conference on Antennas and Propagation (EuCAP), pp. 1–5. IEEE, April (2010)
7. Garcia Rubia, J.M., Riera, J.M., Garcia, P., Siles, G.A., Benarroch, A.: Experimental assessment of slant-path rain attenuation variability in the Ka-band. Int. J. Satellite Commun. Netw. **34**(2), 155–170 (2016)
8. Garcia-del-Pino, P., Riera, J.M., Benarroch, A.: Tropospheric scintillation with concurrent rain attenuation at 50 GHz in Madrid. IEEE Trans. Antennas Propag. **60**(3):1578–1583 (2012)
9. Acosta, R.J., Matricciani, E., Riva, C.: Slant path attenuation and microscale site diversity gain measured and predicted in Guam with the synthetic storm technique at 20.7 GHz. In: 7th European Conference on Antennas and Propagation (EuCAP), pp. 61–64. IEEE, April (2013)
10. Matricciani, E.: Probability distributions of rain attenuation obtainable with linear combining techniques in space-to-Earth links using time diversity. Int. J. Satellite Commun. Netw. **36**(2), 220–237 (2018)
11. Ippolito, L.J., Ippolito Jr, L.J.: Satellite communications systems engineering: atmospheric effects, satellite link design and system performance. Wiley, New York (2017)

Impact of Intense Geomagnetic Storm on NavIC Signals Over Indore

Deepthi Ayyagari(✉), Sumanjit Chakraborty, Abhirup Datta, and Saurabh Das

Discipline of Astronomy, Astrophysics and Space Engineering, Indian Institute of Technology Indore, Simrol, Indore 453552, India
nagavijayadeepthi@gmail.com, {phd1701121001,phd1601121006, abhirup.datta,saurabh.das}@iiti.ac.in

Abstract. Intense geomagnetic storms can have a strong impact on the signals (termed as ionospheric scintillations) emitted by any global navigation satellite system (GNSS). The paper reports the first ever scintillations at Indore region on the NavIC signals due to impact of the intense geomagnetic storm event reported on September 8, 2017 at 01:51 and 13:04 UT. The variation of the planetary indices as well as the DST index which dropped to value of −124 nT on September 8, 2017 indicates the occurrence of an intense geomagnetic storm on September 8, 2017. The observations presented are carried out at Indore, which is located at the equatorial anomaly crest. The S4 index measurements of co-located GNSS receiver showed values ≥0.5 on the disturbed day between 15 and 18 UT. The analysis presented clearly signifies the degradation of carrier–noise measurements of NavIC L5 signal during the same time, which in turn affected the positional accuracy of NavIC, an important consideration for performance.

Keywords: NavIC · GPS · Ionospheric scintillation · S_4 index

1 Introduction

Ionosphere, as the name indicates is the ionized layer of the atmosphere roughly stratified between 50 and 1000 km of altitude from the surface of the Earth, formed by the interaction of sunlight with various gases that aid the formation of the atmosphere [1]. With the advent of technology, many global navigation satellite system (GNSS) aid in positioning precision, remote sensing and other applications for the users. Ionospheric delay (iono-delay) induced in the signals transmitted by the GNSS is the major source that affects the accuracy in these applications, and this effect of the ionosphere depends on the frequency of signal transmitted and the refractive index of the medium. The iono-delay measurements are inversely proportional to the frequency of signal transmitted by GNSS along the path from satellite to receiver [2]. In addition, high frequency (HF) signal propagation from GNSS is also affected by ionospheric irregularities. The two most important effects are increase in ionization of E and D regions and the plasma instabilities. The former effect increases the collision frequency, which accelerates radio

© Springer Nature Singapore Pte Ltd. 2021
N. R. Das and S. Sarkar (eds.), *Computers and Devices for Communication*,
Lecture Notes in Networks and Systems 147,
https://doi.org/10.1007/978-981-15-8366-7_21

wave absorption results in loss of signal strength and triggers radio blackouts. In the E and F region, ionosphere irregularities can alter the reflection height, induce scattering and change the direction of propagation of the signal which again results in loss of signal. The change in direction of propagation of signal plays a major role in navigational accuracy [3]. The latter effect is also a source of ionospheric irregularities but has its significance at high latitudes [4, 5].

2 Ionospheric Scintillation

Ionospheric scintillation is a phenomenon which is best described as a rapid or sudden change in phase and amplitude of the GNSS signal when it passes through ionosphere [6]. Studies reveal that effects of ionospheric scintillation are more severe in the equatorial and low-latitudes regions, as well as in the high latitudes regions [7]. Ionospheric scintillation events in high latitudes are usually related to periods of high solar activity and geomagnetic storms. In the equatorial and low-latitudes regions, such events occur mainly due to equatorial ionization anomaly (EIA) and ionospheric bubbles (equatorial plasma bubbles (EPB)), formed in this region post-sunset [8]. Intense EPB formation leads to severe scintillation in the radio wave communications from satellites to ground. This in turn degrades the accuracy in determining the position by global navigation satellite systems (GNSS). For the horse latitudes (45°) and beyond regions, the scintillation irregularities have been well characterized in the literature.

Some quantitative information of the scintillation is measured as S_4 and σ_φ indexes [9], it is possible to characterize and understand the ionospheric irregularities that cause ionospheric scintillations with these indices. The S-value is a way to characterize the power variation of a signal as a function of time [10, 11]. The S_4 is most widely used index among all the S indices and aids in mapping the intensity of ionospheric scintillation. The σ_φ index indicates the carrier phase measurement (φ) and its variation at the receiver during the past 60 seconds, quantizing the standard deviation of the GPS signal phase. The S_4 index is defined as the normalized variance of intensity of the signal [12, 13] and is given as

$$S_4^2 = \frac{\langle I^2 \rangle - \langle I \rangle^2}{\langle I \rangle^2} \tag{1}$$

where I is the intensity of the signal. The $S4$ index can be classified to three categories depending directly on the signal intensity that occurred on the day and place and are given as strong ($S4 \geq 0.6$), moderate ($0.3 \leq S4 \leq 0.6$) and weak ($0.1 \leq S4 \leq 0.3$) [14].

3 Present Study

The Discipline of Astronomy, Astrophysics and Space Engineering (DAASE) of Indian Institute of Technology, Indore (Lat: 22.52° N, Lon: 75.92° E) operates a dual frequency NavIC, from May 2017, provided by Space Applications Centre, ISRO, capable of receiving GPS L1, NavIC L5 and $S1$ signals. Apart from the NavIC receivers, the discipline is well equipped with another multi-frequency and multi-constellation GNSS (GPS (L1,

L2 and L5), GLONASS(G1, G2 and G3) and GALILEO (E1, E5, E5a, E5b,E6)) receiver which is operational from June 2016 whose ionospheric pierce points (IPP) is shown in Fig. 1. The present study utilizes carrier–noise ratio (*C*/No) data from NavIC receiver measured as (dB-Hz), and the scintillation indices estimated for all GPS operational frequencies *S*4 to analyse the signal strength of NavIC during the disturbed period of the ionosphere. As the receivers are located at Indore, geographically located at the crest of equatorial ionization anomaly, analysis based on the data of such locations give the best estimate of the potential of the signal strength of these receivers. An event reported by Space Weather Prediction Centre [15] for the arrival of Coronal Mass Ejection (CME) on September 6, 2017, which lasted till September 7, 2017 triggered a G4 level geomagnetic storms at 23.50 UT, on September 7, 2017 as well as September 8, 2017 at 01:51 UT and at 13:04 UT. The variation of the planetary indices [16] as well as the DST index [17] which dropped to value of −124 nT on September 8 2017 indicates the occurrence of an intense geomagnetic storm on September 8, 2017. However, on any geomagnetically quiet day measurements of *C*/No of NavIC L5 signal and the S_4 index would remain as represented in Fig. 2. The values of *C*/No (as shown in Fig. 2) of all the satellites of NavIC range from 43 to 52 dB-Hz on any geomagnetically quiet day. The corresponding S_4index values given by GPS are clearly below the value of 0.2 which indicate a very weak scintillation, not strong enough to disturb the strength of signal.

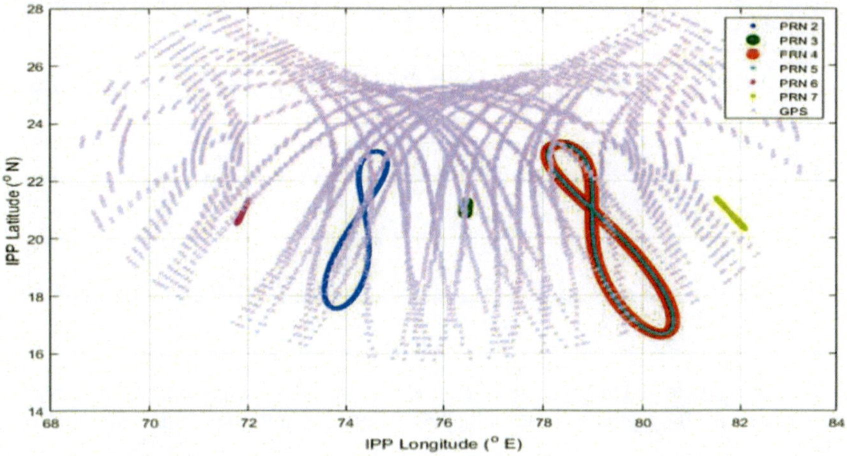

Fig. 1 IPP (tracks) of the satellite vehicles of NavIC and GPS

4 Results and Conclusion

On September 8, 2017 due to the effect of intense geomagnetic storm, the *C*/No value of NavIC L5 signal of PRN 5 and 6 shrunk below 40 dB-Hz during the time of storm *TS* around 15–18 UT and the corresponding *S*4 value peaked beyond the mark of 0.6 (area marked in ellipse of Fig. 3) during the same time significantly indicating the storm impact

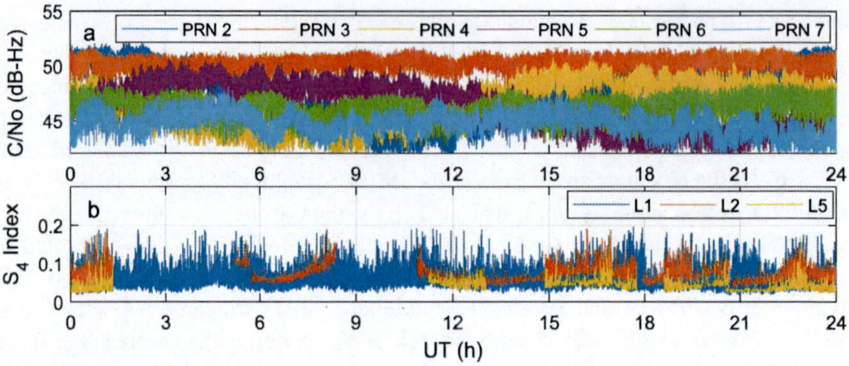

Fig. 2 **a** *C*/No values of all the NavIC satellites and **b** *S*4 index from GPS satellites as a function of time for one day

over the signal strength of NavIC satellite system. Moreover, the accuracy in estimating the position coordinates (latitude, longitude and altitude) has degraded significantly and is shown in Fig. 4, and the data of position coordinates has been sampled to two parts during the *Ts*, i.e., 15–18.

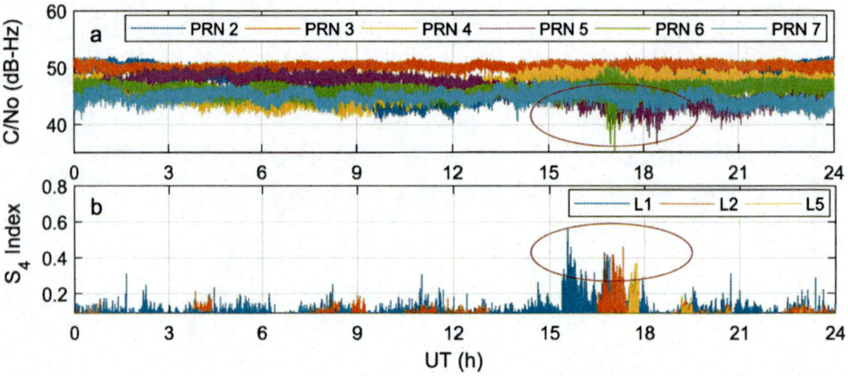

Fig. 3 **a** *C*/No values of all the NavIC satellites and **b** *S*4 index from GPS satellites as a function of time for one day. The region marked with ellipses clearly signifies the impact of scintillation over NavIC L5 signal

UT and before the *Ts* during 3–6 UT, then the distribution of the sample points is shown in Fig. 5a–c. The error in the position accuracy of latitude, longitude and altitude is observed to be 1.0273 m, 1.5789 m and 1.9905 m, respectively. The spread of points is clear proof that the scintillation-induced storm clearly affects the accuracy of position estimates, which in turn signifies the impact of geomagnetic storms on NavIC signals. Furthermore, such studies should be carried out throughout the Indian region in order to investigate the effects of scintillation triggered due to impact of intense storm. Such studies would help formulating modelling techniques to detect such scintillation events on signals and on the methods to mitigate these effects.

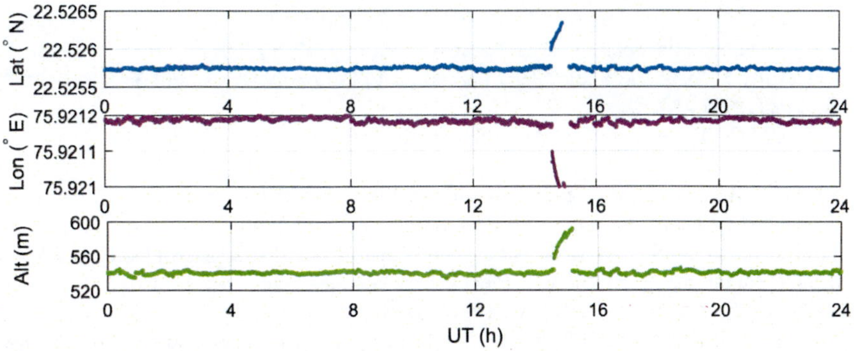

Fig. 4 Plots represents the degradation in the estimates of position coordinates, i.e., latitude (Lat), longitude (Lon) and altitude (Alt) on September 8, 2017

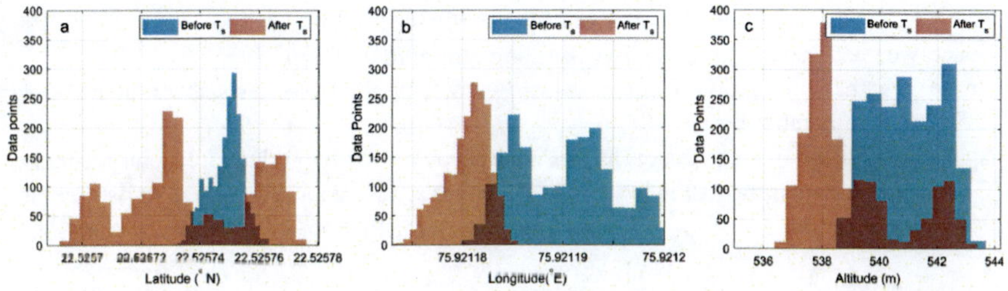

Fig. 5 Each of the plot (**a** latitude, **b** longitude, **c** altitude) represents the distribution of the samples of position estimates before and after the affect of scintillation induced in the signals of NavIC system due to the impact of storm

Acknowledgements. Deepthi acknowledges Department of Science and Technology (DST) for providing her the DST-Inspire fellowship grant to pursue her research. The authors acknowledge Space Applications Centre (SAC), ISRO for providing the NavIC data (ACCORD receiver) under the project number: NGP-17 to Discipline of Astronomy, Astrophysics and Space Engineering, IIT Indore.

References

1. McNamara, L.F.: The ionosphere: communications, surveillance, and direction finding. Krieger Publishing Company, Malabar (1991)
2. Camargo, P.O., Monico, J.F.G., Ferreira, L.D.D.: Application of ionospheric corrections in the equatorial region for L1 GPS users. Earth Planets Space **52**(11), 10831089 (2000)
3. Horne, R.B.: Benefits of a space weather programme, WP1100. ESA Space Weather Programme Study Alcatel Consortium (2001)
4. Kintner, P.M., Ledvina, B.M., de Paula, E.R.: GPS and ionospheric scintillations. Space Weather**5**, S09003. https://doi.org/10.1029/2006SW000260 (2007)
5. Pirjola, R., Kauristie, K., Lappalainen, H., Viljanen, A.: Space weather risk. Space Weather **3**, S02A02, https://doi.org/10.1029/2004SW000112 (2005)

6. Kintner, P.M., Kil, H., Beach, T.L., de Paula, E.R.: Fading timescales associated with GPS signals and potential consequences. Radio Sci. **36**(4), 731743 (2001)
7. Davies, K.: Ionospheric Radio. Peter Peregrinus Ltd., London. https://doi.org/10.1049/PBEW031E (1990)
8. Kelley, M.C.: The earths ionosphere: plasma physics and electrodynamics. Academic Press, San Diego (1989)
9. Streets, R.B.J.: Variation of radio star and satellite scintillations with sunspot number and geomagnetic latitude. J. Can. Soc. Expl. Geophys. **5**, 3552 (1969)
10. Tiwari, R., Skone, S., Tiwari, S., Strangeways, H.J.: 3WBMod assisted PLL GPS software receiver for mitigating scintillation affect in high latitude region. In: IEEE General Assembly and Scientific Symposium, 2011 XXXth URSI, pp. 14 (2011)
11. Tiwari, R., Strangeways, H.J., Tiwari, S., Ahmed, A.: Investigation of ionospheric irregularities and scintillation using TEC at high latitude. Adv Space Res. **6**, 11111124 (2013)
12. Briggs, B.H., Parkin, I.A.: On the variation of radio star and satellite scintillations with zenith angle. J. Atmos. Terr. Phys. **25**(6), 339366 (1963)
13. Kung, C.Y., Chao, H.L.: Radio wave scintillations in the ionosphere. Proc. IEEE **70**(4) (1982)
14. Space Weather Services, https://www.sws.bom.gov.au/Satellite/6/3/1
15. Space Weather Prediction Center, https://www.swpc.noaa.gov/news/shock-arrival-6-sep-2308-utc-4-september-cme
16. Space Weather Prediction Center, https://www.swpc.noaa.gov/products/planetary-k-index
17. Geomagnetic Equatorial Dst index Home Page, https://wdc.kugi.kyoto-u.ac.jp/dst_realtime/201709/index.html

Characteristics of Raindrop Size Distribution Over a Tropical Location, Kolkata

Arijit De[1](✉), Arpita Adhikari[2], and Animesh Maitra[3]

[1] Netaji Subhash Engineering College, Kolkata, India
arijitde33@gmail.com
[2] Techno Main, Salt Lake, Kolkata, India
[3] Institute of Radio Physics and Electronics, University of Calcutta, Kolkata, India

Abstract. Raindrop size distribution (DSD) characteristics of monsoon and pre-monsoon seasons over Kolkata are analyzed by using three years raindrop size data from Joss–Waldvogel disdrometer located at Kolkata. The probability of occurrence DSD for different rain rate cluster has been observed for monsoon and pre-monsoon season. The mean raindrop concentration at a particular drop diameter and rain rate has been observed for both the seasons. The variations of number concentration with drop diameter and time have also been shown for monsoon and pre-monsoon season. The diurnal variations of DSD have also been investigated.

Keywords: Raindrop size distribution · Disdrometer · Mean raindrop concentration · Diurnal variation

1 Introduction

The knowledge about the raindrop size distribution (DSD) is useful to realize rain integral parameters and to understand precipitation microphysics. DSD study is important to establish radar reflectivity (Z) and rain rate (R) relations [1] and play a significant role to deduce rain integral parameter for modeling studies [2–5] also. DSD characteristics variability play a vital role in earth to space radio links operating at frequencies above 10 GHz [6, 7]. It has already been reported that DSD shows distinct seasonal variations [8, 9]. A relatively higher mass-weighted mean diameter value in northeast Asian monsoon (December–March) has been observed than southwest monsoon (June–September) over a maritime continental station (Sumatra) [10]. Suh et al. [11] found different seasonal and diurnal variations in DSD characteristic for Busan, Republic of Korea. Kolkata, a tropical location, shows different climatic behavior for different seasons. The rainfall mainly occurs in monsoon (June–September) and pre-monsoon (March–May) seasons. So it is necessary to observe the characteristics of DSD here. In the present study, PDF of DSD has been observed for different rain rate cluster for monsoon and pre-monsoon seasons. The diurnal variations of number of drops and mass weight mean diameter have been investigated in the present study.

© Springer Nature Singapore Pte Ltd. 2021
N. R. Das and S. Sarkar (eds.), *Computers and Devices for Communication*,
Lecture Notes in Networks and Systems 147,
https://doi.org/10.1007/978-981-15-8366-7_22

2 Data and Methodology

The rainfall rate and DSD have been measured with a Joss–Waldvogel type disdrometer at Kolkata (22°34′ N, 88°29′ E). DSD data has been obtained from disdrometer. The disdrometer measures raindrop size distributions of different drop diameters that are sorted into 20 individual size intervals (starting from 0.359 to 5.5 mm) with 30 s sampling interval. The observations during the monsoon and pre-monsoon period of three years (2007, 2009, and 2010) have been used for this study.

3 DSD Characteristics

The rain rate exceedance shows higher values at higher rain rates for ORG than disdrometer. Disdrometer, due to its electromechanical working principle, considers drops larger than 5.3 mm into the same size class. So there is a possibility that the mean drop size may be underestimated during heavy convective rain (Fig. 1).

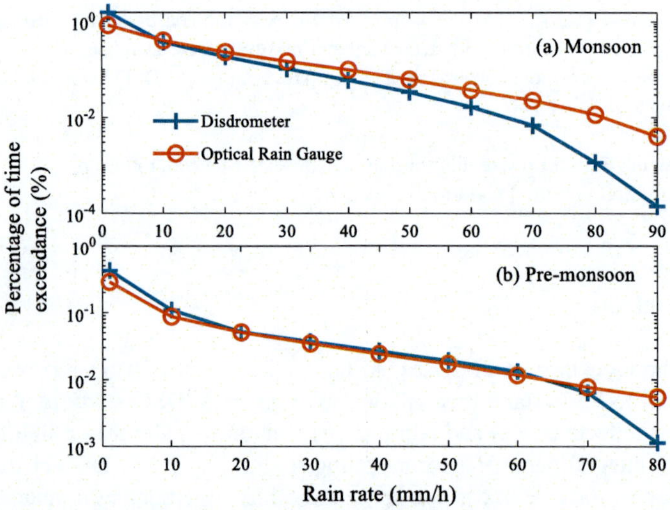

Fig. 1 Comparison of rain rate exceedance between disdrometer and optical rain gauge. **a**) Monsoon, **b**) pre-monsoon

The number of drops for same rain rate (1 mm/h, 5 mm/h, 20 mm/h, 50 mm/h, 80 mm/h) has been compared for pre-monsoon and monsoon seasons. The number of drops shows different characteristics for pre-monsoon and monsoon seasons. At lower rain rate (1 mm/h), a number of smaller drops (<2.5 mm) are more in pre-monsoon season whereas a number of medium and larger drops (>2.5 mm) are more in monsoon season. For 20 mm/h rain rate, lower drops are abundant in monsoon than pre-monsoon season. The two peaks of drop size distribution have been observed prominently at 50 mm/h rain rate for both monsoon and pre-monsoon seasons. For pre-monsoon, the

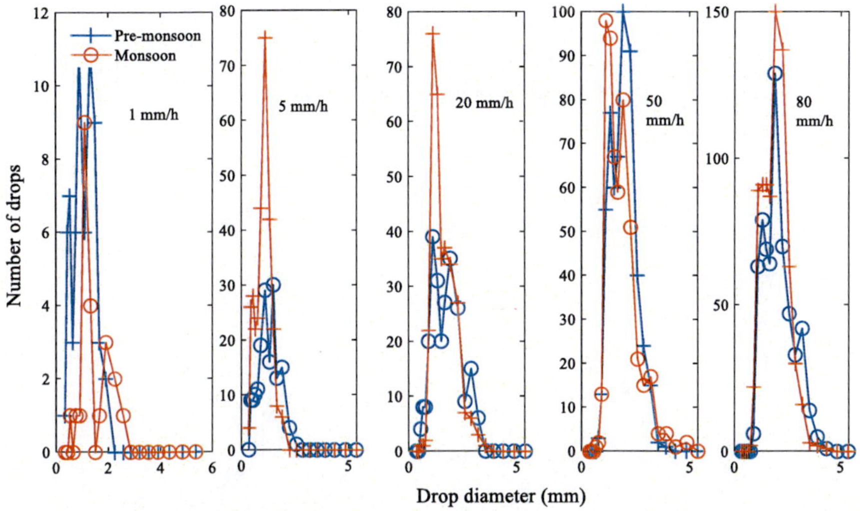

Fig. 2 Drop concentration for various rain rates in pre-monsoon and monsoon seasons

peaks for number of drops are observed at 1.33 and 1.91 mm diameter. For monsoon, the peaks for number of drops are observed at 1.116 and 1.912 mm diameter (Fig. 2).

The probability of occurrence of DSD for two different seasons has been observed with respect to rainfall. The rainfall in both seasons is divided into four rain rate class (C1-(0–20 mm/h), C2-(20–50 mm/h), C3-(50–80 mm/h), and C4 (>80 mm/h)). For each rain rate class, probability of occurrence of raindrop concentration with raindrop size (D, mm) for monsoon and pre-monsoon is depicted in Fig. 3. For low, medium, and high rain rate (0–20 mm/h, 20–50 mm/h and 50–80 mm/h), the probability of occurrences of larger drops (>2.5 mm) is more in pre-monsoon season than in monsoon season. However, for higher rain rates (>80 mm/h), the probability of all the drops is more in pre-monsoon than monsoon season.

To clearly understand it, a normalization method has been used. A total number of drops for monsoon and pre-monsoon are summed up and denoted by $X1$ and $X2$, respectively. The DSD has been normalized with respect to $X1$ and $X2$. It actually signifies the contribution of different drops for different rain classes (0–20 mm/h, 20–50 mm/h, 50–80 mm/h and >80 mm/h) with respect to total number of drops in pre-monsoon and monsoon. For moderate (50–80 mm/h) and high (>80 mm/h) rain rates, contribution of medium to high raindrops is more than lower raindrops. However, for low (0–20 mm/h) and medium (20–50 mm/h) rain rates, contribution of lower raindrops is more than medium and higher raindrops. From Fig. 4, it can be seen that medium and higher drops are dominant in pre-monsoon whereas lower and medium drops are dominant in monsoon season (Fig. 5).

4 Diurnal Variation

To examine the diurnal differences between monsoon and pre-monsoon DSD, 6 hourly drop size distributions have been observed (0–6, 6–12, 12–18, and 18–24 IST) in Fig. 6.

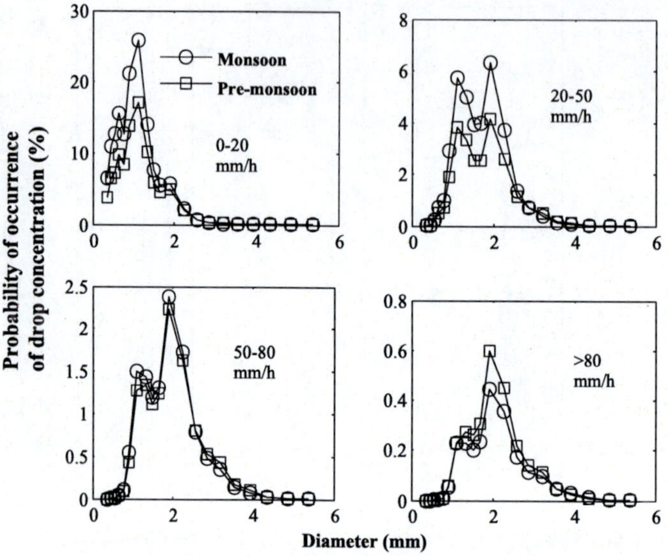

Fig. 3 Probability of occurrence of DSD for different rain rate

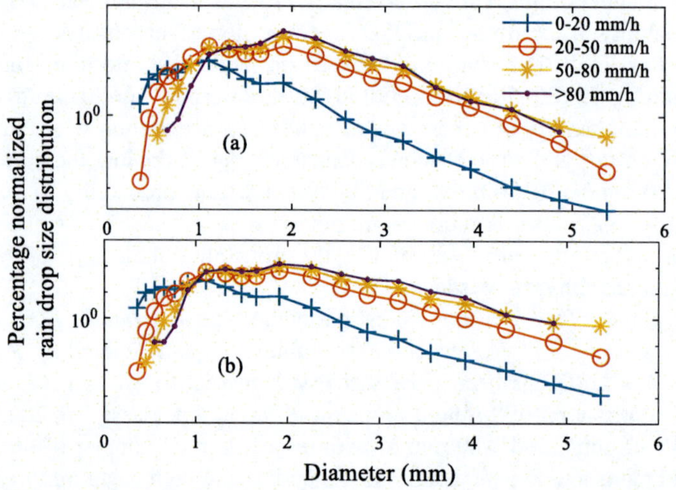

Fig. 4 Normalized raindrop size distribution. **a** Pre-monsoon, **b** monsoon

In the monsoon season, the higher drops (>2.5 mm) are more during 6–12 IST and 12–18 IST whereas the lower drops are more in 0–6 IST. However, in the pre-monsoon season, the major rain occurs in the 12–18 IST and 18–24 IST. In pre-monsoon season, intense thunderstorm associated with heavy rainfall occurs at this time over Kolkata due to strong local convection. The hourly variation of mass-weighted mean diameter shows no significant variation for monsoon season (higher all time), but in pre-monsoon season

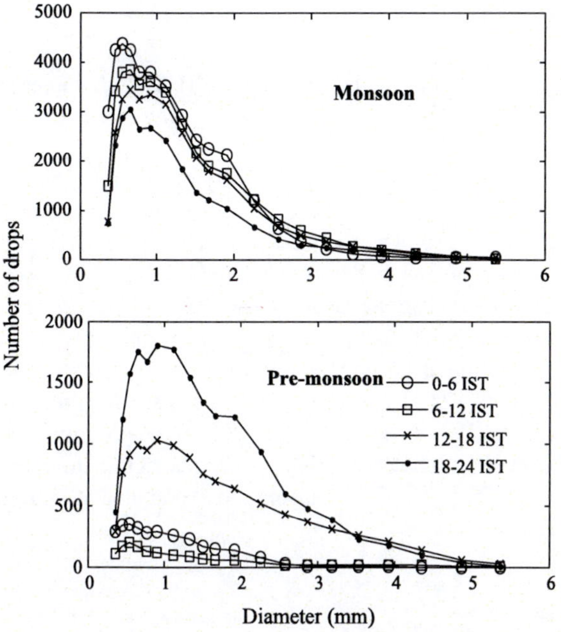

Fig. 5 Diurnal variation of number of drops for monsoon and pre-monsoon seasons

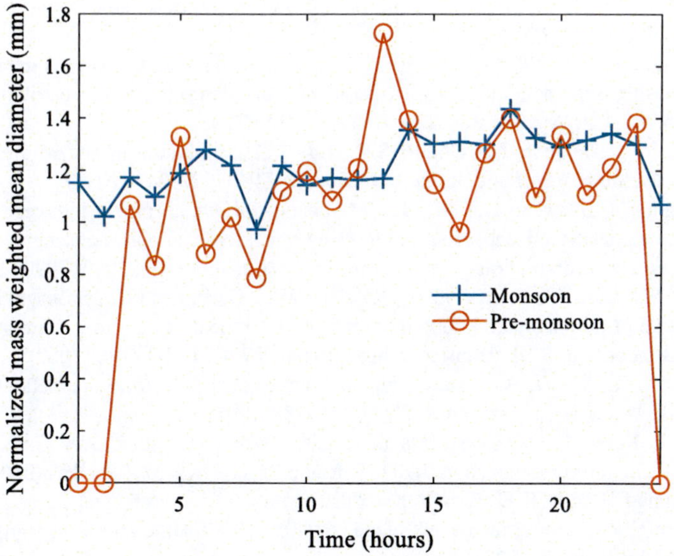

Fig. 6 Hourly mass-weighted mean diameter for monsoon and pre-monsoon

it shows higher values in the afternoon to late night (4–12 pm) indicating the dominance of convective phenomena at that time.

5 Conclusion

In the present paper, the raindrop size distribution (DSD) characteristics of monsoon and pre-monsoon of Kolkata have been analyzed by using three years of DSD data from Joss–Waldvogel disdrometer deployed at Institute of Radio Physics and Electronics, University of Calcutta. The exceedance probability of rain rate shows higher value for ORG than disdrometer at higher rain rates as disdrometer consider drops larger than 5.3 mm into the same size class. The probability of occurrences of larger drops is more in pre-monsoon than monsoon season. Double peak of DSD has been observed prominently for both monsoon and pre-monsoon seasons for moderate to high rain rate (20–50 mm/h). Due to the prevalence of convective phenomena, large raindrops are more abundant in the pre-monsoon than in the monsoon season. In the monsoon season, the larger drops (>2.5 mm) are more in 6–12 IST and 12–18 IST whereas in the pre-monsoon season the large drops occurs in the 12–18 IST and 18–24 IST. In the pre-monsoon season, intense thunderstorm associated with heavy rainfall occurrence at this time over Kolkata due to strong local convection plays a role in causing the abundance of large raindrops.

References

1. Seliga, T.A., Bringi, V.N.: Potential use of the radar differential reflectivity measurements at orthogonal polarizations for measuring precipitation. J. Appl. Meteorol. **15**(1), 69–76 (1976)
2. Gilmore, M.S., Straka, J.M., Rasmussen, E.N.: Precipitation uncertainty due to variations in precipitation particle parameters with in a simple microphysics scheme. Mon. Weather Rev. **132**(11), 2610–2627 (2004)
3. Cohen, C., McCaul, E.W., Jr.: The sensitivity of simulated convective storms to variations in prescribed single-moment microphysics parameters that describe particle distributions, sizes, and numbers. Mon. Weather Rev. **134**(9), 2547–2565 (2006)
4. Fadnavis, S., Deshpande, M., Ghude, S.D., Raj, P.E.: Simulation of severe thunder storm event: A case study over Pune, India. Nat. Hazards **72**(2), 927–943 (2014)
5. Wainwright, C.E., Dawson, D.T., II., Xue, M., Zhang, G.: Diagnosing the intercept parameters of the exponential drop size distributions in a single-moment microphysics scheme and impact on supercell storm simulations. J. Appl. Meteorol. Climatol. **53**(8), 2072–2090 (2014)
6. Marzuki, M., Kozu, T., Shimomai, T., Randeu, W.L., Hashiguchi, H., Shibagaki, Y.: Diurnal variation of rain attenuation obtained from measurement of raindrop size distribution in equatorial Indonesia. IEEE Trans. Antennas Propag. **57**(4), 1191–1196 (2009)
7. Chakravarty, K., Maitra, A.: Rain attenuation studies over an earth space path at a tropical location. J. Atmos. Solar Terr. Phys. **72**(1), 135–138 (2010)
8. Reddy, K., Kozu, T.: Measurements of raindrop size distribution over Gadanki during southwest and northeast monsoon. Indian J. Radio Space Phys. **32**, 286–295 (2003)
9. Jayalakshmi, J., Reddy, K.K.: Raindrop size distributions of southwest and northeast monsoon heavy precipitations observed over Kadapa (14°40N, 78°820E), a semi-arid region of India. Curr. Sci. **107**(8), 1312–1320 (2014)
10. Marzuki, Randeu, W. L., Kozu, T., Shimomai, T., Hashiguchi, H., Schönhuber, M.: Raindrop axis ratios, fall velocities and size distribution over Sumatra from 2D-video disdrometer measurement. Atmos. Res. **119**, 23–37 (2013)
11. Suh, S.H., You, C.H., Lee, D.I.: Climatological characteristics of raindrop size distributions in Busan, Republic of Korea. Hydrol. Earth Syst. Sci. **20**, 193–207 (2016)

A Novel Handoff Algorithm for 5G

Prithwijit Mukherjee[1](✉), Sanchita Ghosh[2], and Anisha Halder Roy[1]

[1] Institute of Radio Physics and Electronics, University of Calcutta, Kolkata, India
mukherjeprithwijit@gmail.com, halder.anisha@gmail.com
[2] Institute of Engineering and Management, Kolkata, India

Abstract. The main challenges of 5G cellular network are capacity increase of the existing cellular system, providing seamless and ubiquitous communication with very less delay. In future, femtocells and mobile femtocells will be deployed with macrocell base stations. The handoff algorithms which are available cannot be implemented efficiently in femtocell/mobile femtocell/macrocell network scenario. The essence of this paper is to propose a new handoff algorithm for 5G cellular network and evaluate its performance in 5G network scenarios. The performance of the proposed handoff algorithm is compared with a conventional handoff algorithm in terms of number of handoff, number of disconnection and efficiency for 10,000, 20,000, 30,000 randomly generated instances of 5G network scenarios.

Keywords: 5G · Handoff algorithm

1 Introduction

Cellular technology has evolved from 1 to 2G, 2G to 3G, and 3G to 4G. Researchers have found that in future number of cell phone users will increase 100 times and traffic volume will increase 1000 times by 2020. 5G cellular network will support future wireless scenarios and client demands.

Mobile femtocell is the one of the proposed technologies for 5G cellular network. Mobile femtocell is a low power cell which moves from place to place. The legacy handoff algorithms, i.e., handoff algorithms which are used in 2G, 3G networks, cannot be used efficiently in a cellular network where femtocell base stations, mobile femto base stations, and macrocell base stations are deployed simultaneously. So, a new handoff scheme is needed for macrocell, femtocell, and mobile femtocell deployed cellular network.

In our experiment, we generate different 5G network scenarios using MATLAB 13b. At first we randomly generate number of available macrocells, femtocells, and mobile femtocells for each instance of the generated 5G network scenarios. After that we generate values of various parameters such as (i) remaining channel of the available macrocells and femtocells, (ii) remaining channel, velocity of the available mobile femtocells, (iii) velocity of the user, and (iv) SNR of the received signals from all of the available macrocells, femtocells, and mobile femtocells. We propose a handoff algorithm which will

© Springer Nature Singapore Pte Ltd. 2021
N. R. Das and S. Sarkar (eds.), *Computers and Devices for Communication*,
Lecture Notes in Networks and Systems 147,
https://doi.org/10.1007/978-981-15-8366-7_23

perform efficiently in 5G network scenarios. Performance of the proposed handoff algorithm [1] is also compared with a conventional handoff algorithm in 10,000, 20,000, and 30,000 instances of randomly generated 5G scenarios.

The paper is ordered as follows. Section 2 describes our proposed work in details. Section 3 illustrates the proposed handoff algorithm. Section 4 illustrates the experimental details and results. Section 5 concludes the paper.

2 Proposed Work

In our experiment, we wish to propose a handoff algorithm for 5G cellular network. In fifth generation cellular network, macrocells, femtocells, and mobile femtocells will be deployed simultaneously in the cellular network. Large number of small cells will be used in 5G cellular network for improving the spatial frequency reuse [2, 3]. One mobile user will be connected to the network through any of the available base stations, i.e., macrocell base station or femtocell base station or mobile femto base station. This will increase the number of handoffs in 5G network. So an efficient handoff algorithm is needed which will reduce the total number of handoffs by eliminating unnecessary handoffs of the system.

Parameters such as number of free channels, traffic increase rate, signal-to-noise ratio of the received signal of the serving base station, signal-to-noise ratio of the signals of the available base stations, speed of the user are used for deciding handoff in our handoff algorithm.

At first we generate different network scenarios where macrocells, femtocells and mobile femtocells are present simultaneously using MATLAB 13b. Then, values of different handoff decision parameters, i.e., (i) user's velocity, (ii) number of free channels of the serving base station and available base stations, (iii) traffic increase rate of serving base station, (iv) signal-to-noise ratio of the received signal from the serving base station, and (v) signal-to-noise ratio of the received signals from the available base stations are generated in a random manner for each instance of previously generated 5G network scenarios.

We obtain total number of the handoff, total number of disconnection and efficiency of the proposed handoff algorithm, and a conventional handoff algorithm in the randomly generated 5G network scenarios using previously generated handoff decision parameters. We analyze the performance the proposed handoff algorithm and compare the performance of the proposed handoff algorithm with a conventional handoff algorithm [1] in terms of number of handoff, number of disconnection, and efficiency.

3 Proposed Handoff Algorithm

Our proposed handoff algorithm is illustrated in details in this section. We use different notations such as m, f, mf, th, δ, θ, ψ, ξ, τ, V_{user} for illustrating the flowchart of our proposed handoff algorithm. Here, m = macrocell, f = femtocell, mobile femtocell = mf, th = threshold value, δ = traffic increase rate of serving base station, θ = remaining channel of the serving base station, ψ = SNR of the serving base station, ρ = SNR of received signals of available base stations, Δ = velocity threshold, ξ = relative velocity of serving mobile femto base station and user, τ = relative velocity of available mobile femto base stations and user, V_{user} = velocity of the user. Figure 1 illustrates the flowchart of our proposed handoff algorithm.

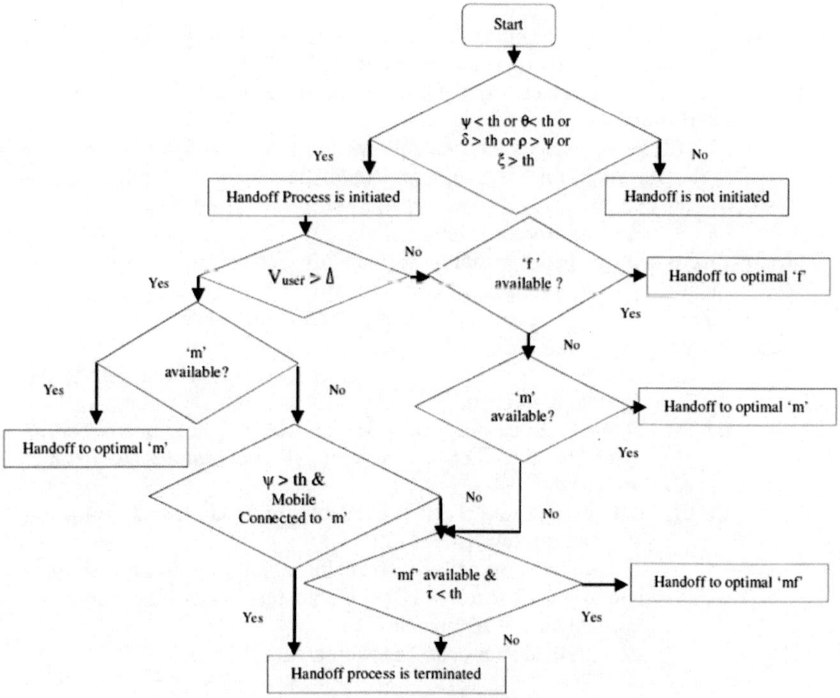

Fig. 1 Flowchart of the proposed handoff algorithm

The proposed handoff algorithm is illustrated below.

- **If ,** [(SNR of the serving base station (ψ) < threshold (th)) or (Remaining channel of the serving base station (θ) < th) or (Traffic increase rate of the serving base station (δ) > th) or (SNR of received signals of available Base Stations (ρ) > SNR of the serving base station (ψ)) or (Relative velocity of Serving mobile femto base station and user (ξ) > th)] is 'True'

  Then, handoff process is initiated.

  Else, Handoff process is not initiated.
- **If ,** [(Velocity of the user (V_{user}) > Velocity threshold (Δ))]

  Then, handoff is only possible to macrocell or mobile femto cell.

  Else, Handoff is possible to macro cell, femto cell and mobile femtocell.
- **When,** [(Velocity of the user (V_{user}) > Velocity threshold (Δ))]

  i. Handoff request will be sent to the optimal macro cell if macro cell is available for handoff.

  ii. Handoff request is terminated if any macro cell base station or mobile femtocell base station is not available for handoff.

  iii. If only mobile femtocell base station is available then following steps are performed -

  If [(Relative velocity of available mobile femto base stations and user (τ) < threshold (th)] is true, then handoff is done to the optimal mobile femto cell.

  Else, Handoff process is terminated.

  iv. **If** [macro cell (m) is not available for handoff && serving base station is a macro cell (m) base station && SNR of the serving Base Station (ψ) < threshold (th)] is true, then handoff process is terminated.
- **When,** [(Velocity of the user (V_{user}) < Velocity threshold (Δ))]

  i. If, femto cell is available then handoff request will be sent to the optimal femto cell.

  ii. If, femto cell is not available for handoff but macro cell is available, then handoff request will be sent to the optimal macro cell.

  iii. If, only mobile femto cell is available for handoff, then following steps are followed-

  If [(Relative velocity of available mobile femto base stations and user (τ) < threshold (th)] is true, then handoff is done to the optimal mobile femto cell.

  Else, Handoff process is terminated.

4 Experimental Details and Results

In our experiment, we have compared the performance of the proposed handoff algorithm with a conventional handoff algorithm [1] in terms of number of handoff, disconnection, and efficiency for 10,000, 20,000, 30,000 randomly generated instances of the 5G network in MATLAB 13b.

The number of disconnections of the proposed handoff algorithm is 11, 35, and 46 for 10,000, 20,000, and 30,000 randomly generated instances of 5G network, respectively. The number of disconnections of the conventional handoff algorithm is 139, 290, and 397 for 10,000, 20,000, and 30,000 randomly generated instances of the 5G network, respectively. Figure 2 shows the performance comparison between proposed and conventional handoff algorithm in terms of number of disconnection for 30,000 randomly generated instances of 5G network.

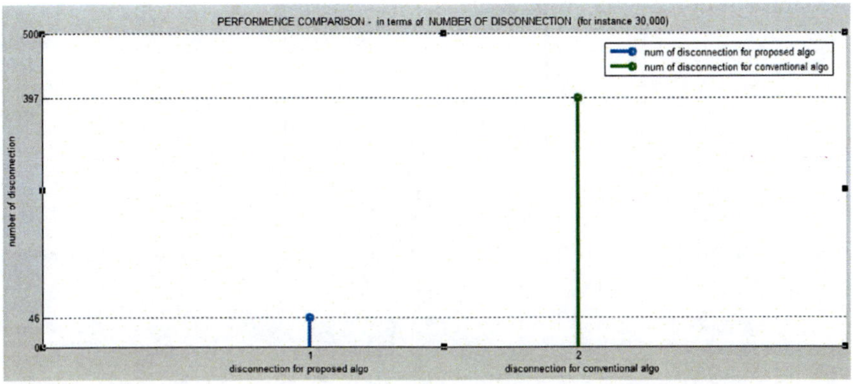

Fig. 2 Performance comparison—in terms of number of disconnection (for 30,000 instances)

The number of handoff of the proposed handoff algorithm is 8789, 17,544, and 26,343 for 10,000, 20,000, and 30,000 randomly generated instances of 5G network, respectively. The number of disconnections of the conventional handoff algorithm is 8994, 18,087, and 27,468 for 10,000, 20,000, and 30,000 randomly generated instances of 5G network, respectively.

Figure 5 illustrates the performance comparison between proposed and conventional handoff algorithm in terms of efficiency for 1000, 10,000, 20,000, 30,000 randomly generated instances in the 5G network. Figure 4 shows the plot between efficiency of the handoff algorithm and velocity of the user for 1000 randomly generated instances of the 5G network. Figure 3 shows the performance comparison between proposed and conventional handoff algorithm in terms of number of handoff for 30,000 randomly generated instances of 5G network (Figs. 4 and 5).

5 Conclusion

In our research work, we have proposed a handoff algorithm which will work efficiently in a cellular network where macrocells, femtocells, and mobile femtocells are deployed simultaneously. We have found that the total number of handoff and disconnection of proposed handoff algorithm is lesser than the total number of handoff and disconnection of conventional handoff algorithm for 10,000, 20,000, and 30,000 randomly generated instances of 5G cellular network. It has been found that efficiency of the proposed handoff algorithm is higher than efficiency of the conventional handoff algorithm. It can be concluded from the obtained results of the experiment that proposed handoff algorithm will perform better than the conventional handoff algorithm in 5G network scenarios.

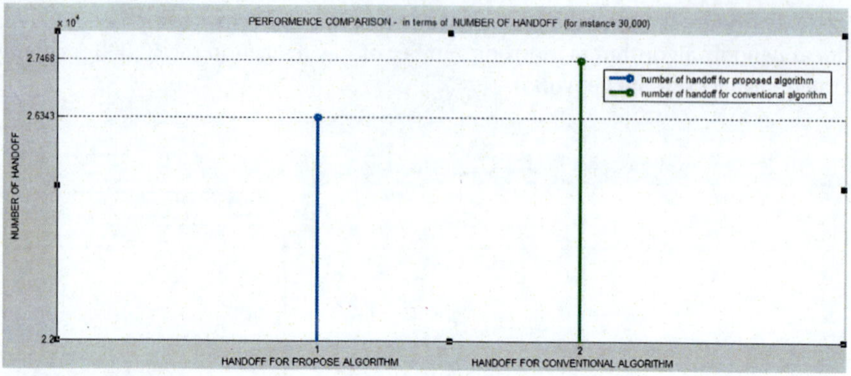

Fig. 3 Performance comparison—in terms of number of handoff (for 30,000 instances)

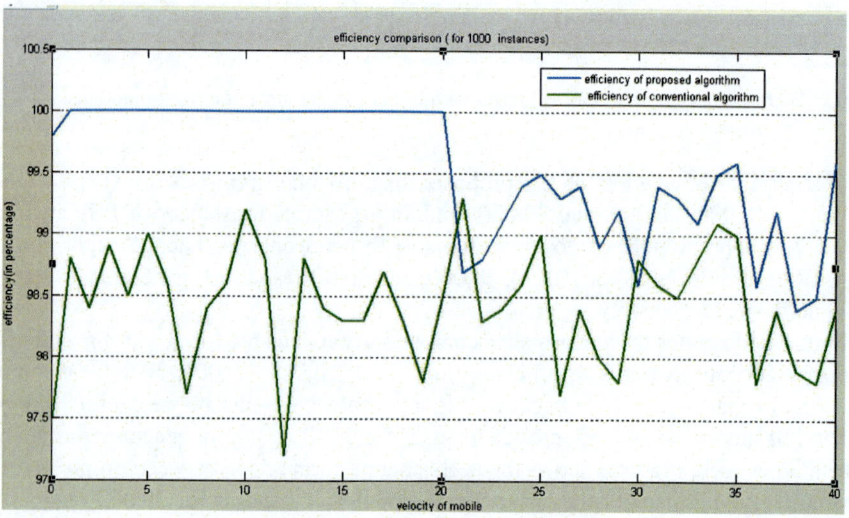

Fig. 4 Efficiency versus velocity plot

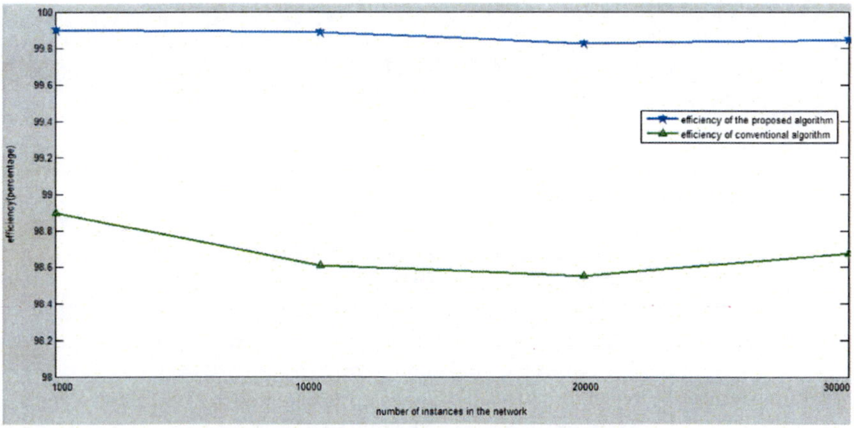

Fig. 5 Performance comparison—in terms of efficiency

References

1. Bhoite, K.S., Gengaje, S.: Handover Management in Two-Tier Femtocell—Macrocell Network. Springer, Berlin (2017)
2. Taha, M., Parra, L., Garcia, L., Lloret, J.: An Intelligent handover process algorithm in 5G networks: The use case of mobile cameras for environmental surveillance. SCPA (2017)
3. Chowdhury, M., Zaman, J., Yeong, M.: Call admission control and traffic modeling for integrated macrocell/femtocell networks. ICUFN (2012)

OFDM-SIM with Adaptive Modulation Through Fuzzy Interface

Susmita Chaki[1](✉), Ipsita Sengupta[2], and Shounak Dasgupta[3]

[1] ECE Department, University of Engineering and Management, Kolkata University Area, Kolkata 700160, West Bengal, India
susmitachaki1991@gmail.com

[2] Electronics Department, Government College of Engineering and Leather Technology, Block-LB 11, Sector-III, Salt Lake, Kolkata 7000106, West Bengal, India
ipsen1980@yahoo.co.in

[3] ECE Department, Heritage Institute of Technology, Kolkata 700107, West Bengal, India
shounak.dasgupta@heritageit.edu

Abstract. OFDM-SIM offers the advantage of reduced peak-to-average power ratio (PAPR) and improved spectral efficiency in comparison with classic OFDM. Due to frequency selective channel fading, OFDM frames undergo non-uniform attenuation and this results in different SNR values for different frames. Consequently, BER of the system is degraded. This problem can be mitigated by using adaptive modulation scheme depending on SNR values estimated by channel estimator. In this paper, a simple model of OFDM-SIM with adaptive modulation is proposed. Switching decision is implemented by fuzzy interface system. The output BER performance of the OFDM-SIM has been investigated for two types of modulations: 8-PSK and 8-QAM with 10 dB SNR as threshold level.

Keywords: OFDM · OFDM-SIM · Adaptive modulation · Fuzzy interface

1 Introduction

Orthogonal frequency division multiplexing (OFDM) with sub-carrier index modulation (OFDM-SIM) is a promising scheme for the high-speed wireless communication due to its high spectrum and energy efficiency [1, 2] along with the inherent ability of OFDM to combat inter-symbol interference caused by multipath effects. In OFDM, the total frequency bandwidth is split into many narrowband sub-channels to transmit parallel data streams. In OFDM-SIM, the information bits are also segregated into several sub-blocks. The bits in each sub-block are further split into two parts, i.e., the index bits and the symbol bits. The symbol bits are transmitted by M-ary signal constellations and the carriers are activated according to the index bits [3]. Effective selection of the number of active carriers results in improved spectral efficiency, compared to conventional OFDM [4]. Apart from this, OFDM-SIM also has many noteworthy features, including enhanced bit error rate (BER) performance, reduced peak-to-average power ratio (PAPR), relatively

© Springer Nature Singapore Pte Ltd. 2021
N. R. Das and S. Sarkar (eds.), *Computers and Devices for Communication*,
Lecture Notes in Networks and Systems 147,
https://doi.org/10.1007/978-981-15-8366-7_24

higher robustness against the inter-carrier interference (ICI), etc., compared to classic OFDM [1, 5].

In an OFDM transmission system, individual sub-carrier is subjected to attenuation caused by frequency selective channel fading. If same modulation scheme is used for all the transmitted symbols, individual OFDM frame undergoes through different power attenuation, so also varying SNR values, which in turn induces to symbol dependent BER degradation. This problem can be mitigated if different modulation schemes can be employed for every OFDM frame. Though many adaptive techniques of modulation combined with classic OFDM have been proposed in the literature [6–9], performance of OFDM-SIM with adaptive modulation is yet to be explored. In this paper, we have proposed an adaptive modulation scheme combined with OFDM-SIM. The switching decision for different modulations is implemented by fuzzy interface system.

2 Theory

2.1 Description of the System Model

In this section, OFDM-SIM scheme with adaptive modulation is proposed. The block diagram is depicted in Fig. 1. Unlike classic OFDM [10], OFDM-SIM does not use all the available sub-carriers for transmission of data symbols [5]. Here, input data stream is segmented into multiple sub-blocks or packets of bit length p. Each sub-block is further divided into two subgroups p_x and p_y, p_y for data symbol mapping, and p_x for index selection of sub-carriers used in the subsequent packet.

In this scheme, the available sub-carriers are also distributed among the segments. However, not all the available sub-carriers in a segment are used. A lookup table, available both at the transmitter and the receiver, is defined to assign sub-carriers for the data symbol bits (p_y), as per the index selection bits (p_x). Such sub-carriers are known as active sub-carriers. The positions of active sub-carriers in each sub-block are decided by the pattern of index selecting bits and the look up table (Table 1) is configured accordingly.

The index bits p_x are determined by possible combinations of k active sub-carriers out of n sub-carriers in each sub-block. So, $2^{p_x} \leq {}^nC_k$.

For *M-QAM* modulation, $p_y = k \times \log_2 M$.

Total number of bits in any sub-block is $p = p_x + p_y$.

In Fig. 1, the OFDM sub-block is configured by choosing the proper modulation scheme by the adaptive modulator and the active carriers are assigned by the index selector. In this communication, the modulation scheme applied to the data symbols is decided according to the signal-to-noise ratio (SNR) at channel, as obtained from the channel estimator. The fuzzy logic decision device suggests the modulation scheme to the adaptive modulator and shares the same information to the adaptive demodulator. At the receiver end, the corresponding received OFDM sub-block is fed to both the adaptive demodulator and the index selector. The adaptive demodulator recovers the data using the appropriate demodulation process as per the applied modulation scheme. The index bits are identified by the index selector consulting the lookup table (as presented in Table 1). The data symbol bits and index selection bits are then combined to form the packet of data.

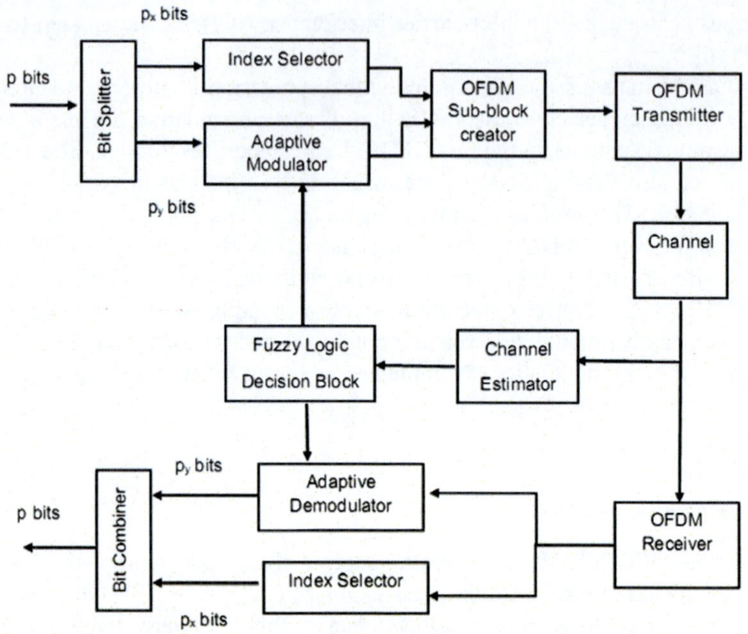

Fig. 1 Block diagram of OFDM-SIM with adaptive modulation

Table 1 Lookup table for OFDM-SIM Tx and Rx

$n = 4$, $k = 2$, $M = 8$, sub-carriers in each OFDM frame $N = 256$
Input bit sequence: $p_1\ p_2\ S_{y1}\ S_{y2}$;
$p_x = p_1\ p_2$ = Index selecting bits
$p_y = S_{y1}\ S_{y2}$ = M-PSK/M-QAM modulated symbols

Index bits p_x		Index combination I_β	M-PSK/M-QAM symbol sub-block S_β
p_1	p_2		
0	1	[1 2]	$[S_{y1}\ S_{y2}\ 0\ 0]$
0	0	[2 3]	$[0\ S_{y1}\ S_{y2}\ 0]$
0	1	[3 4]	$[0\ 0\ S_{y1}\ S_{y2}]$
1	1	[1 4]	$[S_{y1}\ 0\ 0\ S_{y2}]$

OFDM-SIM offers the advantage of presenting lower PAPR value of the transmitted signal in comparison with classic OFDM, thereby had already established its acceptability over the later one [5, 11]. In conventional OFDM-SIM scheme, only one type of modulation is used throughout the entire process. In this paper, we have proposed to adapt variable modulation scheme depending on instantaneous SNR values of the received signal estimated by the channel estimator on frame by frame basis. BER performance of the received signal is subjected to variation of SNR at the front end of the receiver and the modulation scheme used for transmission of data symbols at the transmitter. Usually, higher-order modulation is preferred to achieve improved bandwidth efficiency, on the contrary it degrades the BER performance. In our model, instead of using fixed modulation scheme adaptive modulation technique has been employed to cater better bandwidth efficiency without causing much degradation to BER values. Two types of modulations, M-PSK and M-QAM, have been used adaptively depending on the estimated SNR value at the receiver. Order of modulation schemes selected for adaptive allocation to a particular frame should be kept unchanged so as to maintain constant input packet or sub-block length. For a particular frame, initial modulation scheme used is M-PSK due to its less complexity from implementation point of view. In this work, we have chosen 8-PSK and 8-QAM as two modulation options for every frame. If the SNR value for a particular frame evaluated by the channel estimator drops below a threshold, 10 dB, modulation scheme is supposed to switch automatically to 8-QAM in order to maintain an acceptable range of BER. The decision making for choosing the right modulation scheme for every frame is done by fuzzy interface system (FIS), which also governs the use of proper demodulation scheme at the receiver on frame-by-frame basis.

2.2 Decision Making with Fuzzy Interface System (FIS)

In this paper, fuzzy logic interface is used for making decision for adaption of modulation scheme. The modulation scheme will change depending on the SNR of the signal estimated from channel using a channel estimator.

The fuzzy logic interface system is modeled in MATLAB. In this communication, the fuzzy logic designer has two inputs, first one is SNR and the second one is present modulation scheme. The output of the FIS will control the modulation scheme of adaptive modulator and demodulator blocks. By comparing with the threshold value, the SNR input of FIS have two membership functions—low SNR (if SNR is with the range 0–9.8 dB) and High SNR (if SNR is with the range 9.9–20 dB).

The other input that is the present modulation scheme also has two membership functions: for 8-QAM and for 8-PSK. The logic for assigning the modulation scheme is added in the rule editor.

3 Results

The logic that was implemented is as follows: for high SNR, modulation scheme is 8-PSK; and for low SNR, modulation scheme will be 8-QAM. Simulation has been done based on the rules in the rule editor and after simulation FIS will produce a rule viewer (Fig. 2) and a surface viewer (Fig. 3). In the 3D surface viewer plot, three axes represent SNR, present modulation scheme, and the adapted modulation scheme.

Fuzzy logic system is used in this paper for decision making because of two reasons. Non-fuzzy system is controlled using if-else, where fuzzy system is controlled by membership values. For example, if the range of low SNR is 0–9 in non-fuzzy system, 9.1 will not be considered as low SNR (but it is low). But in fuzzy logic system, 9.1 will also be considered as low SNR. So, fuzzy logic interface will improve the performance of

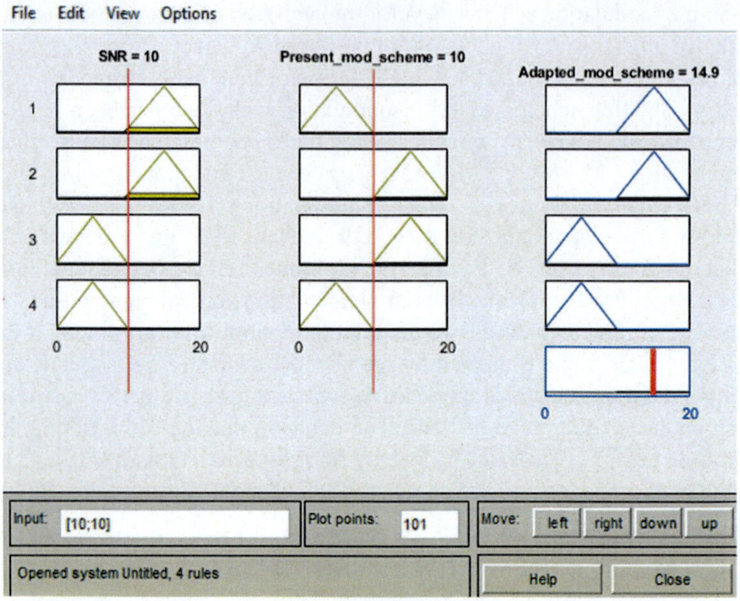

Fig. 2 FIS rule viewer

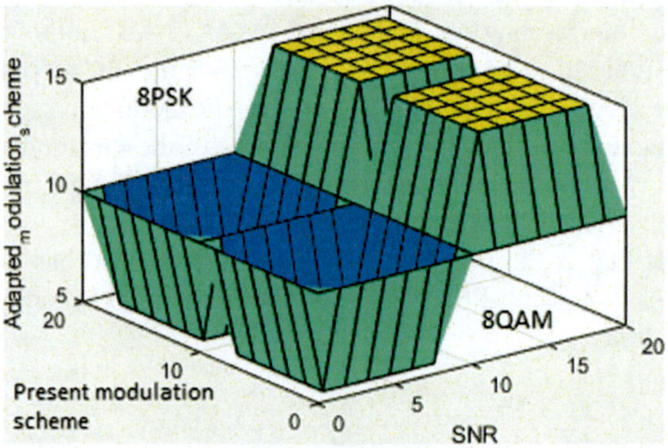

Fig. 3 FIS surface viewer

adaptive modulation and demodulation. Besides, fuzzy logic system controlled decision making offers faster processing speed.

BER plot for OFDM-SIM with the entire adaptive modulation process is shown in Fig. 4. SNR has been estimated considering additive white Gaussian noise. The solid line curve indicates switching of modulation scheme from 8-PSK to 8-QAM when the SNR drops below 10 dB.

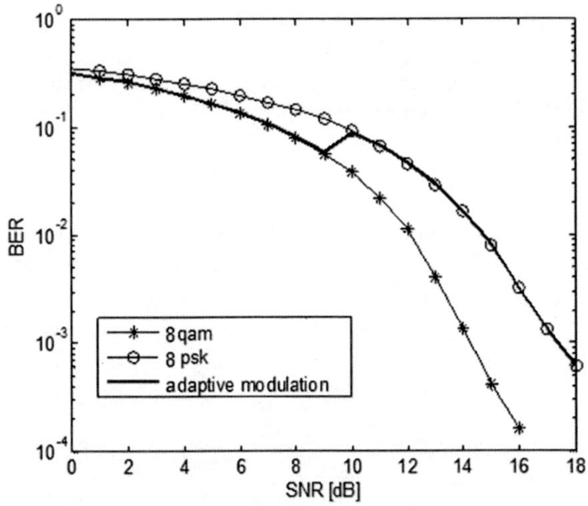

Fig. 4 BER plot for OFDM-SIM with 8-PSK and 8-QAM modulation schemes employed adaptively

4 Conclusions

In this paper, we have proposed an adaptive modulation scheme for SIM-OFDM. The variation in modulation for individual data packets mitigates the variable BER and SNR due to frequency selective fading. In this communication, the primary modulation scheme is set to 8-PSK. The SNR of the system for each packet is monitored. As the SNR becomes less than 10 dB, the modulation scheme is changed to 8-QAM to maintain the BER within the acceptable range. The decision is taken by a fuzzy logic interface which implements this adaptive modulation scheme in a practical approach.

References

1. Basar, E., Wen, M., Mesleh, R., Di Renzo, M., Xiao, Y., Haas, H.: Index modulation techniques for next-generation wireless networks. IEEE Access **5**, 16693–16746 (2017)
2. Wen, M., Cheng, X., Yang, L.: Index Modulation for 5G Wireless Communications. Springer, Berlin (2017)
3. Basar, E., Aygölü, U., Panayırcı, E., Poor, H.V.: Orthogonal frequency division multiplexing with index modulation. IEEE Trans. Signal Process. **61**(22): 5536–5549 (2013)
4. Fan, R., Yu, Y.J., Guan, Y.L.: Orthogonal frequency division multiplexing with generalized index modulation. In: IEEE Global Communications Conference, pp. 3880–3885, Dec 2014
5. Basar, E., Aygölü, Ü., Panayrc, E., Poor, H.V.: Orthogonal frequency division multiplexing with index modulation. IEEE Trans. Signal Process. **61**(22), 5536 5549 (2013)
6. Chow, P.S., Cioffi, J.M., Bingham, J.A.C.: A practical discrete multi-tone transceiver loading algorithm for data transmission over spectrally shaped channels. IEEE Trans. Commun. **38**, 772–775 (1995)
7. Keller, T., Hanzo, L.: Adaptive modulation techniques for duplex OFDM transmission. IEEE Trans. Veh. Technol. **49**(5), 1893–1906 (2000)

8. Lei, Y., Burr, A.: Adaptive modulation and code rate for turbo coded OFDM transmissions. In: Vehicular Technology Conference VTC2007, pp. 2702–2706, 22–25 Apr 2007
9. Tsugi, T., Itami, M. : A study on adaptive modulation of OFDM under impulsive power line channel. In: IEEE International Symposium on Power Line Communications and Its Applications, ISPLC, pp. 304–309, 2–4 Apr 2008
10. Armstrong, J.: OFDM for optical communications. J. Lightwave Technol. **27**(3), 189–204 (2009)
11. Fasn, R., Yu, Y.J., Guan, Y.L.: Generalization of orthogonal frequency division multiplexing with index modulation. IEEE Trans. Wireless Commun. **14**(10), 5350–5359 (2015)

Lower Atmospheric Wind Profile Studies and Validation of VHF Doppler Radar of University of Calcutta

Tanmay Das[1,2](✉), Debyendu Jana[1,2], Arpan Mitra[1], P. Nandakumar[1,2], Sudipto Datta[1], Jawad Y. Siddiqui[1,2], Ashik Paul[1,2], Gopal Singh[1,2], Arnam Ghosh[1], and Souvik Majumder[1,2]

[1] Institute of Radio Physics and Electronics, University of Calcutta, 92, Acharya Prafulla Chandra Road, Kolkata 700009, India
stanmaydas@gmail.com

[2] Stratosphere Troposphere (ST) Radar Facility, Ionosphere Field Station, Institute of Radio Physics and Electronics, University of Calcutta, Mondouri, North 24 Parganas 743145, India

Abstract. An active phased-array VHF Doppler radar operated at 53 MHz is being established by University of Calcutta, Kolkata, at Ionosphere Field Station (latitude: 22.93 °N, longitude: 88.50 °E), Haringhata of the University near the transition between tropical-to-sub-tropical regions. A Pilot Array is presently operational. When completed, the main radar will be a unique facility to study the lower atmospheric dynamics over the eastern and north-eastern parts of India. Initial results of three components of wind profiles, i.e. zonal wind (u), meridional wind (v), and vertical wind (w) and their validation through balloon-borne GPS Radiosonde measurements made during July 2019 have been presented in this paper.

Keywords: Three component wind velocities · GPS Radiosonde · Doppler · SNR

1 Introduction

A fully active phased-array Stratosphere Troposphere (ST) Doppler radar of University of Calcutta (CU-STR) at 53 MHz is being established by University of Calcutta, Kolkata, at Ionosphere Field Station (IFS) (latitude: 22.93 °N, longitude: 88.50 °E), Haringhata, near the transition regions between tropics-to-sub-tropics regions as shown in Fig. 1. A 19-element Pilot Array is presently operational located about 50 km north-east of Kolkata in a relatively less radio frequency disturbances. The main radar will be a unique facility to study the lower atmospheric dynamics over the eastern and north-eastern parts of India as well as the land–sea interactions around Bay of Bengal. This radar outputs three components of wind velocity, signal-to-noise ratio (SNR), range–time–intensity plots (RTI) plots and Doppler width. The main objectives of the ST radar will be to facilitate studies on lower atmospheric dynamics in the form of Stratosphere–Troposphere coupling

© Springer Nature Singapore Pte Ltd. 2021
N. R. Das and S. Sarkar (eds.), *Computers and Devices for Communication*,
Lecture Notes in Networks and Systems 147,
https://doi.org/10.1007/978-981-15-8366-7_25

processes, atmospheric turbulence, and role of atmospheric gravity waves in development of weather systems. Validation of the radar wind measurements has been done in collaboration with Space Physics Laboratory (SPL), VSSC of Indian Space Research Organization (ISRO) during the months of July and August 2019. Initial results of zonal wind (u), meridional wind (v), vertical wind (w), wind speed and wind direction and their validation through balloon-borne GPS-sonde made during July 2019 have been presented in this paper.

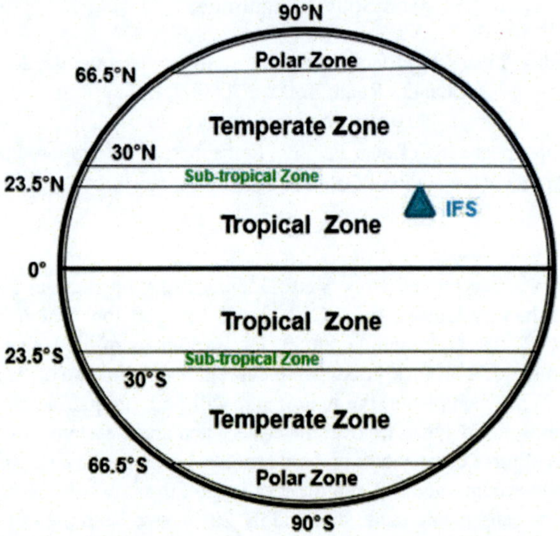

Fig. 1 Geographical location of Pilot Sub-Array of University of Calcutta at IFS, Haringhata

2 Data

The ST Radar at University of Calcutta (CU-STR) has the following technical specifications and outputs:

- Phased-array radar, operating at 53 MHz with a bandwidth of 3 MHz.
- Planar antenna array—475 three-element Yagi antennas
- Beam steering up to an off-zenith angle of 30°
- Organized into 25 sub-array groups each with 19 Yagi elements.
- Height coverage: 0.5–20 km for 90% of time
- Height resolution: 50 m up to 3 km
  150 m from 3 to 20 km
- Horizontal wind velocity: 1–70 m/s
- Vertical wind velocity: 0.1–30 m/s
- Time resolution: 5–15 min for full profile
- Average power aperture product: 3×10^8 Wm^2

Three-component wind velocities along with Doppler and signal-to-noise ratios (SNR) are being routinely measured and range-time-intensity (RTI) maps corresponding to backscatter are generated. Validation of the wind velocities with balloon-borne GPS-sonde measurements have been performed at IFS, Haringhata, during 10–19 July 2019.

A total number of 50 balloons were launched with GPS-sonde during the period of ten days in July 2019. The balloon-borne GPS-sonde were launched every three hours during the first three days and at intervals of six hours for the remaining seven days. Validation has been performed with 41 balloon-borne GPS-sonde data which provide the components of zonal wind (u) and meridional wind (v). During the validation period, the radar has been operated with five beams with an off-zenith angle of 15° along east, west, zenith, north, and south directions, respectively, to complete one scan cycle within 2.8 min. During the validation period, radar was operated with the following specifications mentioned in Table 1.

Table 1 Radar specifications during operation

Number of range bins	70
Number of FFT points	256
Number of coherent integrations	128
Number of in-coherent integrations	4
Inter-pulse period	250 μs
Pulse width	8 μs
Number of beams	5
Duty ratio	3.2%

3 Results

Figure 2 shows the Pilot Array operational at Ionosphere Field Station, Haringhata, of the University of Calcutta which consists of 19-element active phased array. Figure 3 shows a sample of Doppler profile of five beams along east, west, zenith, north, and south directions with an off-zenith angle of 15° on 11 July 2019 at 17:00 IST. It is clearly observed that echoes have been received up to 8 km of height without any appreciable atmospheric disturbances or interference which is often a precursor to the existence of upper level cloudy weather. The x-axis represents the Doppler frequency in Hz and y-axis their corresponding height in kilometre (km).

In Fig. 4, the left section represents the east-west component, i.e. zonal wind (u) and right section shows the north-south component, i.e. meridional wind (v). The x-axis represents the velocity in m/s and y-axis, the corresponding height in km. The red-coloured line represents the balloon-borne GPS-sonde and blue-coloured line represents the CU-ST radar wind components. It could be observed that zonal wind and meridional

Fig. 2 Present status of Pilot Array of CU-STR at IFS, Haringhata

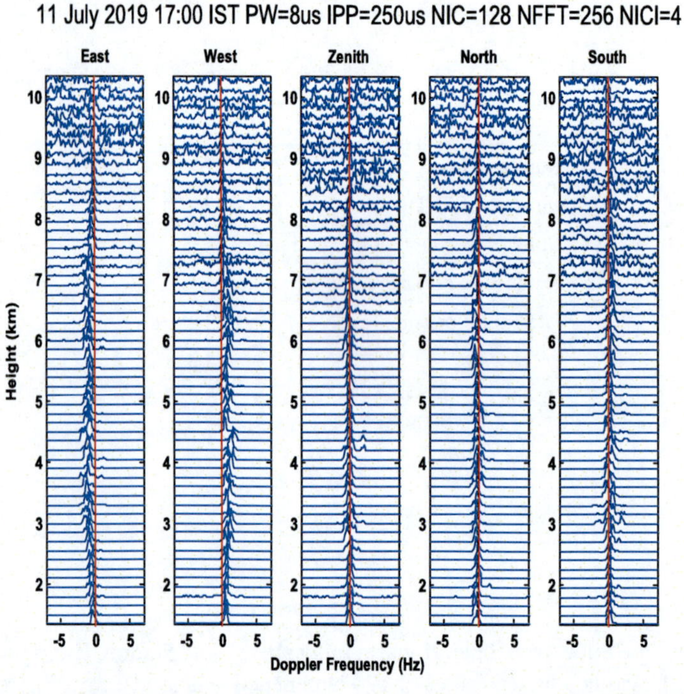

Fig. 3 Doppler profile of five beams along east, west, zenith, north, and south directions at 17:00 IST on 11 July 2019

wind matches very well up to 11 km of height on 11 July 2019 at 08:30 IST with a maximum velocity of 13.1 km/s and 9.6 km/s, respectively, as shown in Fig. 4.

Figure 5 shows the comparison of wind speed and wind directions between CU-ST Radar and balloon-borne GPS-sonde on 11 July 2019 at 08:30 IST. The red-coloured line represents the balloon-borne GPS-sonde wind component and blue-coloured line plots that for CU-ST Radar.

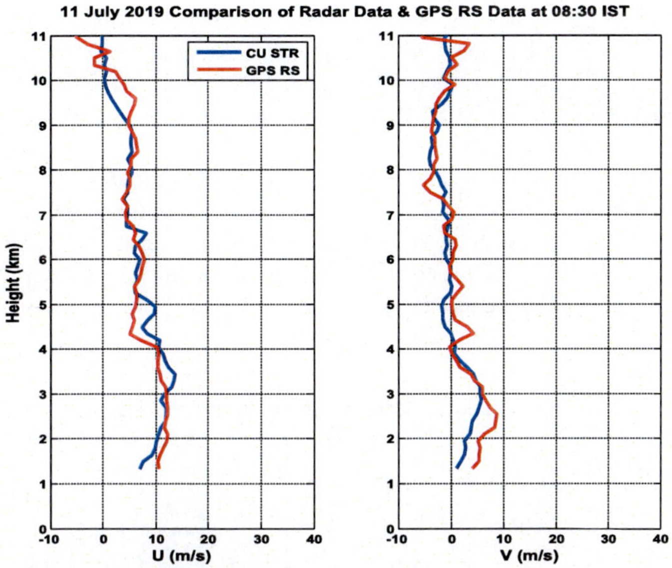

Fig. 4 Comparison of zonal and meridional winds between CU-STR and GPS radiosonde data on 11 July 2019 at 08:30 IST

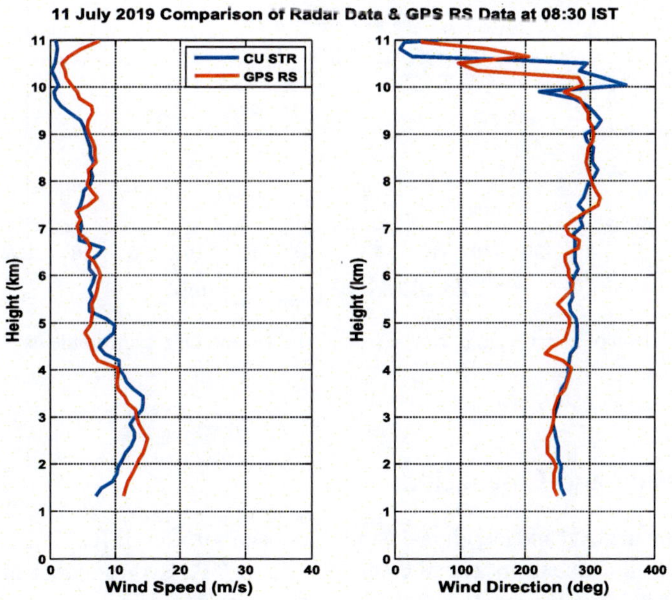

Fig. 5 Comparison of wind speed and wind direction between CU-STR and GPS Radiosonde data on 11 July 2019 at 08:30 IST

It have been noted that wind speed as well as wind direction shows good correspondence up to 10 km, beyond which some changes in direction have been observed.

The comparison between CU-ST Radar retrieved zonal wind, meridional wind, wind speed and wind directions has been made for entire ten days of 10–19 July 2019 with 41 balloon-borne GPS-sonde wind measurements. Figures 6 and 7 represent the comparison of zonal and meridional winds between CU-STR data and balloon-borne GPS radiosonde data respectively for these ten days. The x-axis represents the wind components measured by GPS radiosonde and y-axis, the corresponding wind components measured by the ST radar. The correlation coefficients have been noted 0.94 and 0.79 for zonal and meridional winds, respectively.

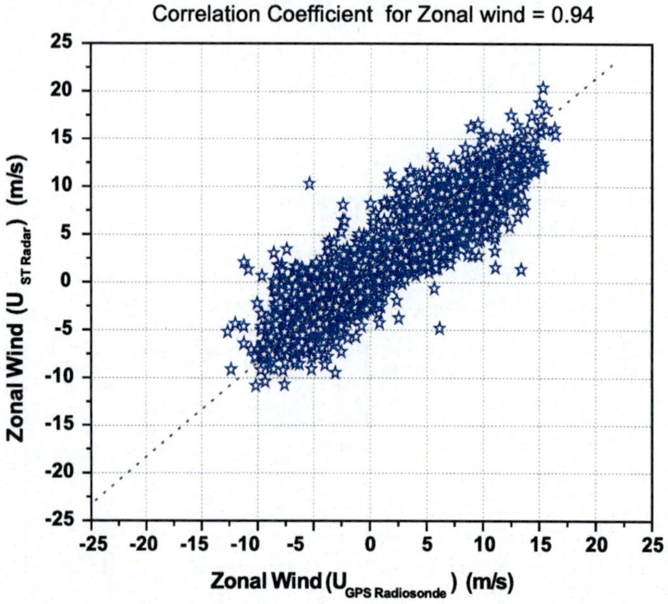

Fig. 6 Comparison of zonal wind retrieved from CU-STR and GPS radiosonde data during 10–19 July 2019

4 Discussion and Conclusions

The validation of zonal and meridional wind components of CU-STR shows good match with the wind components measured by balloon-borne GPS Radiosonde launched from IFS, Haringhata. It is important to note that these measurements are unique being the first from this geophysically sensitive region made using an active phased array and may form the benchmark for future wind profiles. The high resolution data from ST Radar is expected to provide significant information about the highly dynamic Stratosphere–Troposphere exchange processes [1] and the main drivers and their relative roles in generation of atmospheric gravity waves.

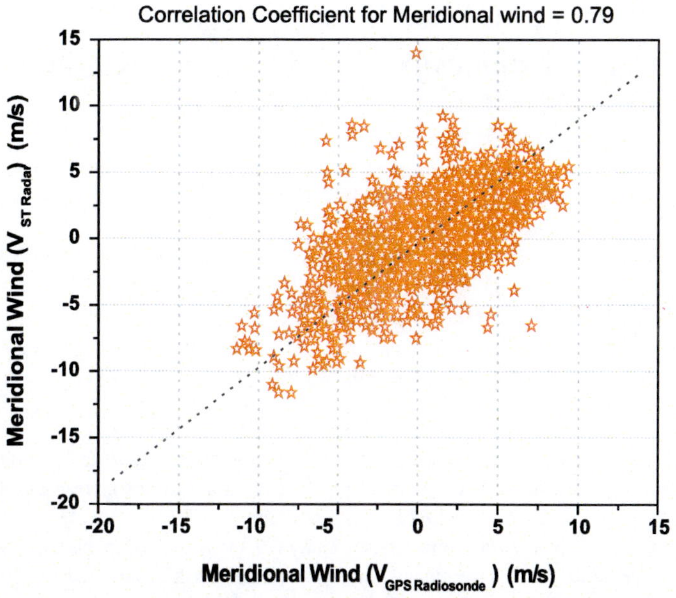

Fig. 7 Comparison of meridional wind retrieved from CU-STR and GPS radiosonde data during 10–19 July 2019

These waves could affect development of regional weather systems. A detailed study may give some interesting clue to understand the large-scale spatio-temporal variability of weather systems [2, 3], effect of global changes on climate, formation and development of cyclones and depressions over Bay of Bengal, etc. Extreme weather events could be identified and characterized by understanding the underlying the physical mechanism.

Acknowledgements. The VHF Radar at University of Calcutta is being implemented with financial support from the Science and Engineering Research Board (SERB), Department of Science and Technology (DST), Govt. of India. The authors thank Space Physics Laboratory (SPL), VSSC, ISRO for their support in validating wind measurements from this radar through balloon-borne GPS Radiosonde.

References

1. Mohanakumar, K.: Stratosphere troposphere interactions: an introduction. Springer Science & Business Media (2010)
2. Ackley, M., Chadwick, R., Cogan, J., Crosiar, C., Eaton, F., Gage, K., Gossard, E., Lucci, R., Merceret, F., Neff, W., Ralph, M., Strauch, R., van de Kamp, D., and an Allen White, B.W.: U.S. Wind profilers: a review. Tech. Rep. FCM-R14-1998, U.S. DoC/ NOAA/OFCM, Washington, DC (1998)
3. Andersson, E., Garcia-Mendez, A.: Assessment of European wind profiler data, in an NWP context. Technical Memorandum 372, ECMWF (2002)

Summer Night-Time E-Layer Echoes Observed Using University of Calcutta ST Radar

Tanmay Das[1,2](✉), Ashik Paul[1,2], P. NandaKumar[1,2], Gopal Singh[1,2], Debyendu Jana[1,2], Jawad Y. Siddiqui[1,2], and Souvik Majumder[1,2]

[1] Institute of Radio Physics and Electronics, University of Calcutta, 92, Acharya Prafulla Chandra Road, Kolkata 700009, India
stanmaydas@gmail.com

[2] Stratosphere Troposphere (ST) Radar Facility, Ionosphere Field Station, Institute of Radio Physics and Electronics, University of Calcutta, Mondouri, North 24 Parganas 743145, India

Abstract. A fully active phased-array Stratosphere Troposphere (ST) radar at 53 MHz is being established by University of Calcutta at Ionosphere Field Station (22.93 °N, 88.50 °E geographic; magnetic dip: 34 °N), Haringhata, near the northern crest of equatorial ionization anomaly (EIA). When completed, this will be a unique facility at this frequency in the eastern and north-eastern parts of India and also in the south-east Asian longitude sector. Initial results of ionospheric backscatter of irregularities observed at the range of 120–140 km during early evening hours to midnight of the months of May through June 2019 from a Pilot Array are presented in this paper.

Keywords: Doppler · SNR · Ionospheric backscatter

1 Introduction

An indigenously developed, state-of-the-art 53 MHz VHF ST radar is being established at Ionosphere Field Station (IFS), University of Calcutta, Haringhata, with the support of Science and Engineering Research Board, DST, Government of India, under its scheme—**Intensification of Research in High Priority Areas (IRHPA)**. It is conceived as a National Facility, open to all academicians and researchers across the country, and those in the east and north-east India, in particular. Once fully functional, it is expected to provide three components of wind velocity, Doppler width, signal-to-noise ratio (SNR), range–time–intensity (RTI) plots, range–time–velocity (RTV) plots and range–time Doppler spread plots. The ST radar will facilitate research on Stratosphere–Troposphere exchange processes, atmospheric dynamics, turbulence, role of atmospheric gravity waves in development of weather systems and ionosphere. Presently, a pilot version of the radar is operational. The ST Radar at University of Calcutta (CU-STR) (22.93 °N, 88.50 °E geographic; magnetic dip: 34 °N) is situated near the northern crest of the equatorial ionization anomaly (EIA), about 50 km north-east of Kolkata, in an area of relatively low radio frequency interference. This paper reports some initial results of ionospheric backscatter from irregularities observed at the range of 120–140 km during early evening hours to midnight of the months of May through June 2019.

© Springer Nature Singapore Pte Ltd. 2021
N. R. Das and S. Sarkar (eds.), *Computers and Devices for Communication*,
Lecture Notes in Networks and Systems 147,
https://doi.org/10.1007/978-981-15-8366-7_26

2 Data

The CU-STR is a fully phased-array radar, operating at 53 MHz with a bandwidth of 3 MHz. Main array will consist of 475 three-element Yagi–Uda antennas organized into 25 sub-arrays. Each array consists of 19 Yagi–Uda antennas. The beam can be steered up to an off-zenith angle of 30°.

The radar will provide the height coverage from 0.5 to 20 km for 90% of time with height resolution of 50 m up to 3 km and 150 m from 3 to 20 km with a time resolution of 5-15 min for full profile with average power aperture product 3×10^8 Wm^2.

The profiles with Doppler and signal-to-noise ratio (SNR) are being routinely measured and range–time–intensity (RTI) and range–time–velocity (RTV) maps corresponding to ionospheric backscatter generated. At present, one sub-array has been operational since April 2018.

In the present observations, only the beams towards the geomagnetic north direction have been taken into consideration with off-zenith angle of 23°, 25°, 27°, 28° and 30° with a resolution of ~1.2 km. This orientation has been made to ensure quasi-transverse (QT) mode of propagation. The radar was operated during post-sunset to early morning hours of the months of May through June 2019.

3 Results

Figure 1 shows the present status of Pilot Sub-Array operational at Ionosphere Field Station (IFS), University of Calcutta, Haringhata. It is a 19-element active phased array, and each TRM transmits a peak power of 2 kW. The schematic diagram of the main array of 475 antennas is shown in Fig. 2. Each red dot represents one antenna. There are 19 hexagonal sub-arrays, and the remaining 6 sub-arrays are distributed around the sub-arrays to form a circular array.

ST Radar Pilot Sub-array at IFS, Haringhata

Fig. 1 Pilot Sub-Array of University of Calcutta at IFS, Haringhata

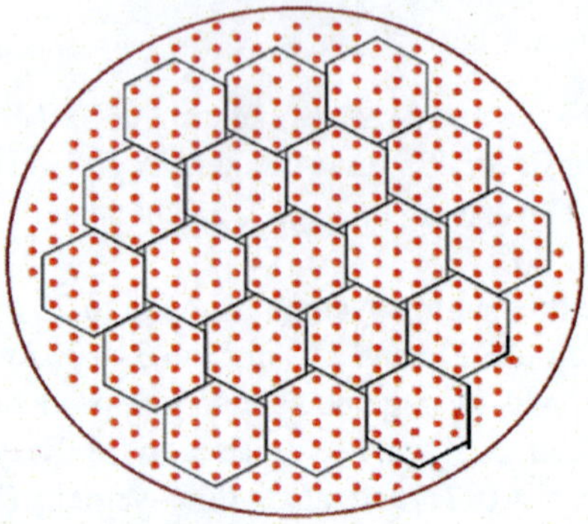

Diagram of Fully active phased array of 475 antennas at IFS, Haringhata

Fig. 2 Schematic diagram of Main Array of University of Calcutta Stratosphere Troposphere Radar (CU-STR)

Figure 3 shows a representative three-dimensional (3D) power spectra with 32 μs coded pulse observed on 31 May 2019 at 19:22 IST towards geomagnetic north direction with 28° off-zenith angle. In the diagram, right-hand green side of x-axis denotes the positive parts of Doppler frequencies in Hz and orange regions represent the negative Doppler frequencies. The y-axis represents the range in Km.

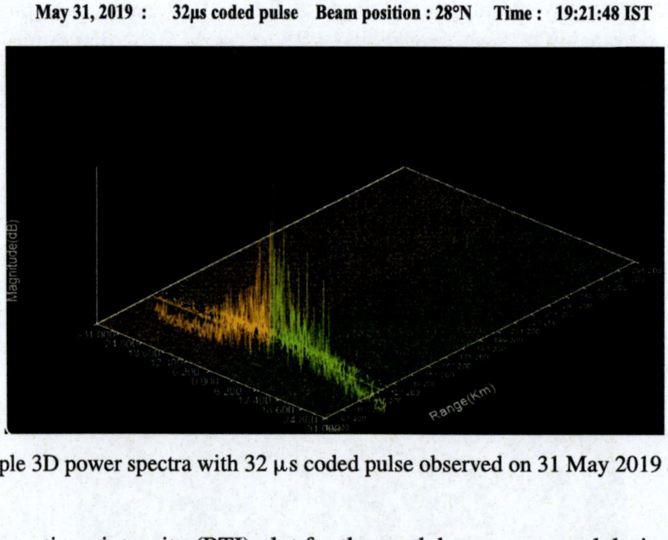

Fig. 3 Sample 3D power spectra with 32 μs coded pulse observed on 31 May 2019 at 19:22 IST

The range–time–intensity (RTI) plot for the north beam measured during 19:12 IST of 31 May 2019 through 03:58 IST of 1 June 2019 is shown in Fig. 4. The x-axis represents time in Indian Standard Time (IST), and y-axis represents the corresponding range in km. It is clearly evident that ionospheric backscatter from ionization density irregularities had occurred over the range of 120–140 km with power of −11 to −5 dB during early evening (19:12 IST) to pre-midnight hours (22:42 IST).

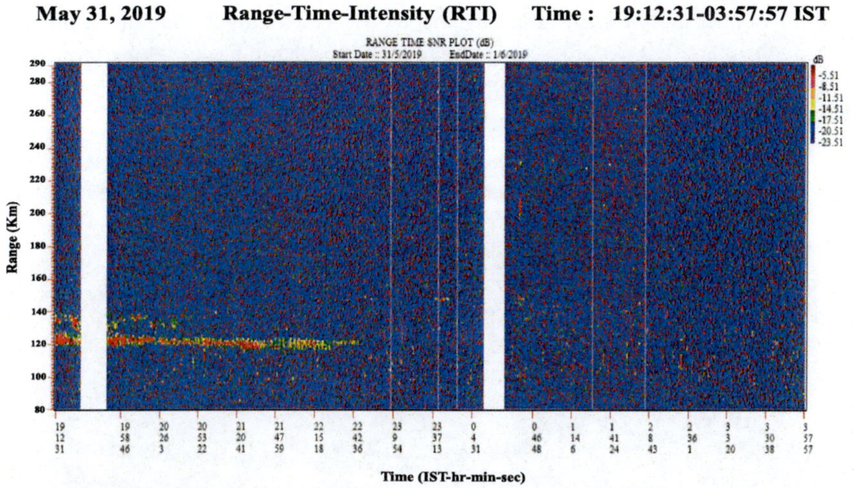

Fig. 4 Range–time–intensity (RTI) plot for the north beam measured during 19:12 IST on 31 May 2019 through 03:58 IST of 1 June 2019

The corresponding range–time–velocity (RTV) plot of the same ionospheric backscatter is shown in Fig. 5. The time (IST) represents along x-axis, and range (Km) represents along y-axis. The velocity of the irregularities is found to be ~−32 m/s during early evening hours.

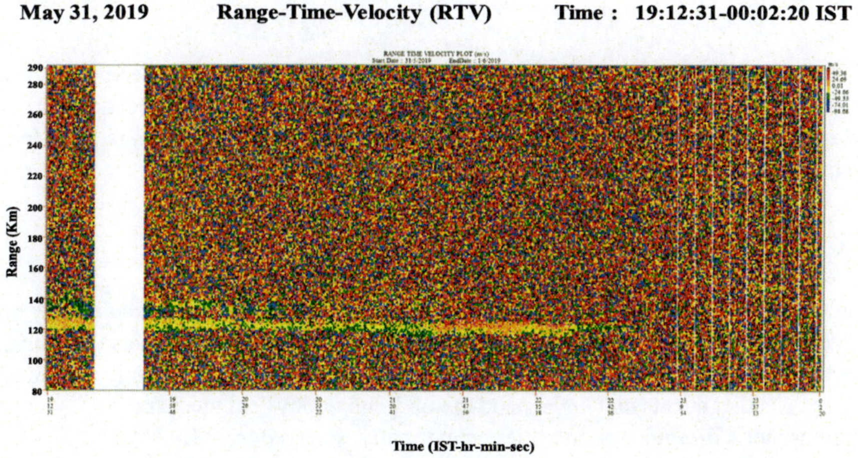

Fig. 5 Range–time–velocity (RTV) for the north beam measured during 19:12 IST of 31 May 2019 through 00:02 IST of 1 June 2019

A total number of nine such cases have been observed during the summer months of May through June 2019. Another sample of such ionospheric backscatter from irregularities observed by the Pilot Array from a range of 120–130 km on 6 June 2019 during 18:45-20:30 IST is shown in Fig. 6.

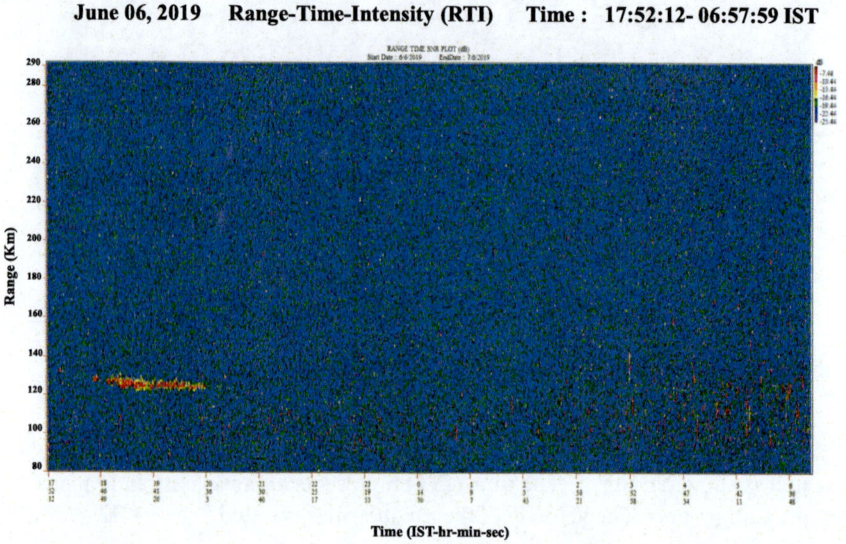

Fig. 6 Range–time–intensity (RTI) plot for the north beam during 17:52 IST of 6 June 2019 through 06:58 IST of 7 June 2019

There are extensive reports of ionospheric E-region backscatter observations from such heights from measurements near the magnetic equator and at intermediate locations in the Indian longitude sector [1]. However, such observations are non-existent from regions around the northern crest of EIA.

4 Conclusions

Results from the ST Radar at University of Calcutta (CU-STR) are unique as they are situated in a region where there have not been any previous measurements done using an active phased array. Such observations could effectively be used to study the impact of space weather events and understand the underlying physical mechanism.

Ionospheric E-region irregularities are normally observed from Kolkata during summer daytime [2, 3] and equinox night-time [4, 5]. This paper reports, perhaps for the first time using radar, occurrences of E-region night-time irregularities at 53 MHz during the summer months (May–June) from a station like Kolkata, situated near the northern crest of equatorial ionization anomaly (EIA).

Acknowledgements. The ST Radar at University of Calcutta (CU-STR) is being implemented with financial support from the Science and Engineering Research Board (SERB), Department of

Science and Technology (DST), Government of India. The authors are also thankful to NARL, Gadanki, for helpful research discussions.

References

1. Patra, A.K., Rao, N.V.: Low-latitude valley region irregularities studied using the Gadanki radar. J. Geophys. Res. (Space Phys.) **112**, A03303 (2007)
2. Das, S.K., Chakraborty, S.K., Das, A.: Daytime amplitude scintillations at a low latitude station, Calcutta. IJRSP **25**(3) (1996)
3. Das, T., Roy, B., Das Gupta, A., Paul, A.: Impact of equatorial ionospheric irregularities on transionospheric satellite links observed from a low-latitude station during the minima of solar cycle 24". IJRSP **41**, 247–257 (2012)
4. Paul, A., Ray, S., DasGupta, A., Chandra, H.: Radio signatures of November 1998 Leonid meteor on transionospheric VHF satellite signal. Planet. Space Sci. **49** (2001)
5. Das, T., Roy, B., Paul, A.: Effects of transionospheric signal decorrelation on Global Navigation Satellite Systems (GNSS) performance studied from irregularity dynamics around the northern crest of the EIA. Radio Sci. **49**(10), 851–860 (2014)

An Approach to Reduce Power Consumption and Delay of Single Error Correction Codes in WSNs for IoT Applications

Jhilam Jana[1](✉), Sayan Tripathi[1], Jagannath Samanta[2], Jaydeb Bhaumik[1], and Soma Barman (Mandal)[3]

[1] Department of ETCE, Jadavpur University, Kolkata, India
jhilamjana2014@gmail.com, tripathysayan@gmail.com, bhaumik.jaydeb@gmail.com

[2] Haldia Institute of Technology, Haldia, India
jagannath19060@gmail.com

[3] Department of Radio Physics and Electronics, University of Calcutta, Kolkata, India
barmanmandal@gmail.com

Abstract. Single error correction (SEC) codes have been employed to protect the data bits as well as control bits in wireless sensor networks (WSNs) for Internet of Things (IoT) applications. In these systems, soft error typically single event upset (SEU) has been occurred. A low delay and power efficient error correcting codes are desirable in most of the wireless sensor networks. In this paper, three schemes have been proposed to construct the *H*-matrix of SEC codes. Based on these schemes, a low power and fast SEC codec has been designed and implemented for data width of 128 bits. Performances of the proposed SEC codes are observed in terms of area, power and delay. The maximum improvements in terms of number of LUTs and delay are 21.43% and 5.47%, respectively, compared to Reviriego et al. codes in FPGA platform. ASIC-based synthesis result shows that maximum power reduction of three proposed schemes are 16.06%, 14.8% and 6.06%, respectively, and delay reduction are 14.34%, 21.70% and 39.63%, respectively, compared to existing designs. The proposed codes may be used in WSNs for IoT applications such as smart cities, smart home and smart health care.

Keywords: Internet of Things (IoT) · Wireless sensor networks (WSNs) · Soft errors · Single error correction (SEC) codes

1 Introduction

Internet of Things (IoT) is a platform that is connected with physical objects via cloud server with the help of Internet [1]. In wireless communication system, the wireless sensor network is a group of sensors which is used to collaborate, detect, communicate and process the desirable information within its network coverages [2]. The reliability of these received data depends on types of channel and noise. The major issues in WSNs are as follows:

© Springer Nature Singapore Pte Ltd. 2021
N. R. Das and S. Sarkar (eds.), *Computers and Devices for Communication*, Lecture Notes in Networks and Systems 147,
https://doi.org/10.1007/978-981-15-8366-7_27

1. Soft errors are generally occurred due to alpha particles radiation, cosmic rays which correspond to loss of signal integrity.
2. Significant power consumption in wireless communication networks.

Reliable communication between wireless sensor networks (WSNs) is obtained by employing different types of error detection and correction schemes which reduce the probability error of the received signal. There are number of single error correction (SEC) codes and multi-bit error correction codes such as Hamming code [3], Hsiao code [4], Reviriego et al. code [5] have been presented in the literature. Memory cells are affected by soft error which can also be protected by employing SEC codes and single error correction-double error detection (SEC-DED) codes [3, 5–7]. Low delay and power efficient codes are always amenable for wireless networks.

Reviriego et al. [8] proposed a technique to implement error correction code that has low delay and low complexity. Reviriego et al. [9] presented a SEC code that is able to protect data bits and a few additional control bits. Hamming et al. also have described the approach of error correction codes to recover the original message bits with some redundant bits [3]. Cha and Yoon also proposed a novel SEC code which is tested on their developed model by facilitating simultaneous data bit and check bit arrays [10].

To overcome these challenges, this paper presents a unique approach of low delay and power efficient single error correction (SEC) code for data words of 128 bits. The main contributions of this paper are as follows:

(i) Propose a new method to construct the parity check matrices (H) for SEC codes. (ii) The SEC codes with message length of 128 bit have been designed and implemented. (iii) These codes have least delay and also power efficient compared to the existing designs. (iv) These codes have simpler encoding and decoding processes.

The rest of the paper is organized as follows. Section 2 contains the proposed SEC codes. Section 3 presents an estimation for number of logic gates. Section 4 describes the implementation of proposed SEC codes. Section 5 states the conclusion.

2 Proposed SEC Codes

Proposed SEC codes are discussed in this section. H-matrix is initialized with $(n - k)$ number of rows and (n) number of columns; where n is the codeword length and $(n - k)$ is the number of parity bits. H-matrix consists of $(n - k)$ numbers of parity columns having identity property and k numbers of data columns. Construction procedures of proposed H-matrix of SEC codes are as follows:

Step 1: In method 1, the k number of data columns is filled with element of minimum total weight and balanced row weight. In method 2, three control bits are added at the first and then method 1 is repeated. In method 3, the H-matrix is separated into two parts using masking techniques.

Step 2: H-matrices of these three methods are modified by increasing the number of parity columns as well as minimum weight data columns.

To minimize the power and delay, the parity bits of these H-matrices are increased which is the main advantage of proposed schemes over Reviriego et al. schemes. To reduce delay in proposed designs, the maximum number of 1s in a row is minimized.

The H-matrix of (137, 128) SEC code is shown in Fig. 1. This H-matrix consists of 128 data columns and nine parity columns.

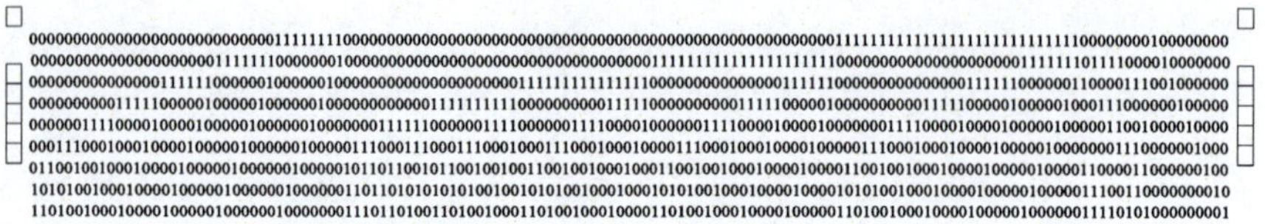

Fig. 1 H-matrix of proposed (137, 128) SEC code

2.1 Design of Proposed SEC Codes

In this section, proposed SEC codes are designed and implemented. The encoding and decoding processes are also described. In the encoding process, the codeword is generated with the combination of data and parity bits. The codeword is stored in memory to protect data bits and control bits in some cases against soft error.

In decoding technique, error detection and correction are done using the syndrome values. Syndromes (S) are calculated employing the following equation.

$$S = r * H^T \tag{1}$$

In received codeword, if the syndrome values are zero ($S = 0$), it means the received codeword has no error. On the other hand, if the syndrome value is nonzero ($S \neq 0$), it means there are some bit errors. Finally, the error associated with each bit is corrected by error correction block. The error correction logic for the first data bit of proposed (137, 128) SEC code is expressed in Eq. (2):

$$dc_1 = r_1 \oplus \left(s_1' s_2' s_3' s_4' s_5' s_6' s_7' s_8 s_9\right) \tag{2}$$

The error correction logic is used to protect the data bits as well as recover the original codeword of the SEC code.

3 Estimation of Logic Gates

In this section, the area in terms of logic gates and critical path delay have been also described. The area in terms of 2-input XOR gates (XOR2), 2-input AND gates (AND2) and NOT gates for proposed schemes and existing schemes is determined. Critical path is calculated by identifying the longest path. The critical path analysis of the proposed SEC codes in terms of number of logic levels present in the longest path from the input to output of the decoder has been also presented in Table 1.

The proposed SEC codes exhibit the maximum improvement of 29.31% in critical path delay compared to the related codes. The synthesis results of proposed codes and existing codes are described in the following section.

Table 1 Area in terms of logic gates of proposed and existing SEC codes

Scheme	SEC Code	Area in terms of logic gates					Critical path delay				
		XOR2	AND2	NOT	Equivalent NAND2	Impro (%)	XOR2	AND2	NOT	Equivalent NAND2	Impro (%)
128 bits	Reviriego et al. (136, 128) [8]	920	896	620	6092	–	102	7	1	423	–
	Proposed (137, 128)	840	1024	790	6198	−1.73	80	8	1	337	20.33
	Proposed (138, 128)	796	1152	934	6422	−5.41	70	9	1	299	29.31
	Proposed (139, 128)	775	1280	1071	6731	−10.48	70	10	1	301	28.84
128 bits + 3 control bits	Reviriego et al. (139, 131) [9]	947	917	636	6258	–	104	7	1	431	–
	Proposed (140, 131)	858	1048	811	6339	−1.29	84	8	1	353	18.09
	Proposed (141, 131)	817	1179	962	6588	−5.27	76	9	1	323	25.05
	Proposed (142, 131)	796	1310	1103	6907	−10.37	72	10	1	309	28.30
128 bits + 3 control bits (Masking)	Reviriego et al. (139, 131) [9]	1121	917	534	6852	–	146	7	1	599	–
	Proposed (141, 131)	1039	1048	851	7103	−3.66	102	9	1	427	28.71

4 Implementation of Proposed SEC Codes

The SEC encoder and decoder blocks have been represented in Verilog Hardware Description Language. Proposed SEC codecs are simulated and synthesized on FPGA and ASIC platforms. The performance of existing and proposed codecs have been measured in terms of lookup tables (LUTs) and delay for FPGA-based design and in terms of area, delay, power, power area product (PAP), power delay product (PDP) and product of area, delay and power (Cost) for ASIC-based design.

4.1 Result of FPGA-Based Design

All the functional blocks are simulated and synthesized using FPGA-based virtex 6 device (XC6VCX75T). The FPGA-based synthesis results have been shown in Table 2. The maximum improvements over Reviriego et al. codes in terms of LUTs and delay for SEC codes are 21.43% and 5.47%, respectively.

4.2 Result of ASIC-Based Design

The encoder and decoder of proposed SEC codes and existing codes have been further synthesized in ASIC platform using Cadence-based Genus Synthesis Solution (TSMC18) tool. The ASIC-based synthesis results have been presented in Table 3. The ASIC-based synthesis results show the overall reduction in terms of power, delay, PDP, PAP and cost but area of the proposed codes is slightly higher compared to existing designs. ASIC-based designs provide maximum reductions in power of these schemes are 16.06%, 14.8% and 6.06%, respectively, and also maximum reductions in delay are 14.34%, 21.70% and 39.63% compared to existing related designs. The area of the proposed schemes is slightly higher compared to the existing codes but the maximum improvements in terms of PDP, PAP and cost are 43.29%, 13.07% and 39.92%, respectively.

5 Conclusion

In this paper, a new method for designing SEC codecs has been proposed by increasing number of parity bits. These proposed codes and existing codes have been designed and implemented on both FPGA and ASIC platforms. Theoretically, maximum reduction in critical path delay of these three methods are 29.31%, 28.30% and 28.71%, respectively. Three methods of proposed SEC codes require lesser power, delay, PDP, PAP and cost for 128 bit. In ASIC platform, the synthesis results also indicate the reductions in both power (maximum of 16.06%) and delay (maximum of 39.63%). The maximum improvements in terms of LUTs and delay are 21.43% and 5.47%, respectively. The proposed codes can be employed to protect data bits with the advantage of low delay and power in WSNs for IoT applications at the cost of additional control bits.

Table 2 FPGA-based synthesis results of proposed and existing SEC codecs

Scheme	SEC code	Encoder				Decoder			
		LUTs	Delay (ns)	Impro Over LUTs (%)	Impro Over Delay (%)	LUTs	Delay (ns)	Impro Over LUTs (%)	Impro Over Delay (%)
128 bits	Reviriego et al. (136, 128) [8]	84	2.50	–	–	240	4.44	–	–
	Proposed (137, 128)	72	2.49	14.29	0.40	235	4.36	2.08	1.82
	Proposed (138, 128)	69	2.44	17.86	2.44	234	4.39	2.50	1.13
	Proposed (139, 128)	66	2.44	21.43	2.44	235	4.60	2.08	−3.51
128 bits + 3 control bits	Reviriego et al. (139, 131) [9]	86	2.50	–	–	238	4.50	–	–
	Proposed (140, 131)	76	2.50	11.63	0.32	242	4.60	−1.68	−2.16
	Proposed (141, 131)	71	2.44	17.44	2.44	236	4.32	0.84	3.96
	Proposed (142, 131)	70	2.44	18.60	2.44	238	4.62	0.00	−2.56
128 bits + 3 control bits (Masking)	Reviriego et al. (139, 131) [9]	87	2.78	–	–	238	4.34	–	–
	Proposed (141, 131)	93	2.63	−6.90	5.47	253	4.51	−6.30	−3.89

Table 3 ASIC-based synthesis results of proposed and existing SEC codecs

Scheme	SEC code	Block	Area (μm^2)	Power (mW)	Delay (ps)	PAP (μm^2 mW)	PDP (mW ps)	Cost (μm^2 mW ps)
128 bits	Reviriego et al. (136, 128) [8]	Enc	5941	1.41	361.6	8376.8	509.9	302.91
		Dec	11,014	2.75	831.4	30,288.5	2286.4	2518.19
	Proposed (137, 128)	Enc	5994	1.25	363.6	7492.5	454.5	272.43
		Dec	11,263	2.55	743.8	28,720.7	1896.7	2136.24
	Proposed (138, 128)	Enc	6104	1.23	361.6	7507.9	444.8	271.49
		Dec	11,323	2.33	819.3	26,382.6	1909.0	2161.53
	Proposed (139, 128)	Enc	6131	1.21	428.2	7418.5	518.1	317.66
		Dec	11,410	2.31	712.2	26,357.1	1645.2	1877.15
128 bits + 3 control bits	Reviriego et al. (139, 131) [9]	Enc	5954	1.46	392.4	8692.8	572.9	341.11
		Dec	11,090	2.76	974.3	30,608.4	2689.1	2982.18
	Proposed (140, 131)	Enc	6190	1.31	410.5	8108.9	537.8	332.87
		Dec	11,486	2.65	861.1	30,437.9	2281.9	2621.01
	Proposed (141, 131)	Enc	6204	1.28	361.6	7941.1	462.8	287.15
		Dec	11,543	2.39	825.1	27,586.9	1972.0	2276.19
	Proposed (142, 131)	Enc	6280	1.24	363.6	7787.5	450.9	283.15
		Dec	11,779	2.36	762.9	27,798.0	1800.4	2120.71
128 bits + 3 control bits (Masking)	Reviriego et al. (139, 131) [9]	Enc	6473	1.66	361.6	10,745.5	600.3	388.56
		Dec	11,436	2.95	754.4	33,736.7	2225.5	2545.10
	Proposed (141, 131)	Enc	6686	1.59	363.6	10,630.7	578.1	386.53
		Dec	12,115	2.77	455.4	33,557.8	1261.5	1528.22

References

1. Farahat, I.S., Tolba, A.S., Elhoseny, M., Eladrosy, W.: Data security and challenges in smart cities. In: Security in Smart Cities: Models, Applications, and Challenges, pp. 117–142. Springer, Cham (2019)
2. Zhang, S., Zhang, H.: August. A review of wireless sensor networks and its applications. In: 2012 IEEE International Conference on Automation and Logistics, pp. 386–389. IEEE (2012)
3. Hamming, R.W.: Error detecting and error correcting codes. Bell Syst. Tech. J. **29**(2), 147–160 (1950)
4. Hsiao, M.Y.: A class of optimal minimum odd-weight-column SEC-DED codes. IBM J. Res. Dev. **14**(4), 395–401 (1970)
5. Reviriego, P., Maestro, J.A., Baeg, S., Wen, S., Wong, R.: Protection of memories suffering MCUs through the selection of the optimal interleaving distance. IEEE Trans. Nucl. Sci. **57**(4), 2124–2128 (2010)
6. Samanta, J., Tripathi, S.: Comments on "A novel approach of error detection and correction for efficient energy in wireless networks." Multimedia Tools Appl. **78**(6), 7579–7584 (2018). https://doi.org/10.1007/s11042-018-6481-8
7. Saiz-Adalid, L.J., Gracia, J., Gil-Toms, D., Baraza-Calvo, J.C., Gil-Vicente, P.J.: Ultrafast codes for multiple adjacent error correction and double error detection. IEEE Access **7**, 151131–151143 (2019)
8. Reviriego, P., Pontarelli, S., Maestro, J.A., Ottavi, M.: A method to construct low delay single error correction codes for protecting data bits only. IEEE Trans. Comput. Aided Des. Integr. Circuits Syst. **32**(3), 479–483 (2013)
9. Reviriego, P., Demirci, M., Evans, A., Maestro, J.A.: A method to design single error correction codes with fast decoding for a subset of critical bits. IEEE Trans. Circuits Syst. II Express Briefs **63**(2), 171–175 (2015)
10. Cha, S., Yoon, H.: Single-error-correction code for simultaneous testing of data bit and check bit arrays for word-oriented memories. IEEE Trans. Device Mater. Reliab. **13**(1), 266–271 (2013)

Microwave and Lightwave Technology (MLT)

Design and Modelling of a FSS-Based Wideband Absorber

Priyanka Das[1(✉)] and Kaushik Mandal[2]

[1] University of Engineering and Management, Kolkata, India
priyankadas131986@gmail.com
[2] Institute of Radio-Physics and Electronics, University of Calcutta, Kolkata, India

Abstract. This paper aims at designing a thin low profile frequency-selective surface (FSS)-based absorber without an air gap in the frequency range of 6.1–12.5 GHz. A resistive FSS-based impedance surface of 11 Ω/square is designed on a two-layered FR4 substrate backed by a metallic plate. The return loss exhibited by the absorber is below −10 dB in the operating frequency range, yielding more than 90% absorption. The absorber shows an angular stability upto 50° for both TE and TM polarizations. The absorber is polarization insensitive due to its fourfold symmetric structure. Parametric studies of different impedance surfaces have been conducted. Finally, an equivalent circuit model is proposed which is used for deriving various circuit parameters.

Keywords: FSS · Resistive surface · Absorber · Wideband

1 Introduction

Frequency-selective surface (FSS) is used for electromagnetic shielding, meta-materials fabrication and radar cross section (RCS) reduction according to its spatial filter characteristics such as bandpass or bandstop response to electromagnetic (EM) waves [1]. Electromagnetic absorbers find many important applications throughout the frequency spectrum: RAM (radar absorbing materials), electromagnetic interference at microwave range, thermo detector, microbolometer, solar cell at IR, visible range and so forth. In recent days, most of the practical applications demand the characteristics like reduced thickness, light weight, conformability, wide bandwidth and so forth all in a single absorber. Scientists and researchers have done several works to achieve these characteristics. The use of parallel *LC* resonators in conjunction with the lumped resistors [2] exhibits absorption from 4.5 to 7.5 GHz and from 9.1 to 11.3 GHz. A Swastika shaped metamaterial absorber is designed in [3] which gives an absorption bandwidth of 0.68 GHz but at the expense of increased unit cell dimension. A periodic array of square rings [4] with Giusepe Peano fractal arms on one side and a metallic ground on another side of a lossy dielectric layer yield a bandwidth of 7.5% (0.35 GHz) with respect to the centre frequency (4.68 GHz). The absorber [5] comprising two-dimensional array of conductive crossed dipoles with lumped resistor elements on the top of a single-layer

© Springer Nature Singapore Pte Ltd. 2021
N. R. Das and S. Sarkar (eds.), *Computers and Devices for Communication*,
Lecture Notes in Networks and Systems 147,
https://doi.org/10.1007/978-981-15-8366-7_28

FR4 substrate, backed by continuous metallic layer offers a bandwidth of 70.7% (5.3–11.2 GHz). The design [6] is based on a two-layer capacitive circuit absorber with the back-metal layer replaced with a polarization-sensitive resistive frequency-selective surface. The upper layer comprising lumped resistors assists wide absorption band (above 90% absorption) from 3.8 to 10.3 GHz (entire C and partial X band); whereas, PIN diodes are mounted in the bottom layer to enable the switching operation of the rasorber [7] at S-Band. Three resonances within the operating frequency range of a single-layer circuit analogue (CA) absorber consisting of double-square-loop (DSL) resistor-loaded arrays are realized in [8]. Two-dimensional periodic arrangements of incurved square loop (ISL) loaded with lumped resistor [9] perform as a bandpass filter at operation band around 0.92 GHz and act as an absorber over a wide out of band 3–9 GHz. Use of lumped resistors increases the cost of fabrication. The earlier designs have incorporated an air gap between the two dielectric layers which increases the overall thickness of the absorber.

The behaviour of absorbing instead of reflecting makes it impossible to realize a low RCS. The standard objective of designing an absorber is to maximize the absorption bandwidth and minimize the thickness of the absorber above a perfect electric conducting (PEC) surface. The novelty of the absorber presented in this paper is twofold. Firstly, there is no air gap between the two FR4 substrate layers which makes the design low profile in nature. Secondly, surface impedance of 11 Ω/square is chosen which is much low as compared to the reported designs. The proposed structure is a potential candidate for RCS reduction in the aircraft stealth technology for the application of battlefield airborne and radar.

2 Design of the Unit Cell

The unit cell in Fig. 1 comprises modified cross-shaped impedance structure both on the top layers of FR4 substrates having thickness 1.58 mm connected back to back. The bottom layer of the lower FR4 substrate is backed by a metallic plate. No air gap exists between the top FR4 layer and the bottom FR4 layer. Compared with earlier reported structures, the designed FSS demonstrates a better miniaturization performance having unit cell dimension of 7.5 mm ($0.15\lambda_0$) where λ_0 corresponds to the lowest frequency of operation. The thickness of the proposed absorber is $0.0632\lambda_L$ where λ_L is the wavelength corresponding to the lowest cut-off frequency.

For oblique angle of incidence, angular stability up to 50° is observed in Figs. 2 and 3 providing absorption above 80% for both TE and TM modes. At 60°, the absorption characteristics degrade and fall below 75%. For normal angle of incidence, high level of polarization stability is observed in Fig. 4, due to fourfold symmetric structure of the proposed absorber. The absorption performance of the absorber changes with variation of resistivity of the surface as seen in Fig. 5. An impedance surface of 11 Ω/square gives optimum absorption performance and hence chosen. Identical resistive FSS layers are placed on both FR4 substrate layers of thickness 1.58 mm connected back to back. Frequencies from 6.1–12.5 GHz are absorbed, and hence, they are not reflected back. Frequencies lower than 6.1 GHz and higher than 12.5 GHz are reflected back from the absorber surface.

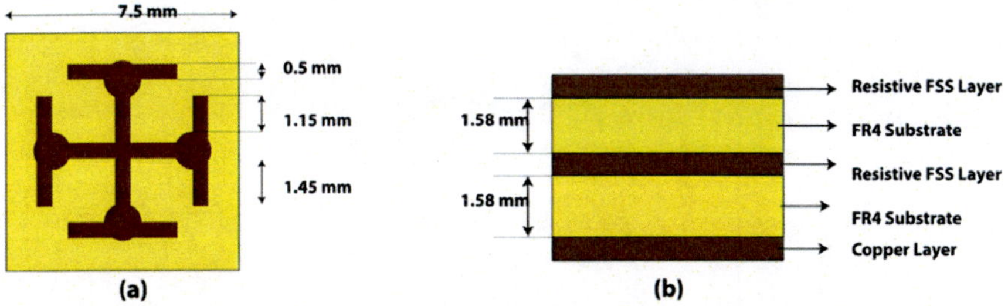

Fig. 1 Unit cell geometry of the proposed absorber **a** Top view. **b** Side view

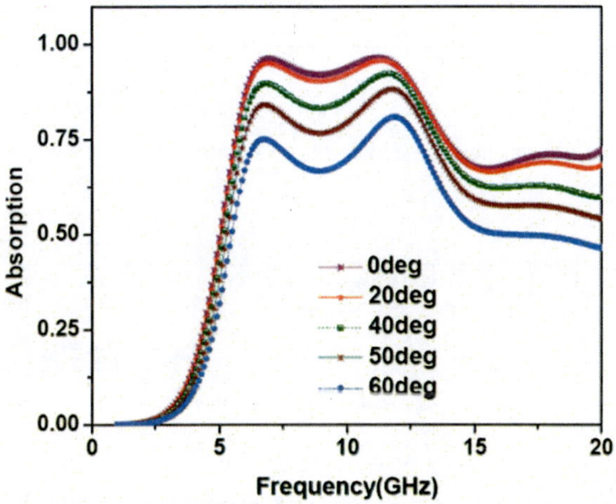

Fig. 2 Changes in absorption characteristics due to incidence angle variation in TE mode

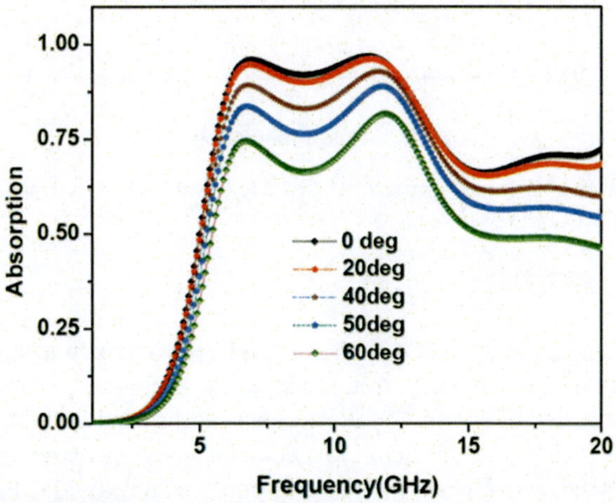

Fig. 3 Changes in absorption characteristics due to incidence angle variation in TM Mode

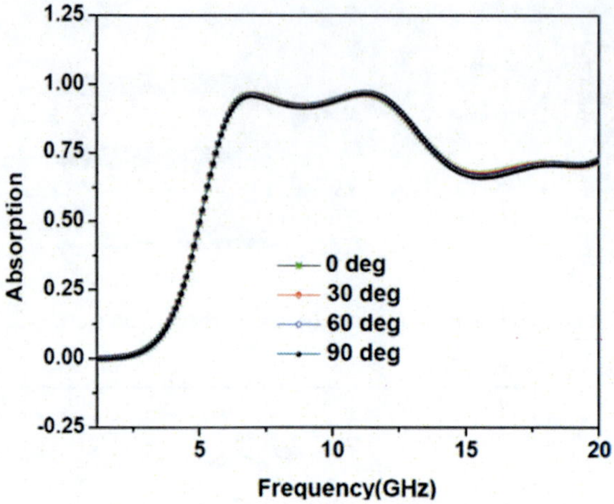

Fig. 4 Changes in absorption characteristics due to polarization angle variation under normal incidence

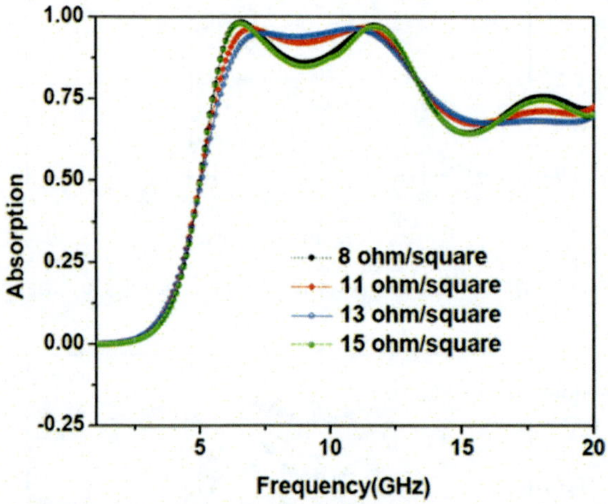

Fig. 5 Changes in absorption characteristics due to variation of surface resistivity

3 Wave Port Analysis

For wave port analysis, an array of 4 × 4 is considered as shown in Fig. 6. Wave port analysis shows identical return loss performance as that of Floquet port analysis as shown in Fig. 7. The introduction of a periodic pattern enhances the bandwidth–thickness ratio of the absorbers since it adds a reactive component to the former resistive sheet impedance. This is particularly attractive since the reactive component provides the possibility to localize more absorption of electromagnetic energy within prescribed finite bandwidth. The EM waves incident on the absorber are either reflected, transmitted or absorbed. Absorption coefficient (A) is defined as the fraction of incident power absorbed by the frequency-selective surface. It is expressed by (1). Since there is a copper layer at the bottom of the absorber, a negligible amount of power is transmitted through the absorber which makes $S_{21} = 0$.

$$A = 1 - \|S_{11}\|^2 - \|S_{21}\|^2 \tag{1}$$

$$A = 1 - \|S_{11}\|^2 \tag{2}$$

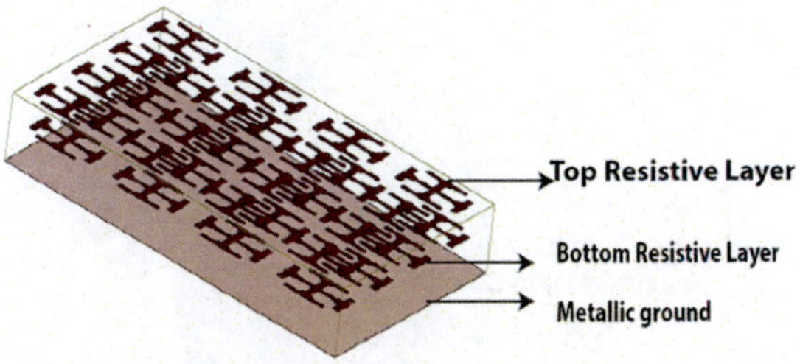

Fig. 6 4×4 array configuration of the proposed absorber for the wave port analysis

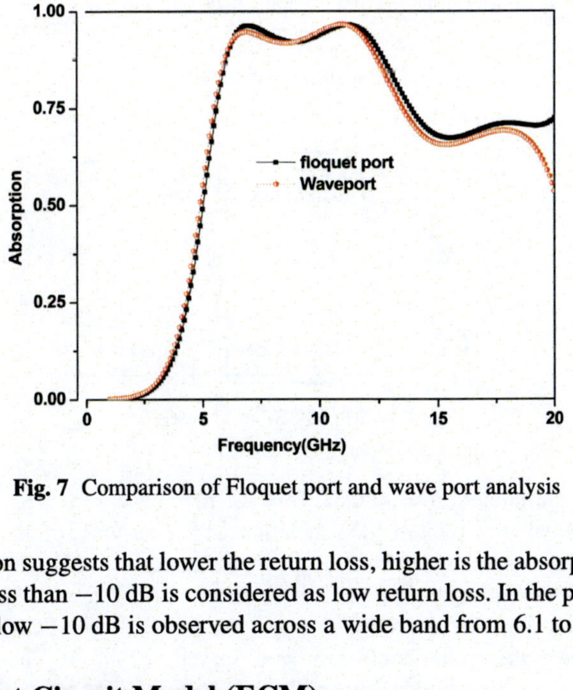

Fig. 7 Comparison of Floquet port and wave port analysis

This equation suggests that lower the return loss, higher is the absorption. Usually, a return loss of less than −10 dB is considered as low return loss. In the proposed design, a return loss below −10 dB is observed across a wide band from 6.1 to 12.5 GHz.

4 Equivalent Circuit Model (ECM)

The impedance contour plot in Fig. 8 shows the variation of return loss of the absorber with that of the real and imaginary values of the surface impedance. The impedance contour plot helps in determining the surface resistance and reactance required for matching with the intrinsic impedance of air. From the impedance contour plot, it can be inferred that below −10 dB return loss can be obtained when the real value of impedance is above 240 Ω. The FSS-based impedance surface is modelled as a series resonant circuit comprising resistance (R), inductance (L) and capacitance (C) as depicted in Fig. 9. The FR4 substrate below the impedance surface can be modelled as a transmission line of characteristic impedance $Z_0/\sqrt{\epsilon_r}$ where ϵ_r is the relative permittivity of the substrate.

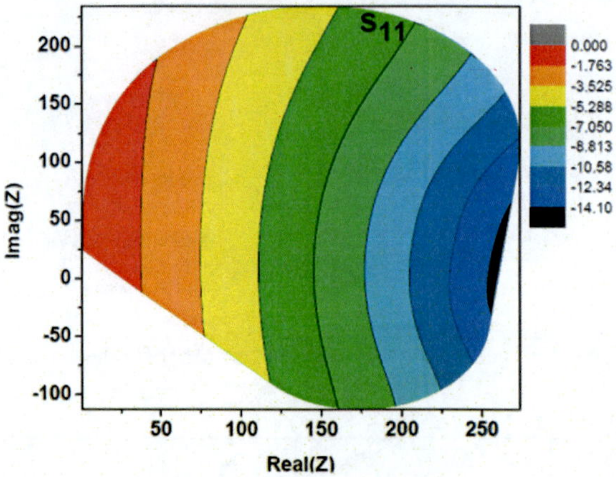

Fig. 8 Impedance contour plot

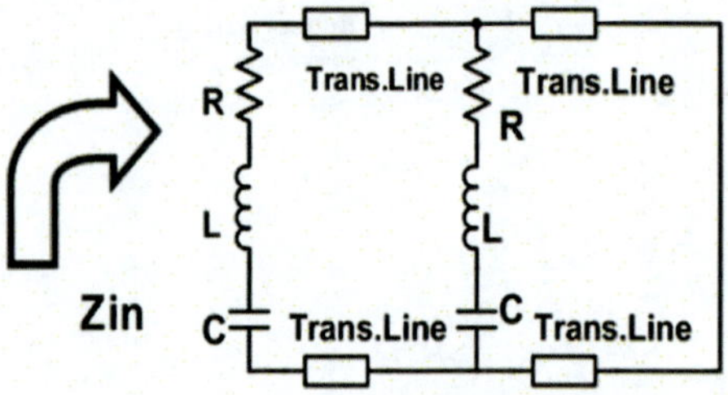

Fig. 9 ECM of FSS-based absorber

The circuital parameters $R = 480\ \Omega$, $L = 0.1$ nH and $C = 0.5$ pF are derived by simulating the equivalent circuit in ADS such that the S_{11} response of the ECM emulates the full-wave simulation results in HFSS. The input impedance of ECM(Z_{in}) is matched to 377 Ω which is the intrinsic impedance of the medium (air) surrounding the absorber. The proposed equivalent circuit provides close agreement between ECM calculated and full-wave simulated absorption coefficient as observed in Fig. 10. The comparison of return loss in Fig. 11 shows close resemblance between ECM analysis in ADS and full-wave simulation in HFSS.

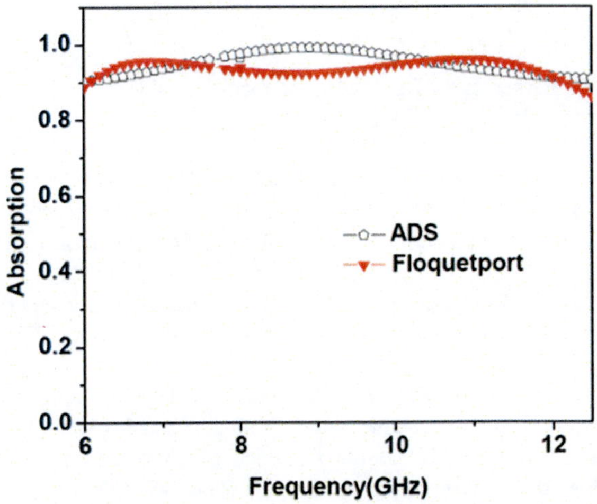

Fig. 10 Comparison of absorption coefficient between circuit analysis and full-wave simulation

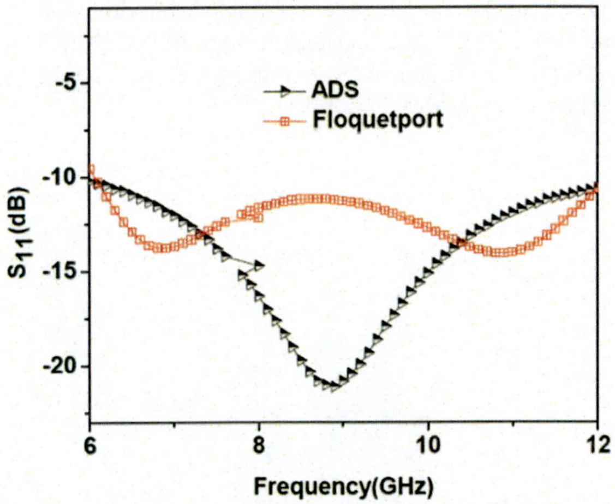

Fig. 11 Comparison of Return Loss between circuit analysis and full-wave simulation

5 Conclusion

A low profile wide band-absorptive FSS has been proposed in this paper exhibiting a bandwidth of more than 90%. The impedance surface has been realized using modified cross loop structure, which achieves high-frequency absorption without using lumped resistors. The absorptive FSS is a potential candidate for military applications such as stealth radome since it reduces the platform's radar cross section (RCS) at high-frequency bands.

References

1. Sanz-Izquierdo, B., Parker, E.A.: Dual polarized reconfigurable frequency selective surfaces. IEEE Trans. Antennas Propag. **62**(2), 764–771 (2014)
2. Zhang, K., Jiang, W., Ren, J., Gong, S.: Design of frequency selective absorber based on parallel *LC* resonators. Progress Electromagnet. Res. M **65**, 91–100 (2018)
3. Ghosh, S., Bhattacharyya, S., Srivastava, K.V.: Bandwidth-enhancement of an ultrathin polarization insensitive metamaterial absorber. Microw. Opt. Technol. Lett. **56**(2), 350–355 (2014)
4. Kundu, D., Mohan, A., Chakraborty, A.: Ultrathin polarization independent absorber with enhanced bandwidth by incorporating Giusepe Peano fractal in square ring. Microw. Opt. Technol. Lett. **57**(5), 1072–1078 (2015)
5. Kundu, D., Mohan, A., Chakraborty, A.: Single layer wideband microwave absorber using array of crossed dipoles. IEEE Antennas Wirel. Propag. Lett. **15**, 1589–1592 (2016)
6. Motevasselian, A., Jonsson, B.L.G.: Design of a wideband rasorber with a polarization sensitive transparent window. IET Microw. Antennas Propag. **6**(7), 747–755 (2012)
7. Bakshi, S.C., Mitra, D., Ghosh, S.: A frequency selective surface based reconfigurable rasorber with switchable transmission/reflection band. IEEE Antennas Wirel. Propag. Lett. **18**(1), 29–33 (2019)
8. Shang, Y., Shen, Z., Xiao, S.: On the design of single-layer circuit analog absorber using double-square-loop array. IEEE Trans. Antennas Propag. **61**(12), 6022–6029 (2013)
9. Chen, Q., Bai, J., Chen, L., Fu, Y.: A miniaturized absorptive frequency selective surface. IEEE Antennas Wirel. Propag. Lett. **14**, 80–83 (2014)

RF Energy Harvesting Circuits and Designs

Joydeep Banerjee(✉) and Subhasish Banerjee

Department of ECE, MCKV Institute of Engineering, Liluah 711204, India
idjoydeepbu81@gmail.com, subha_banerjee12@rediffmail.com

Abstract. The RF energy harvesting, a new booming research area during last decade for generating a small amount of electrical power, and its application in low-power electronics. A novel approach of Schottky diode-based Villard voltage multiplier circuit for energy harvesting application is proposed. A two-port Wilkinson power combiner circuit followed by Villard voltage multiplier circuit to combine output power from different energy sources is also studied. A Monte Carlo simulation has also been carried out for 10% tolerance value of the circuit components and results show proximity to the nominal or actual outcome. 30 mW power is achieved for Wilkinson power combiner circuit for 1 V input voltage.

Keywords: RF energy · Energy harvesting · Voltage multiplier · Power combiner · Monte Carlo simulation

1 Introduction

RF energy harvesting is a promising area of research during this decade. One of the applications of energy harvesting in electronics is the production of enough electrical power which is the essential requirement to drive the low power electronic devices [1]. Recharge of the battery is a major problem for the low power electronic devices in situation where the power source is not available. Energy harvesting [2, 3] is the solution of this problem. The energy harvesting technology also takes crucial role in biomedical field where the output power delivered by the biomedical devices needs to be constant to decrease the patient's risk of death. There are wide range of frequencies (3 kHz to 300 GHz) of RF energy sources [4, 5] available in nature. RF signal is available in our daily lives in form of signals transmission from TV, radio, wireless LAN, mobile phone, etc. [6, 7]. Though most commonly used popular frequencies are 900 MHz and 2.4 GHz ISM bands but also comparatively low frequency like radio, TV signals can take important role as energy sources, since the attenuation of the RF waves is inversely proportional to the frequency.

In this work 909 kHz frequency (AM radio) is taken as input RF source. This work first addresses the design, implementation, and characterization of a Villard voltage multiplier circuit including rectifier and latter transmission line-based Wilkinson power combiner [8] connected voltage multiplier circuit to harvest RF energy. The voltage

© Springer Nature Singapore Pte Ltd. 2021
N. R. Das and S. Sarkar (eds.), *Computers and Devices for Communication*,
Lecture Notes in Networks and Systems 147,
https://doi.org/10.1007/978-981-15-8366-7_29

output of an n-stage voltage multiplier circuit, with stage capacitance (C) can be given by Eq. 1.

$$Vo = n.(2Vin) - \frac{n-1}{fC} I_{Load} \tag{1}$$

V_O, V_{in}, I_{Load} are the output voltage, input voltage, load current, respectively, and C, f, n are the value of capacitance, operating frequency, and number of stage of voltage multiplier, respectively.

2 Results and Discussions

A two-stage voltage doubler circuit to convert RF to DC signal is shown in Fig. 1. A Schottky diode-based voltage multiplier circuit for RF energy harvesting system [9] has been proposed by authors. Schottky diode shows low forward voltage, high switching speed, low noise and can be considered as an ideal component for RF energy harvesting. After studying different Schottky diode from literature it is found that HSMS series of Schottky diode (Ex: HSMS 2850) is most suitable for almost zero bias voltage.

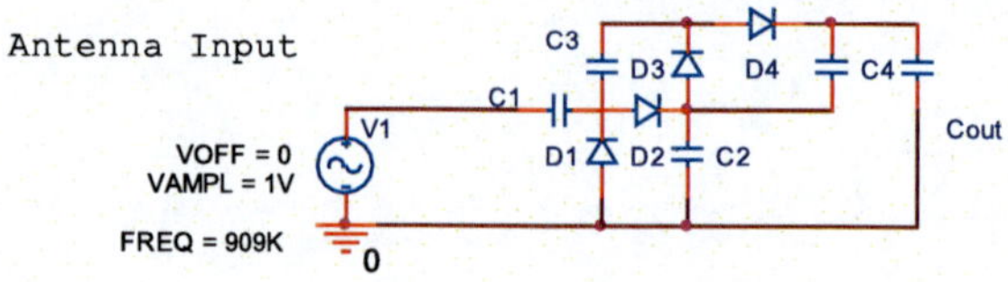

Fig. 1 Two Stage Voltage doubler Circuit

Figure 2 shows the output voltages for two stage voltage doubler circuit obtained as 1.75 and 3.493 V for first stage and second stage, respectively, after 20 ms.

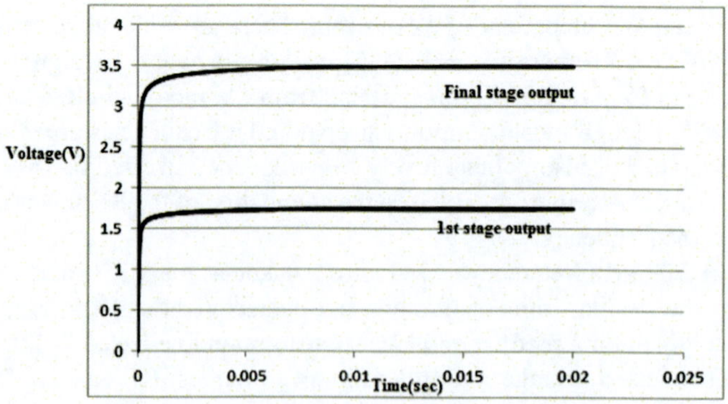

Fig. 2 Output voltage of voltage doubler circuit

Monte Carlo Simulation is also carried out for final output voltage for 100 samples as shown in Fig. 3. It is observed that output voltage varies between 3.490 and 3.498 V for 10% tolerance of circuit components and results show maximum 0.143% variation only from nominal or actual value of output voltage.

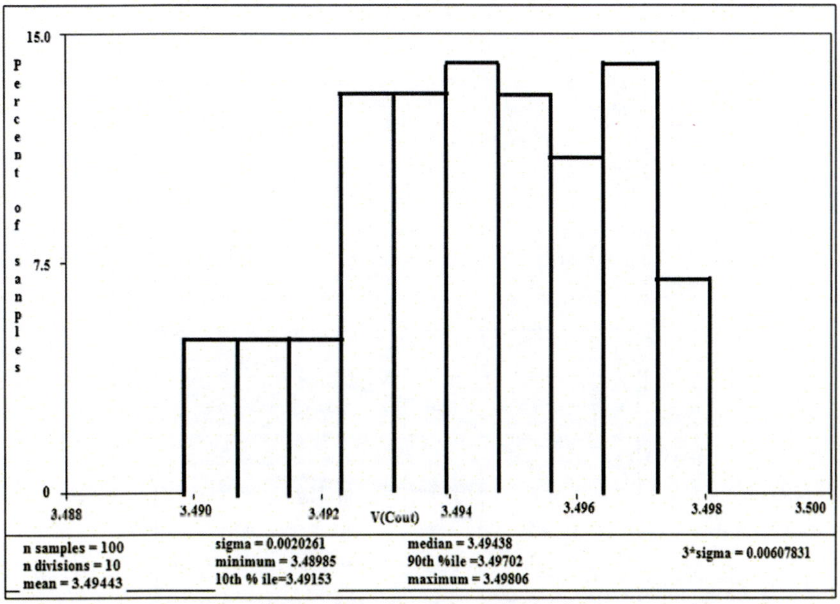

Fig. 3 Monte Carlo simulation output for final stage voltage

As captured RF power by antenna [7] is limited, authors have proposed to insert a Wilkinson power combiner circuit between antenna and voltage doubler circuit as shown in Fig. 4. Wilkinson power combiner is a good choice due to its low loss and good isolation characteristics.

The output voltages for first stage and second stage of Wilkinson power combiner connected voltage multiplier circuit are 3.51 and 7.01 V, respectively, after 20 ms as shown in Fig. 5. Monte Carlo simulation of the circuit for final stage as shown in Fig. 4 has also been carried out for 20 samples, and the results are shown in Fig. 6.

Monte Carlo results of Fig. 6 show the variation of output voltage between 6.6 and 7.5 V range for 10% tolerance of circuit components. It is observed from Monte Carlo simulation that there is maximum 7% variation at output voltage comparing with nominal or actual output voltage.

The output power obtained from the output capacitor for conventional voltage multiplier circuit and Wilkinson power combiner connected circuit are shown in Figs. 7 and 8, respectively. The improvement is observed in Wilkinson power combiner connected circuit increasing output power by 85.41% comparing with conventional voltage multiplier circuit.

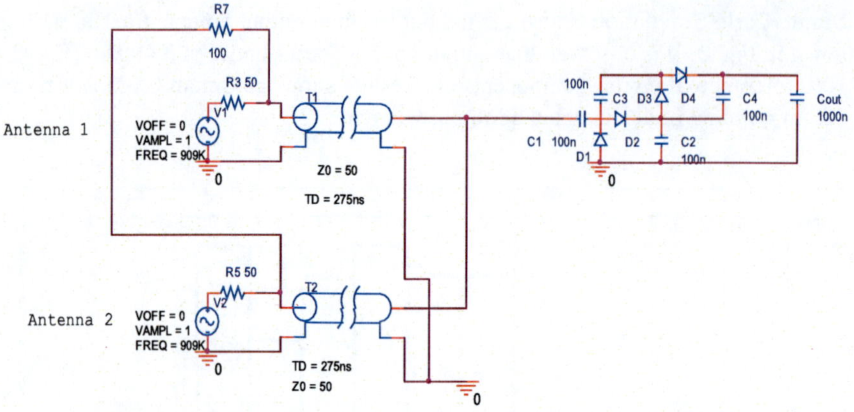

Fig. 4 Wilkinson power combiner connected voltage multiplier circuit

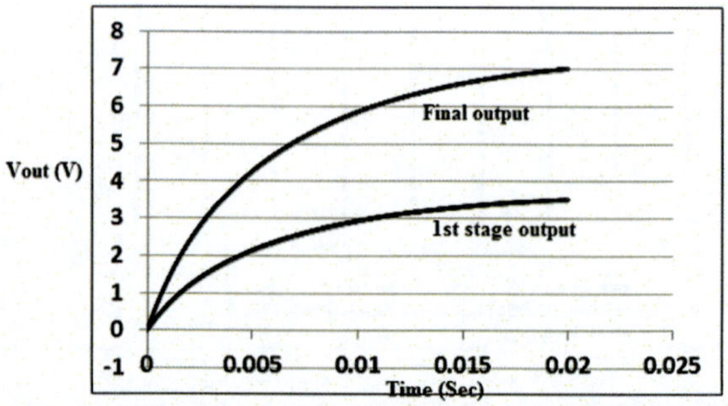

Fig. 5 Output voltage of Wilkinson power combiner

In this proposed design, the Wilkinson power combiner is basically composition of transmission lines, each having one quarter-wavelength long. The fabrication of this component can be easily achieved by using microstrip line through photolithography or milling techniques.

Authors have measured the RF input voltage captured by single dipole antenna at different frequency like 93.3, 104 MHz. The RF voltage is obtained at dBuV range at this frequency. The captured RF power can be improved if proper rectenna is designed for this energy harvesting system.

3 Conclusions

Results obtained from the simulation show higher output voltage and output power generation from Wilkinson power combiner connected voltage doubler circuit as compared with conventional voltage doubler circuit. Also, the output voltage obtained from power combiner followed by two-stage voltage doubler has been found double in comparison with conventional voltage multiplier. The maximum output voltages of final stages were

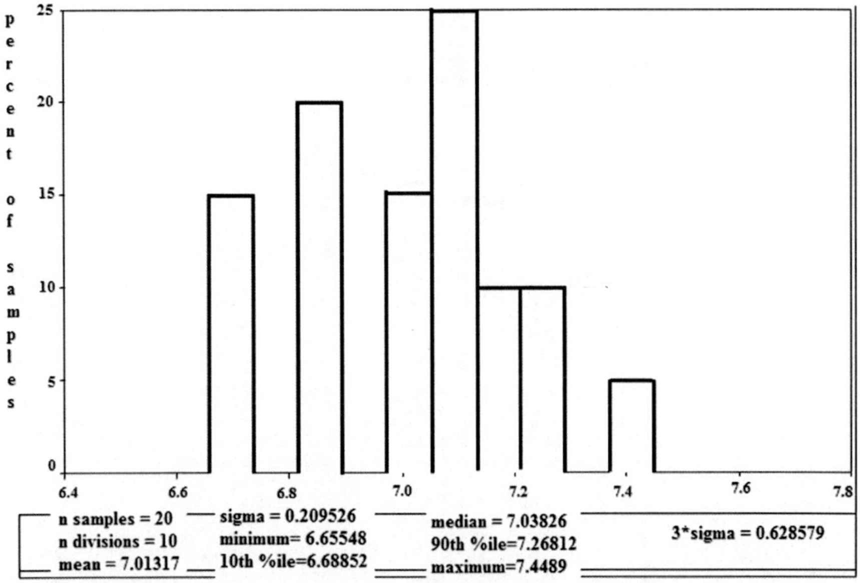

Fig. 6 Monte Carlo simulation output of Wilkinson combiner connected circuit

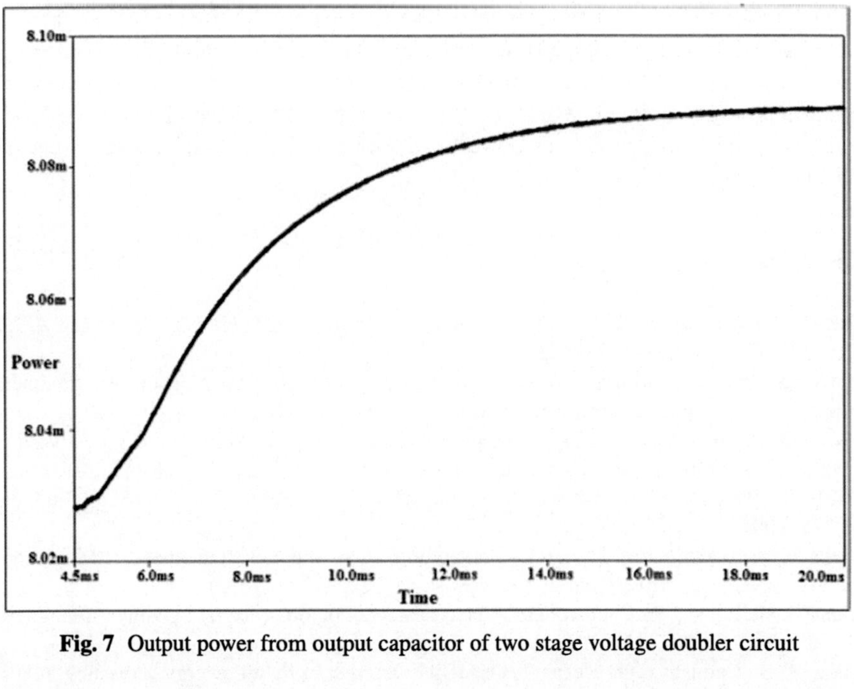

Fig. 7 Output power from output capacitor of two stage voltage doubler circuit

obtained as 3.493 and 7 V for the conventional two-stage voltage doubler and Wilkinson

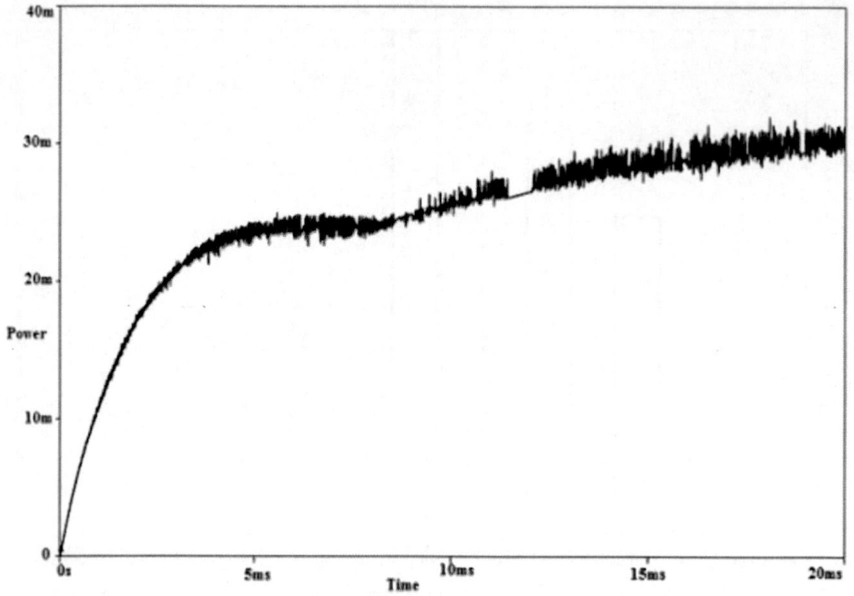

Fig. 8 Output power from output capacitor of Wilkinson combiner connected circuit

power combiner connected voltage doubler circuit, respectively. Monte Carlo simulation for the final stage output voltage shows close proximity to the actual output voltage in both circuits. The 85.41% improvement is observed at the output power generation of Wilkinson power combiner connected circuit as compared to conventional diode-based circuit. These circuits promise as powerful techniques for the application of energy harvesting technology.

References

1. Paradiso, J.A., Starner, T.: Energy scavenging for mobile and wireless electronics. IEEE Pervasive Comput. **4**, 18–27 (2005)
2. Ando, B., Baglio, S., Bulsara, A.R., Marletta, V., Pistorio, A.: Investigation of a nonlinear energy harvester. IEEE Trans. Instrum. Meas. **66**, 1067–1075 (2017)
3. Alvarado, U., Juanicorena, A., Adin, I., Sedano, B., Gutierrez, I., No, J.D.: Energy harvesting technologies for low-power electronics. Trans. Emerging Tel. Tech. **23**, 728–741 (2012)
4. Yildiz, F.: Potential ambient energy-harvesting sources and techniques. J. Technol. Stud. **35**, 40–48 (2009)
5. Collado, A., Georgiadis, A.: Conformal hybrid solar and electromagnetic (EM) energy harvesting rectenna. IEEE Trans. Circuits Syst. **60**, 2225–2234 (2013)
6. Danesh, M., Long, J.R.: Photovoltaic antennas for autonomous wireless systems. IEEE Trans. Circuits Syst. **58**, 807–811 (2011)
7. Akan, O.B., Cetinkaya, O., Koca, C., Ozger, M.: Internet of hybrid energy harvesting things. IEEE Internet Things J. **5**, 736–746 (2017)

8. Pozar, D.M.: Power Dividers and Directional Couplers. Microwave Engineering, 3rd edn. WILEY Press, pp 318–323 (2005)
9. Jabbar, H., Song, Y.S., Jeong, T.T.: RF Energy harvesting system and circuits for charging of mobile devices. IEEE Trans. Consum. Electron. **56**, 247–253 (2010)

Impacts of Emitter Layer Thickness on the Cutoff Frequency of GeSn/Ge Heterojunction Phototransistors

Harshvardhan Kumar(✉) and Rikmantra Basu

Department of Electronics and Communication Engineering, National Institute of Technology, Delhi 110040, New Delhi, India
{harshvardhan,rikmantrabasu}@nitdelhi.ac.in

Abstract. In this work, the effect of emitter layer thickness variation on the frequency performance of GeSn heterojunction phototransistors (HPTs) is studied. Various characteristics parameters such as capacitance and transit time with respect to the emitter layer thickness (t_E) are calculated over the surface of the GeSn/Ge HPT. In addition, we have also calculated the cut-off frequency (f_T) and maximum frequency (f_{max}) with respect to the t_E of GeSn/Ge HPT. The calculated results show that the proposed device exhibits excellent $f_T > 16$ GHz and $> f_{max}90$ GHz for $t_E = 100$nm.

Keywords: Cut-off frequency · GeSn HPT · Junction capacitance · Transit time

1 Introduction

For long-wavelength photodetection, low bandgap, high spectral responsivity, low noise [1], and high-speed heterojunction phototransistors (HPTs) are desired. The device design of HPT is based on the bipolar junction transistor, which also consists of photodiode in it. Therefore, HPT possesses high internal gain and low noise as compared to the conventional photodetectors (PDs). Therefore, it does not require additional trans-impedance amplifier (TIA)/ preamplifier as studied with SiGe HPTs [2] and III-V compound-based HPTs [3].

Silicon (Si) and germanium (Ge) are the preferred group IV elements for the development of photonics, optoelectronics and electronic components on the Si substrate. However, they cannot be suitable for the efficient photodetection for the longer wavelengths such as C- and L-bands of fibre-optic telecommunication and mid-infrared (MIR) applications because of their indirect bandgap nature and large energy bandgap. In support of that, a novel material, germanium-tin ($Ge_{1-x}Sn_x$) alloy has appeared as the potential candidate for photodetection in the longer wavelengths region ($1550 - 3000$nm) depending on the Sn concentration. Another unique advantage of GeSn alloy includes direct bandgap nature for Sn $> 6\%$, high absorption coefficient, CMOS compatibility. Various GeSn-based PDs and HPTs have been reported earlier theoretically and experimentally [4, 5]. The current gain as a function of emitter and base layer thickness [6]

© Springer Nature Singapore Pte Ltd. 2021
N. R. Das and S. Sarkar (eds.), *Computers and Devices for Communication*,
Lecture Notes in Networks and Systems 147,
https://doi.org/10.1007/978-981-15-8366-7_30

and signal-to-noise ratio (SNR) as a function of emitter layer thickness (t_E) of GeSn HPT has been studied [7]. Recently, the cut-off frequency (f_T) and maximum operating frequency (f_{max}) of GeSn HPT have been reported by Kumar et al. [8]. However, the impact of t_E on f_T and f_{max} of GeSn/Ge HPT has not been studied.

Therefore, the design of HPTs, moreover their optimization with respect to maximum 3-dB bandwidth (f_T), requires the flexibility to vary t_E. The optimization of t_E is crucial in order to reduce the respective transit time delay. The main purpose of this paper is to envisage the impacts of t_E of GeSn/Ge HPT in order to increase their frequency performance.

This paper is organized as follows. Section 2 describes the device structure and theory, and Sect. 3 describes the results and discussions. Then, the final Sect. 4 draws a conclusion and future scope on HPT design aspects to improve its frequency performance.

2 Ge/$Ge_{1-x}Sn_x$/Ge HPT Structure and Theory

The proposed GeSn/Ge HPT's structure is shown in Fig. 1. The optimized emitter size of the proposed device is $75 \times 1000\text{nm}^2$, and the doping concentration of emitter region is $1 \times 10^{18}\text{cm}^{-3}$ (*p*-doped). The base layer is 50nm thin abrupt $Ge_{1-x}Sn_x$ layer with Sn composition in the vicinity of 3%, and the doping concentration is $1 \times 10^{18}\text{cm}^{-3}$ (*n*-type). The collector is typical ~ 400nm thick and low *p*-doping concentration of $1 \times 10^{17}\text{cm}^{-3}$. HPT of this technology attains f_T up to ~17 GHz and f_{max} up to ~90 GHz.

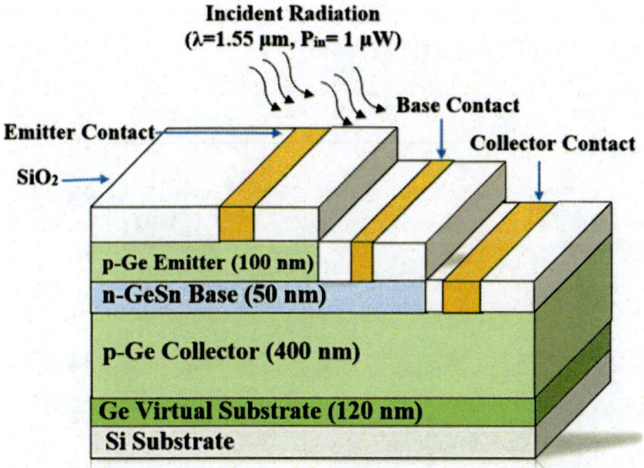

Fig. 1 3-D schematic of the proposed GeSn/Ge HPTs

The variation of t_E changes the HPT's speed. Thus, the optimization of t_E is essential to achieve high f_T. The transit time and capacitance of phototransistors can effectively characterize the effect of the t_E on f_T and f_{max}. Theoretically, f_T and f_{max} can be calculated using Eqs. (1) and (2) [9].

$$f_T = \frac{1}{2\pi \tau_{EC}} = \frac{1}{2\pi \left[\frac{V_T}{i_{cop}} C_{ecop} + \tau_F \right]} \tag{1}$$

$$f_{max} = \sqrt{\frac{f_T}{8\pi C_{bcop} R_{bop}}} \quad (2)$$

where V_T represents the thermal voltage and i_{cop} is the collector current. C_{ecop} is the total capacitance (which include base-emitter (B-E) capacitance (C_{beop}) and base–collector (B-C) capacitance (C_{bcop})) and τ_F represents the transit time delay (also known as forward transit time) due base layer and B-C depletion region. R_{bop} is the base resistance. The subscript 'op' represents under illumination. The expressions for various capacitances and resistance are given in [10]. The first parameter in Eq. (1) is the emitter transit time delay, is highly affected by the t_E. However, base and B-C transit time are independent of the t_E.

3 Results and Discussions

This section explains the effect of t_E variation on the various transit time, capacitances, and f_T and f_{max} of the GeSn HPT. For this purpose, we have estimated all the parameters at 300 K. We have considered the incident light of power 1 μW at 1550nm.

Figure 2 shows the calculated value of f_T and f_{max} with respect to t_E for Sn concentration of 3% inserted in the $Ge_{1-x}Sn_x$ base. The base-emitter voltage is fixed $v_{be} = -0.4$V in the simulation. A significant improvement in both f_T and f_{max} can be achieved through variation of t_E, as it leads to a decrease in emitter transit time. The maximum value of f_T appears for t_E of 75nm. In HPT, the increase in t_E leads to a significant increase in the electron path from the emitter to the collector. Therefore, f_T and f_{max} of HPT decrease with increasing t_E of the GeSn/Ge HPT.

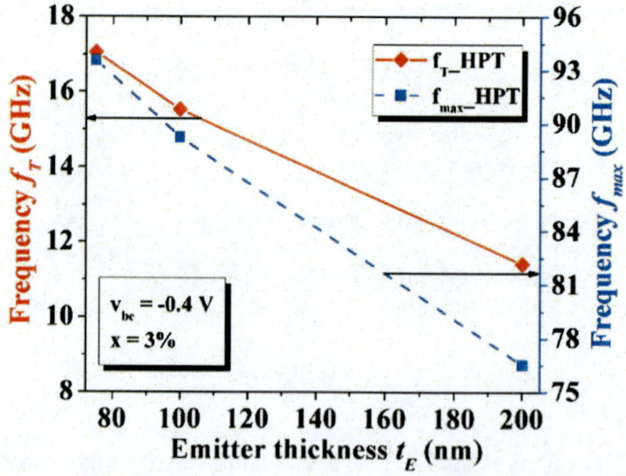

Fig. 2 f_T and f_{max} with respect to the t_E of GeSn HPT

Figure 3 shows the variation of various transit time delays of GeSn/Ge HPT with respect to t_E. It is evident that base and base-collector transit time delays are independent of the emitter thickness. However, emitter transit time delay increases with increasing t_E. The increase in emitter transit time delays can be explained on the basis of collector-to-emitter capacitance, which increases with increasing t_E of GeSn/Ge HPT.

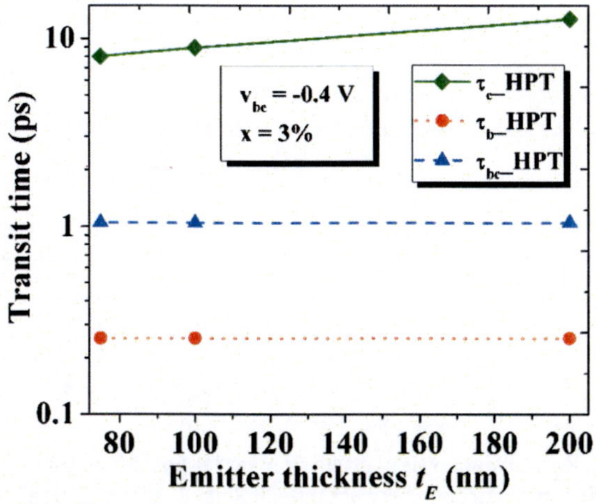

Fig. 3 Various transit time with respect to the t_E of GeSn HPT

From Fig. 4, we draw the observation that the transit time (τ_{EC_HPT}) and junction capacitance (C_{EC_HPT}) of HPT increase with increasing t_E. The increase in transit time with increasing t_E can be explained by the $R_C C_{BC}$ terms become predominant. Whereas, an increase in junction capacitance is attributed to the increase in the effective emitter area due to an increase in the t_E.

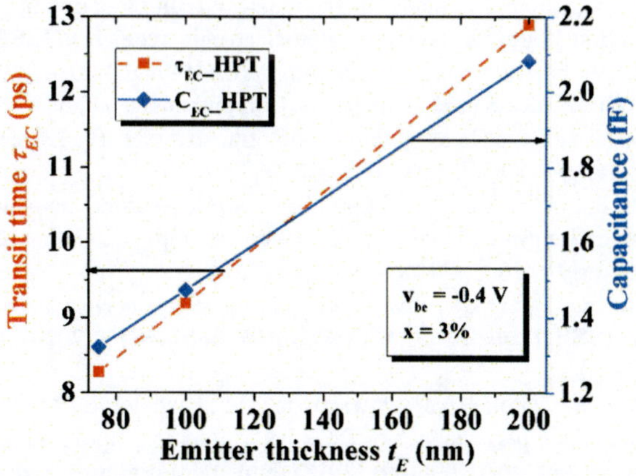

Fig. 4 Total transit time (emitter-to-collector) and junction capacitance with respect to the t_E of GeSn HPT

4 Conclusion and Future Scope

The influence of the emitter thickness on the important characteristics such as capacitance and transit time of GeSn/Ge HPT has been studied. The calculated results show that the HPT's cutoff frequency f_T and maximum frequency f_{max} decrease as the emitter thickness increases from 75nm to 200nm due to the increase in the transit time. The proposed device attains the cutoff frequency f_T up to ~17 GHz and the maximum frequency f_{max} up to ~90 GHz for the Sn concentration of 3%. Further improvements in f_T and f_{max} of the GeSn/Ge HPT can be made by higher Sn concentration in the base and reducing base transit time.

Acknowledgement. Harshvardhan Kumar is thankful to DST-SERB (Under Early Career Research Award Scheme 2017) Project (File No. ECR/2017/000794), India for providing SRF fellowship.

References

1. Kumar, H., Basu, R.: Noise analysis of group IV material-based heterojunction phototransistor for fiber-optic telecommunication networks. IEEE Sens. J. **18**, 9180–9187 (2018). https://doi.org/10.1109/JSEN.2018.2869975
2. Egels, M., Delacressonniere, B., Sahabun, Y., Lecoy, P.: Design of an optically frequency or phase-controlled oscillator for hybrid fiber- radio Lan at 5.2 GHz. Microw. Opt. Technol. Lett. **45**, 104–107 (2005). https://doi.org/https://doi.org/10.1002/mop.20737
3. Kim, J., Kanakaraju, S., Johnson, W.B., Lee, C.: InP/InGaAs uni-travelling carrier heterojunction phototransistors. Electron. Lett. **45**, 649–651 (2009). https://doi.org/10.1049/el.2009.0243
4. Oehme, M., Schmid, M., Kaschel, M., Gollhofer, M., Widmann, D., Kasper, E., Schulze, J.: GeSn p-i-n detectors integrated on Si with up to 4% Sn. Appl. Phys. Lett. **101**, 141110(1–4) (2012). https://doi.org/https://doi.org/10.1063/1.4757124
5. Kumar, H., Basu, R., Gupta, J.: Small-signal compact circuit modeling of group IV material-based heterojunction phototransistors for optoelectronic receivers. IEEE Trans. Electron Devices, 1–7 (2019). https://doi.org/https://doi.org/10.1109/TED.2019.2896068
6. Pandey, A.K., Basu, R., Kumar, H., Chang, G.: Comprehensive analysis and optimal design of Ge/GeSn/Ge p-n-p infrared heterojunction phototransistors. IEEE J. Electron Devices Soc. (2018). https://doi.org/10.1109/JEDS.2018.2884253
7. Basu, R., Kumar, H.: Noise analysis of optimized Ge/Ge1−xSnx/Ge p–n–p heterojunction phototransistors for long-wavelength optical receivers. Opt. Quantum Electron. **51**, 1–12 (2019). https://doi.org/10.1007/s11082-019-1765-4
8. Kumar, H., Basu, R.: Effect of active layer scaling on the performance of Ge 1–x Sn x phototransistors. IEEE Trans. Electron Devices. **66**, 3867–3873 (2019). https://doi.org/10.1109/ted.2019.2925892
9. Tegegne, ZG., Viana, C., Polleux, J., Grzeskowiak, M., Richalot, E.: Study of lateral scaling impact on the frequency performance of SiGe heterojunction bipolar phototransistor. IEEE J. Quantum Electron. **54**, 4600109(1–9) (2018). https://doi.org/https://doi.org/10.1109/JQE.2018.2822179
10. Chakrabarti, P., Kumar Agrawal, N., Kalra, P., Agrawal, S., Gupta, G.: Noise modeling of an InP/InGaAs heterojunction bipolar phototransistor. Opt. Eng. **42**, 939–947 (2003). https://doi.org/10.1117/1.1557693

Error Probability Analysis of Hexagonal 16QAM

Satyabrata Singha(✉), Bishanka Brata Bhowmik, and Nitish Sinha

Department of Electronics & Communication Engineering, Tripura University, Tripura, India
{satyabrata.ece,bishankabhowmik}@tripurauniv.in

Abstract. High-dimensional modulation schemes have become the preferred choice for high data rate transmission. Compared to the conventional square 16QAM, hexagonal 16QAM configuration is less vulnerable to noise. This paper presents a theoretical error performance of uncoded hexagonal 16QAM using a lower computational complex approach.

Keywords: Phase shift keying · 16QAM · Spectral efficiency · IQ-modulators · Mach–Zehnder modulators or phase

1 Introduction

With the constantly growing demand for higher data rates, research on higher-order modulation formats has increased rapidly. Modulation formats like M-ary phase shift keying and M-ary quadrature amplitude modulation offer higher spectral efficiency and bit rate [1, 2]. With double spectral efficiency, 16QAM has found its place as a potential candidate for next-generation high data rate optical transmission system. Research on 16-Ary design was started in the late 1960s. Early work on this type of constellation was circular type consisting of signal points in two concentric rings [3, 4]. The conventional square 16QAM was first proposed in 1962 [5]. Some of the recent works on different variations of 16QAM such as square 16QAM, circular 16QAM and hexagonal 16QAM are [6–8]. Square constellation with limited boundary and uneven distance between the constellation points severely limits the symbol detection margin, thus making it difficult to recover the transmitted signal. Whereas, in circular constellation case, with an increase in M value plotting of signal points becomes complicated and also with an increase in distance between constellation points increases the complexity in detection. However, these structural limitations can be overcome by employing hexagonal 16QAM configuration. The hexagonal variant has equal Euclidean distance between its neighboring constellation points, thus allowing a larger symbol detection region.

Several methods to generate 16QAM are in practice, cascade or parallel configuration using IQ-modulators, Mach–Zehnder modulators or Phase modulators. But complex structures add more complexity to fabrication. In general, hexagonal 16QAM is difficult to generate because of its complex arrangement of signal points. Some of the earlier

© Springer Nature Singapore Pte Ltd. 2021
N. R. Das and S. Sarkar (eds.), *Computers and Devices for Communication*,
Lecture Notes in Networks and Systems 147,
https://doi.org/10.1007/978-981-15-8366-7_31

works on hexagonal 16QAM are monolithic InP square/hexagonal modulator with 28-Gbaud using four integrated amplitude/phase modulators with calculated BER of 2.7 × 10^{-3} and 3.8 × 10^{-3} for square and hexagonal structures, respectively [9]. 25-Gbaud hexagonal 16-QAM with offline BER of 3.7 × 10^{-3} at 32 dB OSNR was generated using dual-drive IQ modulator. 0.5 dB improvement of BER of hexagonal 16QAM over square 16QAM was achieved [8].

In this paper, we have analyzed the error probability of hexagonal 16QAM by defining decision boundary (shaped as a hexagon) and calculated the probability of error for each boundary of a particular signal point.

2 Mathematical Analysis of Hexagonal 16QAM

In hexagonal 16QAM, the neighboring constellation points are placed at equal Euclidian distance. The constellations in the signal space are divided into decision regions, and the received signal can be demodulated by tracking points in the particular decision region. Hexagonal-shaped decision boundary maximizes the minimum distance between constellation points and thus provides an energy-efficient system with minimum probability of symbol error. For calculating the probability of symbol error, we first calculate the average symbol energy and then probability of error for each defined boundary regions and finally adding all the error probabilities.

where,

$$\mathrm{qo} = \frac{\sqrt{19}}{2}x \quad \mathrm{lo} = \frac{\sqrt{19}}{2}x \quad \mathrm{jo} = \frac{\sqrt{7}}{2}x \quad \mathrm{no} = \frac{\sqrt{7}}{2}x \quad \mathrm{eo} = \frac{\sqrt{7}}{2}x$$

$$\mathrm{go} = \frac{\sqrt{7}}{2}x \quad \mathrm{io} = \frac{\sqrt{43}}{4}x \quad \mathrm{mo} = \frac{\sqrt{43}}{4}x \quad \mathrm{po} = \frac{3}{2}x \quad \mathrm{ko} = \frac{3}{2}x$$

$$\mathrm{ho} = \frac{\sqrt{51}}{4}x \quad \mathrm{fo} = \frac{\sqrt{51}}{4}x \quad \mathrm{ao} = \frac{1}{2}x \quad \mathrm{co} = \frac{1}{2}x \quad \mathrm{do} = \frac{\sqrt{3}}{2}x$$

$$\mathrm{bo} = \frac{\sqrt{3}}{2}x$$

2.1 Average Symbol Energy

Average symbol energy E_{savg} is given by

$$E_{savg} = \frac{1}{M}\left[\left(N_1 * r_1^2\right) + \left(N_2 * r_2^2\right) + \left(N_3 * r_3^2\right) + \ldots\ldots\ldots\right] \tag{1}$$

where $N_1, N_2, N_3, \ldots.$ are the numbers of signal points at distances $r_1, r_2, r_3, \ldots.$ from origin, respectively. If 'x' is the Euclidian distance between two signal points, then

$$E_{savg} = \frac{1}{16}\left[2\left(\frac{\sqrt{19}}{2}x\right)^2 + 4\left(\frac{\sqrt{7}}{2}x\right)^2 + 2\left(\frac{\sqrt{43}}{4}x\right)^2 + 2\left(\frac{3}{2}x\right)^2 + 2\left(\frac{\sqrt{51}}{4}x\right)^2 + 2\left(\frac{1}{2}x\right)^2 + 2\left(\frac{\sqrt{3}}{2}x\right)^2\right] \tag{2}$$

$$E_{savg} = 2.171x^2 \tag{3}$$

$$x = \sqrt{\frac{E_{savg}}{2.171}} \tag{4}$$

2.2 Probability of Error

If 'x' is the distance between 2 symbols, 'n' is the number of bits per symbol and 'σ' is the noise parameter given by $\sigma = \sqrt{\frac{N_0}{2}}$. Then, BER can be calculated by

$$P_b = Q\left(\frac{\frac{x}{2} * \sqrt{n}}{\sigma}\right) \tag{5}$$

Figures 2 and 3 show the boundary regions and dimensions of the symbols. 'Boundary 1' is the region representing four symbols on the top and bottom of the hexagonal structure. 'Boundary 2' represents four symbols in the corners of the structure. 'Boundary 3' represents six symbols on the inner side of the structure. 'Boundary 4' represents two symbols on left and right side of the structure.

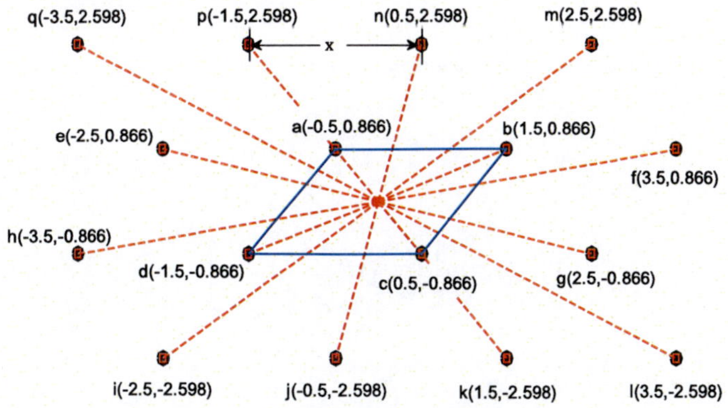

Fig. 1 Constellation diagram

Probability of correct decision is the product of all probabilities in each direction that causes the symbol to be detected inside the specified boundary. And probability of error is the product of all the probability of noise components which forces the symbol to be received outside the specified boundary [10–12].

Boundary 1

Probability of correct decision,

$$P_{c1} = \left[1 - Q\left(\frac{0.5x\sqrt{n}}{\sigma}\right)\right]\left[1 - Q\left(\frac{0.5x\sqrt{n}}{\sigma}\right)\right]\left[1 - Q\left(\frac{0.5x\sqrt{n}}{\sigma}\right)\right]\left[1 - Q\left(\frac{0.5x\sqrt{n}}{\sigma}\right)\right] \tag{6}$$

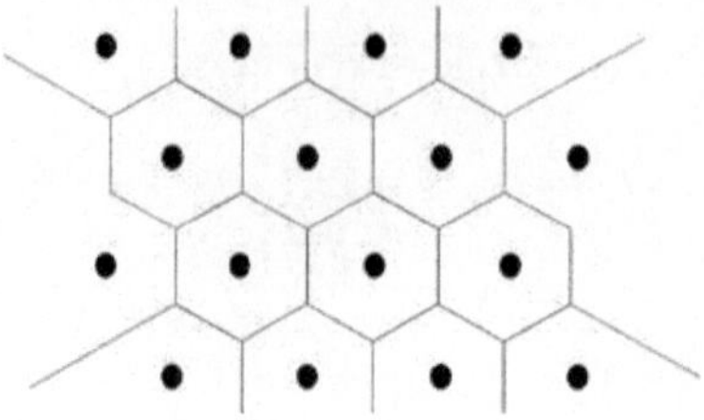

Fig. 2 Constellation with decision boundary

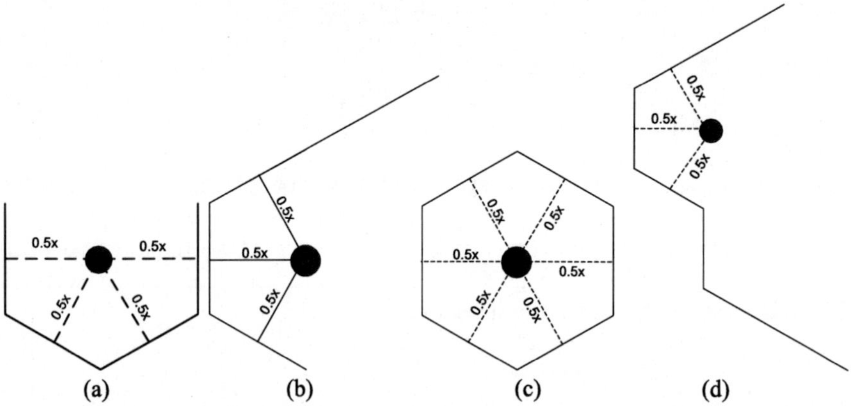

Fig. 3 **a** Decision boundary 1. **b** Decision boundary 2. **c** Decision boundary 3. **d** Decision boundary 4

Probability of error,

$$P_{e1} = 1 - \left[1 - Q\left(\frac{0.5x\sqrt{n}}{\sigma}\right)\right]^4 \tag{7}$$

Boundary 2
Probability of correct decision,

$$P_{c2} = \left[1 - Q\left(\frac{0.5x\sqrt{n}}{\sigma}\right)\right]\left[1 - Q\left(\frac{0.5x\sqrt{n}}{\sigma}\right)\right]\left[1 - Q\left(\frac{0.5x\sqrt{n}}{\sigma}\right)\right] \tag{8}$$

Probability of error,

$$P_{e2} = 1 - \left[1 - Q\left(\frac{0.5x}{\sigma}\sqrt{n}\right)\right]^3 \tag{9}$$

Boundary 3
Probability of correct decision,

$$P_{c3} = \left[1 - Q\left(\frac{0.5x\sqrt{n}}{\sigma}\right)\right]\left[1 - Q\left(\frac{0.5x\sqrt{n}}{\sigma}\right)\right]\left[1 - Q\left(\frac{0.5x\sqrt{n}}{\sigma}\right)\right]\left[1 - Q\left(\frac{0.5x\sqrt{n}}{\sigma}\right)\right]$$

$$\left[1 - Q\left(\frac{0.5x\sqrt{n}}{\sigma}\right)\right]\left[1 - Q\left(\frac{0.5x\sqrt{n}}{\sigma}\right)\right] \tag{10}$$

Probability of error,

$$P_{e3} = 1 - \left[1 - Q\left(\frac{0.5x}{\sigma}\sqrt{n}\right)\right]^6 \tag{11}$$

Boundary 4
Probability of correct decision,

$$P_{c4} = \left[1 - Q\left(\frac{0.5x\sqrt{n}}{\sigma}\right)\right]\left[1 - Q\left(\frac{0.5x\sqrt{n}}{\sigma}\right)\right]\left[1 - Q\left(\frac{0.5x\sqrt{n}}{\sigma}\right)\right] \tag{12}$$

Probability of error,

$$P_{e4} = 1 - \left[1 - Q\left(\frac{0.5x}{\sigma}\sqrt{n}\right)\right]^3 \tag{13}$$

Putting the value $\sigma = \sqrt{\frac{N_0}{2}}$ and Eq. (4) in Eqs. (7), (9), (11) and (13),

$$P_{e1} = 1 - \left[1 - Q\left(\frac{0.5x\sqrt{n}}{1.0855}\sqrt{\frac{E_{savg}}{N_0}}\right)\right]^4 \tag{14}$$

$$P_{e2} = 1 - \left[1 - Q\left(\frac{0.5x\sqrt{n}}{1.0855}\sqrt{\frac{E_{savg}}{N_0}}\right)\right]^3 \tag{15}$$

$$P_{e3} = 1 - \left[1 - Q\left(\frac{0.5x\sqrt{n}}{1.0855}\sqrt{\frac{E_{savg}}{N_0}}\right)\right]^6 \tag{16}$$

$$P_{e4} = 1 - \left[1 - Q\left(\frac{0.5x\sqrt{n}}{1.0855}\sqrt{\frac{E_{savg}}{N_0}}\right)\right]^3 \tag{17}$$

Probability of error for hexagonal 16QAM will be combining the error probabilities of all the defined boundaries, given by

$$P_e = \frac{4}{16}P_{e1} + \frac{4}{16}P_{e2} + \frac{6}{16}P_{e3} + \frac{2}{16}P_{e4} \tag{18}$$

2.3 Bit Error Rate

See Fig. 4.

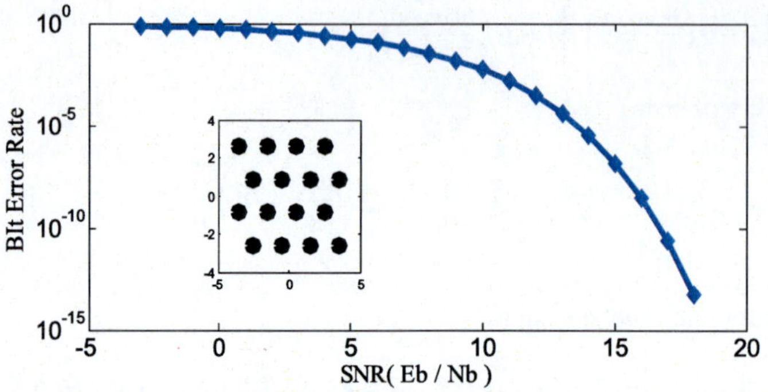

Fig. 4 BER versus SNR curve

3 Conclusion

We have calculated theoretical BER for hexagonal 16QAM that can reduce the detection error probability. The hexagonal-shaped decision boundary provides less error detection probability which is believed to provide better performance than conventional square 16QAM.

References

1. Gao, J., et al.: 40-Gb/s star 16-QAM transmitter based on single dual-drive Mach-Zehnder modulator. Chin. Opt. Lett. **7**(2), 109–111 (2009)
2. Yu, K., Bao, J. -q, Yin, J.J.: All-optical modulation format conversion from PSK to ASK based on phase-sensitive amplification. Optoelectron. Lett. **15**(1), 35–38 (2019)
3. Cahn, C.: Combined digital phase and amplitude modulation communication systems. IRE Trans. Commun. Syst. **8**(3), 150–155 (1960)
4. Je, H., Lucky, R.: Performance of combined amplitude and phase-modulated communication systems. IRE Trans. Commun. Syst. **8**(4), 232–237 (1960)
5. Campopiano, C., Glazer, B.: A coherent digital amplitude and phase modulation scheme. IRE Trans. Commun. Syst. **10**(1), 90–95 (1962)
6. Winzer, P.J., et al.: Spectrally efficient long-haul optical networking using 112-Gb/s polarization-multiplexed 16-QAM. J. Lightwave Technol. **28**(4), 547–556 (2009)
7. Dong, J., Zou, Y., Li, D.: Optimal 2-circular 16QAM constellation design. In: 14th IEEE Proceedings on Personal, Indoor and Mobile Radio Communications, 2003. PIMRC 2003, vol. 3, IEEE (2003)
8. Yan, S., et al.: Generation of square or hexagonal 16-QAM signals using a dual-drive IQ modulator driven by binary signals. Opt. Express **20**(27), 29023–29034 (2012)
9. Doerr, C.R., et al.: 28-Gbaud InP square or hexagonal 16-QAM modulator. In: Optical Fiber Communication Conference. Optical Society of America (2011)
10. Iyamabo, P.E.: Exact BER calculation of TCM-MAPSK using pairwise probability of product trellis algorithm for DVB applications. Diss. University of Toledo (2016)
11. Rugini, L.: Symbol error probability of hexagonal QAM. IEEE Commun. Lett. **20**(8), 1523–1526 (2016)
12. Abdullah, K., Mahmoud, S.S., Hussain, Z.M.: Performance analysis of an optimal circular 16-QAM for wavelet based ofdm systems. Int. J. Commun. Netw. Syst. Sci. **2**(09), 836 (2009)

A Wideband Transmittive-Type Cross Polarization Converter for Terahertz Waves

Meghna Mishra[1], Lavesh Nama[2], Sambit Kumar Ghosh[2], and Somak Bhattacharyya[2(✉)]

[1] Department of Electronics & Communication Engineering, Shri Mata Vaishno Devi University, Katra 182320, Jammu and Kashmir, India
16bec033@smvdu.ac.in

[2] Department of Electronics Engineering, Indian Institute of Technology, (BHU), Varanasi 221005, U.P.-, India
{laveshn.ece16,sambitkrghosh.rs.ece17, somakbhattacharyya.ece}@iitbhu.ac.in

Abstract. A bilayer metasurface constituting of asymmetric split-ring resonator (ASRR) array as top layer and an E-Shaped resonator (ER) array as a bottom layer has been discussed in this paper. The former is aimed to make the waves passing through it go unaltered while the latter behaving as a bandstop filter through which polarization selectivity feature has been achieved. The measured transmission is maximized to 0.68 at 1.15 THz leading to realization of polarization conversion ratio (PCR) up to 99.9%. The PCR is above 90% in the range of 1.07–1.21 THz. The proposed structure has been compared with the other conventionally used optical devices. The polarization conversion of the structure has been carried out for various substrate materials used in the terahertz region like silicon dioxide and preperm L270(lossy).

Keywords: Metasurfaces · Cross polarization converter (CPC) · Wideband · Waveplates · Birefringence · Bandstop resonator

1 Introduction

Metamaterials have recently created huge interest among the research communities owing to their physically non-existent electromagnetic properties [1–2]. Due to the miniaturized dimensions, metamaterials have been proposed for various potential applications [3–5]. Metasurfaces, a two-dimensional representation of metamaterial designs, have been found to be of great interests for several applications like filter, absorber, polarization conversion, etc. [6]. Various devices including photoelastic modulators, optical grating, and birefringence effects-based nonlinear optical components were reported to control the polarization of the incident wave [7].

In this article, we propose an anisotropic unit cell structure made of a bi-layered metasurface for transmittive-type cross-polarization converter in terahertz region consisting of a front array of joint asymmetric split-ring resonators (ASRRs) for unaltered

© Springer Nature Singapore Pte Ltd. 2021
N. R. Das and S. Sarkar (eds.), *Computers and Devices for Communication*, Lecture Notes in Networks and Systems 147,
https://doi.org/10.1007/978-981-15-8366-7_32

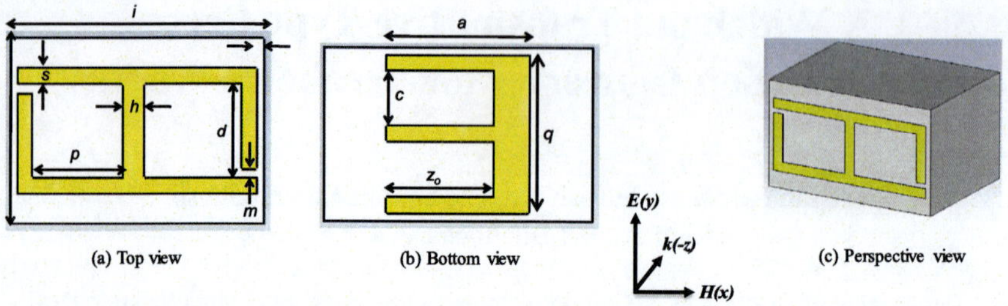

Fig. 1 **a** Top view (ASSRs), **b** bottom view (ER) and **c** perspective view of the proposed structure, respective dimensions are ($i = 66$ μm, $o = 44$ μm, $p = 26$ μm, $d = 22$ μm, $h = 6$ μm, $s = 4$ μm, $e = 2$ μm, $c = 14$ μm, $q = 40$ μm, $a = 36$ μm, $z_o = 9$ μm)

traveling of electromagnetic incident wave and a rear array of E-shaped resonators (ER) for polarization selectivity.

2 Design of the Structure

The structure comprises two layers of 0.2 μm thick gold imprinted over 50 μm thick polycarbonate substrate on both the sides. The top layer of the unit cell has been designed in the form of ASRR, while the bottom layer is made of ER. The top and bottom views of the unit cell are shown in Fig. 1a, b, respectively. The three-dimensional perspective view has been shown in Fig. 1c. The optimized geometrical dimensions of the unit cell are also shown in Fig. 1. The unit cell under periodic boundary conditions has been simulated using CST microwave studio based on finite integration technique (FIT). The metallic patterns are designed along the *x*–*y* plane while the electromagnetic wave is incident along—*z*-direction as illustrated in Fig. 1.

3 Simulation and Results

When the *y*-polarized THz wave is incident onto the ASSR array on the top surface, it passes unharmed while it gets rotated by an angle of; 90° i.e., the incident wave gets cross polarized while it gets passed through the rear layer. This yields that E-shaped patterned bottom metallic layer acts as a notch filter, allowing *x*-polarized component to pass through while blocking the *y*-polarized component of the wave. The co-polarized and cross-polarized transmission coefficients are shown in Fig. 2a. It is observed that no co-polarized transmission occurs at 1.15 THz while significant amount of cross-polarized components have been received at the output end. This results in the polarization conversion ratio (PCR) of near unity value at about 1.15 THz. It is further observed that the structure offers PCR greater than 90% over the frequency range 1.07–1.21 THz as evident from Fig. 2b. In the frequency band 0.6–1.4 THz, the PCR achieved is greater than 50%.

The surface currents in the top layer consisting ASRR array have been studied. It is clear from Fig. 3a that at 1.15 THz, the currents on the two sides around the central strip are parallel while that at 1.164 THz are anti-parallel. This results in formation of

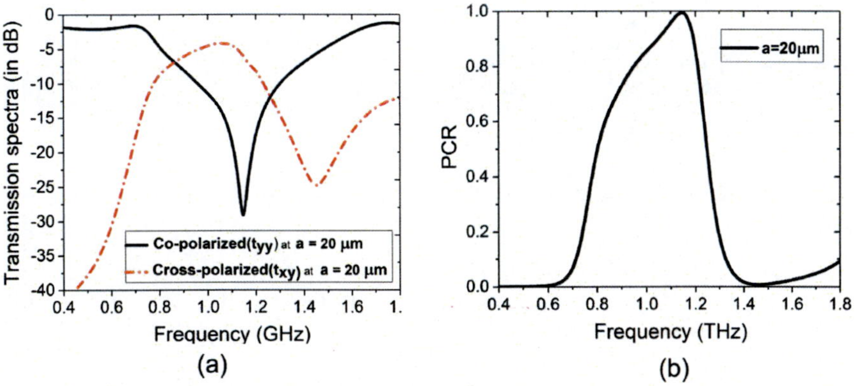

Fig. 2 Frequency responses of **a** co & cross-polarized transmitted components of the EM wave along with **b** PCR

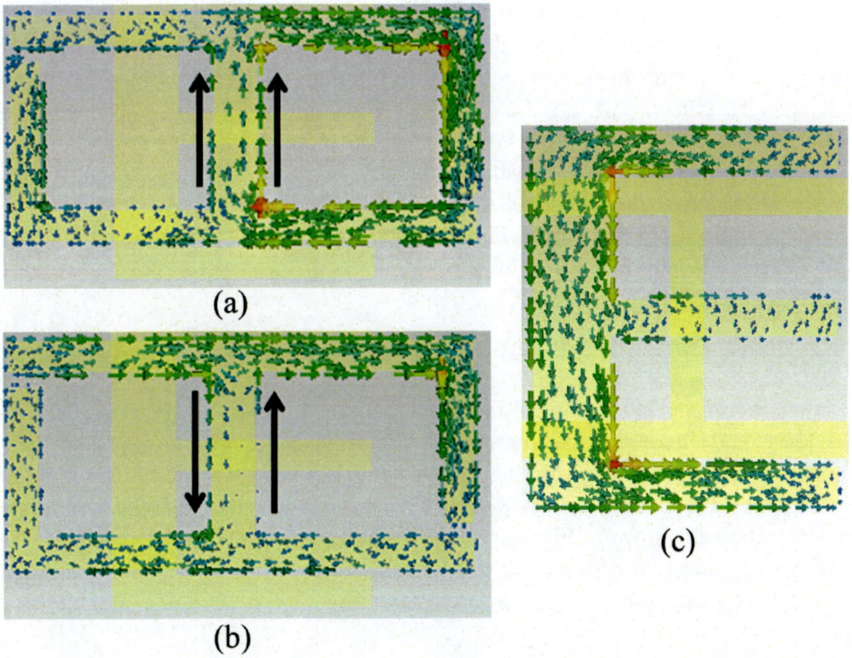

Fig. 3 Responses of **a** parallel surface current distribution at 1.15 THz. **b** Anti-parallel surface current distribution at 1.164 THz. **c** Surface current distributions within E-shaped resonator

electric and magnetic coupling, respectively, at the two strips. The formation of parallel and anti-parallel current distributions at two different frequencies introduces chirality in the ASRR array [9–10].

The PCR response has been studied with the effect of the length of the E-shaped resonator at the bottom side (*a*). The co- and cross-polarized transmission coefficients

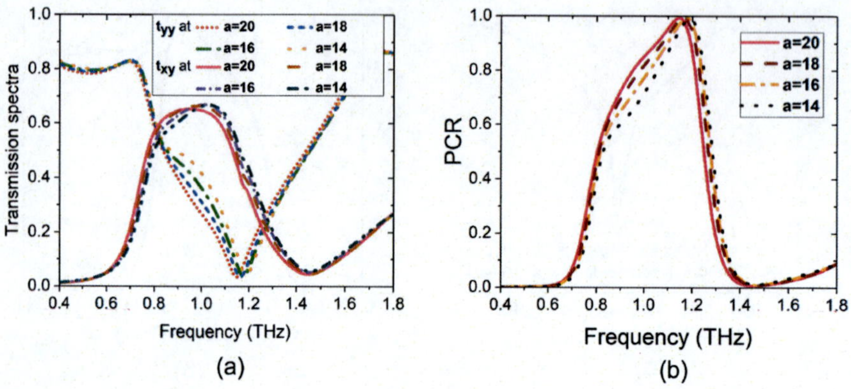

Fig. 4 Frequency responses with increasing values of length *a* of E-shaped resonator **a** co-and cross-polarized transmission coefficients and **b** PCR

of the structure are shown in Fig. 4a for various *a*. It is evident from Fig. 3c that the surface current end arms of ER are densely populated with surface current or in other words, the surface current density is not much high but considerably large at the end of the extended arms of ER. So, the increment in *a* contributes to the enhancement of the effective inductance generated. Furthermore, the increase in *a* also produces rise in the width of the engraved metallic patterns; thereby, enhancing the effective capacitance [11]. These, in combination, produce the red-shift of the PCR response as evident from Fig. 4b. It has been observed from Fig. 4b that the structure exhibits maximum PCR bandwidth for side length of ER of 20 μm.

4 Response Under Different Substrate Materials

The structure has also been examined with another two different substrates, PREPERM L270 (lossy) and silicon dioxide having dielectric constants of 2.7 and 3.9, respectively. A short comparison table regarding the above study has been provided in Table 1. The cross component reaches to 0.70 with the PCR narrower in range than the former, i.e., with polycarbonate shown in Fig. 4a, b. The optimized results have been achieved for silicon dioxide substrate with $a = 18$ μm while by using preperm, maximum PCR has been realized by having $a = 17.33$ μm.

5 Conclusion

The proposed structure with enhanced transmission component reaching 0.68 at 1.15 THz leading to polarization conversion ratio upto 99.99% in terahertz domain has been investigated in details in this paper. The PCR is above 90% in the range of 1.07–1.21 THz. The structure has also been studied for variations of geometrical parameter to realize the maximum PCR bandwidth. Furthermore, a brief simulation study has been successfully carried out by making variation of substrates used in the terahertz domain to optimize the results in terahertz gap.

Table 1 Comparison of existing polycarbonate-based metasurface with silicon dioxide and preperm L270

Substrate material under consideration	Dielectric constant of the substrate	PCR > 0.5 (THz)	Cross-polarized transmission coefficient (t_{xy})
Polycarbonate ($a =$ 20 μm)	2.9	0.6–1.4	0.67532
Preperm L270 Lossy ($a =$ 17.33 μm)	2.7	1.25–1.5	0.69326
Silicon dioxide ($a =$ 18 μm)	3.9	0.8–1.2	0.70566

References

1. Chiang, Y.J., Yen, T.J.: A composite-metamaterial-based terahertz-wave polarization rotator with an ultrathin thickness an excellent conversion ratio, and enhanced transmission. Appl. Phys. Lett. **102**, 011129 (2013)
2. Chen, H.T., Taylor, A.J. Yu, N.: A review of metasurfaces: physics and applications review article, **3** (2016)
3. Smith, D.R., Padilla, W.J., Vier, D.C., Nemat-Nasser, S.C., Schultz, S.: Composite medium with simultaneously negative permeability and permittivity. Phys. Rev. Lett. **84**, 4184–4187 (2000)
4. Pendry, J.B.: Negative Refraction Makes a Perfect Lens. Phys. Rev. Lett. **85**, 3966–3969 (2000)
5. Fang, N., Lee, H., Sun, C., Zhang, X.: Sub-diffraction-limited optical imaging with a silver superlens. Science **308**, 534–537 (2005)
6. Bhattacharyya, S., Ghosh, S., Srivastava, K.V.: A wideband cross polarization conversion using metasurface. Radio Sci. **52**, 1395–1404 (2017)
7. Yadav, V.S., Ghosh, S.K., Das, S., Bhattacharyya, S.: Graphene based metasurface for a tunable broadband terahertz cross polarization converter over a wide angle of incidence. Appl. Opt. **57**, 8720–8726 (2018)
8. Chen, H., Ran, L., Zhang, X., Chen, K.: Left-handed materials composed of only S shaped resonators. Phys. Rev. E **70**, 057605 (2004)
9. Dincer, F., Sabah, C., Karaaslan, M., Unal, E., Bakir, M., Erdiven, U.: Asymmetric transmission of linearly polarized waves and dynamically wave rotation using of chiral metamaterial. Progress Electromagnet. Res. **140**, 227–239 (2013)
10. Wang, B., Zhou, J., Koschny, T., Kafesaki, M., Soukoulis, C.M.: Chiral metamaterials: simulations and experiments. J. Opt. A: Pure Appl. Opt. **11**, 114003 (2009)
11. Huang, S.Y., Lee, Y.H.: A compact E-shaped patterned ground structure and its applications to tunable bandstop resonator. IEEE Trans. Microw. Theory Tech. **57**, 657–666 (2009)

An Ultra-Thin X-band Metasurface-Based Transmittive-Type Linear to Circular Polarization Converter

Lavesh Nama[1], Nilotpal[1], Somak Bhattacharyya[1](✉), and P. K. Jain[1,2]

[1] Department of Electronics Engineering, Indian Institute of Technology (BHU), Varanasi 221005, Uttar Pradesh, India
{laveshn.ece16,nilotpal.rs.ece17,somakbhattacharyya.ece, pkjain.ece}@iitbhu.ac.in

[2] Department of Electronics & Communication Engineering, National Institute of Technology, Patna 800005, Bihar, India

Abstract. In this manuscript, an ultra-thin (~λ/18.75), transmittive-type linear to circular (LTC) electromagnetic wave polarization converter in X-band employing metasurface (MS) has been reported. The proposed MS-based LTC polarization converter consists of two mirrored L-shaped structure and a thin strip designed over the top surface of 1.6 mm thick FR4 substrate while meander structure has been designed on the bottom side of the substrate. Numerical simulation results show the conversion of 45° linearly polarized incident wave into circularly polarized transmitted wave having axial ratio (AR) less than 3-dB over the frequency range 7.42–12.00 GHz; thereby yielding fractional bandwidth of 47.02% under normal incidence.

Keywords: Metasurface · Linear to circular polarization converter · Wideband

1 Introduction

In recent years, researches on metamaterials (MTM) have been explored to uncover unique electromagnetic properties like invisible cloaking, holograms, negative refractive index, perfect lenses, etc. [1, 2]. Several types of metasurface (MS) designs, a two-dimensional realization of MTMs, have been introduced for diverse applications, viz absorber, filter, phase-shifter, etc., in high-frequency regime [3–5]. MS structures have also been used to convert the state of polarization of electromagnetic waves. These are preferred over the conventional polarization controlling mechanisms utilizing birefringence effect, chiral activity, wave plates, etc., where the structures suffer from bulky volume, low adaptability for practical applications, narrow bandwidth, high loss, etc. [6, 7].

Circularly polarized waves are preferred over the linearly polarized ones due to their superior signal propagation characteristics and better response in adverse weather condition, phasing issues, polarization mismatch problem, etc. [8–10]. MS-based polarization

© Springer Nature Singapore Pte Ltd. 2021
N. R. Das and S. Sarkar (eds.), *Computers and Devices for Communication*,
Lecture Notes in Networks and Systems 147,
https://doi.org/10.1007/978-981-15-8366-7_33

converting structures have been designed either by incorporating anisotropy (using birefringence effect) [11–13] or chirality (using optical activity) [14, 15]. Reflective-type MS polarization converting structures have been designed more efficiently exploiting anisotropy while chirality has been exposed to realize the transmittive type of the same. Several transmittive-type linear to circular (LTC) polarization converting MS structures have been proposed till date [16–19], but they are limited by thickness, bandwidth, periodicity, or angle of incidence.

In this article, a transmittive-type LTC polarization converter MS in X-band has been proposed. The proposed MS offers LTC polarization conversion of the incident electromagnetic wave with 3-dB fractional axial ratio (AR) bandwidth of 47.02% under normal incidence. The unit cell of the MS comprises two layers of metallic patches printed on both sides of the dielectric substrate. The bottom surface of the unit cell consists of meander line structure, while a pair of mirrored L-type structure along with rectangular strip has been printed on the top surface of the dielectric. The dielectric substrate is 1.6 mm thick and made of FR4. In the frequency range 7.42–12.00 GHz, the structure offers left-hand circular polarized (LHCP) transmitted wave for a 45° linearly polarized incident wave. The polarization ellipses have been presented for transmitted wave at several frequencies within X-band to visualize the circular polarization.

2 Design of the Structure

The schematic diagram of top and bottom layers consisting of 0.035 mm thick copper layer imprinted on FR4 ($\varepsilon_r = 4.2$ and $\tan\delta = 0.02$) substrate of the proposed MS unit cell are illustrated in Fig. 1a, b, respectively. The optimized values of various geometrical parameters are also depicted in Fig. 1. The geometrical dimensions of the structure have been chosen, so that the magnitudes of co- and cross-polarized transmitted coefficients become almost equal in X-band. The strip in the top surface is responsible to obtain a phase quadrature among the two orthogonally transmitted electromagnetic waves, so that the transmitted wave exhibits circular polarization. The geometrical dimension of the strip has been optimized too to satisfy the phase condition in X-band. The complete unit cell comprising top and bottom layers is shown in Fig. 1c. The linearly polarized electromagnetic wave is incident on top surface of the designed MS along—z-direction. The electric field vector of the incident wave makes an angle 45° with y-axis along $\hat{v}$ direction, whereas the incident magnetic field is oriented along $\hat{u}$ direction; thereby subtending an angle 45° with x-axis. The incident v-polarized wave after transmission through the proposed MS structure generates two orthogonal components of electric field in $\hat{u}$ and $\hat{v}$ directions in X-band.

3 Simulated Results

The unit cell of the proposed MS structure has been simulated using CST MicroWave Studio simulator. The co-polarized (t_{vv}) and cross-polarized (t_{uv}) transmission coefficients for incident v-polarized electromagnetic wave on the proposed MS structure are shown in Fig. 2a. The frequency responses of the ratio of the magnitudes of co-polarized transmission co-efficient with respect to the cross-polarized one (t_{vv}/t_{uv}) along with

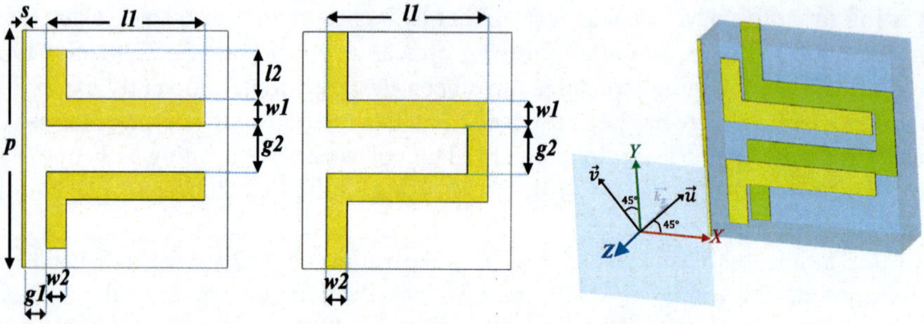

Fig. 1 Schematic of **a** top side and **b** bottom side of the unit cell of transmittive-type LTC polarization converter using MS ($p = 6$ mm, $l1 = 4.7$ mm, $w1 = g1 = 0.7$ mm, $S = 0.1$ mm, $l2 = g2 = 1.2$ mm, $w2 = 0.6$ mm). **c** Incidence of 45° linearly polarized (v-polarized) wave

the phase difference between them ($\psi = \phi_u - \phi_v$) are shown in Fig. 2b. It has been observed from Fig. 3b that the ratio is nearly unity over X-band by maintaining a near phase quadrature; thereby producing circularly polarized wave.

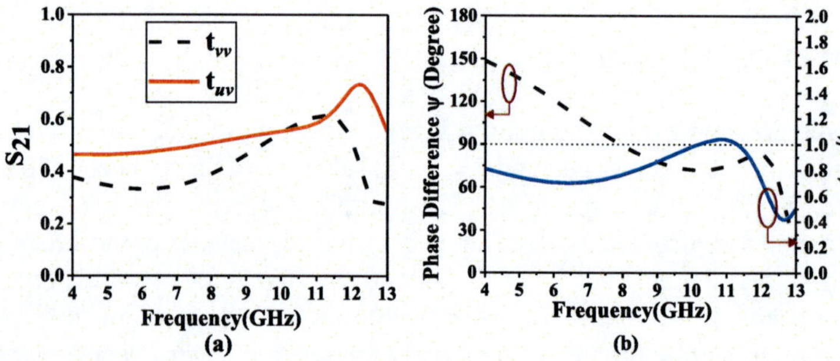

Fig. 2 **a** Co-polarized and cross-polarized transmission coefficients of the v-polarized wave passing through the proposed structure whose unit cell is shown in Fig. 1. **b** Phase difference between co- and cross-polarized transmitted waves along with their respective amplitude ratio

The nature of polarization of the transmitted wave has been further investigated from the axial ratio (AR) and ellipticity [20]. Figure 3a, b depicts the respective computed AR and ellipticity. It has been revealed in Fig. 3a that the AR is less than 3 dB over the range 7.42–12 GHz. Simultaneously, the ellipticity is above +0.9 over the same range as evident from Fig. 3b; implying that the transmitted wave is left-handed circularly polarized in nature.

For better visualization of the transmitted circularly polarized waves, the instantaneous electric field has been studied at various X-band frequencies in the plane perpendicular to the direction of wave propagation as shown in Fig. 4. It has been observed that near-circular paths have been traced at all frequencies in X-band.

The surface current distributions have been studied at 11.7 GHz, where LTC polarization conversion occurs. The surface currents at the top and bottom surface of the unit cell are shown in Fig. 5a. It has been found that the surface currents are anti-parallel

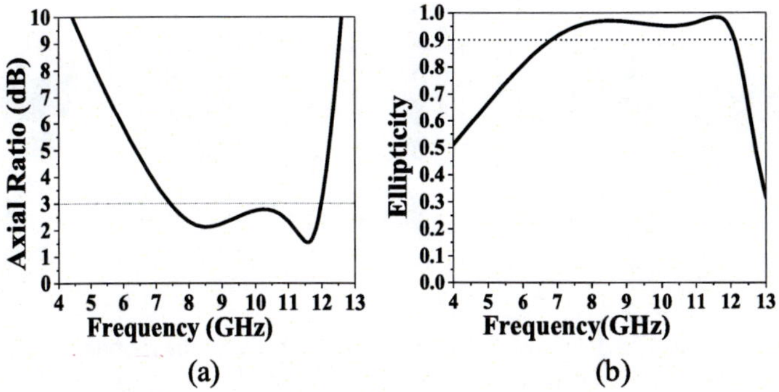

Fig. 3 Frequency responses of computed **a** AR (dB) and **b** ellipticity for v-polarized wave, propagating along—z direction

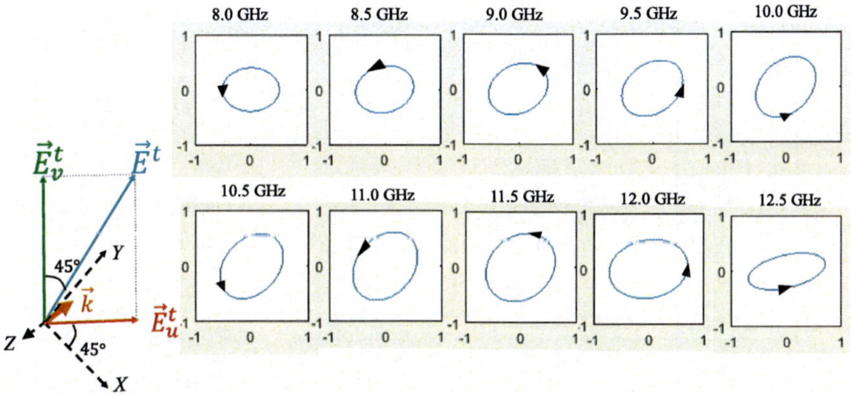

Fig. 4 Polarization ellipses for the transmitted wave in the plane perpendicular to wave propagation direction at various frequencies in X-band

between the top mirrored L-shaped structure and bottom meander line geometry; thereby creating magnetic resonance which yields the induced magnetic field perpendicular to the plane of the current loop [21]. This, in turn, results in induced electric field perpendicular to the direction of induced magnetic field. The electric field distributions within the structure has also been studied as illustrated in Fig. 5b. It has been clearly observed from Fig. 5b that the incident linearly polarized electric field gets converted into the circularly polarized one. The performance of the proposed LTC structure has been compared with existing LTC designs as listed in Table 1 where it has been revealed that the proposed LTC design offers wideband LTC conversion maintaining the compactness in periodicity and ultra-thin nature.

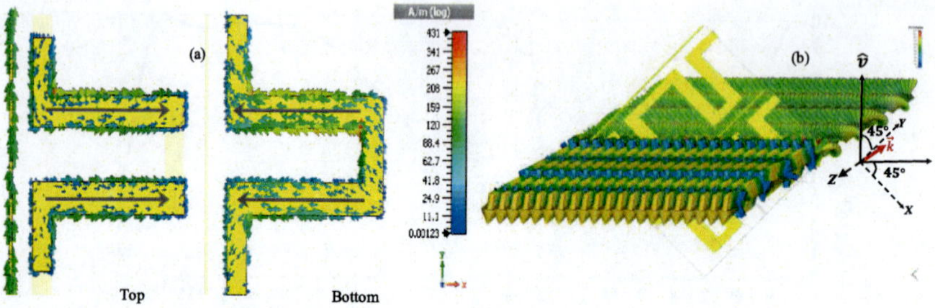

Fig. 5 **a** Surface current distributions at top and bottom surfaces of the proposed metasurface unit cell and **b** electric field representation on the design at 11.7 GHz

Table 1 Comparison with existing transmittive-type LTC polarization converter structures using metasurface

Transmittive-type LTC polarizer	3-dB AR range (Bandwidth in GHz)	Periodicity	Thickness	3-dB AR fractional bandwidth (%)
Akgol et al. [16]	3.87–4.38 GHz (0.51)	~ λ/5.2	~ λ/46	12.36
Lin et al. [17]	12.21–18.39 GHz (6.18)	~ λ/3.2	~ λ/4.8	40.4
Ma et al. [18]	13.25–15.7 GHz (2.45)	~ λ/2.15	~ λ/41	17
Lin et al. [19]	7.63–11.13 GHz (3.5)	~ λ/4	~ λ/13	37.4
Proposed structure	7.42–12.00 GHz (4.58)	~ λ/5.14	~ λ/19	47.02

4 Conclusions

A X-band transmittive-type linear to circular polarization converter has been realized. The structure converts incident linearly polarized wave to left-handed circularly polarized one as transmitted wave within the frequency range 7.42–12.00 GHz. The computed 3-dB AR has been found to be less than 3 dB over the complete X-band. The positive ellipticity implies that the transmitted wave is left-handed circularly polarized in nature. The polarization ellipses at distinct frequencies in X-band reveal the better visualization of the transmitted polarized wave. The proposed wide-band ultra-thin LTC polarization conversion metasurface may find potential applications in satellite communication and military domain.

Acknowledgement. SB acknowledges Science and Engineering Research Board (SERB), Government of India, for partial funding.

References

1. Caloz, C., Itoh, T.: Electromagnetic Metamaterials: Transmission Line Theory And Microwave Applications. Wiley-Interscience, Hoboken, NJ (2006)

2. Chen, H.T., Taylor, A.J., Yu, N.: A review on metasurfaces: physics and applications Rep. Prog. Phys. **79**, 076401 (2016)
3. Ghosh, S., Bhattacharyya, S., Kaiprath, Y., Srivastava, K.V.: Bandwidth-enhanced polarization-insensitive microwave metamaterial absorber and its equivalent circuit model. J. Appl. Phys. **115**, 104503 (2014)
4. Varuna, A.B., Ghosh, S., Bhattacharyya, S., Srivastava, K.V.: Design of a dual-band polarization-insensitive and angular-stable frequency selective surface. In: IEEE Applied Electromagnetics Conference (AEMC), IIT Guwahati, India, 18–21 Dec 2015
5. Antoniades, M., Eleftheriades, G.: Compact linear lead/lag metamaterial phase shifters for broadband applications. IEEE Antennas Wirel. Propag. Lett. **2**, 103–106 (2003)
6. Young, L., Robinson, L., Hacking, C.: Meander-line polarizer. IEEE Trans. Antennas Propag. **21**, 376–378 (1973)
7. Huang, Y., Zhou, Y., Wu, S.T.: Broadband circular polarizer using stacked chiral polymer films. Opt. Express **15**, 6414 (2007)
8. Balanis, C.: Antenna Theory, Analysis, and Design, 2nd edn. Wiley, New York (1997)
9. Kajiwara, A.: Line-of-sight indoor radio communication using circularly polarized waves. IEEE Trans. Veh. Technol. **44**, 487–493 (1995)
10. Rafii, V., Nourinia, J., Ghobadi, C., Pourahmadazar, J., Virdee, B.S.: Broadband circularly polarized slot antenna array using sequentially rotated technique forC-band applications. IEEE Antennas Wirel. Propag. Lett. **12**, 128–131 (2013)
11. Nama, L., Bhattacharyya, S., Jain, P.K.: An ultra-thin wideband linear to circular polarization converter using metasurface. In: IEEE International Symposium on Antennas and Propagation and USNC/URSI National Radio Science Meeting, 777-778, Boston, USA, 8–13 July (2018)
12. Bhattacharyya, S., Ghosh, S., Srivastava, K.V.: A wideband cross polarization conversion using metasurface. Radio Sci. **52**, 1395–1404 (2017)
13. Yadav, V.S., Ghosh, S.K., Bhattacharyya, S., Das, S.: Graphene-based metasurface for a tunable broadband terahertz cross-polarization converter over a wide angle of incidence. Appl. Opt. **57**, 8720 (2018)
14. Yan, S., Vandenbosch, G.A.E.: Compact circular polarizer based on chiral twisted double split-ring resonator. Appl. Phys. Lett. **102**, 103503 (2013)
15. Mutlu, M., Akosman, A.E., Serebryannikov, A.E., Ozbay, E.: Asymmetric chiral metamaterial circular polarizer based on four U-shaped split ring resonators. Opt. Lett. **36**, 1653 (2011)
16. Akgol, O., Altintas, O., Unal, E., Karaaslan, M., Karadag, F.: Linear to left- and right-hand circular polarization conversion by using a metasurface structure. Int. J. Microw. Wirel. Technol. **10**, 133–138 (2017)
17. Lin, B.Q., Guo, J.X., Huang, B.G., Fang, L.B., Chu, P., Liu, X.W.: Wideband linear-to-circular polarization conversion realized by a transmissive anisotropic metasurface. Chin. Phys. B **27**, 054204 (2018)
18. Ma, X., Huang, C., Pu, M., Hu, C., Feng, Q., Luo, X.: Single-layer circular polarizer using metamaterial and its application in antenna. Microw. Opt. Technol. Lett. **54**, 1770–1774 (2012)
19. Lin, B.Q., Guo, J., Wang, Y., Wang, Z., Huang, B., Liu, X.: A wide-angle and wide-band circular polarizer using a bi-layer metasurface. Progress Electromagnet. Res. **161**, 125–133 (2018)
20. Goldstein, D., Goldstein, D.H.: Polarized Light, Revised and Expanded, 2nd edn. CRC Press, Florida (2003)
21. Bhattacharyya, S., Srivastava, K.V.: Triple band polarization-independent ultra-thin metamaterial absorber using electric field-driven LC resonator. J. Appl. Phys. **115**, 064508 (2014)

Energy-Efficient Frequency Octupling Using Mach–Zehnder Optical Modulator

Abhirup Das Barman[1(✉)], Arnav Mukhopadhyay[1], and Antonella Bogoni[2]

[1] Institute of Radio Physics and Electronics, University of Calcutta, Kolkata, India
abhirup1.rpe@gmail.com

[2] Inter-University National Consortium for Telecommunications (CNIT), Pisa, Italy

Abstract. Mach–Zehnder modulator (MZM)-based energy-efficient scheme for 25 GHz RF generation, using frequency octupling technique is proposed. By employing a lower phase modulation index (PMI) of RF drive to MZM and utilizing phase cancelation of optical carrier by symmetric X-couplers, the ±fourth-order optical sidebands are generated efficiently. With the proposed scheme, a good optical and RF sideband suppression ratios of 26.2 dB and 25 dB, respectively, are achieved along with a reasonable output RF carrier power of 0 dBm.

Keywords: Mach–Zehnder modulator · mmWave generation · OSSR · RFSSR

1 Introduction

Rise in volume of traffic and frequency congestions below 6 GHz demands 5G mobile to operate in the new radio mmWave bands (24.25–27.5 GHz) [1]. mmWave (MMW) generation using photonic technology is not only cost effective [2] but also provide superior quality signal in terms of spectral purity, stability and the advantage of tunability [3, 4].

Among several methods of mmWave generation, frequency multiplication using external optical modulation has become popular due to its simplicity and low cost. Frequency octupling is one such technique, which have been used to generate MMW using parallel MZM arrangements [2, 4]. In these methods, the fourth-order optical sideband is retained via suppression of second-order optical sideband using quadrature RF signal applied to parallel MZMs. Then, the optical carrier is eliminated by increasing the phase modulation index (PMI) of the RF drive applied to the MZM [4], leading to higher RF power requirements, making the system energy inefficient. PMI is defined mathematically as the phase modulation of light that occurs due to applied electrical voltage. Hence, PMI is proportional to the electrical power consumed by the system. Alternatively, in [2], optical carrier is suppressed with biasing the cascaded MZMs across the null transmission point. The bias deviation, however, depends on applied RF power, hence, RF fluctuations can affect the system performance and to prevent this, complicated bias control circuit is required, which adds cost to the system. Single MZM-based octupling requires a high value for RF drive frequency, so as to allow the

© Springer Nature Singapore Pte Ltd. 2021
N. R. Das and S. Sarkar (eds.), *Computers and Devices for Communication*,
Lecture Notes in Networks and Systems 147,
https://doi.org/10.1007/978-981-15-8366-7_34

optical filters to isolate the fourth-order optical sidebands [5], which is again inflexible in the environment of dynamic tuning.

In the proposed method, energy efficiency is achieved with cascaded MZMs biased at maximum transmission point (MATP), with a low power RF drive. In this case, the optical carrier, second- and fourth-order optical sidebands will exist. Adding a MZM in parallel to the cascade, and driving it with phase quadrature RF, will lead to cancelation of the second-order sidebands with insignificant effect on the fourth-order sideband. Finally, the optical carrier is eliminated in the output, using interferometry with an opposite phase shifted optical carrier derived from the laser source using symmetric X-couplers. Optical attenuators are used to further improve on amplitude matching, sideband cancelation and remove the effect of asymmetry due to component aging.

2 System Description and Mathematical Analysis

The detailed scheme for frequency octupling is shown in Fig. 1. The light from a laser source is split equally along two optical path, E_1 and E_2, using a symmetric X-coupler X_1. The light along E_1 is again split equally into optical fields E_3 and E_4, using optical splitter S_1. Along the path of E_3, two cascaded MZMs, M_1 and M_2, both biased at MATP, are driven by a RF source, represented in phasor form, $\beta_1(t) = \overline{\beta}_1 \cos[\omega_m t + \phi_m(t)]$. The output optical field (E_6) from the cascaded MZM is then attenuated by an optical attenuator L_1. The resultant optical spectrum E_7 contains optical carrier, second- and fourth-order sidebands. Along the optical path E_4, another MZM M_3, biased at MATP and driven by a phase quadrature RF, $\beta_3(t) = \overline{\beta}_3 \cos\left(\omega_m t + \phi_m(t) - \frac{\pi}{2}\right)$ is placed. Its optical output, E_8 is then attenuated using an optical attenuator, L_2, and combined with the attenuated optical field (E_7) from MZM cascade. The resultant spectrum E_{10}, after proper adjustments of optical attenuators, L_1 and L_2, contains optical carrier and the fourth-order optical sidebands, only then, is applied to another symmetric X-coupler X_2. An optical carrier derived from another output of X-coupler X_1, denoted by E_2 is applied to the second input port of X-coupler X_2 after proper attenuation by the attenuator L_3. The combined output from coupler X_2 will contain only the fourth-order optical sidebands, because the optical carrier from E_{10} will destructively interfere with the phase-inverted carrier along E_{11}. The phase-inversion of the optical carrier is achieved through the combination of the X-coupler, X_1 and X_2, where each coupler introduces a quadrature phase shift to its optical path.

The optical signal dominated by the fourth-order sidebands is remotely delivered using a combination of optical amplifier and optical fiber, providing an optical path gain G, onto a photodiode (PD). Assuming that the PMI applied to the three MZMs are same, the optical field incident on the PD is given:

$$E_p = \frac{E_0}{4} G \left[\left\{ \frac{J_0^2(\overline{\beta}_1)}{L_1'} + \frac{2J_2^2(\overline{\beta}_1)}{L_1'} + \frac{J_0(\overline{\beta}_1)}{L_1'} - \frac{2}{L_3'} \right\} \right.$$
$$- 2 \left\{ \frac{2J_0(\overline{\beta}_1) J_2(\overline{\beta}_1)}{L_1'} - \frac{J_2(\overline{\beta}_1)}{L_2'} \right\} \cos 2(\omega_m t + \phi_m(t))$$

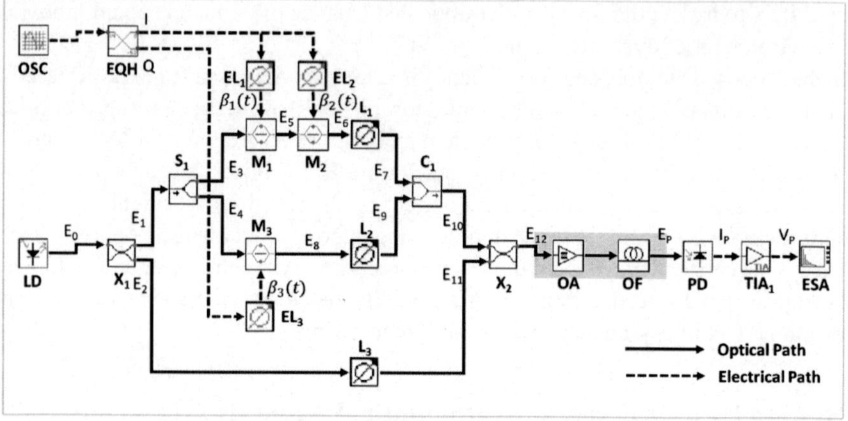

Fig. 1 Block diagram of a frequency ocutupling photonic circuit. OSC: RF signal source, EQH: electrical quadrature hybrid, EL: electrical attenuator, TIA: trans-impedance amplifier, ESA: electrical spectrum analyzer, LD: laser diode, X: symmetric optical X-coupler, S: symmetric optical power splitter, C: symmetric optical power combiner, M: single input balanced MZM (with high extinction ratio), OA: optical amplifier, OF: optical fiber, PD: PIN photodiode

$$+2\left\{\frac{J_2^2(\overline{\beta}_1)}{L_1'}\right\}\cos 4(\omega_m t + \phi_m(t))\Bigg] \qquad (1)$$

In Eq. 1, the square of the respective *L'* denotes the respective path losses along the three parallel paths, L_1, L_2 and L_3; their values are given by $L_1' = \frac{L_{X1}L_{S1}L_{M1}L_{M2}L_1L_{C1}L_{X2}}{1+\frac{1}{2}(\Delta c_{X1}+\Delta c_{X2})}$, $L_2' = \frac{L_{X1}L_{S1}L_{M3}L_2L_{C1}L_{X2}}{1+\frac{1}{2}(\Delta c_{X1}+\Delta c_{X2})}$ and $L_3' = \frac{L_{X1}L_1L_{X2}}{1+\frac{1}{2}(\Delta c_{X1}+\Delta c_{X2})}$. The square of the factor L_k represents the insertion loss corresponding to the *k*th component. The factors Δc_{X1} and Δc_{X2} represent the coupling ratio deviation for symmetric X-couplers X_1 and X_2, respectively.

The optical spectrum containing the fourth-order sidebands, as shown in Eq. 1, heterodyned at the PD to produce octupled RF, which is then amplified by a trans-impedance amplifier (TIA).

3 Results and Discussion

The simulation of the proposed scheme as shown in Fig. 1 is implemented in OptiSystem®16.0. The CW tunable laser source is centered at 193.1 THz with a linewidth of 10 MHz. A RF signal source at 3.125 GHz with PMI $\beta_1 = 1.2$ is applied to the MZMs, such that the total RF power consumed by all MZM ($V_\pi = 5.5$ V) is 27 dBm. Optical attenuators L_1 and L_2 are adjusted, such that $L_1 = 22.568$ dB and $L_2 = 20$ dB, which minimizes the second-order optical sidebands. Then, the optical attenuator L_3 is adjusted to 25.644 dB, in order to suppress optical carrier. A line amplifier amplifies the optical signal by 25 dB. Then, the amplified signal is carried to the remote unit through a 10 km ITU-T recommended G.652.D complaint single mode optical fiber, with an attenuation of 0.2 dB/km. Dispersion effect on the signal for a 10 km length of fiber is negligible.

The optical signal received at the remote unit is optically preamplified before applying to the photodiode PD. The spectrum of the signal incident on photodiode is shown in Fig. 2. The spectrum exhibits an optical suppression ratio (OSSR) of 26.2 dB, and the fourth-order optical sideband power is found to be −20.6 dBm. The photodiode considered here is an InGaAs PIN photodiode. It is followed by a TIA [6], providing a total optical-to-electrical power conversion gain of 19.5 dB. The output RF spectrum is shown in Fig. 2b shows a 25 GHz RF carrier with 0 dBm power and RF spurious suppression ratio (RFSSR) is obtained as 27.8 dB.

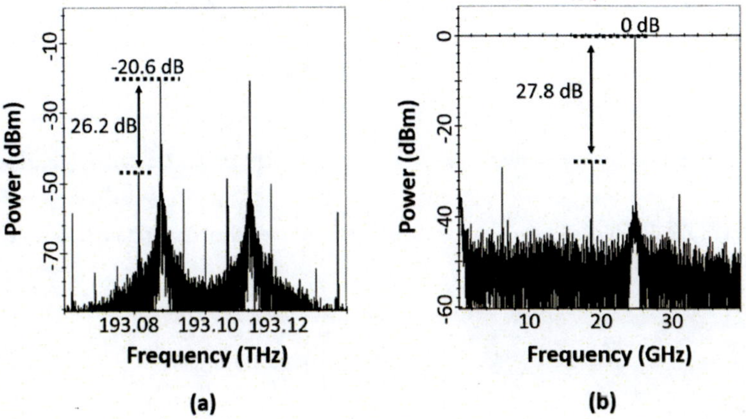

Fig. 2 **a** Optical spectrum incident on the photodiode (PD), **b** output electrical spectrum after O/E conversion using PD-TIA

Practically, there may be some component inaccuracy in electrical quadrature hybrid which may affect spurious suppression ratio (SSR), as shown in Fig. 3. The degradation of SSR is due to RF phase imbalance, which causes unequal suppression of second-order sidebands, resulting in a spur in RF spectrum. The asymmetry of X-couplers X_1 and X_2 can also result in SSR degradation, due to unmitigated optical carrier, is shown in Fig. 4. This imbalance can be corrected by properly re-adjusting the optical attenuators. The adjustment of optical attenuation versus the variation in coupling ratio deviation is shown in Fig. 5.

4 Conclusion

Frequency octupling technique is proposed for generation of 25 GHz RF signal. The proposed scheme requires lower PMI, and consequently low power RF drive to MZM. The scheme can be used to generate any other mmWave frequency as desired, as no optical filters are used. Other than energy efficiency and tunability, a good OSSR of 26.2 dB and RFSSR of 27.8 dB are also obtained.

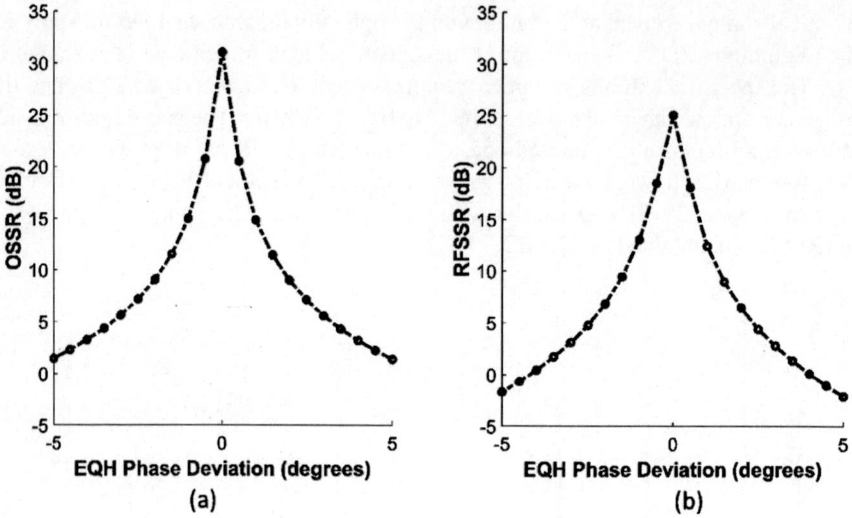

Fig. 3 Plot of the effect of phase deviation from 90°[52] of electrical quadrature hybrid (EQH) on OSSR and (b) RFSSR

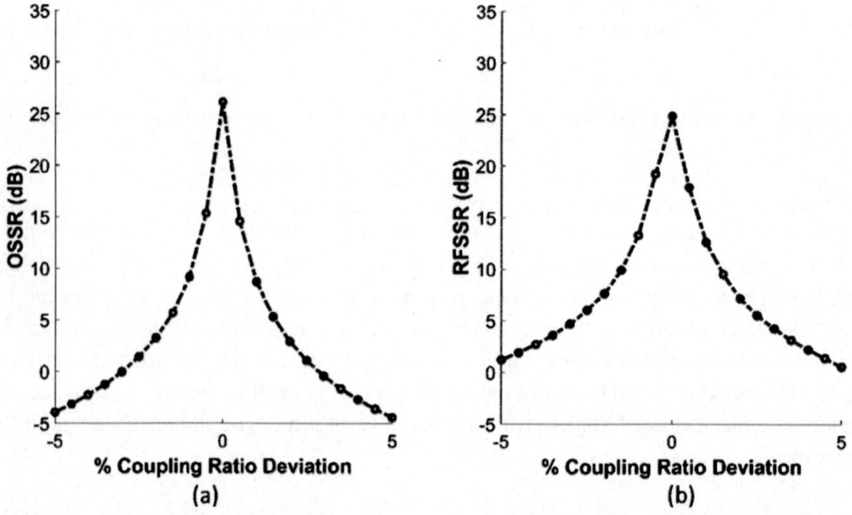

Fig. 4 **a** Plot of the effect of coupling ratio deviation of symmetric X-coupler X_1 on OSSR and **b** RFSSR

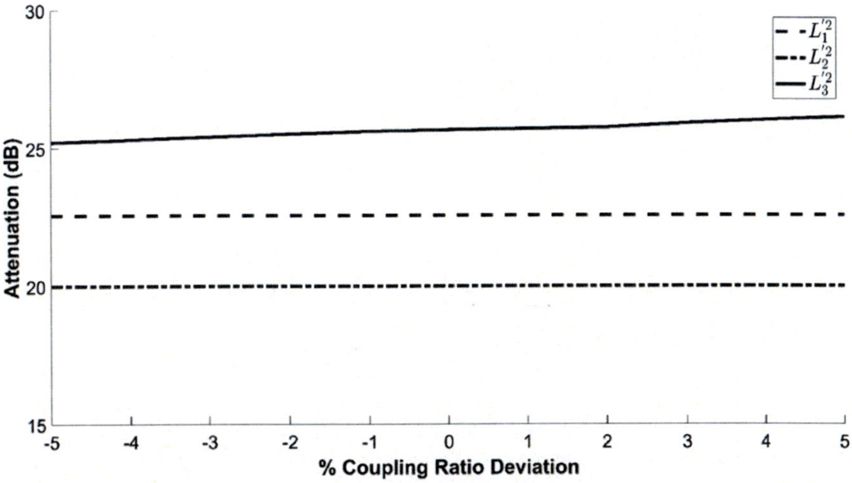

Fig. 5 Plot of variation in optical power attenuation for $L_1'^2$, $L_2'^2$ and $L_3'^2$ against variation in coupling factor deviation of X-coupler X_1

Acknowledgement. This work is supported by Department of Science and Technology (DST) (Project No.: INT/Italy/P-16/2016 (SP)).

References

1. Dahlman, E., Parkvall, S., Skold, J.: 5G NR: The Next Generation Wireless Access Technology. Academic Press (Elsevier) (2018)
2. Shang, L., et al.: A filterless optical millimeter-wave generation based on frequency octupling. Optik **123**(13), 1183–1186 (2012). https://doi.org/10.1016/j.ijleo.2011.07.047
3. Li, W., Yao, J.: Microwave generation based on optical domain microwave frequency octupling IEEE Photonics Technol. Lett. **22**(2), 24–26. 10.1109/LPT.2009.2035332 (2010)
4. Ma, J., et al.: Optical millimeter wave generated by octupling the frequency of the local oscillator. J. Opt. Netw. **7**(10), 837–845 (2008). https://doi.org/10.1364/JON.7.000837
5. Hu, J., Xu, Z., Huang, G.: Optical Millimeter-wave generation based on an external modulator Microw. Opt. Technol. Lett. **53**(12), 2902–2907. 10.1002/mop (2011)
6. Wu, C.Q., Sovero, E.A, Massey, B.: 40 GHz transimpedance amplifier with differential outputs using InGaAs heterojunction bipolar transistors. In: 24th Annual Technical Digest Gallium Arsenide Integrated Circuit (GaAs IC) Symposium (2002). doi: https://doi.org/10.1109/GAAS.2002.1049030

Check for updates

Design of a Bident-Shaped Metamaterial-Embedded Triple Band Microstrip-Printed Antenna with Defected Ground Structure

Apratim Chatterjee[1], Dweepayan Sen Sharma[1], Diptiranjan Samantaray[2], Chittajit Sarkar[1], Chinmoy Saha[3], and Somak Bhattacharyya[2](✉)

[1] Department of Electronics and Communication Engineering, Swami Vivekananda Institute of Science and Technology, Kolkata 700145, India
apratimchatterjee6@gmail.com, dssharma090298@gmail.com, chittajit.sarkar.in@ieee.org

[2] Department of Electronics Engineering, Indian Institute of Technology (BHU), Varanasi 221005, India
{drsamantaray.rs.ece17,somakbhattacharyya.ece}@iitbhu.ac.in

[3] Department of Avionics, Indian Institute of Space Science and Technology, Trivandrum 695547, India
csaha@ieee.org

Abstract. A compact high-gain-printed antenna with triple band characteristics has been proposed in this paper. The design consists of microstrip-printed antenna embedded with bident-shaped metamaterial unit cells in the top side along with defected ground structure in which Rogers RT/duroid 6006 has been used as the substrate. The antenna resonates at three frequencies viz., 6.34, 9.79, and 10.30 GHz making the design versatile with a high gain of 7.9 dBi. A bandwidth of 1.08 GHz lying between the frequency range from 10.11 to 11.19 GHz has also been realized at 10.30 GHz. The proposed antenna finds diverse application in the fields of radar engineering, satellite communication, and defence tracking. It is also suitable for future 5G mobile communication services.

Keywords: Bident-shaped metamaterials · Wideband · Defected ground structure · Gain · Triple band antenna

1 Introduction

Due to extremely thin profile and high conformality to the main system, printed antennas have been extensively used in wireless communication both commercial and scientific applications [1, 2]. The performance characteristics of such an antenna are crucial in wireless devices to serve ubiquitously in several frequency bands of interest. A number of design attempts have been made to tailor the efficiency of printed antenna [3, 4]. Artificially engineered metamaterials have been incorporated in the antenna design to

© Springer Nature Singapore Pte Ltd. 2021
N. R. Das and S. Sarkar (eds.), *Computers and Devices for Communication*,
Lecture Notes in Networks and Systems 147,
https://doi.org/10.1007/978-981-15-8366-7_35

exhibit unusual electromagnetic properties [5–8]. Due to the exotic properties of metamaterials, significant enhancement of antenna parameters have been realized over the last few decades [9–12].

This paper proposes a new and simple technique of designing triple band microstrip-printed antenna. The proposed design introduces a novel bident-shaped periodic metasurface into a conventionally designed square-printed antenna. In addition to the novel-shaped periodic elements, it employs two conventional techniques called defected ground structure and defected microstrip structure to yield a high-gain multi-band radiator [13, 14]. Major losses in the form of 'surface waves' in printed antenna have been drastically reduced as explained in [15]. The modification of RF signals with bident-shaped metamaterials placed on the printed antenna body is the major attraction of this paper. The proposed antenna is capable of resonating at three frequencies viz., 6.34 GHz, 9.79 GHz, and 10.30 GHz having fractional bandwidths of 2.2%, 2.7%, and 10.48%, respectively. In addition, broadband characteristics at 10.30 GHz have been achieved with a bandwidth of 1.08 GHz. It covers a wide range of frequency bands without compromising the gain and radiation characteristics of the antenna.

2 Design of the Bident-Shaped Metamaterial-Embedded Antenna

The schematic diagram of the proposed bident-shaped metamaterial-based antenna is shown in Fig. 1a. The enlarged view of the single bident-shaped metamaterials and the bottom view of the proposed structure are shown in Fig. 1b, c, respectively. The optimized design parameters of the proposed prototype antenna are shown in Table 1.

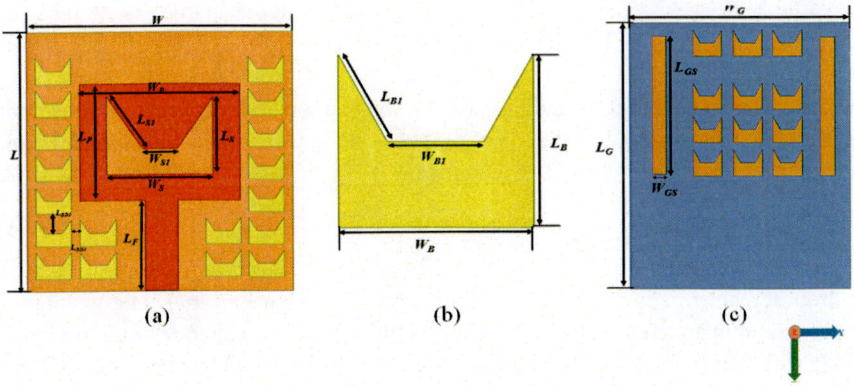

Fig. 1 **a** Top view of the proposed antenna with **b** enlarged view of the single metamaterial unit cell and **c** bottom view of the proposed antenna

Figure 2 shows the evolution of the bident-shaped embedded microstrip-printed antenna. The conventional patch antenna has been initially considered as shown as Case-I in Fig. 2 while the top view of the bident-shaped defected microstrip structure (DMS) has been considered as Case-II. Case III indicates patch with bident-shaped DMS (top view) as well as bident-shaped and rectangular-defected ground structure (DGS) (bottom

Table 1 Optimized geometrical dimensions of the proposed bident-shaped metamaterial-based antenna designed on Rogers RT/duroid 6006 ($\varepsilon_R = 6.16$) with thickness $h = 2$ mm

Antenna dimensions	Dimension (mm)	Antenna dimensions	Dimension (mm)
Length of substrate (L)	40	Length of the bident metamaterial (L_B)	4
Width of substrate (W)	30	Width of the bident metamaterial (W_B)	4
Length of patch (L_P)	18	Length of the two bident metamaterial prongs (L_{B1})	2
Width of Patch (W_P)	18	Distance between the two bident metamaterials prongs (W_{B1})	2
Length of slot in patch (L_S)	12	Distance between each bident metamaterial (L_{BD1})	4
Width of slot in patch (W_S)	12	Length of ground (L_G)	4
Length of the two bident prongs (L_{S1})	8	Width of Ground (W_G)	2
Distance between the two bident prongs (W_{S1})	4	Length of rectangular slot in ground (L_{GS})	2
Length of the feed line (L_F)	14	Width of rectangular slot in ground (W_{GS})	4

view). In the final design depicted as Case-IV, the patch with bident-shaped DMS and bident-shaped metamaterial on the top of the substrate (top view) and bident-shaped and rectangular DGS (bottom view) has been revealed.

3 Results and Discussion

The proposed antenna is designed and simulated by using electromagnetic simulation tool [16]. A comparative study of return loss with respect to frequency of the proposed antenna with four different configurations has extensively been illustrated in Fig. 3. It has been observed that Case-IV offers triple band characteristics of the proposed antenna. The antenna resonates in three frequencies viz., 6.34 GHz, 9.79 GHz, and 10.30 GHz with respective return losses of 32 dB, 20 dB, and 38 dB. From Fig. 3, it has also been observed that the proposed antenna is broadband in nature at 10.30 GHz offering a bandwidth of 1.08 GHz and thereby a fractional bandwidth of 10.48%.

The E-plane and H-plane co-polarized and cross-polarized radiation patterns are illustrated in Fig. 4a, b, respectively, at 6.34 GHz, 9.79 GHz, and 10.30 GHz. It has been observed from Fig. 4 that nearly omni-directional radiation patterns with reduced back lobes and side lobes have been achieved by incorporating the proposed design.

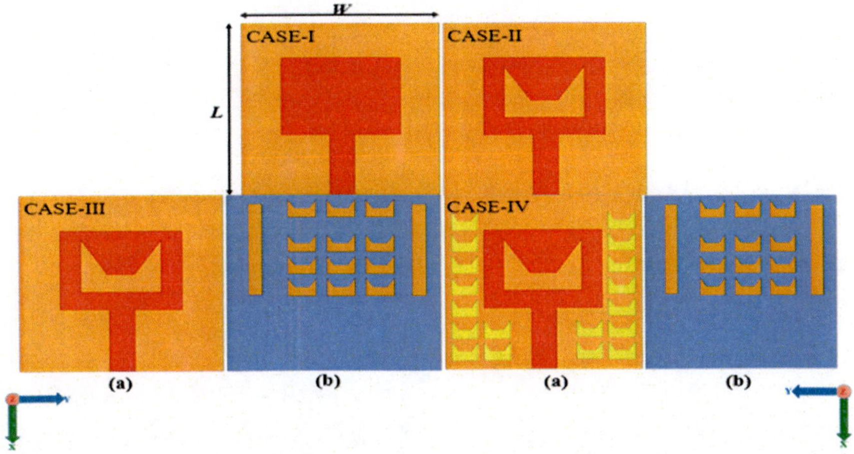

Fig. 2 Evolution of the proposed bident-shaped embedded microstrip-printed antenna. Case-I: Conventional printed antenna (top view); Case-II: Antenna with bident-shaped DMS (top view); Case-III: **a** Antenna (top view) with bident-shaped DMS. **b** Antenna (bottom view) with bident and rectangle-shaped DGS; Case-IV: **a** Antenna (top view) with bident-shaped metamaterial unit cells and DMS. **b** Antenna (bottom view) with bident- and rectangle-shaped DGS

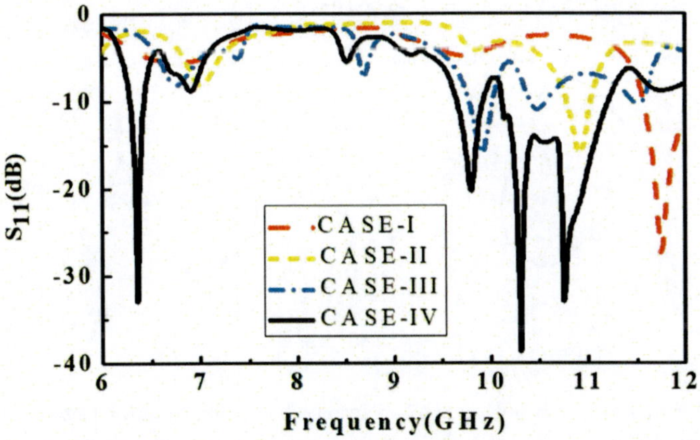

Fig. 3 Frequency responses of return loss characteristics for various cases depicted in Fig. 2

The electric field and surface current distributions at 6.34 GHz for both microstrip patch and defected ground structure are shown in Fig. 5a, b, respectively. From these figures, it is clear that the antenna is in radiating mode at this frequency.

The three-dimensional realized gains have also been studied for the proposed design at the three resonating frequencies as shown in Fig. 6. It has been observed that the maximum gain of 7.9 dBi has been achieved at 6.34 GHz. Further, the maximum realized gains of 6.5 dBi and 5.17 dBi have been obtained at 9.79 GHz and 10.30 GHz, respectively.

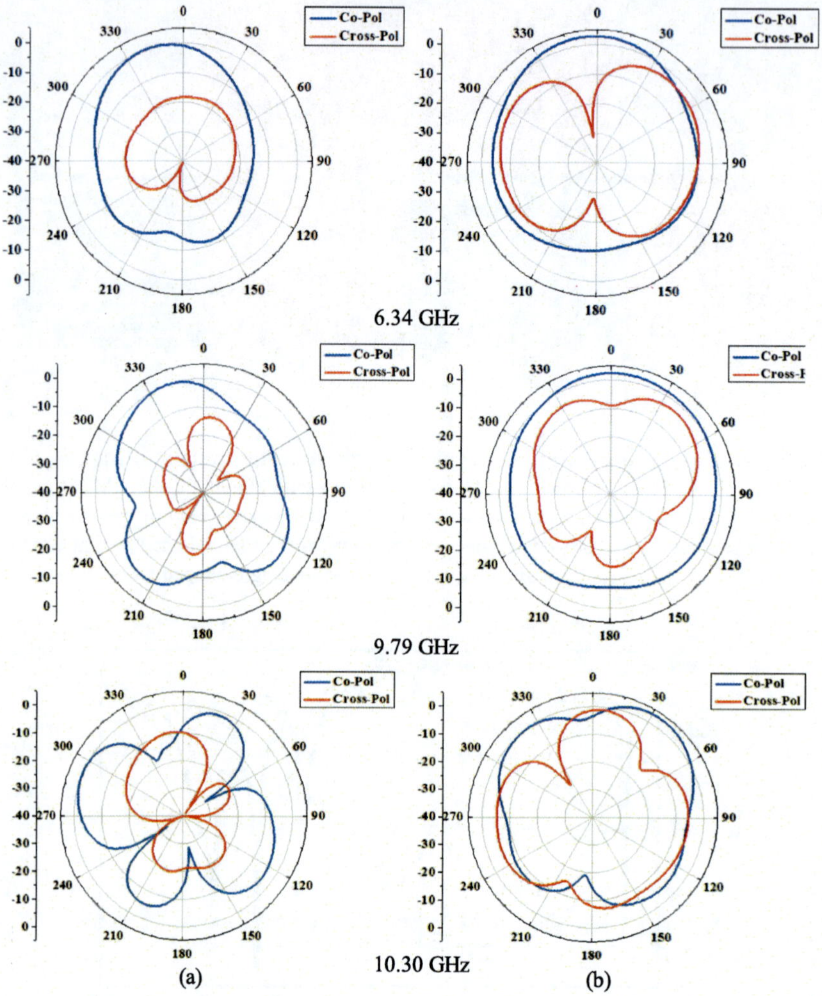

Fig. 4 Co-polarized and cross-polarized **a** E-plane and **b** H-plane radiation patterns at 6.34, 9.79, and 10.30 GHz

4 Conclusion

An improved performance of a high-gain triple band microstrip-printed antenna has been realized by embedding novel bident-shaped metamaterial unit cells along with defected ground structure. A bandwidth of 1.08 GHz has been achieved at 10.30 GHz with fractional bandwidth of 10.48% making the antenna broadband at this frequency. A maximum return loss of 38 dB has also achieved at 10.30 GHz. The bident geometry effectively reduces the back lobe and side lobe radiations from the antenna radiator to enhance the gain of the antenna. The proposed configuration is found to be suitable for 5G mobile communication application.

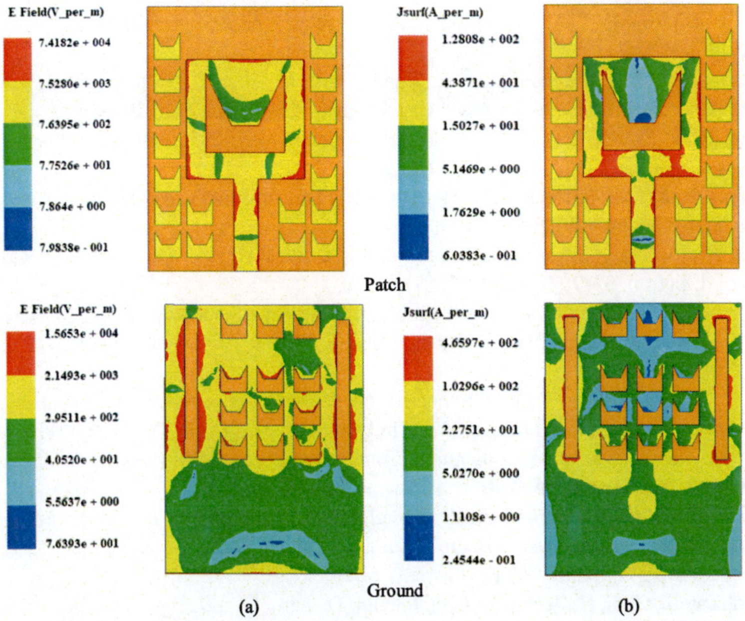

Fig. 5 **a** Electric field and **b** surface current distributions at 6.34 GHz

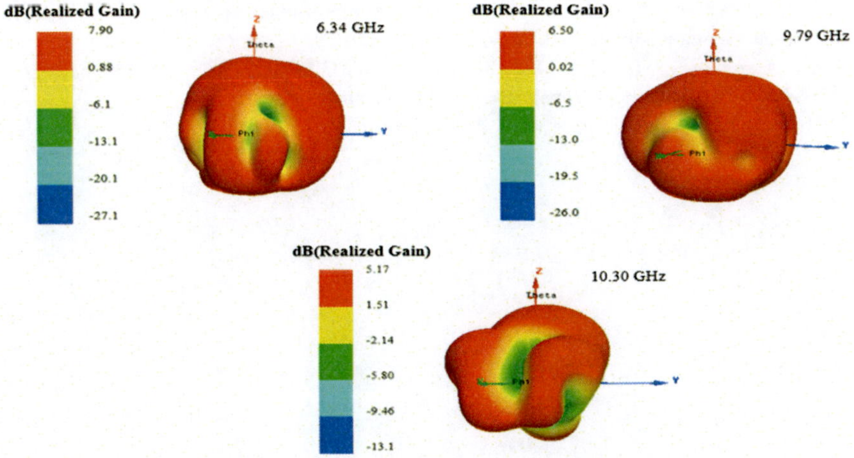

Fig. 6 Three-dimensional gain patterns at 6.34, 9.79, and 10.30 GHz

References

1. Stutzman, W.L., Thiele, G.: Antenna Theory and Design, IEEE antenna definition (2012)
2. Das, S., Choudhary, S.K.: Rectangular microstrip antenna of ferrite substrate. IEEE Trans. Antenna Propag. AP-**13**(3), 499–502 May (1982)
3. Balanis, C.A.: Antenna Theory Analysis & Design, 3rd edn. Wiley, Hoboken, New Jersey (2005)

4. Garg, R., Bhartia, P., Bahl, I., Ittipiboon, A.: Microstrip Antenna Design Handbook. Artech House (2000)
5. Veselago, V.G: Electrodynamics of substances with simultaneously negative electrical and magnetic permeabilities. Soviet Phys. Uspekhi **10**(4), 5–13 Jan–Feb (1968)
6. Shelby, R.A., Smith, D.R., Schultz, S.: Experimental verifications of a negative index of refraction. Science **292**, 77–79 (2001) (April)
7. Caloz, C., Itoh, T.: Electromagnetic Metamaterials Transmission Line Theory and Microwave Applications. Wiley—IEEE Press (2006)
8. Mittra, R.: A critical look at metamaterials for antenna-related applications. J. Commun. Technol. Electron. **52**(9), 972–978 (2007)
9. Ziolkowski, R.W., Kipple, A.: Application of double negative metamaterials to increase the power radiated by electrically small antennas. IEEE Trans. Antennas Propag. **51**(10), 2626–2640 (2003) (Oct)
10. Lim, S., Caloz, C., Itoh, T.: Electronically scanned composite right/left handed microstrip leaky-wave antenna. . IEEE Microwave Wirel Comp. Lett. **14**, 277–279 (2004) (June)
11. Yang, F., Samii, Y.R.: Electromagnetic Band Gap Structures in Antenna Engineering. Cambridge University Press (2009)
12. Mostafa, B.M., Abdel Rehman, A.B., Hamed, H.F.A.: Gain and bandwidth improvement of microstrip patch antenna using complementary G shape split ring resonator. IEEE Trans. Antennas Propag. **67**, 250–255 (2014)
13. Chaudhary, G., Choi, H., Jeong, Y., Lim, J., Kim, D., Kim, J.C.: Design of dual-band bandpass filter using DGS with controllable second passband. IEEE Microwave Wirel. Compon. Lett. **21**(11), 589–591 (2011)
14. Shi, S., Choi, W.-W., Che, W., Tam, K.-W., Xue, Q.: Ultrawideband differential bandpass filter with narrow notched band and improved common-mode suppression by DGS. IEEE Microwave Wirel. Compon. Lett. **22**(4), 185–187 (2012)
15. Lee, Y., Ganguly, S., Mittra, R.: Multi-band L5-capable GPS antenna with reduced backlobes. IEEE Int. Symp. Antennas Propag. **1**A, 3–8 July (2005)
16. High Frequency Simulation Software, Ansoft corp. v.14.

Planar Waveguide-based Optofluidic Refractive Index Sensors for Real-time Biomedical Sensing

Devesh Barshilia and Guo-En Chang(✉)

Department of Mechanical Engineering and Advanced Institute of Manufacturing With High-Tech Innovations (AIM-HI), National Chung Cheng University, Chia-Yi County 62102, Taiwan

imegec@ccu.edu.tw

Abstract. A low-cost intensity detection-based refractive index (RI) sensor is proposed and developed for rapid RI sensing. The sensor is composed of a planar waveguide structure on a low-cost glass substrate with an integrated microfluidic channel, which was fabricated using a simple, low-cost, vacuum-less and lithography-less process, making it suitable for mass production. Variation in the output light intensity, proportional to the RI of input solutions, is used as the simple and real-time mechanism for RI detection. A good RI resolution of 4.65×10^{-4}RIU is achieved through RI experiments. These results suggest that the developed waveguide optofluidic sensors are promising for rapid and sensitive bio-medical detection.

Keywords: Intensity detection · RI sensing · Microfluidic · Biomedical detection

1 Introduction

Optofluidics based sensing approaches have gained significant popularity as they not only meet the requirements of next-generation sensor platforms of increased sensitivity, specificity and parallelity, but also have the advantage of being simple in structure [1]. Primarily distinguished as being label-free, optofluidic refractive index (RI) sensors are particularly attractive for detection of ultralow quantities of molecules, making them useful in a multitude of applications ranging from biomedical detection and chemical analysis to food safety. A variety of optofluidic RI sensors, including Mach-Zender interferometers sensors [2], waveguide (WG) sensors [3], guided mode resonance (GMR) sensors [4] and surface plasmon resonance (SPR) sensors [5], have been developed to explore the possibilities of maximizing the light–analyte interaction for high-performance RI detection. Although successful in exhibiting high-performance RI sensing, limitations like complicated and time-consuming fabrication process, costly and bulky readout systems and delayed RI detection make these technologies less likely to be commercialized for practical applications.

The abovementioned problems can be addressed by using an intensity detection-based refractive index (RI) sensor. Recently, a polymer bent ridge waveguide (BRWG)

© Springer Nature Singapore Pte Ltd. 2021
N. R. Das and S. Sarkar (eds.), *Computers and Devices for Communication*,
Lecture Notes in Networks and Systems 147,
https://doi.org/10.1007/978-981-15-8366-7_36

structure on a low-cost glass substrate has been developed for low-cost, rapid RI sensing [6]. In this study, we propose and develop another WG-type optofluidic RI sensor to further improve sensitivity and reduce the cost and mass production time. The proposed WG-type sensor consists of a boro-silicate cover slip glass planar waveguide structure as the sensing region integrated with a microfluidic module. On the injection of sample solutions into the sensor, the output light intensity of the WG is modified due to the variations in RI on the surface of the planar sensing region. By monitoring the variations in output light intensity, sensitive and real-time RI detection can be achieved. In addition, the glass WG RI sensors are fabricated by simple, rapid, vacuum-less and lithography-less processes using UV glue, thus providing the unique advantages of high throughput and low cost for successful mass production. The detection system employs a low-cost, high-stability light-emitting diode (LED) as the light source and a photodetector (PD) as the optical receiver to precisely record the output light power, thereby enabling simple and real-time RI detections.

2 Sensor Design and Sensing Principle

Our designed RI sensor based on planar waveguide structures is schematically shown in Fig. 1. A low-cost glass slab is utilized as the low-RI substrate ($n = 1.513$). Borosilicate glass cover slip having a slightly higher RI ($n = 1.525$) than the substrate glass with a thickness of t is chosen as the WG material. The glass waveguide has a planar structure with a width, w_{in}, large enough to match the beam size of the input light from the far field to minimize the coupling loss between the WG and input light [6]. A microfluidic module is also integrated with the sensor, with the fluidic channel being centred in the planar waveguide sensing region to permit the light and flow interaction. The use of a microfluidic module substantially improves the stability and measuring accuracy of the RI sensor, resulting in enhanced sensor performance. The coupled light with an intensity of I_{in} propagates through the planar sensing region and reaches the output of the waveguide with significant loss due to RI variation in the cladding layer of WG. The magnitude of this received output light intensity, I_{out}, is a function of RI of the injected solution. Therefore, changes in RI of the solutions can be reflected as corresponding variations in the output light intensity which can be used as a measure to accurately determine the RI of the injected solution.

3 Fabrication of Waveguide Refractive Index Sensors

The optofluidic WG RI sensors were fabricated by vacuum-less and lithography-less fabrication processes. The fabrication process flow is shown in Fig. 2. Low-cost glass (25.4 × 76.2mm) was used as the substrate. First the substrate was cleaned with DI water and air dried through a nitrogen gun followed by an oven bake of 1 h. The same cleaning process was followed for the cover slip waveguide (24 × 50mm) and bonding cover. After the cleaning process, the borosilicate glass cover slip was glued to the substrate with UV glue and illuminated to an UV lamp for 15 min to allow the glue to harden. Extreme care was taken while gluing the cover slip to the substrate to avoid spreading of the glue towards the sensing region. This was followed by gluing of the bonding cover

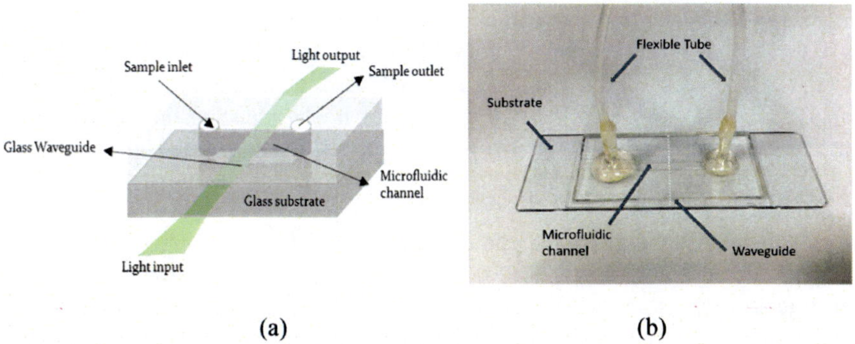

Fig. 1 **a** Schematics and **b** optical image of the proposed optofluidic waveguide refractive index sensor

to the substrate using the same steps. The sensor was then completed by connecting two flexible tubes to the bonding cover for the inlet and outlet of the fluids with the help of AB glue. The sensor was then left for air drying for 2–3 h. The fabrication processes of the sensors do not involve time-consuming vacuum and lithography processes, indicating significantly reduced production time. Also, the estimated cost of the waveguide RI sensors is less than one USD per chip, which is much lower than those of optofluidic RI sensors (typically tens of USD or higher) making the waveguide RI sensors highly suitable for high-throughput mass production.

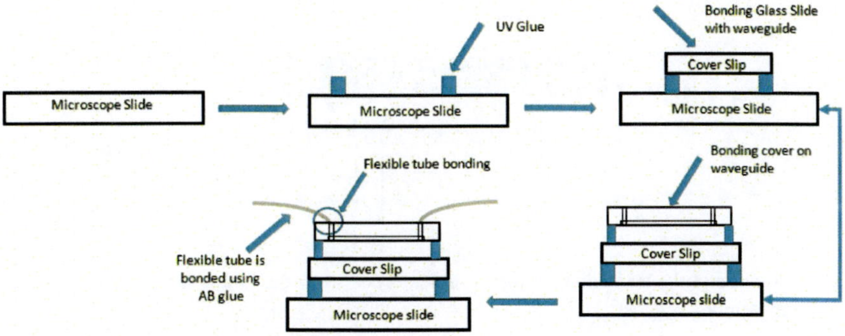

Fig. 2 Fabrication process flow of the optofluidic RI sensor

4 Characterization of Waveguide Refractive-Index Sensors

The arrangement of the optical detection system for the fabricated waveguide RI sensor is shown in Fig. 3. To procure better RI resolutions, a low-cost, commercially available, 532-nm green LED was employed as the light source because of their better power stability than typical lasers. To further enhance the signal-to-noise ratio, a lock-in technique was employed by using a 1-kHz square wave with a 50% duty cycle, generated by a homemade LED driver, to drive and directly modulate the LED [6]. The emitted light was passed through a collimator and coupled to one of the facet of the WG sensor chip through a 20X objective lens. To ensure minimal losses while coupling, the sensor chip was mounted on a three-axis translation stage for finely adjusting the position of the chip. The transmitted light through the WG chip, as in [6], was filtered by an adjustable iris, focused by a lens, and then directed to a Si PD to convert the output light intensity into photocurrent. Subsequently, a homemade current amplifier with a band-pass filter was used to amplify the generated photocurrent followed by the analog-to-digital conversion of the signals. Lastly, the recorded real-time digital signals were demodulated using a lock-in programme. This compact and low-cost detection system eliminates the requirement of any bulky and costly components, making the system suitable for practical applications.

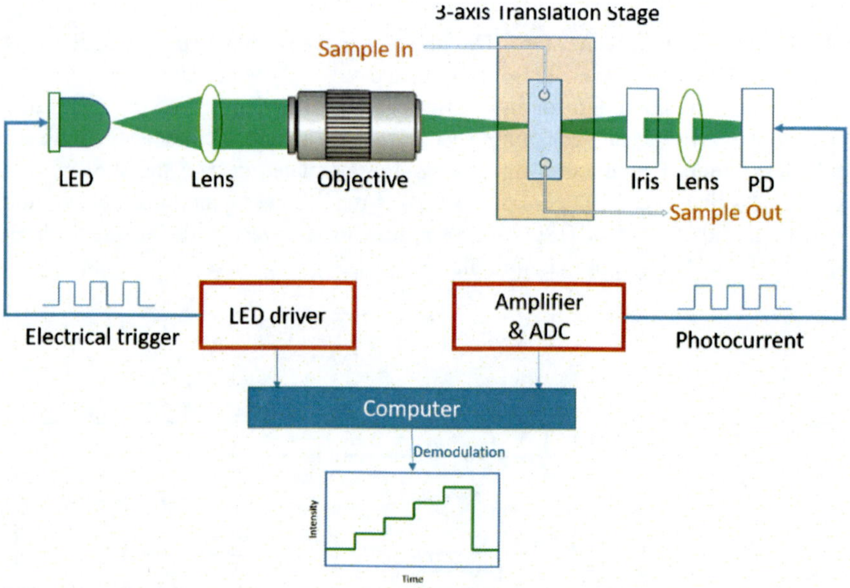

Fig. 3 Schematic of the transmission measurement system for the fabricated optofluidic waveguide refractive index sensors

For characterizing RI sensing performance of the developed system, RI experiments were performed. Deionized (DI) water and sucrose solutions with varying concentrations having RIs in the range $n = 1.333 - 1.373$ were used. First, DI water was injected into the sensor chip as blank solution to make the output light intensity reach a steady-state condition. Sucrose solutions with different RIs were then injected successively followed by the final injection of DI water. The intensity of the transmitted light (I) was synchronously recorded using the data acquisition system described above. The normalized real-time optical responses of the RI sensors are shown in Fig. 4a. As the RI of the sample solution increases, we observe increment in the output light intensity due to the decreasing optical loss in the planar sensing region. From the real-time optical responses, the average output light intensities at each RI resolution (I_{avg}) are extracted, and the normalized average output light intensity (I_{avg}/I_0, with I_0 being the average optical power measured from the blank solution) as a function of the solution's RI is presented in Fig. 4b. From the results, the normalized sensitivity (S_n) and sensor RI resolution (R_s), which represents the minimal detectable change in the RI of the solution, can be evaluated as

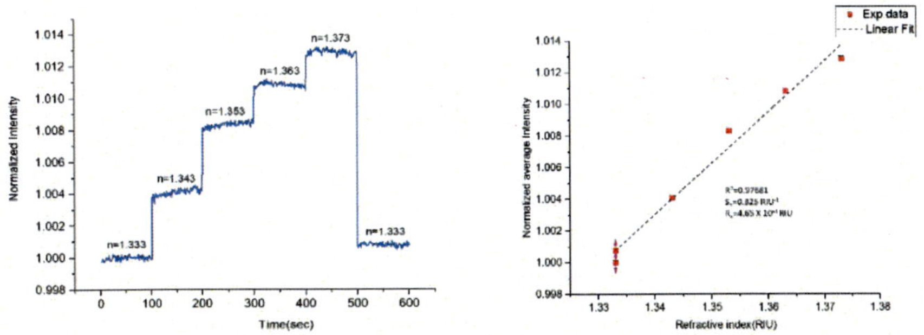

(a) (b)

Fig. 4 a Real-time optical response of the RI sensor with different solutions. **b** Normalized averaged real-time optical responses of the RI sensor as a function of refractive index of the solution

$$S_n = \frac{\mathrm{d}}{\mathrm{d}n}\left[\frac{I_{avg}(n)}{I_0}\right] \tag{1}$$

$$R_s = \left|\frac{\sigma}{S_n}\right| \tag{2}$$

Linear fitting of the experimental results gives the normalized sensitivity $S_n = 0.325 RIU^{-1}$ and linear correction coefficient $R^2 = 0.97681$, demonstrating the excellent linearity of this sensor over a wide dynamic RI range of 0.04 RIU. The RI resolution of this sensing system is determined to be 4.65×10^{-4}RIU. The results obtained are in close agreement with the results of [6] and demonstrate the feasibility of the proposed sensor. Further improvements in the RI resolution of the proposed sensor are possible by tweaking the width of the sensing region, which will be discussed later.

5 Conclusion and Future Work

In this paper, we present glass-based WG sensors for low-cost, rapid RI sensing. The preliminary results exhibit good performance in terms of sensitivity. The sensing principle consists of injecting solutions of different RI onto the planar sensing region that allows the variation of RIs to be manifested in terms of change in output light intensity. In addition to the simple sensing principle, these sensors are fabricated using simple, rapid, vacuum-less and lithography-less processes, offering a further reduction in manufacturing time, as compared with [6], yielding higher throughput and low cost. The detection system employs a low-cost, high-stability LED as the light source and a PD as the optical receiver, making the system cost-effective and compact, suitable for portable detection systems [6]. The achieved RI resolution of 4.65×10^{-4}RIU, over a wide range of 0.04 RIU, through RI simulations, is in close agreement with the results obtained in [6]. These results can be further enhanced by improving the LED light intensity to enhance the signal-to-noise ratio and adjusting the width of sensing region of the sensor to increase the sensitivity. Deeper understanding of the sensing capacity of the proposed RI sensor are also possible from numerical simulations for further optimizing the sensor performance.

Acknowledgements. This work at CCU was supported by Ministry of Science and Technology of Taiwan (MOST) under the grant numbers of MOST 107-2218-E-194 -004 and 108-2218-E-194-001, and Advanced Institute of Manufacturing with High-tech Innovations (AIM-HI) from The Featured Areas Research Centre Program within the framework of the Higher Education Sprout Project by the Ministry of Education (MOE) in Taiwan.

References

1. Psaltis, D., Quake, S.R., Yang, C.: Developing optofluidic technology through the fusion of microfluidics and optics. Nature **442**, 381–386 (2006)
2. Dante, S., Duval, D., Sepúlveda, B., González-Guerrero, A.B., Sendra, J.R., Lechuga, L.M.: All-optical phase modulation for integrated interferometric biosensors. Opt. Express **20**, 7195–7205 (2012)
3. Hu, J., Tarasov, V., Agarwal, A., Kimerling, L., Carlie, N., Petit, L., Richardson, K.: Fabrication and testing of planar chalcogenide waveguide integrated microfluidic sensor. Opt. Express **15**, 2307–2314 (2007)
4. Voros, J., Ramsden, J., Csucs, G., Szendro, I., Paul, S.D., Textor, M., Spencer, N.: Optical grating coupler biosensors. Biomaterials **23**, 3699–3710 (2002)
5. Robelek, R., Wegener, J.: Label-free and time-resolved measurements of cell volume changes by surface plasmon resonance (SPR) spectroscopy. Biosens. Bioelectron. **25**, 1221–1224 (2010)
6. Liu, I.C., Chen, P.C., Chau, L.K., Chang, G.E.: Optofluidic refractive-index sensors employing bent waveguide structures for low-cost, rapid chemical and biomedical sensing. Opt. Express **26**, 273–283 (2018)

Design and Simulation of RF Cavity for Ka-Band Multibeam Klystron

Santigopal Maity[1](✉), M. Santosh Kumar[2], Chaitali Koley[2], Ayan Kumar Bandyopadhyay[3], and Debasish Pal[3]

[1] Department of ECE, Ramkrishna Mahato Government Engineering College, Purulia, WB, India
santigopal.100@gmail.com
[2] Department of ECE, NIT Mizoram, Aizawl, India
[3] Microwave Devices Area, CSIR-CEERI-Pilani, Rajasthan, India

Abstract. This paper presents the computer-aided design of Ka-band RF cavities of a multibeam klystron. The state-of-the-art electromagnetic simulation tool CST microwave studio has been used for the design and optimization of the 28 GHz cavity with four beams. Different cavity parameters such as the quality factor, shunt impedance have been estimated with the help of the simulation tool for comparison among them.

Keywords: Multibeam klystron · Cavity resonator · RF cavity · Coaxial cavity · Radial reentrant cavity · Reentrant cavity

1 Introduction

The Varian brothers invented klystron in 1939. It is a vacuum tube amplifier to amplify microwave frequency signals. The main characteristics of klystron are the amplification of given input RF signal employing the principle of velocity modulation. In Klystron tubes, a DC beam voltage is applied between anode and cathode, from which electrons are emitted. This potential energy accelerates the electron beam, imparting kinetic energy into them. The electrons are either accelerated or decelerated in the microwave interaction region by the microwave field. These electrons are also bunched as they drift down the tube. The current is in turn induced by these bunched electrons in the output structure. These electrons are collected by the collector after they gave up their kinetic energy to the microwave fields [1].

One electron beam is used for the beam-wave interaction in the case of conventional single beam klystron amplifier, but in the case of multiple beam klystrons (MBK), more than one electron beams are used for the same purpose. The output RF power and bandwidth of MBK amplifiers are greater than traditional single-beam klystrons operating with similar DC beam voltage. A number of electron beams propagating in parallel are used by MBK amplifiers to employ higher amount of total beam current interacting with the RF fields of a series of resonant cavities while maintaining lower beam current

© Springer Nature Singapore Pte Ltd. 2021
N. R. Das and S. Sarkar (eds.), *Computers and Devices for Communication*,
Lecture Notes in Networks and Systems 147,
https://doi.org/10.1007/978-981-15-8366-7_37

density. To minimize space charge effects, the separate beams are transported through individual beam tunnels. As a result, the total beam current is high, but individual beam perveances remains low [2].

Achieving enough shunt impedance for effective beam-wave interaction is one of the main challenges in designing RF cavities of a MBK.

In high-frequency operations, ohmic losses are increased and geometrical dimensions of RF cavities are decreased compared to lower frequency operations. This results in lower quality factor *Q*, *R*/*Q* (with *R* being the shunt impedance) and the increased machining difficulties [3]. The gain-bandwidth product of klystrons depends on *R*/*Q* values of the RF cavities. In vacuum microwave, devices either reentrant cavities or slow-wave structures are used to achieve overall high gain over a broad bandwidth. In klystrons, to improve the beam-wave interaction, RF cavities with higher *R*/*Q* values are required. Some solutions to improve the performance using different topologies of cavities to provide higher *R*/*Q* have been proposed in the past [4, 5].

In this paper, we have investigated four types of multibeam reentrant cavities as shown in Fig. 1, for Ka-band frequency (~28 GHz) operation. For all the three cavities, relevant cavity parameters (operating frequency, quality factor, and *R*/*Q*) have been compared. All the electromagnetic simulations have been carried out using commercially available electromagnetic simulation software, CST EM Studio [6].

2 Design Approach

The main parameters involved in the cavity design are the quality factor (*Q*), *R*/*Q*, and resonant frequencies which are dependent on the geometrical parameters like height, gap length, drift tube radius, shunt impedance of the cavity. The optimum model of the cavity also should follow proper choice of electron beam parameters like beam radius and perveance. For proper interaction of electron beam with RF field in the cavity, it is essential that the inter space be short compared to the distance an electron travels per cycle. For our case, in order to be compatible to a predesigned four beam electron gun, the radius of the circle is on which all the beam centers should be placed at 7.5 mm. Therefore, it is essential to design the RF cavities at expected frequency with a dimension sufficiently larger than 15 mm to put up the four beams. The diameter of the electron beam has been taken as 0.7 mm. For a proper beam transmission through the drift tube, normally a beam fill factor of 70% is considered. Therefore, the diameter of the drift tube comes out to be 1 mm.

To achieve efficient beam-wave interaction, RF cavities with high *R*/*Q* values and *Q* factor (unloaded) are desired. To achieve high *Q* (larger than 1000) and *R*/*Q* up to 150 Ω, reentrant cavities are used. In case of a cavity with low *R*/*Q* value, the interaction of the electric field with the beam at the gap decreases. In this case, it may not be enough to set up the proper level of beam modulation to attain the gain and output power specifications. The dimensions of the cavities reduced to a great extent with an increase in operating frequency. For example, the diameter of a cylindrical cavity is in the order of 1 mm for a reentrant cavity operating at 100 GHz [7].

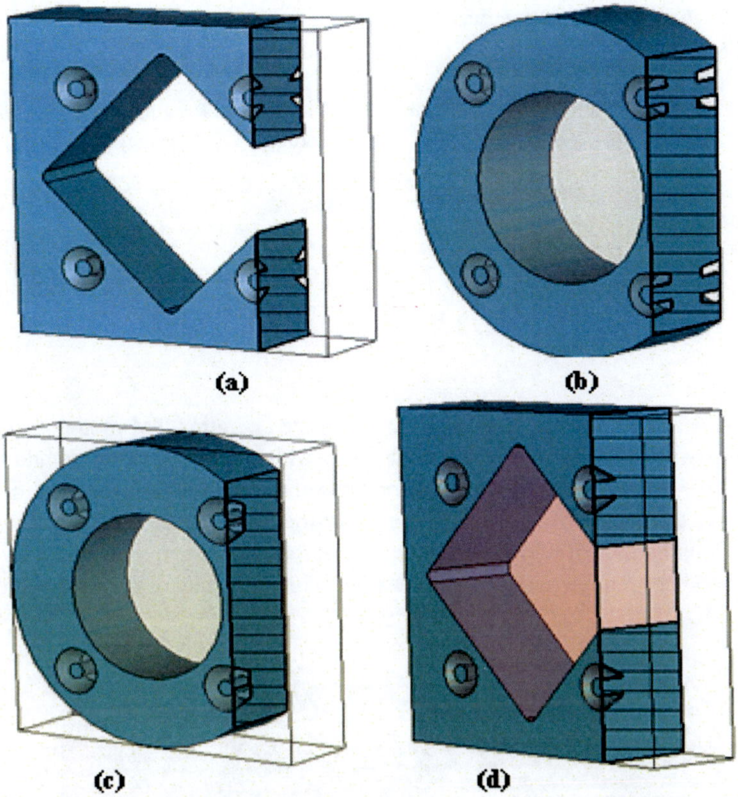

Fig. 1 Cut view of RF cavities (only vacuum) considered for simulation: **a** brick-type multibeam double reentrant vavity (brick-type center conductor), R_1, **b** Coaxial-type multibeam double reentrant cavity R_2 **c** brick-type multibeam single reentrant cavity (brick-type center conductor), R_3 and **d** Coaxial-type multibeam single reentrant cavity R_4

3 Simulation Results

In Table 1, R_1 is brick-type multibeam double reentrant cavity, R_2 is coaxial-type multibeam double reentrant cavity, R_3 is brick-type multibeam single reentrant cavity (brick-type center conductor), and R_4 is coaxial-type multibeam single reentrant cavity.

The geometrical and simulated cavity parameters for R_1, R_2, R_3 and R_4 are presented in Table 1. All the four types of cavity resonators (shown in Fig. 1) are designed to operate at Ka-band frequency. From the cavity parameters, it can be seen that R_1exhibits high electric field intensity as well as high R/Q value. Since the gain-bandwidth product of a klystron is directly proportional to R/Q values of the used RF cavities, higher R/Q value of this cavity offers the possibility of higher output power and/or broad bandwidth.

All the cavities have been designed with the drift tube centers positioned at equidistance on a circle of diameter 15 mm, and diameter of the drift tube has been taken as 1.0 mm.

Table 1 Different cavities geometrical dimensions and simulated cavity parameters (Cavity dimensions in mm)

Cavity	Height	Width/Radius	Gap length	Drift tube Radius	Resonant frequency GHz	Q	R/Q
R_1	5	18.7	3	0.5	28.02	4259	92.9
R_2	5.2	9.3	2.32	0.5	28.03	2567	78.7
R_3	5	18.7	3.675	0.5	28.03	4362	88.38
R_4	5	9.7	3.4	0.5	28.1	3182	66.2

Figures 2 and 3 show the field distribution of the operating TM mode of the brick-type (cubic structure) multibeam reentrant cavity with cubic-type center conductor and the nose cones. Figure 2 shows the cut plane view of the electric field distribution, while Fig. 3 shows the details of the cut plane view of the magnetic field distribution across the gaps. From both the figures, it can be seen that the electric and magnetic field is well concentrated between the nose cones as desired. The frequency of the mode is found to be 28.02 GHz from the CST simulations.

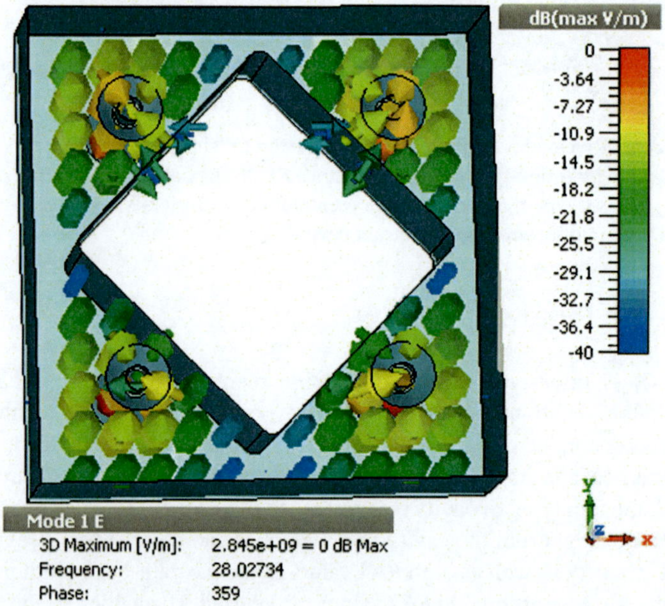

Fig. 2 Electric field distribution of the operating mode of the brick-type (cubic structure) multibeam reentrant cavity with cubic-type center conductor and the nose cones

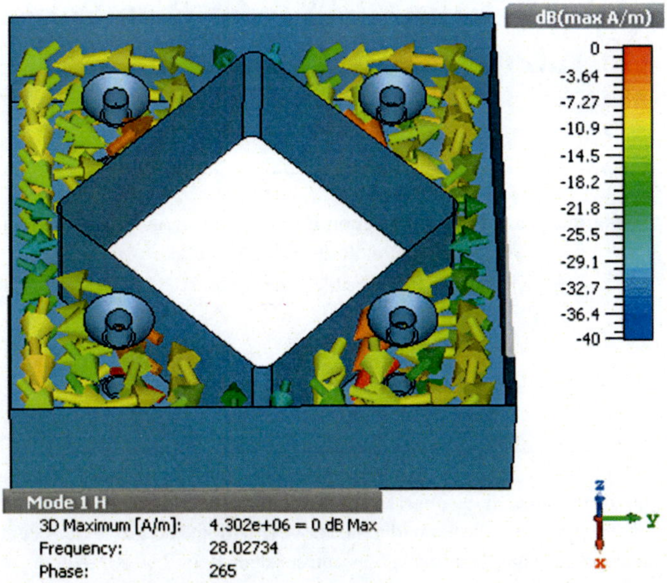

Fig. 3 Cut plane view of the magnetic field distribution of the operating mode of the brick-type (cubic structure) multibeam reentrant cavity with cubic-type center conductor and the nose cones

4 Conclusion

The Ka-band multibeam cavity has been designed and simulated using CST EM Studio. Several cavity parameters like the resonant frequency, quality factor (Q) and R/Q value have been obtained, and the electric and magnetic field distributions of the operating mode have been observed. Brick-type (cubic structure) multibeam reentrant cavity exhibits better performance than the coaxial and radial reentrant cavity; hence, it can be used in klystron amplifiers operating in Ka-band.

References

1. Liao, S.Y.: Microwave Devices and Circuits, pp. 353–363. Pearson Education, Inc. and Dorling Kindersley Pvt. Ltd., India (2003)
2. Maity, S., Bandyopadhyay, A.K., Joshi, L.M.: Design of radio frequency cavities for a J-BAND multibeam klystron. IETE J. Res. **58**(4) (2012)
3. Chang, Z., Meng, L., Yin, Y., Wang, B., Li, H., Rauf, A., Ullah, S., Bi, L., Peng, R.: Circuit design of a compact 5-kV W-band extended interaction klystron. IEEE Trans. Electron. Devices **65**(3), 1179–1184 (2018)
4. Chodorow, M., Wessel-Berg, T.: A high-efficiency klystron with distributed interaction. IRE Trans. Electron. Devices **8**(1), 44–55 (1961)
5. Symons, R.S., Vaughan, R.M.: The linear theory of the clustered cavity klystron. IEEE Trans. Plasma Sci. **22**(5), 713–718 (1994)
6. CST 2018, User's Manual. CSST GmbH. www.cst.com.
7. Paoloni, C.: Periodically allocated reentrant cavity klystron. IEEE Trans. Electron. Devices **61**(6), 1687–1691 (2014)

Check for updates

W-Band InP DDR IMPATTs for High Current Operation Near Avalanche Resonance

S. J. Mukhopadhyay(✉), S. Banerjee, and M. Mitra

Department of Electronics & Telecommunication Engineering, Indian Institute of Engineering Science & Technology, Shibpur, India
jana_sangeeta@yahoo.co.in

Abstract. Investigations are carried out on the space charge dependence of the negative resistances of avalanche and drift layers of double-drift region (DDR), indium phosphide (InP) and impact ionization avalanche transit time (IMPATT) diodes at high bias current levels near avalanche frequency by using computer simulation techniques. It is observed that DDR InP diodes behave like uniformly avalanching p-i-n diodes under the above situation, and the device negative resistance degrades sharply above a bias current density of 3×10^8 A/m^2.

Keywords: InP IMPATTs · Double-drift IMPATT diode · W-band · p-i-n diode

1 Introduction

In the last part of twentieth century, when the development of IMPATT devices gained its momentum, then basically fabrication was done by using germanium (Ge) and silicon (Si) as semiconducting substance. But at present, Ge IMPATTs are no longer in vogue. Due to very fast development of silicon technology, silicon diodes have received appreciable attention at that time, possessing the potency of generating high RF power with high conversion efficiency at mill-metric wave frequencies. According to some published reports, as discussed elsewhere [1–4], 94 GHz silicon DDR IMPATTs generate high RF power at high bias current density in pulsed mode corresponding to device operation near or above avalanche frequency. At very high current density, the mobile charge density is comparable with the immobile charge density and the device operates near avalanche resonance. Operation near avalanche resonance provides high impedance level and low losses due to compensation of capacitive and inductive currents. A space charge compensated p-i-n like Misawa mode of DDR p+pnn+ diode at high current levels near avalanche resonance has been suggested to explain high RF power output from the diode [5]. The IMPATT diodes were fabricated by III–V material, namely indium phosphide (InP) which was conceived as superior substance for high power, high efficiency power resource at millimeter wave frequency. Theoretical studies on InP IMPATT diodes show that these devices are capable of generating high RF power at high current densities in the millimeter wave frequency bands [6, 7]. Here in this paper, the high bias current density operation and the contributions of electron and hole drift layers toward

© Springer Nature Singapore Pte Ltd. 2021
N. R. Das and S. Sarkar (eds.), *Computers and Devices for Communication*,
Lecture Notes in Networks and Systems 147,
https://doi.org/10.1007/978-981-15-8366-7_38

negative resistance (R_{dn}, R_{dp}) have been shown for DDR InP-based IMPATT diode at 94 GHz. At higher bias current density, InP IMPATT behaves as p-i-n like Misawa mode; whereas, at lower current density it behaves as normal read mode.

2 Substance Parameters and Design Strategy

The ionization coefficients in InP have been extracted from the experimental data, as described by Kao and Crowell 1980 [8] for lower field range (2.5×10^7–5.0×110^7 V/m) and Umebu in [9] for the higher field range (5.0×110^7–8.0×110^7 V/m).

The negative differential mobility in the electron drift velocity versus electric field characteristics of InP as discussed elsewhere [10] has been taken in the computer simulation method.

All the other material parameters are obtained from electronic archive [11]. The material parameters of InP IMPATT are listed in Table 1 at 500 K.

Table 1 Material parameters for InP at 500 K

Carrier	Electric field range ($\times 10^7$, Vm^{-1})	$A_{n,p}$ ($\times 10^8$, m^{-1})	$B_{n,p}$ ($\times 10^8$, m^{-1})	$V_{sn,sp}$ ($\times 10^5$ ms^{-1})	$\mu_{n,p}$ (m^2V^{-1} s^{-1})	Diffusion constants $D_{sn,sp}$ ($\times 10^{-2}$ m^2s^{-1})
Electron	3.00–8.00	7.121	4.072	0.6	0.276	0.15
Hole	3.00–8.00	35.02	3.852	0.568	0.009	0.10

Double-drift n+-n-p-p+ IMPATT arrangement initially designed for specific frequency (f_d) from formula furnished by Sze and Ryder [12]. Based on the strategy highlighted by Roy et al. [13], the electric field is obtained. A small-signal simulation technique narrated elsewhere [14] is used to find out the small-signal parameters [15]. The structural and doping parameters of InP IMPATT designed for operation at 94 GHz are listed in Table 2.

Table 2 Design parameters for InP

Design frequency, f_d (GHz)	*n*-epitaxial layer thickness, W_n (μm)	*p*-epitaxial layer thickness, W_p (μm)	*n*-epitaxial layer doping, N_D ($\times 10^{23}$ m^{-3})	*p*-epitaxial layer doping, N_A ($\times 10^{23}$ m^{-3})	Substrate layer doping, N_{Sub} ($\times 10^{26}$ m^{-3})
94	0.339	0.339	1.6000	1.6000	1.000

3 Simulation Technique

For the simulation, one-dimensional DDR IMPATT structure has been taken, exhibited in Fig. 1.

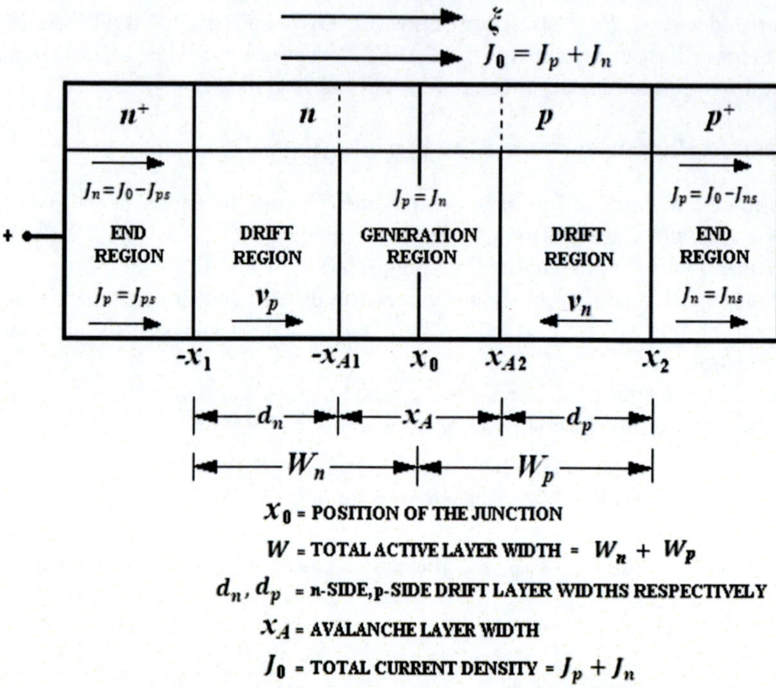

Fig. 1 One-dimensional model of DDR IMPATT device

3.1 DC and High-Frequency Simulation

Solving Poisson's, carrier continuity and current density equations, the electric field and current density profiles are received. DC simulation technique has been narrated elsewhere [13, 14].

DC parameters are used in the high-frequency simulation to obtain the high-frequency admittance profile of the device. The device impedance $Z(x, \omega)$ is given by $Z(x, \omega) = R(x, \omega) + jX(x, \omega)$; real part $R(x, \omega)$ and imaginary part $X(x, \omega)$ [15].

The distribution of negative resistance $R(\mathrm{x})$ and reactance $X(x)$ of DDR InP diode in the space charge layers can thus be obtained. The contributions of the avalanche layer and electron and hole drift layers to negative resistance (R_{a}, R_{dn}, R_{dp}) of a DDR diode are computed from numerical integration of $R(x)$ profiles over the respective layers. Similarly, the negative reactance contributions of different layers X_{a}, X_{dn}, X_{dp} are computed.

The total negative resistance and negative reactance are given by

$$R = R_{\mathrm{a}} + R_{\mathrm{dn}} + R_{\mathrm{dp}} = R_{\mathrm{a}} + R_{\mathrm{d}}$$

and

$$X = X_{\mathrm{a}} + X_{\mathrm{dn}} + X_{\mathrm{dp}} = X_{\mathrm{a}} + X_{\mathrm{d}}$$

4 Result and Discussion

At the initial phase, the bias current density was gradually varied to 1.04×10^8 Am^{-2}, and thereafter it was varied to the maximum limit of 8×10^8 Am^{-2}. The optimum frequency (f_0) for peak conductance is obtained for each current density from admittance studies. It is observed that f_0 is close to the design window frequency at $J_0 = 2 \times 10^8$ Am^{-2}.

The diode negative resistance is an integrated parameter of the device which determines the RF power output of the device. But a physical understanding of generation of RF power from the device can be obtained from layer to layer simulation of negative resistance in the active zone of the diode. The negative resistance of the avalanche zone (R_a) and that of the electron and hole drift zones (R_{dn}, R_{dp}) are computed for different bias current levels. Figure 2 shows the variation of R_a, R_{dn}, R_{dp} and $R_d(R_d = R_{dn} + R_{dp})$ with J_0. It is observed from Fig. 2 that initially both the magnitudes of R_d and R_a increase with the increase of J_0. Beyond $J_0 = 1.2 \times 10^8$ Am^2, magnitude of R_d decreases sharply while magnitude of R_a increases exponentially. It is worthwhile to note that magnitude of R_d is greater than the magnitude of R_a upto $J_0 = 1.4 \times 10^8$ Am^{-2} above which magnitude of R_d is less than that of magnitude of R_a. R_d versus J_0 and R_a versus J_0 curves intersect at a current density which is defined here as critical current density J_c. J_c is found to be 1.4×10^8 Am^{-2} from Fig. 2. Above the critical current density J_c, the contribution of the avalanche zone toward device negative resistance and RF power is more than that of the drift zone. The R versus J_0 curve shows that magnitude of R increases initially to attain a peak value at a current density 1.4×10^8 Am^{-2} close to J_c after which magnitude of R decreases with the increase of J_0.

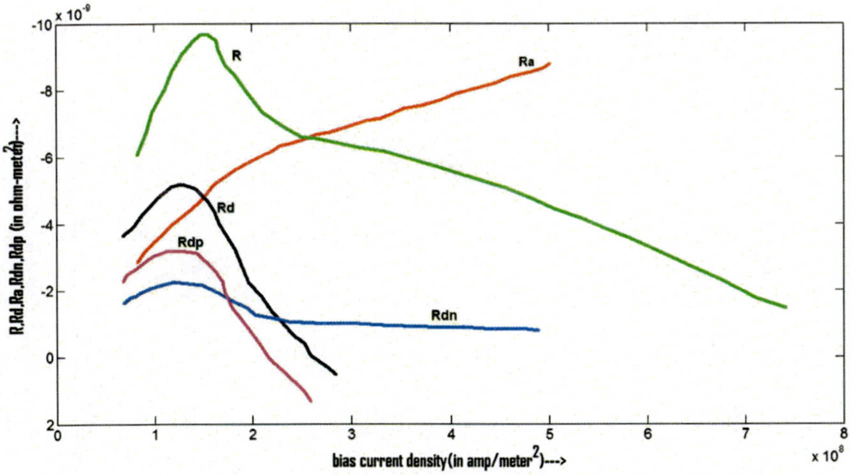

Fig. 2 Variation of R, R_a, R_d, R_{dn} and R_{dp} with bias current density

The contributions of electron and hole drift layers of a DDR InP diode at 94 GHz toward negative resistance (R_{dn}, R_{dp}) and their variations with J_0 have been separately presented in Fig. 2. The negative resistance of the diode at such high current densities degrades to a small value equal to the avalanche zone negative resistance as shown in Fig. 2. At such high current densities, the DDR diodes behave like uniformly avalanching p-i-n diodes since the diode negative resistance is only due to avalanche zone.

The small-signal avalanche resonance frequency (f_a) of the diode increases from 79 to 92 GHz when J_0 increases from 2×10^8 Am^{-2} to 3×10^8 Am^{-2} while the optimum frequency (f_0) increases from 98 to 107 GHz for the same increase of J_0 as shown in Fig. 3. Figure 4 exhibits the behavior of InP IMPATT similar to those of uniformly avalanching Misawa diode at higher bias current density.

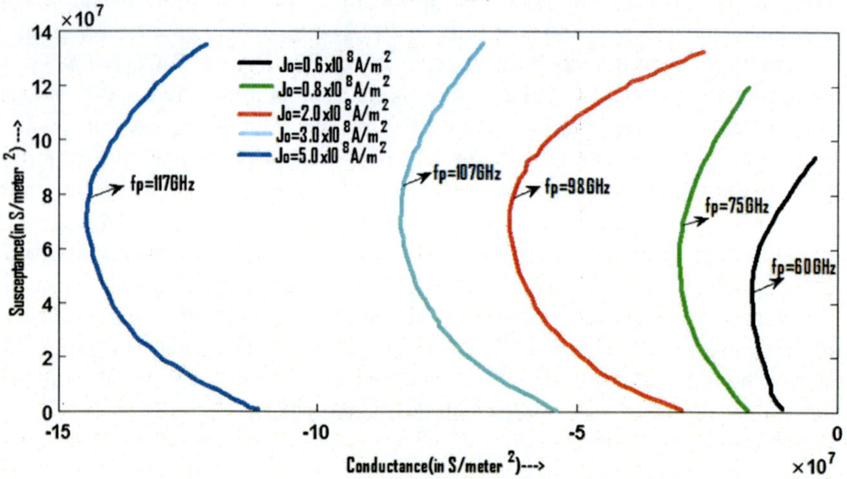

Fig. 3 Admittance characteristics of W-band DDR InP IMPATTs at different bias current density

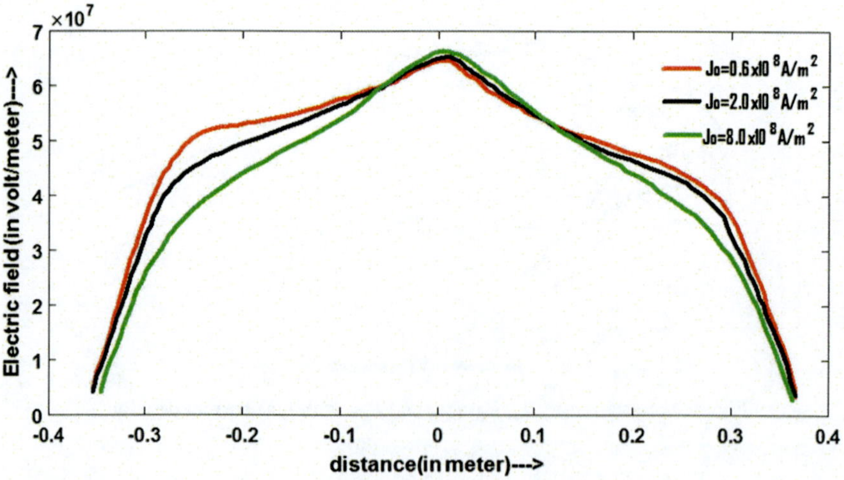

Fig. 4 Electric field profiles of W-band DDR InP IMPATTs at different higher bias current density

5 Conclusion

Here in this paper, space charge dependence of the negative resistances of avalanche and drift layers of double-drift region indium phosphide IMPATT diodes at high bias current density levels have been investigated. It is revealed that space charge compensated p-i-n like Misawa mode of DDR InP diode at high current density near avalanche resonance similar to that of DDR silicon diodes, as focused elsewhere [1, 2]. The device negative resistance degrades sharply above a bias current density of 3×10^8 Am^{-2}. At such high current densities, the avalanche zone becomes the prime contributor to the negative resistance of DDR diode.

Acknowledgements. The authors deeply feel and acknowledge the help and spontaneous support rendered by the authority of IIEST, Shibpur, Howrah, West Bengal, by making a suitable arrangement to perform the research work smoothly.

References

1. Behr, W., Luy, J.F.: High power operation mode of pulsed Impatt diodes. IEEE Electron Device Lett. **11**(5), 206–208 (1990)
2. Luy, J.F.: Impatt operation below the avalanche frequency. Electron Lett. **23**, 1960–1962 (1990)
3. Rolland, P.A., Dalle, C., Friscourt, M.R.: Physical understanding and optimum design of high-power millimeter wave pulsed Impatt diodes. IEEE Electron Device Lett. **12**(5), 221–223 (1991)
4. Chen, C., Mains, R.K., Haddad, G.I.: High power generation in Impatt devices in the 100–200 GHz range. IEEE Trans. Electron Devices **38**(8), 1701–1705 (1991)
5. Gaul, L., Claassen, M.: Pulsed high-power operation of p+pnn+−Avalanche diodes near Avalanche Resonance for mm-wave oscillators. IEEE Trans Electron Devices **41**(8), 1310–1317 (1994)
6. Mukherjee, R., Banerjee, J.P.: Avalanche and drift layer contributions to the negative resistance of millimeter wave p+nn+ InP IMPATT diode for different current densities. Int. J. Electron, **76**(4), pp. 589–600 (1994)
7. Banerjee, J.P., Pati, S.P., Roy, S.K.: High frequency characterization of double drift region InP and GaAs diodes. Appl Phys A **48**(5), 437–443 (1989)
8. Kao, C.W., Crowell, C.R.: Impact ionization by electrons and holes in InP. Solid State Electron **23**(8), 881–891 (1980)
9. Umebu, I., Chowdhury, A.N.M.M., Robson, P.N.: Ionization coefficients measured in abrupt InP junction. Appl. Phys. Lett. **36**(4), 302–303 (1980)
10. Kramer, B., Micrea, A.: Determination of saturated electron velocity in GaAs. Appl. Phys. Lett. **26**(11), 623–624 (1975)
11. Electronic Archive New semiconductor Materials, Characteristics and Properties. https://www.ioffe.ru/SVA/NSM/semicond
12. Sze, S.M., Ryder, R.M.: Microwave avalanche diodes. Proc IEEE Spec Issue Microw Semicond Devices **59**(8), 1140–1154 (1971)
13. Roy, S.K., Sridharan, M., Ghosh, R., Pal, B.B.: Computer method for the dc field and carrier current profiles in the IMPATT device starting from the field extremum in the depletion layer. In: Miller, J.H. (ed.) Proceedings of the 1st Conference on Numerical Analysis of Semiconductor Devices (NASECODE I), Dublin, Ireland, pp. 266–274 (1979)

14. Roy, S.K., Banerjee, J.P., Pati, S.P.: A Computer analysis of the distribution of high frequency negative resistance in the depletion layer of IMPATT diodes. In: Proc. 4th Conference on Numerical Analysis of Semiconductor Devices (NASECODE IV) (Dublin), Ireland, pp. 494–500 (1985)
15. Gummel, H.K., Blue, J.L.: A small-signal theory of avalanche noise in IMPATT diodes. IEEE Trans. Electron Devices **14**(9), 569–580 (1967)

Interaction of a Pair of Parabolic Self-similar Pulses in Nonlinearity Varying Chalcogenide Fibers (NVCFs)

Somen Adhikary[1(✉)], Binoy Krishna Ghosh[1], Roshmi Chatterjee[1], Dipankar Ghosh[2], Navonil Bose[3], and Mousumi Basu[1]

[1] Department of Physics, Indian Institute of Engineering Science and Technology, West Bengal, Shibpur, Howrah 711103, India
adhikary.somen@gmail.com, binoy.phys@gmail.com, roshmic987@gmail.com, mousumi.basu@gmail.com

[2] Department of Basic Science, MCKV Institute of Engineering, Liluah, Howrah 711204, West Bengal, India
ghosh.dip@gmail.com

[3] Department of Physics, Supreme Knowledge Foundation Group of Institutions, 1 Khan Road, Mankundu, Chandannagar, Hooghly 712139, West Bengal, India
navonil05@gmail.com

Abstract. Generation of a self-similar parabolic pulse (PP) in presence of virtually gained normal dispersion tapered chalcogenide fibers is presented in this work. The variations of group velocity dispersion and mainly nonlinearity along the fiber length produce the proposed nonlinearity varying chalcogenide fibers (NVCFs) which can induce sufficient amount of virtual gain leading to obtain desired PPs at smaller optimum length. While studying the interaction among such pair of PPs, it is observed that an oscillating pulse train is created in the overlapping zone. The so generated beat frequency can be tuned in a range of 575– 832 GHz when a lengthwise variation of fiber is considered.

Keywords: Group velocity dispersion (GVD) · Virtual gain · Nonlinearity variation · Interaction · Parabolic pulse (PP) · Chalcogenide fiber

1 Introduction

Parabolic pulse generation and propagation have drawn considerable research interests in recent times due to its potential applications as high-power lasers sources [1], super continuum generation [2] and many more. The interactions of the so generated parabolic pulses are very much fascinating as in the overlapping region of the pair of pulses some new features can be seen during the propagation. Literatures reveal that Peacock et al. have dealt with the amplification of a self-similar parabolic pulse pair [3], while C. Finot and G. Millot have experimentally demonstrated various interaction properties of two overlapping parabolic pulses while transmitting in a Raman fiber amplifier [4]. At the

© Springer Nature Singapore Pte Ltd. 2021
N. R. Das and S. Sarkar (eds.), *Computers and Devices for Communication*,
Lecture Notes in Networks and Systems 147,
https://doi.org/10.1007/978-981-15-8366-7_39

same time, Zhang et al. concentrated on the interaction of parabolic pulses in a dispersion decreasing optical fiber within the normal dispersion regime [5, 6]. The interaction is limited only to their overlapping parts and at the beginning of the interaction it produces a sinusoidal like beating [3, 5]. In our work, we have an aim to produce such beating in a nonlinearity varying chalcogenide fiber (NVCF). We have also shown how this interaction can be helpful for the generation of stable high-frequency (~GHz) oscillating pulse train which can change in presence of nonlinearity induced gain [7].

2 Proposals for Nonlinearity Varying Chalcogenide Fibers (NVCFs) for Parabolic Pulse Generation:

With an aim of designing a NVCF, we first design and optimize a dispersion increasing fiber (DIF) at an operating wavelength of 1.55 μm (λ_0). For this, we consider a step index profile where As_2S_5 chalcogenide glass, used as core and $Ge_{17}Ga_4Sb_{10}S_{69}$ glass as cladding with refractive indices at the core $n_1 = 2.2332$ [8] and cladding $n_2 = 2.2297$ [9], respectively. The relative refractive index difference (Δ) of the so designed fiber is sufficiently small (~0.0015), so that the weakly guiding approximation [10] remains valid and the propagation constant (β) of the fiber can be obtained by solving the scalar wave equation [11]. Typical variations of core radii in single mode regime along the fiber length help us to obtain lengthwise variation of group velocity dispersion (β_2), and as a result, we also obtain typical variations of nonlinear parameter (γ) along the fiber length as depicted in Fig. 1. It is to be mentioned that nonlinear variation rate is relatively large with respect to the lengthwise nominal variation of GVD. We consider three profiles corresponding to the variations of core radii, and we rename the fibers as nonlinearity varying chalcogenide fibers (NVCFs) with suffices 1, 2 and 3.

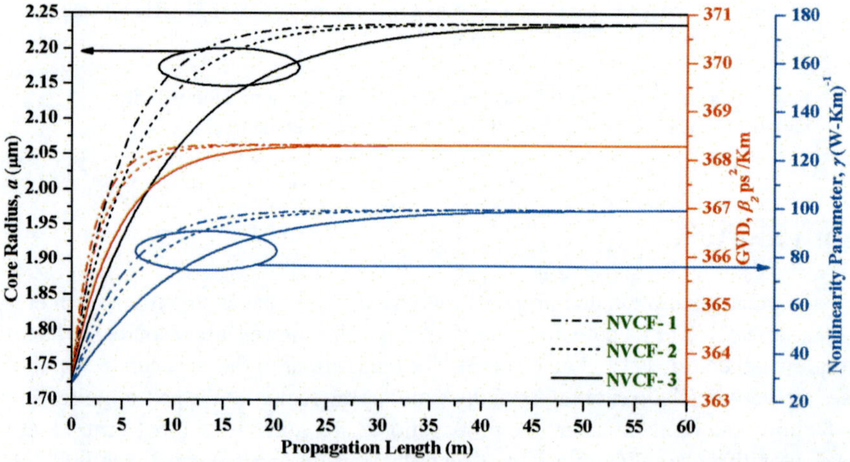

Fig. 1 Variations of core radius (a), GVD (β_2) and nonlinearity (γ) for different profiles of NVCFs along the propagation length (z)

It is well known that the evolution of an optical pulse during its propagation through an optical fiber is governed by the nonlinear effects of the material (γ) and group velocity dispersion (GVD) [10]. Their effects on the pulse envelope while propagating through the fiber can be described by nonlinear Schrodinger's equation (NLSE) [11] which is given by

$$i\frac{\partial A}{\partial z} = \frac{\beta_2(0)D(z)}{2}\frac{\partial^2 A}{\partial T^2} - \gamma(0)\Gamma(z)|A|^2 A + i\frac{(g-\alpha)}{2}A \quad (1)$$

where $A(z, T)$ represents the propagating pulse envelope after time T and at a distance z in propagating frame, $\beta_2(0)$ and $\gamma(0)$ are the initial values of GVD and nonlinear coefficient respectively, $D(z)$ and $\Gamma(z)$ represents the normalized variation of GVD and nonlinearity factor along length, while g represents the amount of the external gain applied to compensate the fiber loss α. It should be mentioned here that the NVCF we consider is a tapered passive fiber and produces virtual gain [7] itself contributing to the factor 'g,' due to lengthwise variation of dispersion as well as nonlinearity and no external gain is provided through the fiber. Here we consider a chirped Gaussian optical pulse as input with the following form as given by

$$A(0, T) = \sqrt{P_0}\exp(-\frac{(1+iC)T^2}{2T_0^2}) \quad (2)$$

where C is the initial chirp parameter. It can be seen that the input pulse evolves into a self-similar parabolic pulse (PP) after propagating through our so designed normal dispersion chalcogenide fiber and the perfection of the produced PP is estimated lengthwise by measuring the misfit parameter (μ) with respect to an exactly shaped parabolic pulse. The misfit parameter (μ) is estimated by the following relation [12]

$$\mu^2 = \frac{\int\left[|A|^2 - |A_{PP}|^2\right]^2 dT}{\int|A|^4 dT} \quad (3)$$

where $|A|^2$ represents the power profile of the propagating pulse and $|A_{PP}|^2$ is that of an ideal parabolic pulse. Now, initially chirped input Gaussian pulse with a peak power (P_0) ~ 8 W, pulse width (T_0) ~ 2 ps and chirp value (C) = 2 is fed into all the proposed fibers and lengthwise variation of the misfit parameter toward the parabolic pulse (PP) generation are estimated. The optimum length for PP generation (L_{Opt}) and its stability within the 60 m of fiber length (L_{St}) in each case are noted down in Table 1.

3 Propagation of Parabolic Pulse Pair Through Nonlinearity Varying Chalcogenide Fibers (NVCFs)

With an aim to study the interaction between a pair of parabolic pulses while propagating through our proposed NVCFs, we consider a pair of input Gaussian pulses separated from each other by an amount $2q_0$. The separation $2q_0$ in time domain is chosen in such a way that each of the individual pulses should achieve the parabolic shape before they interact. Here we take $q_0 = 40$ ps such that the Gaussian pulses transforms into

Table 1 Optimum length of PP generation and its stability through different fibers

Fiber profile	Optimum length (L_{Opt}) (m)	Stability length (L_{St}) (m)
NVCF 1	31.0	12.6
NVCF 2	30.4	29.6
NVCF 3	36.1	23.9

parabolic form before interaction. It can be shown that the pulses evolve independently to produce parabolic profiles and started to interact only when their ends start to overlap. During interaction, the overlapping of the two pulses leads to a beating frequency in the intensity profile [4]. Figure 2 represents the different snapshots of evolution and interaction between the pulses at different positions of the propagation lengths.

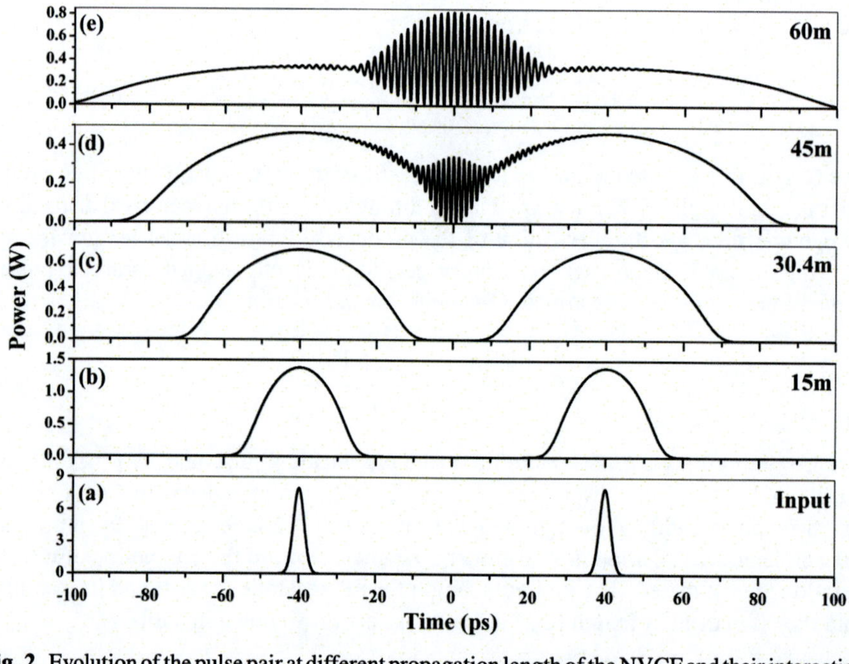

Fig. 2 Evolution of the pulse pair at different propagation length of the NVCF and their interaction: **a** input Gaussian pulses with their centers $q_0 = 40$ ps apart, **b** the pulses at an intermediate stage after traveling $z = 15$ m, **c** the pulses changed to parabolic shape after traveling $L_{opt} \sim 30.4$ m, **d** pulses started interacting when their tails overlapped at $z = 45$ m, **e** the interaction of parabolic pulses at $z = 60$ m

After taking a closer look into the interacting region by zooming into the time scale, we have found that the overlapping part actually produces high frequency beats and the frequencies of those beats have a tendency to change along the NVCF length. By

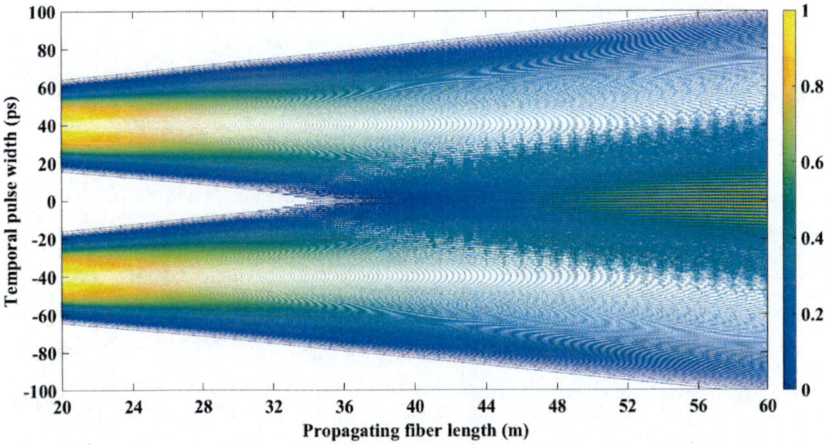

Fig. 3 Lengthwise progression of the pulse pair through the proposed NVCF

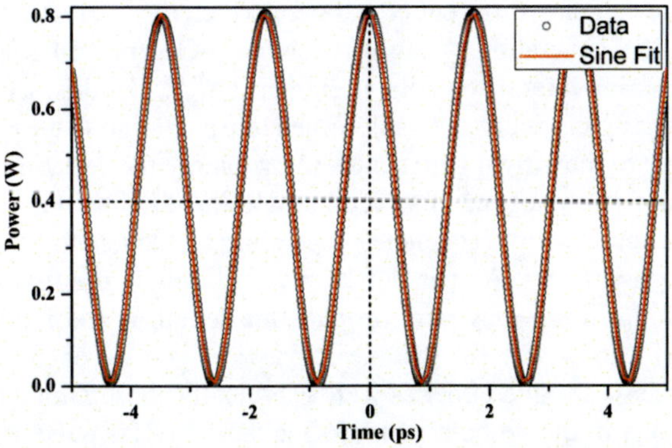

Fig. 4 Temporal power profile of the generated beat frequency and its sinusoidal fit

this technique, we have calculated the frequency of the pulse trains for different NVCF profiles throughout its propagation up to 60 m. After propagating a length of 37.5 m, the interaction becomes prominent and the interaction frequency for different profiles lies within a range of ~832 to 804 GHz. The effect of different values of nonlinearity induced virtual gains in three NVCFs leads to obtain the above change in beat frequency. The frequency decreases as the pulse pair propagates through those fibers, and at 60 m, the frequency of interaction for all three profiles becomes almost same (~575 GHz). This is due to the fact that at ~60 m, all the profiles merge into single values of GVD or nonlinearity. Figure 5 gives a clear idea about the required propagation length for generation of a particular frequency generation within ~832 to 575 GHz. Therefore, one can choose a typical fiber length for a particular NVCF from this optimization for the purpose of generating an oscillating pulse train at a particular frequency. In this aspect, it is to be mentioned that the interaction of parabolic pulse in the NVCFs is studied for the first time.

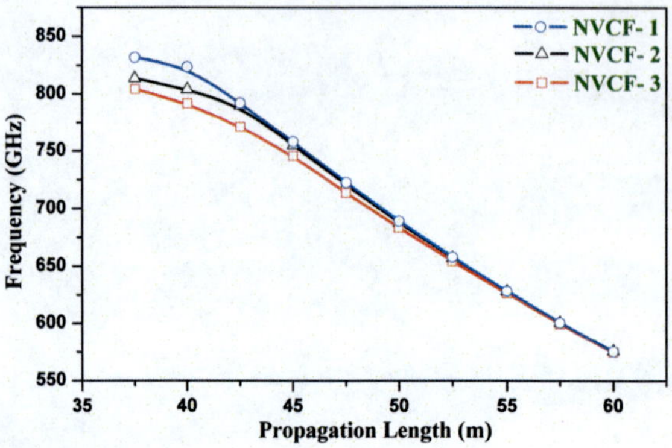

Fig. 5 Variation of oscillation frequency (GHz) with propagation length (m) for three NVCFs

4 Conclusions

Our work presents designing different NVCFs for which the interaction between a pair of PPs is shown. The generation of high-frequency (~GHz) oscillating pulse train in between the overlapping interaction zone is shown as well. For this, we consider a pair of Gaussian pulses as inputs which are transformed into parabolic shape after reaching the optimum length due to the virtual gain achieved by mainly the variation of nonlinearity of the fibers. Interaction among the PPs becomes dominant as their tails starts to overlap and generates different beating frequencies depending on the length of propagation and profile of the designed fibers. As per our knowledge, this is the first time where the variation of interaction frequency with propagation length is studied in such type of NVCFs. The study can be helpful for one in view of generation of a required high frequency. Thus, the so designed fiber as well as typically chosen input pulses can have potential applications in the domain of high-frequency (GHz) pulse generators along with many other applications of PP in the domain of optical 2R regenerators, optical retiming or time-domain Fourier transformation approach, etc.

References

1. Ruehl, A., Prochnow, O., Wandt, D., Kracht, D., Burgoyne, B., Godbout, N., Lacroix, S.: Dynamics of parabolic pulses in ultrafast fiber laser. Opt. Lett. **31**, 2734–2736 (2006)
2. Tamura, K.R., Kubota, H., Nakazawa, M.: Fundamentals of stable continuum generation at high repetition rates. J. Quantum Electron. **36**, 773–779 (2000)
3. Peacock, A.C., Kruglov, V.I., Thomsen, B.C., Harvey, J.D., Fermann, M.E., Sucha, G., Harter, D., Dudley, J.M.: Generation and interaction of parabolic pulses in high gain fiber amplifiers and oscillators. In: Optical Fiber Communication Conference (OFC), WP4, vol. 13 (2000)
4. Finot, C., Millot, G.: Interaction between optical parabolic pulses in a Raman fiber amplifier. Opt. Express. **13**, 5825–5830 (2005)
5. Zhang, Q.F., Wu, L.M., Tang, X.C., Wang, G.T., Deng, Y.H.: Interaction between parabolic pulses in a dispersion-decreasing fiber. Optik. **121**, 517–520 (2010)
6. Zhang, Q.F., Gao, J.: Generation of excellent self-similar pulses in a dispersion -decreasing fiber. Optik. **122**, 1753–1756 (2011)

7. Ghosh, D., Basu, M., Sarkar, S.: Generation of self-similar parabolic pulses by designing normal dispersion decreasing fiber amplifier as well as its staircase substitutes. J. Lightwave Technol. **27**, 3880–3887 (2009)
8. Salem, A.B., Diouf, M., Cherif, R., Wague, A.D., Zghal, M.: Ultraflat-top mid infrared coherent broadband supercontinuum using all normal As_2S_5-borosilicate hybrid photonic crystal fiber. Opt. Eng. **55**, 066109 (2016)
9. Kohoutek, T., Mizuno, S., Suzuki, T., Ohishi, Y., Matsumoto, M., Misumi, T.: Third-harmonic generation measurement of nonlinear optical susceptibility $\chi^{(3)}$ of Ge-Ga-Sb-S chalcogenide glasses proposed for highly nonlinear photonic fibers. J. Opt. Soc. Am. B. **28**, 298–305 (2011)
10. Agrawal, G.P.: Nonlinear Fiber Optics, 3rd edn. Academic Press (2001)
11. Ghatak A, Thyagarajan K (1999) Introduction to Fiber Optics. Cambridge University Press
12. Ghosh, B.K., Ghosh, D., Basu, M.: Prospective use of a normally dispersive step-index chalcogenide fiber in nonlinear pulse reshaping. Appl. Opt. **57**, 3348–3356 (2018)
13. Ghosh, B.K., Ghosh, D., Basu, M.: Potential use of nonlinearity induced virtual gain on parabolic pulse formation in highly nonlinear tapered fiber system. J. Opt. **21**, 045503 (2019)
14. Chowdhury, D., Ghosh, D., Bose, N., Basu. M.: Parabolic pulse regeneration in normal dispersion-decreasing fibers and its equivalent substitutes in presence of third-order dispersion. Appl. Phys. B. **125**, 106 (2019)

Establishment of the Validity of Time Transformation Approach to Study Pulse Compression in Silica-Based Single Mode Optical Fibers

Roshmi Chatterjee(✉), Binoy Krishna Ghosh, Debasruti Chowdhury, Somen Adhikary, and Mousumi Basu

Department of Physics, Indian Institute of Engineering Science and Technology, Shibpur, Howrah, West Bengal 711103, India
{roshmic987,binoy.phys,debasruti.chowdhury,adhikary.somen, Mousumi.basu}@google.com

Abstract. To achieve higher-order soliton pulse compression, single-mode anomalous dispersion silica-based fibers are chosen. Time transformation (TT) approach and symmetrized split step Fourier method (SSFM) are compared here to obtain the compressed optical pulses. The results obtained from TT method match quite well with the results of symmetrized SSFM, which validates the TT method in the domain of pulse compression. Variations of external chirp and input pulse width are also studied to achieve high compression factor in single-mode fibers within a shorter optimum length.

Keywords: Time transformation · Pulse compression

1 Introduction

Ultrashort optical pulses have a number of potential applications in different domains. One of them is in optical communication [1] which is leading to the growth of interest in the generation of high peak power ultrashort pulses through relatively smaller distance. First experimental demonstration of pulse compression was done by Mollenauer et al. [2]. The limitation for the use of nonlinear Schrödinger equation (NLSE) and the finite difference time domain method (FDTD) for ultrashort pulses [3] were controlled using time transformation (TT) approach for propagation of optical pulse. Xiao et al. in their papers presented an approach of pulse propagation through dispersive Kerr nonlinear medium [4] and included the delayed Raman response to study propagation of ultrashort pulses [5]. Similar approach has also been applied here for pulse compression when the nonlinear dispersive as well as attenuating fibers are considered.

© Springer Nature Singapore Pte Ltd. 2021
N. R. Das and S. Sarkar (eds.), *Computers and Devices for Communication*,
Lecture Notes in Networks and Systems 147,
https://doi.org/10.1007/978-981-15-8366-7_40

2 Time Transformation (TT) Method to Study Optical Pulse Compression

The TT approach is used to study the ultrashort pulse propagation in.

dispersive Kerr medium. For ultrashort pulses, the higher-order nonlinear effects such as Raman response function play a vital role ($T_o < 1$ ps) [6]. Involving the medium's nonlinear response function, the refractive index $n(t)$ can be written as [5]

$$n(t) = n_0 + n_2\left((1 - f_R)I(t) + f_R \int_{-\infty}^{t} h_R(t - t_1)I(t_1)\mathrm{d}t_1\right) \tag{1}$$

Here n_0 is the linear part of refractive index and n_2 be the nonlinear index coefficient. The total transit time that arises due to the linear (T_{Linear}) and nonlinear part of the refractive index of the medium as the pulse propagates through the fiber. The nonlinear transit time delay arises due to transit time due to the Kerr-type nonlinearity and due to delayed part of the nonlinearity.

$$t_t = \frac{n(t_1)L}{c} \tag{2}$$

The output electric field (E_{out}) is related to the input electric field (E_{in}) at any instantaneous time t_1 by using the impulse response function $\rho(t_1, t_{\text{out}})$ while the pulse propagates through dispersive Kerr optical medium [4, 5]

$$E_{\text{out}}(t_{\text{out}}) = \int_{-\infty}^{\infty} \rho(t_{\text{out}}, t_1)E_{\text{in}}(t_1)\mathrm{d}t_1 \tag{3}$$

Using the approach of time transformation method, the above equation modifies to

$$E_{\text{out}}(t_{\text{out}}) = \int_{-\infty}^{\infty} \rho(t_{\text{out}} - t_t)E_{\text{in}}{}^{1}(t_t)\frac{\mathrm{d}t_t}{\left(1 + \frac{\mathrm{d}T_K}{\mathrm{d}t_1} + \frac{\mathrm{d}T_R}{dt_1}\right)} \tag{4}$$

where T_K and T_R are delays induced by Kerr effect and Raman. If the GVD (β_2) induced chirp and external chirp (C) is of opposite sign i.e. when product of GVD and external chirp becomes negative, pulse compression is possible [7]. This approach is used to compress pulse using different anomalous and normal dispersion fibers. The pulse compression factor (CF) is defined as

CF = FWHM of the input pulse/FWHM of the output pulse.

The length where the width of the compressed pulse becomes minimum with maximum CF can be defined as optimum fiber length. The well-known nonlinear Schrödinger equation (NLSE) is used in describing the optical pulse propagation [6],

$$\frac{\partial A}{\partial z} + \frac{i\beta_2}{2}\frac{\partial^2 A}{\partial T^2} - \frac{\beta_3}{6}\frac{\partial^3 A}{\partial T^3} + \frac{\alpha}{2}A = i\gamma\left(|A|^2 A - T_R A\frac{\partial |A|^2}{\partial T}\right) \tag{5}$$

where $A(z, T)$ is the propagating pulse envelope, z is the distance of pulse propagation. β_2, β_3, γ, and α are the GVD, third-order dispersion (TOD), nonlinear coefficient and gain over loss of the fiber; T is the time in the co-propagating frame; and T_R is the Raman response function. Soliton number $N = \frac{\gamma P_0 T_0^2}{|\beta_2|}$ plays an important role in the perspective of pulse compression [7].

3 Production of Compressed Pulses by TT as Well as SSFM Method

Here time transformation (TT) approach and symmetrized split step Fourier method (SSFM) both are used to obtain the optical pulses at the output end in its reshaped form. Two silica-based anomalous dispersion fibers (F1, F2) are chosen with different values of GVD, TOD, and γ at the operating wavelength of 1550 nm as depicted in Table 1. A loss of 0.2 dB/km is considered in all the cases.

Table 1 Typical values of fiber parameters at 1550 nm

Choice of silica fiber	Optimized V value	β_2(ps^2/km)	β_3(ps^3/km)	γ (W Km)$^{-1}$
F1	1.351	−10.55	0.11	1.09
F2	1.849	−5.28	0.08	3.34

Here the input is considered as a Gaussian pulse of width 950 fs and power 150 W. A comparative study of pulse compression using TT approach and SSFM for fiber F1 and Fiber F2 is shown in Fig. 1a, b, respectively, by plotting the temporal pulse shape. For the Fiber F1, the FWHM of the output compressed pulse is 145.4 and 144.87 fs as obtained from SSFM and TT approach, respectively, which results to a CF ~ 10.92 for $N = 3.75$ as the pulse propagates through a distance of 21.22 m. A CF ~ 30 at a propagating length 12.22 m is obtained for Fiber F2. It is seen from the figure that result obtained from the above-mentioned two methods matches quite well. Figure 2 shows the variation of CF with propagating length for the two fibers. From Fig. 2, it is clearly observed that the variation of CF is rapid from 9 m in case of Fiber F2, whereas for Fiber F1 the change in CF is quite slow. As obtained from the comparative study, the value of CF for Fiber F2 is the best in respect of higher value of CF and shorter values of optimum fiber length.

Figure 3 shows the variation of CF with input pulse width keeping the input energy of the pulse constant. It is seen from the graph that the CF decreases on decreasing the pulse width though maximum CF is attained at a short optimum length for shorter pulses. A better compression with CF ~ 14.14 at an optimum length of 23.22 m is achieved when the input pulse width is 800 fs and $E_{in} = 70.89$ pJ. Our study reveals that in addition to the GVD and SPM induced chirp, external chirp also plays a vital role in pulse compression. For Fiber F2, the variation of chirp and optimum fiber length is shown for a 950 fs Gaussian input pulse having peak power 150 W in Fig. 4. It is observed from the graph that on applying a positive external chirp of +2, the CF reaches a value of 25.95

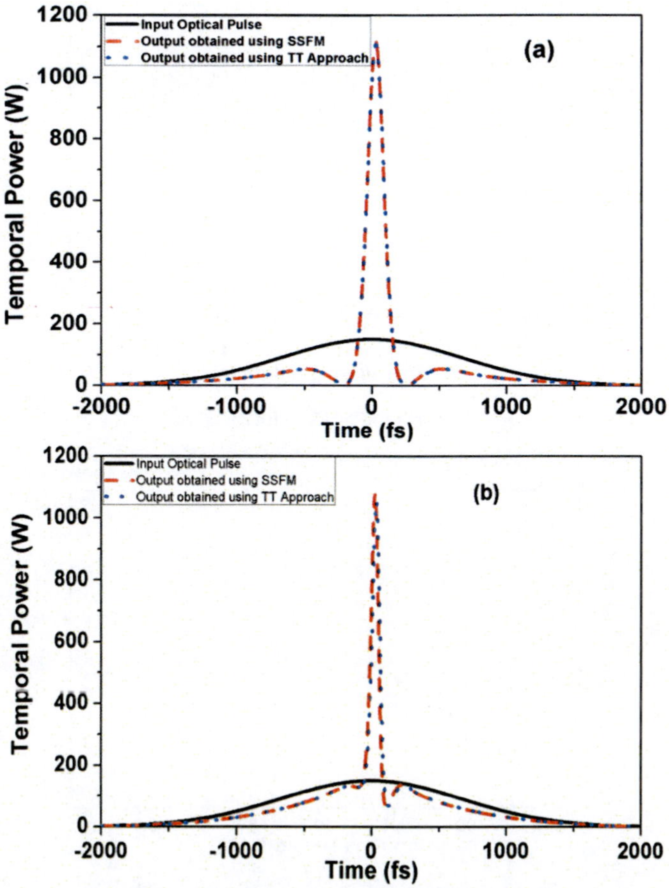

Fig. 1 Temporal power profile for input Gaussian pulse for **a** fiber F1, **b** fiber F2

at a distance of 11.1 m. The CF increases, and the optimum fiber length decreases with the increase of external positive chirp and reach a value of 34.35 at a distance of 9.4 m only when +6 external chirp is applied. Using TT approach, comparable compression factors can be achieved [8, 9].

4 Conclusion

Single-mode silica-based anomalous dispersion optical fibers are chosen with an aim to study pulse compression. In the domain of pulse compression, TT approach is a contemporary approach. The pulse propagation through a dispersive Kerr medium is studied using the symmetrized split step Fourier method (SSFM) as well as time transformation (TT) method. In SSFM, the well-known NLSE can be solved numerically and the results obtained from SSFM matches quite well with that from TT approach. For the chosen anomalous dispersion fibers, pulse compression is possible when no external chirp is

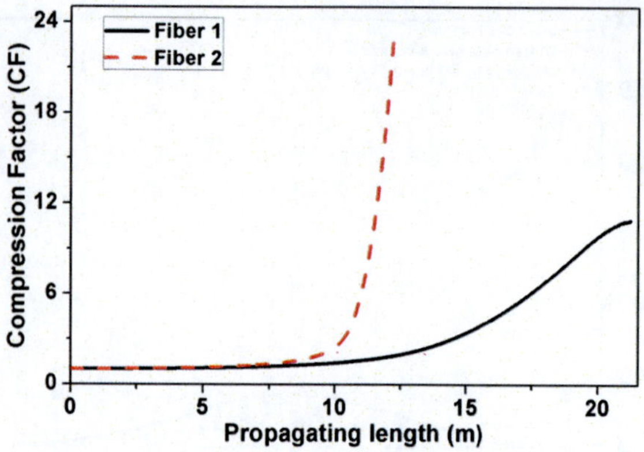

Fig. 2 Variation of CF with distance

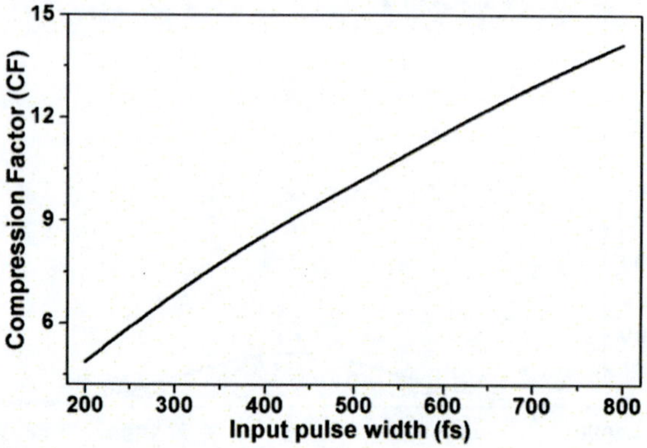

Fig. 3 Variation of CF with input pulse width

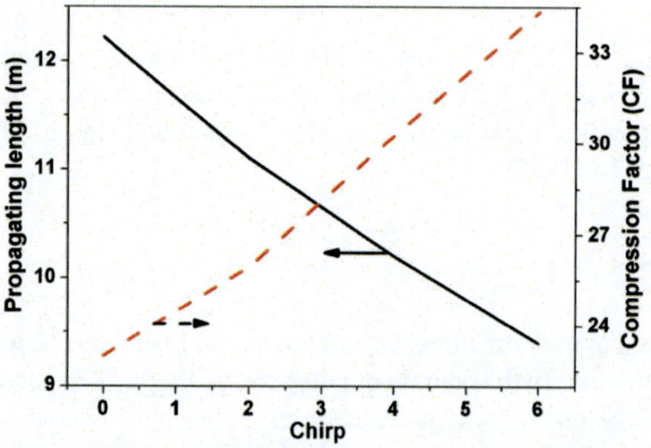

Fig. 4 Variation of CF and optimum fiber length with external chirp

applied and the CF increases with the application of external positive chirp. The TT approach is used for different values of pulse parameters to study the variation of CF with the input pulse width. The role of external chirp in pulse compression is also shown. Relatively high CF ~ 34.35 is attained at an optimum length of 9.4 m using external chirp parameter (C ~ +6). As per our knowledge, this approach is used for the first time in the domain of pulse compression in an anomalous dispersion fiber.

References

1. Nakajima, K., Sillard, P., Richardson, D., Li, M.J., Essiambre, R.J., Matsuo, S.: Transmission media for an SDM-based optical communication system. IEEE Commun. Mag. **53**(2), 44–51 (2015)
2. Mollenauer, L.F., Stolen, R.H., Gordon, J.P.: Experimental observation of picosecond pulse narrowing and solitons in optical fibers. Phys. Rev. Lett. **45**(13), 1095–1098 (1980)
3. Hile, C.V., Kath, W.L.: Numerical solutions of Maxwell's equations for nonlinear-optical pulse propagation. J. Opt. Soc. Am. B **13**(6), 1135–1145 (1996)
4. Xiao, Y., Maywar, D.N., Agrawal, G.P.: New approach to pulse propagation in nonlinear dispersive optical media. J. Opt. Soc. Am. B **29**(10), 2958–2963 (2012)
5. Xiao, Y., Maywar, D.N., Agrawal, G.P.: Time-transformation approach to pulse propagation in nonlinear dispersive media: Inclusion of delayed Raman nonlinearity. Phys. Rev. A **87**(6), 0638161–0638166 (2013)
6. Agrawal, G.P.: Non-linear Fiber Optics. Elsevier Academic Press (2013)
7. Agrawal, G.P.: Applications of Non-linear Fiber Optics. Academic Press (2001)
8. Rehan, M., Kumar, G., Rastogi, V., Korobko, D.A., Sysolyatin, A.A.: Compression of femtosecond pulses in a wide wavelength range using a large mode area tapered fiber. Laser Phys. **29**(2), 0251041–0251049 (2019)
9. Mollenauer, L.F., Stolen, R.H., Gordon, J.P.: Extreme picosecond pulse narrowing by means of soliton effect in single-mode optical fibers. Opt. Lett. **8**(5), 289–291 (1983)

A Tunable Dual-Band Metamaterial Absorber for Terahertz Applications

N. B. Nikhil[1], Bhavana R. Nair[1], Ancilla Philip[1], Nilotpal[2], Anu Mohamed[1], Chinmoy Saha[3], and Somak Bhattacharyya[2(✉)]

[1] Department of Electronics and Communication Engineering, Government Engineering College Bartonhill, Trivandrum 695035, India
nikhilnageshsnpuram@gmail.com,
{bhavananair597,m.anu.in}@ieee.org, ancilla.philip@gmail.com

[2] Department of Electronics Engineeering, Indian Institute of Technology (BHU), Varanasi, Uttar Pradesh 221005, India
{nilotpal.rs.ece17,somakbhattacharyya.ece}@iitbhu.ac.in

[3] Department of Avionics, Indian Institute of Space Science and Technology, Valiamala, Thiruvananthapuram 695547, India
csaha@ieee.org

Abstract. A tunable dual-band metamaterial absorber for terahertz range operation is illustrated in this paper. The tunability can be attained using the variable resistive property of diodes in reverse bias operation. The absorption bands are observed at ~45 μm (6.719 THz) and ~80 μm (3.765 THz). The structure exhibits variation in absorptivity with the variation in the resistance of diode. The proposed structure is with the dimensions having a periodicity of (~λ/4) and the thickness of (~λ/53). The tunable dual-band metamaterial absorber can have potential applications in the area of sensing, stealth technology, etc.

Keywords: Terahertz · Metamaterials · Tunable · Variable resistance

1 Introduction

Metamaterials are artificially engineered effective medium made of metallic inclusions which are not readily available in nature [1]. Though metamaterial research has matured over last decade, thanks to pioneering works by various research groups across the globe [2–6], design and application of metamaterial in the THz band is less explored. Metamaterial absorbers are engineered surface made of planar periodic structures capable of absorbing electromagnetic radiation over a particular band of frequency. Absorbers have the ability to block both reflected and transmitted component of the incident radiation exposed to it. This feature of the absorbers is exploited in various electromagnetic applications, like stealth technology [7], sensing [8], imaging [9]. Metamaterial absorbers usually consist of a three-layered structure [10]. An upper metallic patch designed so that the incident electric field strongly couples with the structure. The middle dielectric layer pairs the top metallic structure with the continuous reflective ground plane.

© Springer Nature Singapore Pte Ltd. 2021
N. R. Das and S. Sarkar (eds.), *Computers and Devices for Communication*, Lecture Notes in Networks and Systems 147,
https://doi.org/10.1007/978-981-15-8366-7_41

Depending on the bandwidth of the absorption band, metamaterial absorbers are classified into (1) narrowband [11], (2) wideband [12], and (3) multi-band [13]. Most of these absorbers reported in open literature works over fixed frequency spectrum. Tunable metamaterial absorbers, having frequency as well as absorption tunability, provide more degree of freedom over passive metamaterial absorbers working on a fixed spectrum [14]. Major techniques that are being conceived to realize such tunable absorbers are use of nematic liquid crystals, embedding active diodes, and incorporating polarization tunable metamaterials [15–17].

In this paper, a dual-band metamaterial absorber with tunable absorptivity is proposed. The proposed design is realized by printing shaped metallization on both sides of a ZnO laminate having $\varepsilon_r = 8.5$. Several appealing features of ZnO, such as high band-gap, high breakdown voltage, and high mobility, make it an attractive choice as the substrate for THz metamaterial applications [18].

2 Design and Simulations

The top and side views of a unit cell of the proposed dual-band metamaterial structure along with the directions of electric field ($E(x)$), magnetic field ($H(x)$), and propagation ($k(z)$) are shown in Fig. 1a, b. The unit cell consists of three layers: (1) the top layer consisting of a gold patch of thickness $t_1 = 0.01$ μm, (2) the middle Zinc Oxide (ZnO) layer ($\varepsilon_r = 8.5$) of thickness $h = 1.5$ μm, and (3) bottom continuous metallic (gold) layer of thickness $t_2 = 0.01$ μm. The diode embedded between the two adjacent metallic patches on the top layer, as shown in Fig. 1a, tunes the structure. The other optimized geometric parameters of the structure in Fig. 1 are: $a = 20$ μm, $b = 19$ μm, $g = 1$ μm, and $w = 2.6$ μm. The equivalent circuit of the diode in reverse biased condition is shown in Fig. 1c in which the variable resistance R is a function of the applied reverse biased voltage applied through the metallic patches itself.

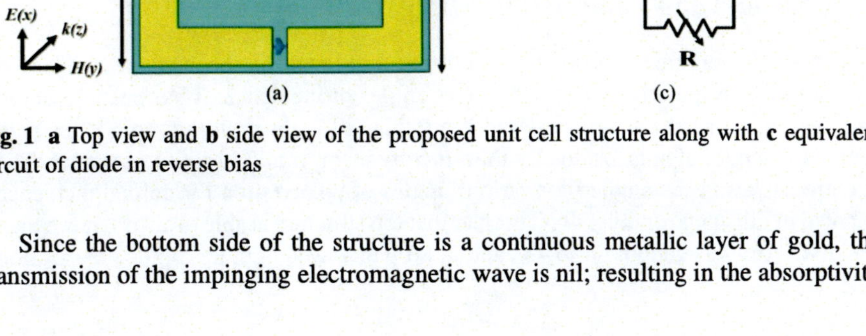

Fig. 1 **a** Top view and **b** side view of the proposed unit cell structure along with **c** equivalent circuit of diode in reverse bias

Since the bottom side of the structure is a continuous metallic layer of gold, the transmission of the impinging electromagnetic wave is nil; resulting in the absorptivity

(*A*) given by Eq. (1). The absorptivity can thus be maximized by minimizing the reflection coefficient (S_{11}).

$$A = 1 - |S_{11}|^2 \tag{1}$$

3 Results

The proposed structure has been simulated and analyzed using the frequency domain solver in CST Microwave Studio software. The reflection coefficient and absorptivity responses of the structure for various values of *R* under normally incident plane wave have been studied and illustrated in Fig. 2a, b, respectively. It has been observed that the absorptivity is greater than 90% for the value $R = 1500\ \Omega$ at two wavelengths of 45 and 80 μm. It has been observed that with the decrease of R, the absorptivity of the structure has been reduced.

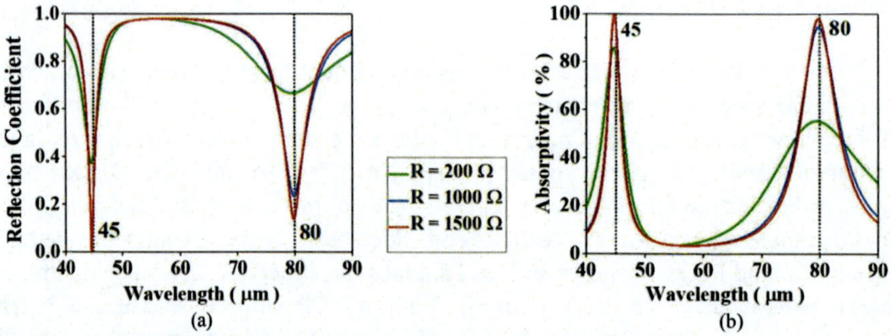

Fig. 2 **a** Simulated reflection coefficient and **b** absorptivity responses of the proposed structure for different resistance (*R*) values

The effective electromagnetic properties have been calculated for different *R* as shown in Table 1. It has been observed that the retrieved effective permittivity and permeability are nearly equal to each other for $R = 1500\ \Omega$, thereby making minimum reflection and enhancing the absorption within the structure itself. For other values of *R*, they are quite separated to each other.

4 Study of the Response of the Proposed Structure Under Oblique Incidences and Polarization Angle Variations

The proposed structure whose unit cell illustrated in Fig. 1 has been studied under TE and TM modes of polarizations for different angles of incidence. It has been observed from Fig. 3a, b that the structure shows more than 80% absorption up to 60° for both TE and TM modes of polarization for the structure incorporating 1500Ω resistance. The structure has also been studied for various angles of polarization for normal incidence as shown in Fig. 4a. From Fig. 4b, it is clear that the structure is able to absorb more than 75% at 45 μm wavelength up to 45° and at 80 μm wavelength up to 30° polarization angle.

Table 1 Comparison of real and imaginary parts of effective permeability and permittivity at the resonant wavelengths for $R = 200\ \Omega$, $R = 1000\ \Omega$, and $R = 1500\ \Omega$

Resistance (Ω)	Wavelength (μm)	Real part		Imaginary part	
		ε_{eff}	μ_{eff}	ε_{eff}	μ_{eff}
200	45	1.5683	1.2658	4.7009	18.7440
	80	26.7558	2.8165	59.5815	3.8994
1000	45	1.0161	0.9806	8.7219	10.2929
	80	0.7934	1.0746	28.0286	10.1247
1500	45	0.996	1.0004	9.5190	1.4310
	80	0.8805	1.0688	22.2929	12.8148

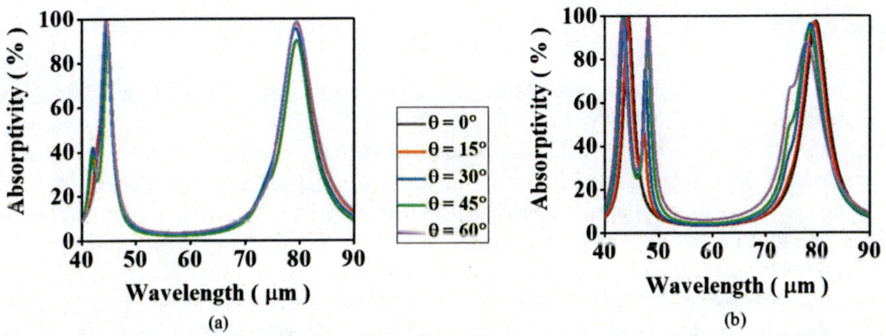

Fig. 3 Simulated results of absorptivity obtained for oblique incidence responses of $R = 1500\ \Omega$ resistance under **a** TE mode of polarization and **b** TM mode of polarization

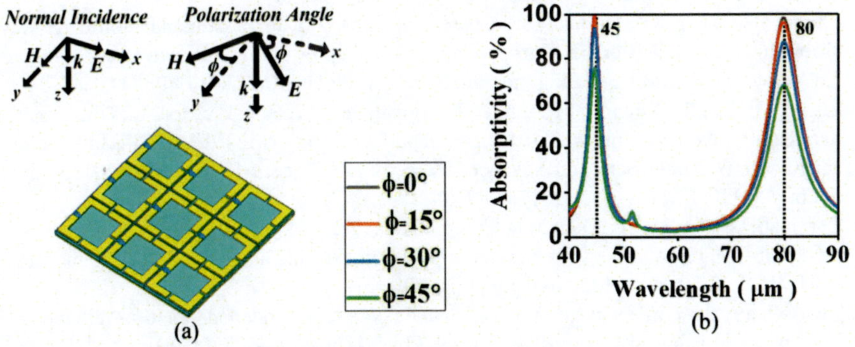

Fig. 4 **a** Setup of the proposed structure and **b** the simulated results of absorptivity obtained for different polarization angles under normal incidence for 1500 Ω resistance

5 Conclusion

A tunable dual-band metamaterial absorber has been proposed in this paper for various applications in terahertz frequency range. The design offers maximum absorptivity of greater than 90% over two bands at 45 μm (6.719 THz) and 80 μm (3.765 THz). The absorptivity of the proposed design can be varied from about 50% to a range of 90% by varying the resistance of the diode which is dependent on the applied bias. The impedance of the structure is perfectly matched with the free space when the value of resistance is high, thereby increasing the absorptivity. As the resistance values get decreased, the mismatch increases, and the absorptivity of the structure gets reduced. The proposed structure offers more than 80% absorptivity up to 60° incident angles under both TE and TM modes of polarizations. The proposed structure is ultra-thin in nature as the thickness is ~λ/53.

References

1. Caloz, C., Itoh, T.: Electromagnetic Metamaterials: Transmission Line Theory and Microwave Applications
2. Ziolkowski, R.W., Kipple, A.D.: Application of double increase the power radiated by electrically small antennas. IEEE Trans. Antennas Propag. **51**, 2626–2640 (2003)
3. Marquez, R., Mesa, F., Martel, J., Medina, F.: Comparative analysis of edge- and broadside-coupled split ring resonators for metamaterial design-theory and experiments. IEEE Trans. Antennas Propag **51**, 2572–2581 (2003)
4. Saha, C., Siddiqui, J.Y.: Theoretical model for estimation of resonance frequency of rotational circular split-ring resonators. Electromagnetics **32**(6), 345–355 (2012)
5. Saha, C., Siddiqui, J.Y., Antar, Y.M.M.: Multifunctional Ultrawideband Antennas: Trends, Techniques and Applications. CRC Press (2019)
6. Marqués, R., Martin, F., Sorolla, M.: Metamaterials with Negative Parameters: Theory, Design, and Microwave Applications, vol. 183. Wiley (2011)
7. Zhang, C., Cheng, Q.: Optically transparent metamaterial for broadband millimeter wave absorption. In: 2017 10th UK-Europe-China workshop on millimetre waves and Terahertz technologies (UCMMT). IEEE. https://doi.org/10.1109/UCMMT.2017.8068475
8. Zhang, W., Lan, F., Xuan, J., Mazumder, P., Aghadjani, M., Yang, Z., Men, L.: Ultrasensitive dual-band terahertz sensing with metamaterial perfect absorber. In: IEEE MTT-S International Microwave Workshop Series on Advanced Materials and Process (IMWS-AMP 2017). https://doi.org/10.1109/IMWS-AMP.2017.8247404
9. Carranzo, I.E., Grant, J.P., Gough, J., Cumming, D.: Terahertz metamaterial absorbers implemented in CMOS technology for imaging applications: scaling large format focal plane arrays. IEEE J. Sel. Top. Quant. Electron. **23**(4) (2017)
10. Bhattacharyya, S., Srivastava, K.V.: Triple band polarization-independent ultra-thin metamaterial absorber using ELC resonator. J. Appl. Phys. **115**(6), 064508 (2014)
11. Nilotpal, S.A.K., Upadhyay, M., Lata, R., Bhattacharyya, S., Chakrabarti, P.: A proposed long wavelength infra-red metamaterial absorber for THz detection. In: IEEE International Symposium on Antennas and Propagation and USNC/URSI National Radio Science Meeting, pp. 2067–2068, 8–13 July 2018 Boston USA. https://doi.org/10.1109/APUSNCURSINRSM.2018.8608291

12. Ghosh, S.K., Yadav, V.S., Bhattacharyya, S., Das, S., A graphene based bandwidth enhanced metamaterial absorber using circular ring. In: IEEE International Symposium on Antennas and Propagation and USNC/URSI National Radio Science Meeting, pp. 1491–1492, 8–13 July 2018, Boston, USA. https://doi.org/10.1109/APUSNCURSINRSM.2018.8608226
13. Wang, B.-X., Wang, G.-Z., Wang, L.-L., Zhai, X.: Design of a five-based Terahertz absorber based on three nested split-ring resonators. IEEE Photon. Technol. Lett. **28**(3) (2016)
14. Zheng, Y., Chen, K., Luo, X., Feng, Y.: Broadband tunable metamaterial absorber with active lumped diodes. In: 2018 International Applied Computational Electromagnetics Society Symposium—China (ACES). https://doi.org/10.23919/ACESS.2018.8669133
15. Zografopaulos, D.C., Ferraro, A., Isić, G., Vasić, B., Gajić, R., Beccherelli, R.: Tuned Terahertz metamaterials based on nematic liquid crystals. In: 2016 41st International Conference on Infrared, Millimeter and Terahertz Waves (IRMMW-THz). https://doi.org/10.1109/IRMMW-THz.2016.7758898
16. Qi, L., Li, C., Fang, G.: Tunable Terahertz metamaterial absorbers using active diodes. Int. J. Electromagn. Appl. **4**(3), 57–60 (2014). https://doi.org/10.5923/j.ijea.20140403.01
17. Wang, B.-X., Wang, G.-Z., Zhai, X., Wang, L.-L.: Polarization tunable Terahertz metamaterial absorber. IEEE Photon. J. **7**(4) (2015). https://doi.org/10.11109/JPHOT.2015.2448718
18. Kim, Y., Ahn, J., Yee, D.-S.: Terahertz birefringence of ZnO. In: 2009 34th International Conference on Infrared, Millimeter and Terahertz Waves. https://doi.org/10.1109/ICIMW.2009.5324942

Dual-Band FSS Backed Printed Antenna with Fractal Geometry for Wearable Applications

M. J. Anand Krishnan[1], Diptiranjan Samantaray[2], Anu Mohamed[2], Chinmoy Saha[3], and Somak Bhattacharyya[2](✉)

[1] Department of Electronics and Communication, Government Engineering College Bartonhill, Trivandrum 695035, India
Anand.k@ieee.org

[2] Department of Electronics Engineering, Indian Institute of Technology (BHU), Varanasi 221005, India
{drsamantaray.rs.ece17,somakbhattacharyya.ece}@iitbhu.ac.in, m.anu.in@ieee.org

[3] Department of Avionics, Indian Institute of Space and Technology, Valiamala, Thiruvananthapuram 695547, India
csaha@ieee.org

Abstract. In this paper, a dual-band fractal-based printed wearable antenna backed with frequency selective surface (FSS) array is proposed for C- and X-band applications. The proposed antenna is designed to operate at 6 and 8 GHz with a high realized gain of 8.5 dBi and 9.4 dBi, respectively. The realized antenna provides a narrow band impedance bandwidth of 70 MHz centered at 6.01 GHz in the first band and UWB bandwidth of 780 MHz centered at 8.05 GHz for the second band, with maximum realized gain of 8.5 dBi and 9.4 dBi, respectively. The ground, FSS unit cell, reflector sheet and the radiating patch of antenna have been made using copper while styrofoam is used as the substrate material. The proposed design can be embedded on fabrics and is a potent candidate for wearable applications.

Keywords: Dual band · FSS · Fractal geometry · Styrofoam substrate

1 Introduction

Wearable antennas have gained a lot of interest in the field of research and industrial applications over the past few years [1–4]. Recently, there has been an increasing trend of using the frequency selective surfaces (FSS) for realizing wearable antennas with better performance. FSS structures which consists of periodic metallization pattern on a dielectric laminate can be easily fabricated at low cost using various techniques, such as (1) etching techniques, (2) inkjet printing on materials, etc. FSS printed on textiles find numerous civilian and defense applications includes (1) telemedicine, (2) security and

© Springer Nature Singapore Pte Ltd. 2021
N. R. Das and S. Sarkar (eds.), *Computers and Devices for Communication*,
Lecture Notes in Networks and Systems 147,
https://doi.org/10.1007/978-981-15-8366-7_42

access control, (3) medical monitoring, (4) fashion/advertising, sports application and (5) emergency services [5, 6]. Frequency selective surfaces provide uninterrupted transmission of electromagnetic waves in specific frequency bands but suppress transmission in other bands. Such structures are composed of unit cells that are placed periodically to form an artificial crystal lattice within the substrate. In this article, an FSS backed microstrip patch with fractal geometry having dual-band radiation is proposed. The designed fractal patch is backed with an FSS array of size 5 × 4 and a partial ground plane [7, 8]. The antenna is designed and simulated using a commercial EM solver. The realized prototype is compared with [9–12], designed for similar applications and reveals better gain performance in both the bands.

2 Antenna and FSS Architecture

2.1 Geometry of Antenna

The overall structure of the prototype is described in Fig. 1a. The detailed dimensions of various design parameters of the radiating patch and the FSS unit cell are provided in Table 1. The proposed structure consists of a fractal antenna printed on a styrofoam substrate of thickness (t_1) of 1 mm, having a dielectric constant of 1.11 and dielectric loss tangent of 0.0018. The fractal structure is formed by cutting a rectangular sheet of area 17 × 22.5 mm^2 from the basic rectangular patch of dimension 34 × 45 mm^2. In the next iteration, three triangular sheets of same area are attached in the three sides of the rectangular patch (excluding the feed side). The antenna is excited by a 50 Ω microstrip line. To obtain the dual-band operation, the first-iterated prototype is used as the radiating element and further improvised. Upon placing the antenna over the FSS array, the resonance frequency of the antenna is detuned due to the mutual coupling between the two. Therefore, to obtain the desired resonant frequency as well as a better reflection coefficient, the height and base length of each triangular sheet are systematically optimized. A rectangular sheet of size 15 × 20 mm^2 is also added in the slot to enhance the gain characteristics. The overall size of the substrate is 127 × 87 mm^2. The partial ground plane is having a length (l) of 10 mm and width (w) of 87 mm and is present in the same plane as that of the FSS layer. The feed to the patch is given with respect to this partial ground plane.

2.2 Geometry of FSS Unit Cell and Ground Plane

The detailed dimensions of the unit cell of the designed FSS structure are described in Fig. 1c. The unit cells are arranged with gap of 1.4 mm and 1.17 mm between the adjacent side cells and top cells, respectively. The width of the ring is 1.55 mm on the left, right and bottom sides, and the two vertical pole like component of the proposed design is having a width (q_2) of 2 mm and length (p_3) of 3.88 mm. The FSS array is placed above the second styrofoam substrate of thickness 2 mm, which is backed by a copper reflector of thickness 0.1 mm. For the characterization of FSS, an infinite array of FSS structure is illuminated with a plane wave, and the associated reflection phase characteristics is shown in Fig. 2. The FSS structure can be considered as an effective

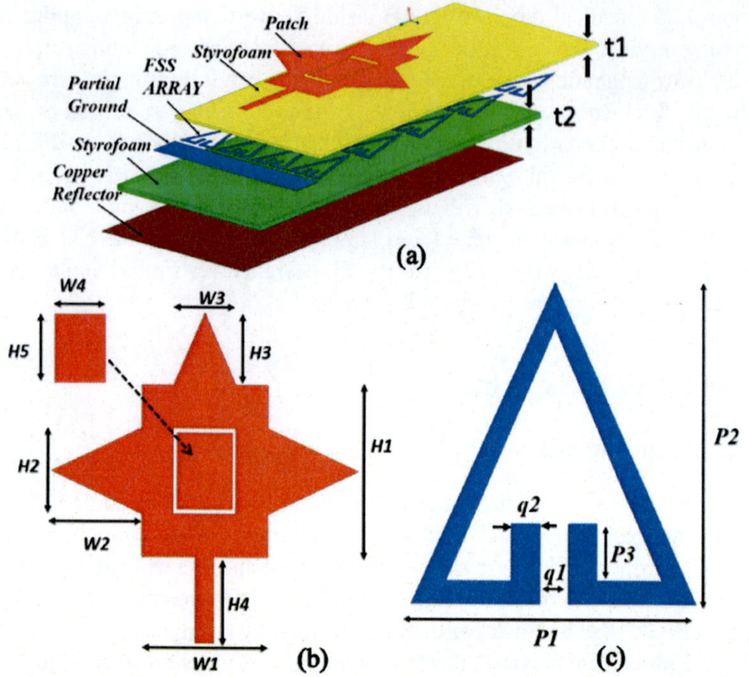

Fig. 1 Structure **a** Three-dimensional view of overall structure, **b** Patch and **c** FSS unit cell

addition to the antenna, if the reflection phase is in the range of [−90° to +90°], so that the reflected waves will be constructively added to the radiated waves and thereby increasing the gain in the desired direction.

3 Simulated Results

3.1 Reflection Coefficient

The proposed antenna without and with the FSS structures is simulated for two cases, named as (a) and (b), respectively. In case (a), a conventional metallic ground plane having same dimensions as that of the substrate is used; whereas, in case (b), the conventional ground plane is replaced with FSS array and partial ground plane lying in the same plane. Figure 3 shows the simulated S_{11} characteristics of the proposed antennas for these two cases. In case (a), the antenna is well matched near the second band, while in the desired first band it is poorly matched. The same antenna with embedded FSS layer yields a good reflection coefficient at both the frequency bands. This is due to the reflection phase characteristics of the FSS structure as depicted in Fig. 2. The reflection phase of the FSS array as shown in Fig. 2 is 0° at 6.01 GHz, which ensures that incident and reflected wave add up constructively. Thus, the antenna integrated with FSS shows good impedance matching with S_{11} less than −25 dB at 6.01 GHz and less than −45 dB at 8.05 GHz.

Table 1 Dimensions of the optimized geometrical parameters

Variables	Values (mm)
H_1	45
H_1	22.5
H_3	19
H_4	22.5
H_5	20
W_1	34
W_2	24
W_3	17
W_4	15
P_1	20
P_2	22
P_3	3.8
a_1	2
a_2	2
t_1	1
t_2	2

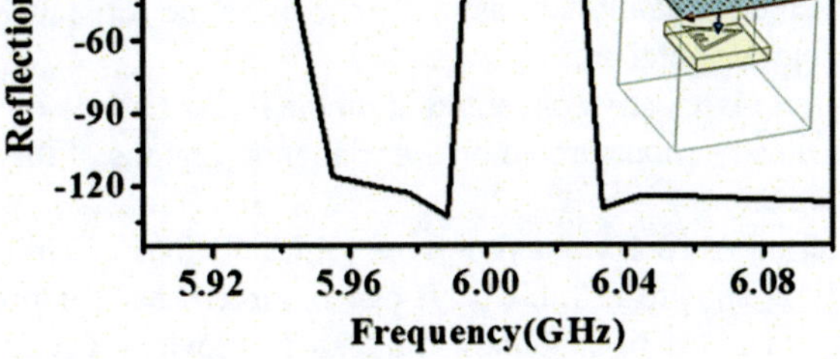

Fig. 2 Reflection phase characteristics of FSS array

3.2 3D Polar Plot and Radiation Pattern

The 3D polar plots of the proposed FSS layer integrated fractal patch antenna are illustrated in Fig. 4a. As revealed from the plots, maximum realized gain obtained is 8.5 and

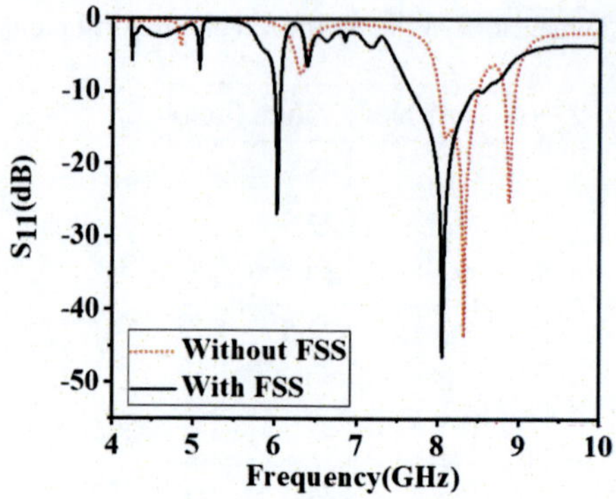

Fig. 3 S_{11} parameter of the proposed antenna with and without FSS Array

9.4 dBi at frequencies 6.01 GHz and 8.05 GHz, respectively, which is nearly 6.8 dBi greater than that of case (a). Also, the back lobe is very less compared to the main lobe as depicted in Fig. 4a, indicating significant reduction in backward radiation. Radiation toward the body is reduced significantly; whereas, the antenna gain is increased in the desired direction. The antenna gives maximum radiation nearly along the positive z-axis (assuming that the antenna is placed in X-Y plane). Figure 4b, c shows the simulated E- and the H-plane 2D radiation pattern of the proposed antenna with FSS backing. It is seen that the antenna with FSS is having very low cross-polarized component in its E-plane and H-plane compared to the co-polarized components. There is a significant difference around 35 dB between co-pol and cross-pol radiation in E-plane at both the frequencies.

3.3 Surface Current Distribution and E-Field Plot

The simulated surface current distribution and E-field distribution of the FSS array loaded antenna are shown in Fig. 5a, b, respectively. Due to the slot in radiating patch and the fractal-based geometry, the current path length is increased leading to a better electrical length compared to the conventional patch. This reduces the resonance frequency and helps in miniaturizing the antenna.

The surface current is mostly concentrated on the linear regions of patch; whereas, the electric field is mostly concentrated on the triangular regions of the patch as shown in Fig. 5a, b, respectively.

The comparison of the proposed prototype with antennas of similar applications reported previously is shown in Table 2. It can be noted that the proposed antenna exhibits superior gain and fractional bandwidth than the previously reported antenna for similar applications.

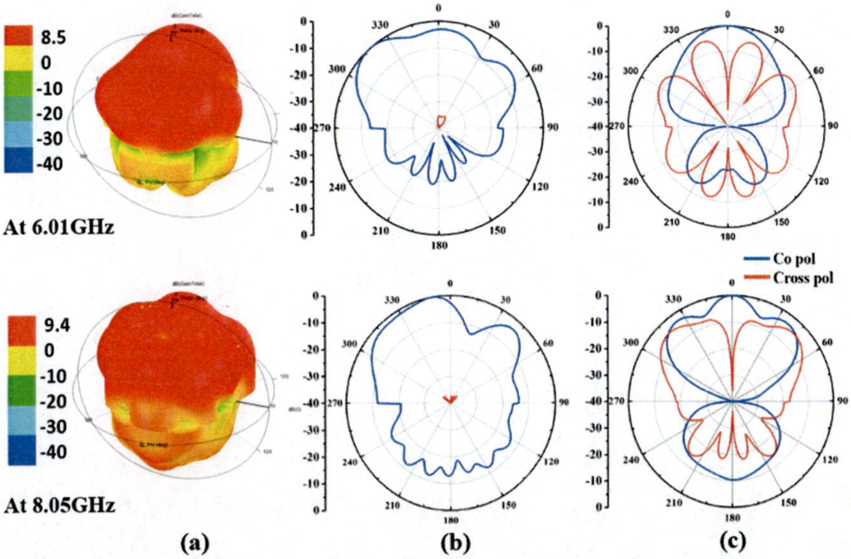

Fig. 4 **a** Realized gain 3D polar plot, **b** Co-pol and cross-pol radiation characteristics in E-plane and **c** Co-pol and cross-pol radiation characteristics in H-plane at 6.01 GHz and 8.05 GHz, respectively

4 Conclusion

This paper presents a new design of a fractal geometry-based FSS layer backed antenna having dual-band operation at 6.01 and 8.05 GHz. A detailed description of the design steps has been presented. The prototype exhibits high gain and enhanced reflection coefficient at the desired frequencies with reduced back lobes. The FSS array is found to be very useful in increasing the gain. The results show that the proposed structure is a good candidate as communicating antenna in wearable applications.

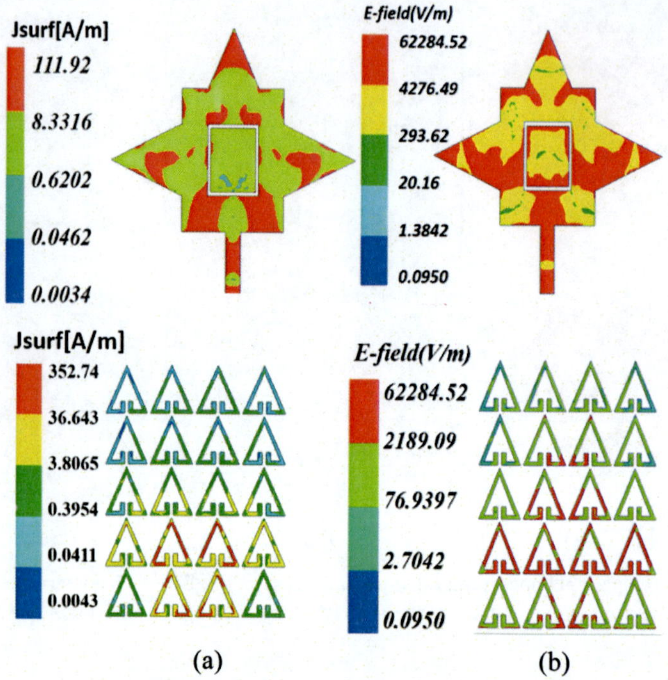

Fig. 5 **a** Surface current and **b** E-field plot of the patch and FSS array

Table 2 Comparison table of the proposed antenna with previously reported antennas

References	Fractional bandwidth (%)	Realized gain (dBi)
Zhu et al. [1]	2.0	3.14
Raad et al. [2]	4.8	4.2
Agarwal et al. [9]	1.88	4.12
Saeed et al. [10]	4.08	6.4
Proposed design	1.17 and 9.69	8.5 and 9.4

Reference

1. Zhu, X.Q., Guo, Y.X., Wu, W.: A compact dual-band antenna for wireless body-area network applications. IEEE Antennas Wirel. Propag. Lett. **15**, 98–101 (2016)
2. Raad, H., Abbosh, A.I., Al-Rizzo, H.M., Rucker, D.G.: Flexible and compact AMC based antenna for telemedicine applications. IEEE Trans. Antennas Propag. **61**(2), 524–531 (2013)
3. Mandal, B., Chatterjee, A.: A wearable button antenna with FSS superstrate for WLAN health care applications. In: IMWS-Bio (2014)
4. Yang F., Rahmat-Samii, Y: Reflection phase characterization of the EBG ground plane for low profile wireless antenna applications. IEEE Trans. Antennas Propag. **51**(10), 2691–2703 (2003)

5. Pirhadi, A., Hosein, A.B.: Wideband high directive aperture coupled microstrip antenna design by using an FSS superstrate layer. IEEE Trans. Antenna Propag. **60**(4), 2101–2108 (2012)
6. Chauraya, A., Seager, R.: Embroidered frequency selective surfaces on textiles for wearable applications. In: Loughborough Antenna & Propagation conference, pp. 388–391(2013)
7. Werner, D.H., Ganguly, S.: An overview of fractal antenna engineering research. IEEE Antennas Propag. Mag. **45**(1), 38–57 (2003)
8. Kim, S., Ren, Y.-J., Member: Monopole antenna with inkjet-printed ebg array on paper substrate for wearable applications. IEEE Antennas Wirel. Propag. Lett. **11**, 663–66 (2012)
9. Agarwal, K., Guo, Y.X., Salam, B.: Wearable AMC backed near end fire antenna for on-body communications on latex substrate. IEEE Trans. Compon. Pack. Manuf. Technol. **6**(3), 346–358 (2016)
10. Saeed, S.M., Student Member, IEEE, Balanis, C.A.: Wearable flexible reconfigurable antenna integrated with artificial magnetic conductor. IEEE Antennas Wirel. Propag. Lett. **16**, 2396–2399 (2017)
11. Chen, H.-Y., Tao, Y.: Performance improvement of a U-slot patch antenna using a dual-band frequency selective surface with modified Jerusalem cross elements. IEEE Trans. Antennas Propag. **59**(9), 3482–3486 (2011)
12. Zhu, S., Langley, R.: Dual-band wearable textile antenna on an EBG substrate. IEEE Trans. Antennas Propag. **57**(4), 926–935 (2009)

Design of a 2.4 GHz Sensor with Low SAR Value for Measuring Vital Signs

Ananya Dey[1](✉), Prapti Ganguly[2], and Jawad Y. Siddiqui[1]

[1] Institute of Radio Physics and Electronics, University of Calcutta, Kolkata, India
18ananyadey@gmail.com
[2] A. K. Choudhury School of Information Technology, University of Calcutta, Kolkata, India
prapti90ganguly@gmail.com

Abstract. A microwave sensor is designed to operate at 2.4 GHz for measuring vital signs in human body. A double-layer patterned ground plane (GP) has been conceived to minimize the sensor size and experimentally verified. More than 360 MHz of 10 dB bandwidth and 2.28 dBi of gain over this band makes this sensor suitable for medical application. The simulated SAR value is found to be 0.5 W/kg for 30 mW of input power, which is much less than maximum limit 1.6 W/kg average over 1 g provided by FCC.

Keywords: Sensor · Vital signs · Pulmonary edema · Specific absorption rate (SAR) · Lung water content (LWC)

1 Introduction

Pulmonary edema is a condition of excess fluid accumulation in the lungs causing difficulty in breathing [1, 2]. Detection of pulmonary edema can help us to detect health conditions such as heart failure [3, 4], blood infections [5], acute lung diseases like pneumonia or injuries [6, 7], burns [8] and dehydration [9]. Since pulmonary edema is usually used to monitor heart failure, a device to detect changes in the lung fluid continuously and noninvasively can be very helpful. The existing techniques for measuring extravascular lung water content (LWC) are body weight scale [10], PAC catheter [11], CT scan [11] and impedance cardiography [12]. These methods are expensive, invasive, indirect and not appropriate for continuous monitoring [13].

To address these needs, a novel, noninvasive, low-cost, microwave-based system for continuous assessment of vital signs such as respiratory rate (RR), heart rate (HR) and changes in lung fluid content is required. The use of microwave methods to measure changes in LWC was first suggested by Susskind [14], and preliminary results were reported by Pederson [15, 16]. Microwave is very safe to use in medicine, particularly in the field of diagnosis because of its completely non-ionizing nature [17] and the ability to penetrate tissues. Researchers have designed microwave sensors to measure different vital signs using the reflection coefficient of the microwave signal transmitted through the lung [18–20].

© Springer Nature Singapore Pte Ltd. 2021
N. R. Das and S. Sarkar (eds.), *Computers and Devices for Communication*,
Lecture Notes in Networks and Systems 147,
https://doi.org/10.1007/978-981-15-8366-7_43

This paper presents a novel cardiovascular sensor at 2.4 GHz for measuring LWC with a better sensitivity. The specific absorption rate (SAR) of the sensor is found to be 0.5 W/kg averaged over 1 g. The 2.3 dBi of gain of the proposed sensor makes it suitable for low-power applications.

2 Design and Characterization

A simple waveguide applicator was designed, Fig. 1a, on a FR4 substrate having dielectric constant $\varepsilon_r = 4.4$ for substrate thickness 0.8 mm using commercially available EM simulator [21] to operate at 2.4 GHz. The sensor consists of a substrate sandwiched between two similar ground structures as shown in Fig. 1b. To reduce the size of the sensor, meander-type slot is incorporated.

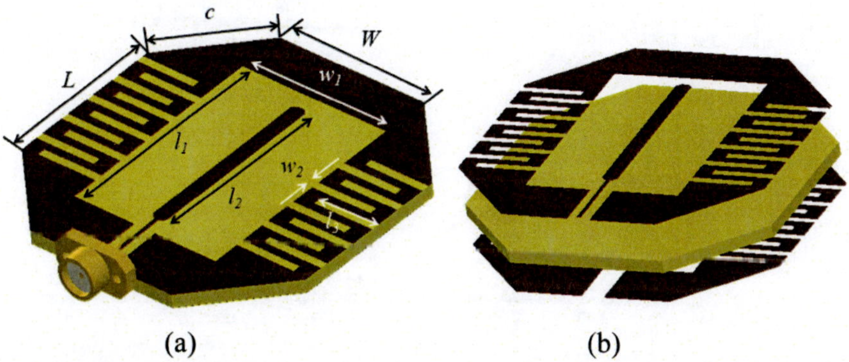

Fig. 1 **a** Optimized microwave sensor at 2.4 GHz simulated on a dielectric substrate with ε_r = 4.4. **b** Different layers of the proposed sensor

The dependence of return loss characteristics on different parameters of the meander is studied through Figs. 2, 3 and 4. Figure 2 characterizes the variation of resonance of the sensor with w_2 as a parameter. A good matching at $w_2 = 1.05$ mm is evident. The shift in resonance frequency with different number of slots is depicted in Fig. 3. It is clear from the figure that as we increase the number of slots, the resonance frequency shifts toward the lower side. Incorporating more slots would have shifted the resonance even more, but that would degrade the antenna gain. So, we have restricted the number of slots up to 5. The variation of reflection coefficient parameter with l_3 is shown in Fig. 4. Here, as we increase the slot length, the current path increases and therefore the resonance frequency shifts toward left. Although a good matching is observed for $l_3 = 10$ mm, but we have chosen $l_3 = 9.5$ mm as it gives a perfect resonance at 2.4 GHz. Therefore, w_2 = 1.05 mm, a number of slots = 5 and $l_3 = 9.5$ mm are chosen as optimum for a good resonance at 2.4 GHz with 10 dB bandwidth being 360 MHz (15%). Figure 5 shows the 3D gain of the antenna at its resonant frequency and two band edge frequencies. A broad side gain of more than 2.2 dBi at the band edges and 2.28 dBi at resonant frequency is seen.

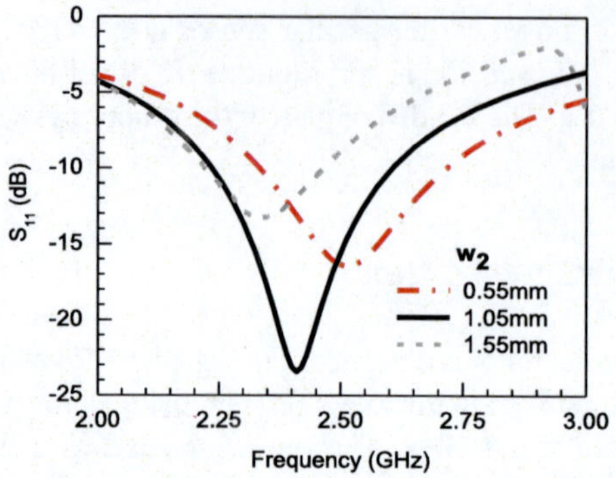

Fig. 2 Comparison of return loss characteristics for different values of w_2; Parameters: $W = 20$ mm, $L = 20$ mm, $l_1 = 27.6$ mm, $l_2 = 23$ mm, $l_3 = 9.5$ mm, $w_1 = 18.4$ mm, $c = 14.14$ mm

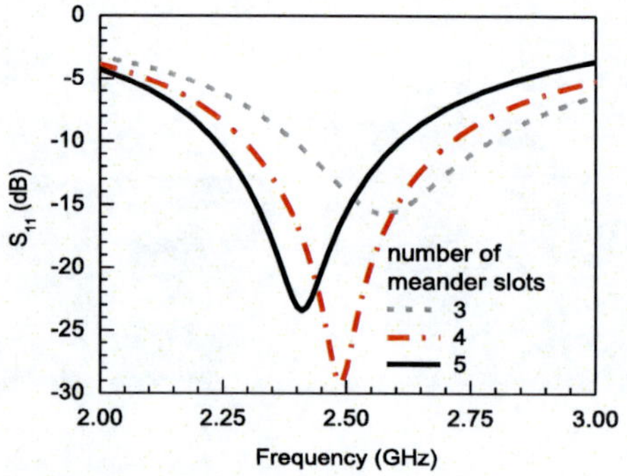

Fig. 3 Comparison of return loss characteristics for different number of meander slots on the ground plane. Other parameters are in Fig. 2.

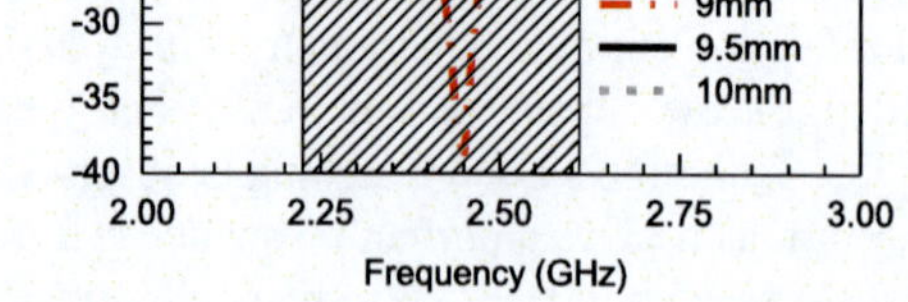

Fig. 4 Comparison of return loss characteristics for different values of l_3. Other parameters are in Fig. 2.

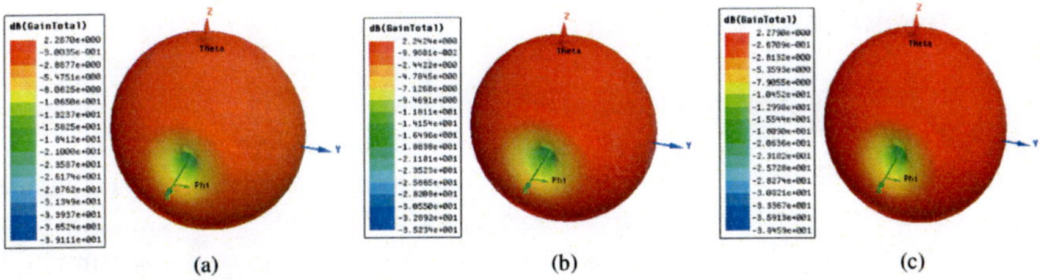

Fig. 5 3D radiation pattern of the sensor shown in Fig. 1 at, **a** Resonant frequency, **b** Lower band edge frequency, **c** Upper band edge frequency

3 SAR Measurement

To measure the specific absorption rate (SAR), we have used CST voxel model. The simulated prototype of the sensor is placed on a human model, and the simulated SAR value is calculated using IEEE/IEC 62,704–1 method. The SAR value is found to be 0.5 W/kg with 30 mW of input power, which is almost 1/3rd of the FCC limit of 1.6 W/kg for 1 g avg.

4 Fabrication and Measurement

The antenna is fabricated on a FR4 substrate having dielectric constant 4.4 and fed with 50 Ω SMA connector as shown in Fig. 6a. The *S*-parameters of the antenna are measured using E8363B series vector network analyzer. The measured result is compared with its simulated version in Fig. 6b. Due to fabrication error, a slight shift in resonant frequency is seen in measured result. But frequency at 2.4 GHz is sufficiently below −10 dB. Therefore, it will not affect the performance of the sensor at 2.4 GHz. Due to lack of proper measurement setup, we were unable to measure the radiation characteristics.

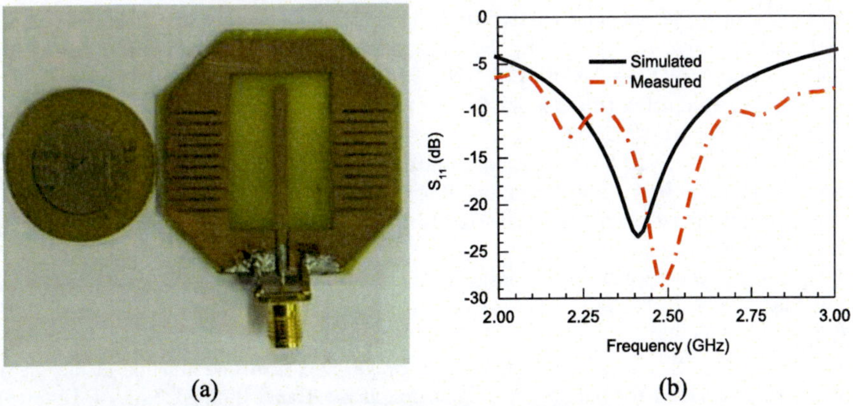

Fig. 6 **a** Fabricated prototype of the proposed sensor; **b** Comparison of simulated and measured return loss characteristics of the proposed sensor

5 Conclusion and Future Work

This investigation thus provides an innovative microwave sensor for detection of changes in lung water content by measuring the changes in the phase of the reflection coefficient. The sensor is flat, lightweight and small in size. The SAR value of the sensor is found to be much less than the permissible SAR limit set by the FCC. But there are certain aspects we could not address. The radiation pattern of the sensor could not be measured due to lack of experimental setup. The SAR value is also simulated SAR using CST voxel model, so it may change a little when measured using conventional SAR measurement systems such as DASY4 system.

In future, we plan to make an integrated continuous monitoring system using the sensor and extract various vital signs such as heart rate, respiration rate, heart amplitude, respiration amplitude along with lung water content from the phase of the reflection coefficient using suitable DSP algorithm. Attempts to achieve these goals are in progress.

References

1. Research, Mayo Foundation for Medical Education: https://www.mayoclinic.org
2. Demling, R.: Pulmonary edema: pathophysiology, methods of measurement, and clinical importance in acute respiratory failure. New Horiz. **1**(2), 371–380 (1993)
3. Grossman, S.: Current Thinking in Acute Congestive Heart Failure and Pulmonary Edema. U.S. Cardiol. (2004)
4. Shamsham, F., Mitchell, J.: Essentials of the diagnosis of heart failure. Am Family Phys **61**(5), 1319–1328 (2000)
5. Miller, E.J., Cohen, A.B., Matthay, M.A.: Increased interleukin-8 concentrations in the pulmonary edema fluid of patients with acute respiratory distress syndrome from sepsis. Crit. Care Med. **24**(9), 1448–1454 (1996)
6. Schuller, D., Schuster, D.P.: Fluid-management strategies in acute lung injury. N. Engl. J. Med. **355**(11), 1175 (2006)
7. Health Scopers, Pneumonia: Causes, Symptoms and Treatment. https://healthscopers.com
8. German, J., Allyn, P., Bartlett, R.: Pulmonary artery pressure monitoring in acute burn management. Arch. Surg. **106**(6), 788–791 (1973)
9. Kalhoff, H.: Mild dehydration: a risk factor of broncho-pulmonary. Eur. J. Clin. Nutr.-Oral Commun. **57**(2), 81–87 (2003)
10. Wireless Body Analysis Scale: https://ihealthlabs.com
11. Swan-Ganz-Right Heart Catheterization, https://www.nlm.nih.gov
12. Medtronic: Clinicians Practical Guide for Using OptiVol® and Other Trends for Managing Heart Failure Patients
13. Lange, N.R., Schuster, D.P.: The measurement of lung water. Crit. Care **3**(2), 19–24 (2009)
14. Susskind, C.: Possible use of microwaves in the management of lung disease. Proc. IEEE **61**(5), 673–674 (1973)
15. Pederson, P.C., Johnson, C.C., Durney, C.H., Bragg, D.G.: An Investigation of the use of microwave radiation for pulmonary diagnostics. IEEE Trans. Biomed. Eng. **BME-23**(5), 410–412 (1976)
16. Pederson, P.C., Johnson, C.C., Durney, C.H., Bragg, D.G.: Microwave reflection and transmission measurements for pulmonary diagnosis and monitoring. IEEE Trans Biomed Eng. **BME-25**(1), 40–48 (1978)

17. Zamanian, A., Hardiman, C.: Electromagnetic radiation and human health: a review of sources and effects. High Freq. Electron. 16–26 (2005) (Summit Technical Media)
18. Iskander, M.F., Durney, C.H., Shoff, D.J., Bragg, D.G.: Diagnosis of Pulmonary edema by a surgically non-invasive microwave technique. Radio Sci. **14**, 265–269 (1979)
19. Gagarin, R., Celik, N., Youn, H.S., Iskander, M.F.: Microwave stethoscope: a new method for measuring human vital signs. In: Proceedings of IEEE International Conference Antennas and Propagation Symposium, Spokane, WA (2011)
20. Celik, N., Gagarin, R., Huang, G.C., Iskander, M.F., Berg, B.W.: Microwave stethoscope: development and benchmarking of a vital signs sensor using computer controlled phantoms and human studies. IEEE Trans. Biomed. Eng. **61**(8) (2014)
21. High Frequency Structure Simulator (HFSS) v. 12

The Scattering Parameter Analysis Using the Circuit Model of UTC-PD

Senjuti Khanra(✉)

Institute of Engineering and Management, Salt Lake, Kolkata, India
senjuti.physics@gmail.com

Abstract. A small signal equivalent circuit model of uni-traveling carrier photodiode (UTC-PD) is developed from integral carrier density rate equation and parasitics are included with it. The technique to obtain scattering parameters from circuit model is given and simulation results are in good agreement with the measurement.

Keywords: Circuit model · Photodiode · Scattering parameter

1 Introduction

Uni-traveling carrier photodiode (UTC-PD) is promising as a key device for millimeter (mm) wave photonic transmitter [1]. Fiber to wireless low power transmitter node can be designed with such device for the access node in next generation 5G pico/femto-cellular wireless transmission in the license-free mmwave band [2]. It can enhance the network capacity for broadband operation. To optimize the device performance, it is essential to develop a suitable model of the device. The existing circuit models of UTC-PD [3, 4] are mostly based on observations and circuit elements are extracted from the measured output reflection coefficient [4].

In this work, the integral carrier density rate equation is extended to develop small signal AC circuit model. The model can be implemented in SPICE or CAD-like circuit simulator. The S_{22} parameter which quantifies the output reflection coefficient is obtained employing small signal circuit model and the results are verified with the experiment results.

The chapter is organized as follows: Sect. 2 provides derivation of the small signal intrinsic and parasitic circuit model of UTC-PD. Scattering parameters are evaluated and verified with the experimental values in Sect. 3 and conclusions are drawn in Sect. 4.

2 Derivation of the Circuit Model

Schematic of vertically illuminated $In_{0.53}Ga_{0.47}As$/InP UTC-PD layer structure and the detail operating principle of UTC-PD can be found in [1, 5], respectively. The following subsection describes small signal circuit model of UTC-PD and how the scattering

© Springer Nature Singapore Pte Ltd. 2021
N. R. Das and S. Sarkar (eds.), *Computers and Devices for Communication*,
Lecture Notes in Networks and Systems 147,
https://doi.org/10.1007/978-981-15-8366-7_44

parameters can be obtained from the circuit model. The continuity equation of the integral photogenerated carrier density $h(t)$ for electrons in the photo-absorption region of UTC-PD can be expressed by [5].

$$\frac{\mathrm{d}h(t)}{\mathrm{d}t} = Q_{\mathrm{in}}(t)\left[\exp\left\{-(m+\beta_1)\frac{h(t)}{A} + (mN_0 - \beta_0)W_{\mathrm{A}}\right\} - 1\right] - \frac{h(t)}{\tau} - \frac{\mu_n E_{\mathrm{indo}} h(t)}{W_{\mathrm{A}}} \tag{1}$$

The description of the parameters in Eq. (1) and their values can be found in [6] which will be used in the simulation.

2.1 Small Signal Circuit Model

When a time varying light (photon flux) with suitable wavelength and energy is incident on UTC-PD then carriers $h(t)$ are generated within it. The modulated integral photogenerated carrier $h(t)$ can be expressed by

$$h(t) = h_{\mathrm{dc}} + \Delta h(t) \tag{2}$$

where h_{dc} is the DC part and $\Delta h(t)$ is the AC part of carrier density variation. Under small signal condition, $\Delta h(t)$ is small. Hence, time varying exponential term ($\mathrm{e}^{\Delta h(t)}$) in Eq. (1) can be approximated as $(1+\Delta h(t))$. Substituting the above approximation and following few mathematical steps, the time varying parts are separated out to obtain Eq. (3).

$$\frac{d(\Delta h(t))}{dt} = \Delta Q_{\mathrm{in}}(t)\left[e^b.e^{-\left\{\frac{(m+\beta_1)}{A}\right\}h_{\mathrm{dc}}} - 1\right] - \left(\frac{1}{\tau} + \frac{\mu_n E_{\mathrm{ind0}}}{W_{\mathrm{A}}}\right)\Delta h(t) \tag{3}$$

2.2 Extraction of Circuit Parameters

Equation (3) can be expressed in the circuit form as follows:

$$C\frac{\mathrm{d}v(t)}{\mathrm{d}t} = i_{\mathrm{ac}}(t) - \frac{v(t)}{R} \tag{4}$$

Comparing Eq. (4) with Eq. (3), the AC current source $i_{\mathrm{ac}}(t)$ at 1550 nm wavelength is given by

$$i_{\mathrm{ac}}(t) = \Delta Q_{\mathrm{in}}(t)\left[e^b.e^{-\left(\frac{(m+\beta_1)}{A}h_{\mathrm{dc}}\right)} - 1\right] \tag{5}$$

To find the voltage v(t) from the dimensionless device intrinsic parameter $\Delta h(t)$, we assume

$$v(t) = \frac{AE_{\mathrm{ind0}}\Delta h(t)}{W_{\mathrm{A}}} \tag{6}$$

The other parameters are

$$R = \left(\frac{W_A}{A\tau E_{indo}} + \frac{\mu_n}{A}\right)^{-1} \text{(in Ohm)},$$

$$C = \frac{W_A}{AE_{indo}} \text{(in Farad) and}$$

$$g = \frac{qW_A}{(\tau_g + \tau_C)AE_{indo}} \text{(in mho)} \quad (7)$$

The Eq. (3) can be implemented using circuit elements given by Fig. 1a.

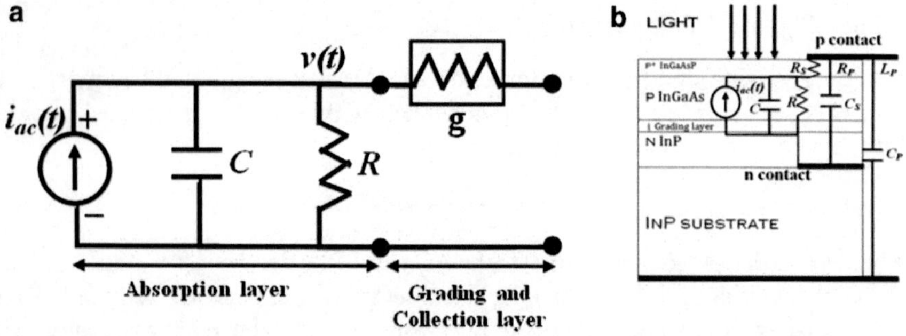

Fig. 1 **a** Small signal circuit model of intrinsic UTC-PD and **b** cross-section of UTC-PD with parasitic element

The circuit element 'g' is included at the output in Fig. 1a in order to incorporate the effect of grading layer in the model.

2.3 Inclusion of Electrical Parasitic

High frequency performance of the device can be significantly affected by the chip and package parasitics. The parasitic elements are included in circuit model as shown in Fig. 1b. R_S is the substrate resistance, C_S is the chip capacitance due to leakage, bonding wires due to packaging causes inductance L_P and provides resistance R_P. The package capacitance C_P arises due to the close proximity of bonding wires which is significant at high frequency.

3 Simulation Results

Circuit as shown in Fig. 1a, b is implemented in Capture CIS OrCAD_10.5 simulation software. The SPICE simulation procedure to extract the S parameters of a two port network is presented briefly. It may be noted that the method is applicable for n-port network also.

S parameters measure the ratio of the powers of the incident and the reflected signals. The incident and corresponding reflected signals are defined as a_1, a_2 and b_1, b_2, respectively, at the input port 1 and output port 2. The scattered waves are related to the incident waves by the following matrix form:

$$\begin{pmatrix} b_1 \\ b_2 \end{pmatrix} = \begin{pmatrix} S_{11} & S_{12} \\ S_{21} & S_{22} \end{pmatrix} \begin{pmatrix} a_1 \\ a_2 \end{pmatrix} \tag{8}$$

The S_{ij} coefficients are dimensionless ratios. S_{22} is the output reflection coefficient can be calculated as

$$S_{22} = \left.\frac{b_2}{a_2}\right|_{a_1=0} = \frac{2Z_L}{Z_L + Z_0} - 1 \tag{9}$$

where Z_L is the device output impedance and Z_0 is 50 Ω. Similarly, S_{21} is the forward transmission coefficient and is defined as the ratio b_2/a_1.

$$S_{21} = \left.\frac{b_{b_2}}{a_1}\right|_{a_2=0} = \frac{2V_{\text{out}}}{V_{\text{in}}} \tag{10}$$

In the simulation, two sub-circuits are required by SPICE to measure the transmitted and reflected powers at the two ports of small signal circuit of Fig. 1. Two S parameters namely S_{21} and S_{22} are important to us. The other two parameters namely S_{12} will be equal to zero for a photodetector [7] and S_{11} is the input reflection coefficient for the optical port which is unimportant as it does not contribute to output photocurrent. To measure the forward transmission coefficient (S_{21}), the output impedance should be matched with the input impedance and the transmission coefficient is the output voltage multiplied by 2. In order to measure S_{21} of the circuit model, a small measuring sub-circuit called "TRANSMIT" is employed. "TRANSMIT" consists of a voltage controlled voltage source (VCVS) (denoted by E) having gain of 2 and associated circuit provided by SPICE as shown in Fig. 2a. During measurement of S_{21}, the hierarchical port "CKT" connects circuit model of UTC-PD given in Fig. 1b and the other port "STR" is declared as "hidden pin". Similarly, the sub-circuit arrangement named "REFLECT" as shown in Fig. 2b is used to measure the output reflection coefficient S_{22}. The reflection coefficients are the input voltage multiplied by 2 minus AC unity. So, VCVS has a gain of 2. Here, also the interface pin CKT is used to connect with the measuring UTC-PD circuit. The hidden pin SRE is left unconnected. There is a provision to apply the bias voltage at V1 for active circuits which is set to zero in our simulation.

S_{22} and S_{21} are measured by connecting two customized hierarchical sub-circuits "TRANSMIT" and "REFLECT", respectively, with UTC-PD circuit port by off-page connectors using SPICE.

The simulated S_{22} parameter is evaluated with frequency from the circuit model in Fig. 1b using SPICE. The magnitude and phase plot of the output reflection coefficient S_{22} versus frequency is shown in Fig. 3a, b, respectively.

Similarly, S_{21} the optical to electrical frequency response is plotted in Fig. 4. The variation of S_{21} versus frequency is agreed well with the experimental result [8] as shown by the arrow in Fig. 4.

In order to plot S_{22} parameter in a conventional way such as in a Smith chart, it is required to evaluate from Fig. 1b by using MATLAB. The output reflection coefficient

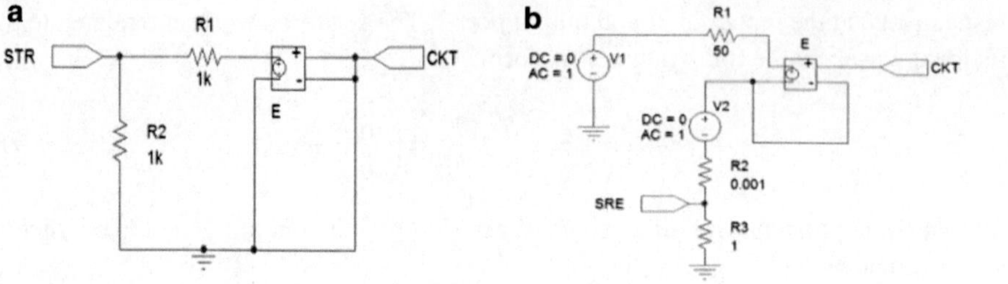

Fig. 2 SPICE sub-circuit model **a** TRANSMIT and **b** REFLECT, respectively, to measure transmission S_{21} and reflection coefficient S_{22} coefficient

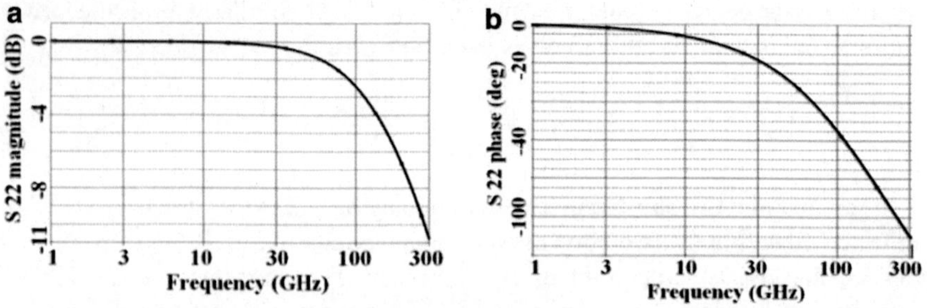

Fig. 3 **a** Magnitude and **b** the phase of output reflection coefficient S_{22} with frequency

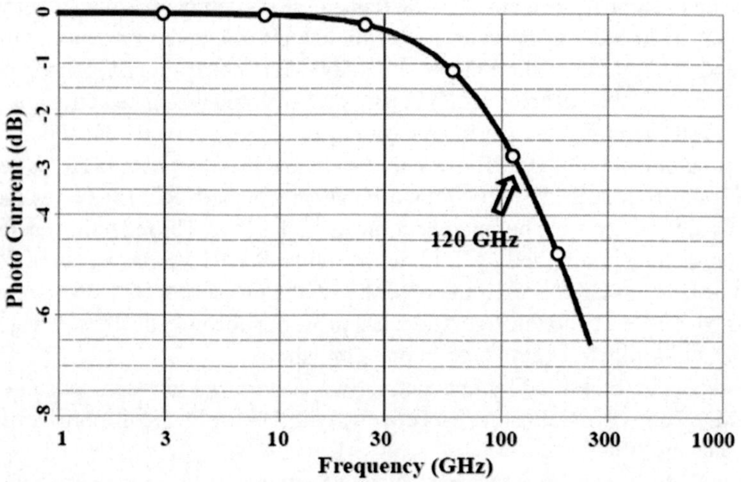

Fig. 4 S_{21}, the optical to electrical frequency response of UTC-PD

(S_{22}) is calculated using the relation

$$S_{22} = \frac{Z_L - Z_0}{Z_L + Z_0} \tag{11}$$

The output reflection coefficient is derived in the frequency range 1–120 GHz at two different bias voltages 0 and −2 V from the small signal circuit model of UTC-PD. The output reflection coefficients with frequencies are shown by the Smith chart in Fig. 5. The simulation results plotted by the dashed line closely matches the experimental data [9] shown by cross symbols. The extracted parasitic values are given in Table 1.

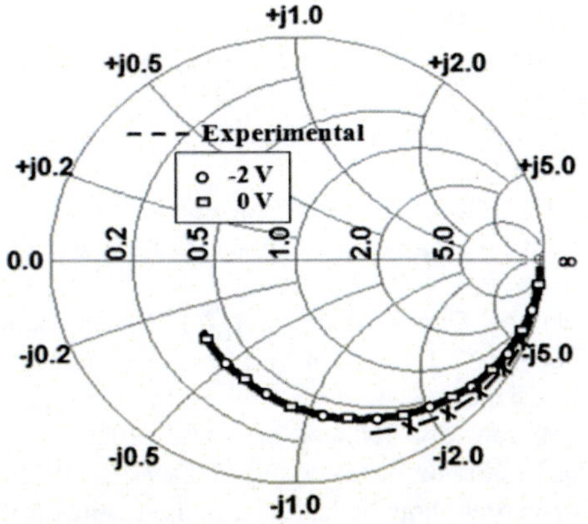

Fig. 5 Output reflection coefficients of $In_{0.53}Ga_{0.47}As/InP$ UTC-PD for the frequency range from 1 to 120 GHz (circles and squares) using the small signal circuit simulation model and that of $GaAs/Al_{0.15}Ga_{0.85}As$ based UTC-PD from 1 to 15 GHz (cross dotted line) using simulation and experiment

Table 1 Extracted values of parasitic from small signal circuit model

Parasitic circuit elements	Values
R_S	1 Ω
L_P	0.015 nH
$(C_S + C_P)$	0.05 pF
R_P	10 Ω

4 Conclusions

We have developed a time varying small signal equivalent circuit model of UTC-PD from carrier density rate equation. The circuits are implemented in a SPICE simulator. Chip and package parasitics of the device are readily incorporated as lumped elements into the model. The scattering parameters of UTC-PD which quantify the input–output transmission and reflection coefficients of the device are obtained from the developed model. Close agreement of the simulation results with the measured values shows that the model can be useful as a tool to optimize UTC-PD performance in mmWave transmissions.

References

1. Ito, H., Kodama, S., Muramoto, Y., Furuta, T., Nagatsuma, T., Ishibashi, T.: High-speed and high-output InP–InGaAs uni-traveling carrier photodiodes. IEEE J. Sel. Top. QE. **10**(4), 709–727 (2004)
2. Ohlen, P., Skubic, B., Rostami, A., Fiorani, M., Monti, P., Ghebretensae, Z., Martensson, J., Wang, K., Wosinska, L.: Data plane and control architectures for 5G transport networks. J. Lightw. Technol. **34**(6), 1501–1508 (2016)
3. Natrella M., Liu C.-P., Graham C., Dijk F.V., Liu H., Renaud C.C., Seeds A.J.: Accurate equivalent circuit model for millimetre-wave UTC photodiodes. Opt. Exp. **24**(5) (2016)
4. Piels, M., Bowers, J.E.: 40 GHz Si/Ge uni-traveling carrier waveguide photodiode. J. Lightw. Technol. **32**(20), 3502–3508 (2014)
5. Khanra S., Barman A.D.: Photoresponse characteristics from computationally efficient dynamic model of uni-traveling carrier photodiode. Opt. Quant. Elecron **48**, 1–11 (2015) (Springer)
6. Khanra, S., Das Barman, A.: Circuit model of UTC-PD with high power and enhanced bandwidth technique. Opt. Quant. Electron. **47**(6), 1397–1405 (2014)
7. Jianjun, G., Baoxin, G., Chunguang, L.: A pin pd microwave equivalent circuit model for optical receiver design. Microw. Opt. Technol. Lett. **38**(2), 102–104 (2003)
8. Shimizu, N., Miyamoto, Y., Hirano, A., Sato, K., Ishibashi, T.: RF saturation mechanism of InP/lnGaAs unitravelling-carrier photodiode. Electron. Lett. **36**(8), 750–751 (2000)
9. Kuo, F-M., Hsu, T-C., Shi, J-W.: Strong bandwidth-enhancement effect in high-speed GaAs/AlGaAs based uni-traveling carrier photodiode under small photocurrent and zero-bias operation. In: Proceedings of LEOS Annual Meeting, TuB3, Belec-Antalya, pp. 141–142 (2009)

A Metasurface Inspired Terahertz Antenna for Multiband Applications

Manikant Jha, Diptiranjan Samantaray, and Somak Bhattacharyya(✉)

Department of Electronics Engineering, Indian Institute of Technology (BHU), Varanasi, Uttar Pradesh 221005, India

manikantjha.ece18@itbhu.ac.in,
{drsamantaray.rs.ece17,somakbhattacharyya.ece}@iitbhu.ac.in

Abstract. In this paper, the performance of a metasurface-based slot antenna in THz region has been studied. A rectangular slot has been designed on the bottom side of the antenna, where a square-shaped patch behaving as radiating element has been designed on the centre of the slot. The antenna offers fractional bandwidths of 6.04%, 12.84% and 2.76% at the frequencies 0.32 THz, 0.35 THz and 0.41 THz, respectively. The maximum realized gain of 7.51 dBi has been achieved at 0.35 THz, and unidirectional radiation pattern has been observed at all the operating frequencies.

Keywords: Terahertz antenna · Multiband · Gain · Bandwidth

1 Introduction

Terahertz region is the least explored electromagnetic spectrum lying in between microwave and infrared regions [1–4]. This band possesses a strong potential for applications in astronomy, security, sensing, imaging and material testing [3–8]. The terahertz spectrum is also being explored for future wireless communication systems [9–11]. Efficient radiators are required for communications in terahertz band of spectrum to avoid path loss [11, 12]. In terahertz region, bandwidth as well as gain of the antenna has been enhanced employing two layers of substrate [9].

In this paper, a multiband terahertz antenna has been designed employing a slot cut in the ground plane in which a square-shaped patch has been introduced at the centre of the slot acting as radiating element. The final design consists of two layers of different substrates, viz. Si_3N_4 and GaAs, to increase the bandwidth and gain of the antenna. A metasurface layer has been designed on the top surface of GaAs substrate. The final proposed optimized geometrical parameters have been evolved from the basic design by optimizing the various parameters. The antenna is designed in such a manner that it operates over the fractional bandwidths of 6.04%, 12.84% and 2.76% at the frequencies 0.32 THz, 0.35 THz and 0.41 THz, respectively. The maximum realized gain of 7.51 dBi at 0.35 THz has been achieved.

© Springer Nature Singapore Pte Ltd. 2021
N. R. Das and S. Sarkar (eds.), *Computers and Devices for Communication*,
Lecture Notes in Networks and Systems 147,
https://doi.org/10.1007/978-981-15-8366-7_45

2 Design of the Antenna

The 3-D view of the designed antenna and metasurface of 5 × 5 order is shown in Fig. 1a, b, respectively. Two different substrates using GaAs ($\varepsilon_r = 12.9$) and Si_3N_4 ($\varepsilon_r = 9.5$) have been used to design the antenna structure. The antenna has been designed using a patch at the bottom side of Si_3N_4 substrate which is acting as a radiating element as shown in Fig. 1c. The gold square-shaped patch of length w_d has been designed at the centre of the slot of width w_s and length l_s cut out from the bottom layer of the antenna. The proposed antenna consists of periodic array of 5 × 5 order of unit cell. The basic unit cell consists of a square gold patch from which a ring-type cut has been extracted as shown in Fig. 2a. The ring possesses inner radius r and outer radius R. The square patch is of dimension d while the periodicity of the patch is p. All the optimized geometrical dimensions are shown in Table 1. The effective permittivity (ε) and effective permeability (μ) of the proposed metasurface are analysed using retrieval extraction algorithm [13], and it is observed that the metasurface shows negative real part permittivity (ε) in the range of 0.317–0.338 THz, 0.383–0.4087 THz and 0.417–0.427 THz as shown in Fig. 2b and also negative real part permeability (μ) in the range of 0.301–0.317 THz, 0.338–0.383 THz, 0.408–0.417 THz and 0.427– 0.45 THz as illustrated in Fig. 2c.

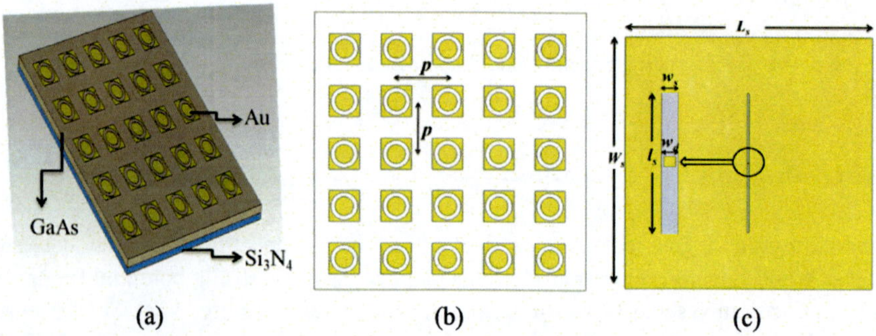

Fig. 1 **a** 3-D view of proposed antenna, **b** top view of antenna, **c** bottom view of antenna

3 Simulated Results

The proposed design shown in Fig. 1 has been simulated in CST microwave studio [14]. The simulated return losses for different cases have been compared and shown in Fig. 3. Initially, the antenna is designed without involving metasurface in which a small patch is placed at the centre of the rectangular slot depicted as case-1. The metasurface has then been introduced as the top layer of the antenna illustrated as case-2. It has been observed from the responses shown in Fig. 3 that case-2 exhibits better performance as compared to case-1. It is evident from the return loss characteristics shown in Fig. 3 of case-2 that the antenna operates at 0.32 THz, 0.35 THz and 0.41 THz with fractional bandwidths of 6.04%, 12.84% and 2.76%, respectively. It is clearly observed that the maximum return

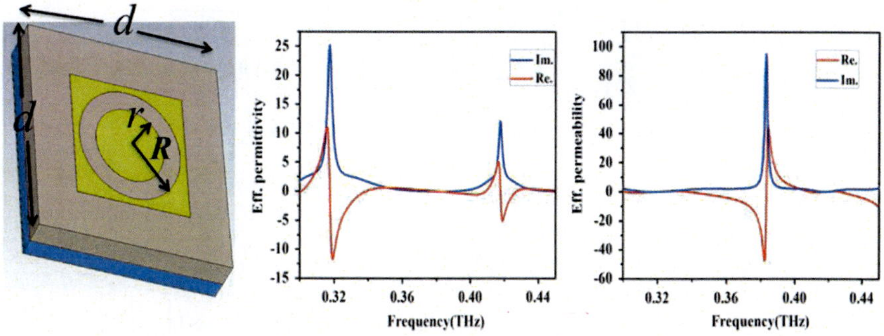

(a) (b) (c)

Fig. 2 **a** 3-D view of unit cell of the metasurface along with **b** extracted effective permittivity and **c** effective permeability

Table 1 Optimized geometrical dimension of parameter used of the proposed antenna

Antenna dimensions	Dimension (μm)	Antenna dimensions	Dimension (μm)
Length of ground (L_s)	1800	Width of patch (d)	200
Width of ground (L_s)	1800	Periodicity of patch (p)	340
Height of ground (t)	0.35	Inner radius of ring (r)	60
Length of substrate (l)	1800	Outer radius of ring (R)	90
Width of substrate (l)	1800	Length of patch on bottom (w_d)	10
Height of substrate (GaAs) (h_1)	40	Width of patch on bottom (w_d)	10
Height of substrate (Si_3N_4) (h_2)	30	Length of slot on bottom (l_s)	500
Length of patch (d)	200	Width of slot on bottom (w_s)	15

loss of 21.25 dB has been realized at 0.32 THz and the maximum fractional bandwidth of 12.84% is obtained at 0.35 THz. The surface current of the proposed antenna is studied over different frequencies. It is observed that surface current is more intense around the aperature of the unit cells, and also, it is extremely high along the slot as shown in Fig. 4a, b, respectively.

The three-dimensional realized gain has been studied at different operating frequencies as shown in Fig. 5a. The realized gain of 5.1 dBi, 7.51 dBi and 6.25 dBi at the frequencies of 0.32 THz, 0.35 THz and 0.41 THz, respectively, has been achieved. The maximum gain of 7.51 dBi occurring at 0.35 THz has been realized. The radiation patterns of the proposed antenna have been studied at different frequencies under E-plane

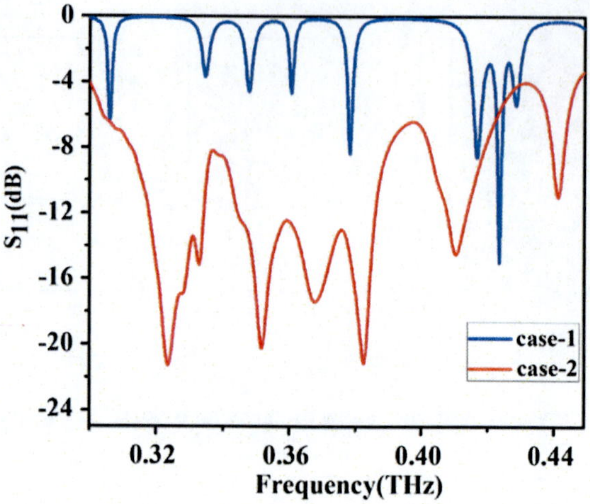

Fig. 3 Plot of S_{11} (dB) with respect to frequency (THz) for different cases

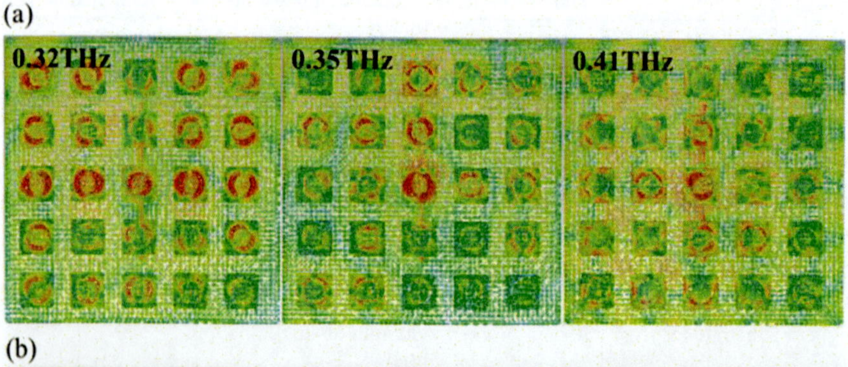

Fig. 4 Surface current distributions at **a** top and **b** bottom layers at 0.32, 0.35 and 0.41 THz

($\varphi = 0$) and H-plane ($\varphi = 90$) as shown in Fig. 5b, c, respectively, in which the unidirectional radiation characteristics have been achieved in both the planes. Furthermore, the co-pol and cross-pol radiation patterns are separated by at least 15 dB.

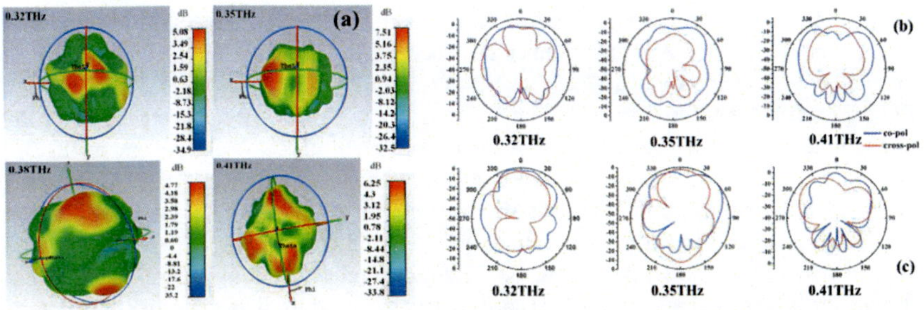

Fig. 5 **a** Three-dimensional plots of realized gain at 0.32, 0.35, 0.38 and 0.41 THz. Co-polarized and cross-polarized radiation patterns of the proposed antenna in (**a**) E-plane (**b**) H-plane at 0.32 THz, 0.35 THz and 0.41 THz, respectively

The proposed antenna has been compared with a few existing terahertz antennas as given in Table 2 from which it is found that the behaviour of the proposed antenna is much superior to the existing ones.

Table 2 Comparison table for the performance of the proposed antenna with existing THz antennas

Terahertz antenna	Fractional bandwidths (%)	Realized gain (dBi)
Nag et al. [6]	7.4	5.95
Prince et al. [15]	8.97	4.25
Saini et al. [16]	4.7	7.36
Proposed design	6.04, 12.84, 2.76	5.08, 7.51, 6.25

4 Conclusions

In this paper, a terahertz antenna has been realized operating at 0.32 THz, 0.35 THz and 0.41 THz, respectively. Two different layers of substrate have been used in designing the antenna structure in which a metasurface layer has been incorporated at the top side of GaAs layer. The maximum realized gain of 7.51 dBi has been obtained at 0.35 THz with fractional bandwidth of 12.84%. Unidirectional radiation pattern has been observed in both E-plane and H-plane. The proposed antenna can be used for various medical and security purposes.

References

1. Hussain, N., Park, I.: Design of a wide-gain-bandwidth metasurface antenna at terahertz frequency. AIP Adv. **7**, 055313-1–12 (2017)

2. Siegel, P.H.: Terahertz technology. IEEE Trans. Microw. Theory Tech. **50**, 910–928 (2002)
3. Dragoman, M., Muller, A.A., Dragoman, D., Coccetti, F., Plana, R.: Terahertz antenna based on grapheme. J. Appl. Phys. **107**(104313), 1–3 (2010)
4. Rabbani, M.S., Ghafouri-Shiraz, H.: Improvement of Microstrip antenna's gain, bandwidth and fabrication tolerance at Terahertz frequency bands. In: IEEE Wideband and Multi-Band Antennas and Arrays for Civil, Security & Military Applications, pp. 1–3 (2015)
5. Kashyap, S. S., Dwivedi, V.: Compact Microstrip patch antennas for Terahertz applications. In: IEEE 9th ASIA Modelling Symposium, pp. 157–163 (2015)
6. Nag, A., Mittal, D., Kaur, A., Sidhu, E.: THz rectangular slitted microstrip patch antenna design for biomedical applications, security purpose and drug detection. In: 2016 International Conference on Automatic Control and Dynamic Optimization Techniques (ICACDOT), pp. 755–758 (2016)
7. Kadoya, Y., Onuma, M., Yanagi, S., Ohkubo, T., Sato, N., Kitagawa, J.: THz wave propagation on strip lines: devices, properties and applications. Radio Eng. **17**(2), 48–55 (2008)
8. Treizebre, A., Bocquet, B., Xu, Y., Bosiso, R.G.: New THz excitation of Planar Goubau Line. Microw. Opt. Technol. Lett. **50**, 2998–3001 (2008)
9. Jha, K.R., Singh, G.: Analysis and design of rectangular microstrip antenna on two-layer substrate materials at terahertz frequency. J. Comput. Electron. **9**, 68–78 (2010)
10. Siegel, P.H.: THz instruments for space. IEEE Trans. Antennas Propag. **55**(11), 2957–2965 (2007)
11. Ostiander, R., Fitch, M.J.: THz waves for communication and sensing. Jhon Hopkins APL Tech. Dig. **25**(4), 348–355 (2004)
12. Azarbar, A., Masouleh, M.S., Behbahani, A.K.: A new Terahertz microstrip rectangular patch array antenna. IEEE Trans. Microw. Theory Tech. **50**, 910–927 (2002)
13. Veselago V.G.: The electrodynamics of substances with simultaneously negative values of ε and μ. Phys.-Uspekhi **10**, 509–514 (1968)
14. CST Studio Suite- User's Manual. Computer Simulation Technology, Darmstadt, Germany (2017)
15. Prince, K.G., Mehta, V., Sidhu, E.: Rectangular Terahertz microstrip patch antenna design for vitamin K2 detection applications. In: IEEE 1st International Conference on Electronics, Material Engineering and Nano- Technology (IEMENTech), Kolkata, India, pp. 1–3 (2017)
16. Saini, S.S., Kaur, G., Rani, N., Kaur, J., Sidhu, E.: High gain reduced ground terahertz microstrip patch antenna design for the detection of trinitrotoluene (TNT) explosives material. In: IEEE Progress in Electromagnetics Research Symposium, pp. 934–938. Spring (PIERS), St. Petersburg, Russia (2017)

Trenched Core Waveguide Structure for Photonic Integrated Circuit

Madhusudan Mishra(✉) and Nikhil Ranjan Das

Institute of Radio Physics and Electronics, University College of Science and Technology, University of Calcutta, 92, A. P. C. Road, Kolkata 700009, India
madhusudandbt@gmail.com, nrd@ieee.org

Abstract. Waveguide structure having a rectangular trench on top of its core has been studied. Variation of optical power overlapping with cladding of the waveguide has been shown as a function of trench dimension. Appropriate dimension for the trench to obtain large amount of optical power in the cladding has been shown, and the reason has been discussed. Usefulness of this study for suitable design of photonic phase actuator and sensor in photonic integrated circuit has also been discussed.

Keywords: Trenched core · Sensor · Phase actuator · Photonic integrated circuit

1 Introduction

Silicon photonics shows a prominent development in integrated circuit domain by introducing photonic integrated circuit (PIC) technology for high-speed and broadband applications [1–4]. Apart from that, it also plays a vital role for sensing and detecting applications [5–7]. Phase change in propagating light is a key factor for realizing devices for such kind of applications. To impose any change in the phase of light traveling through the waveguide, one has to change the refractive index of the core/cladding of the waveguide. Mainly, it is the cladding of which the refractive index is tailored. The rate of the phase change has a direct relation with the presence of light in the cladding. Since major part of light is confined within the core in conventional waveguides, phase change of light using the cladding becomes a very slow process needing a long length of the device. Hence, it is desired to have a large portion of light to be propagating through the cladding for performing phase actuation or sensing function. This is the main aim of this study. Though work has been done for realizing sensors using slot waveguides [8], the present study is based on the trenched waveguide structure in anticipation of improved performance. The study includes a waveguide structure, in which the core has been modified to trenched one and the effect of the trench dimension on the power availability in the cladding has been investigated.

© Springer Nature Singapore Pte Ltd. 2021
N. R. Das and S. Sarkar (eds.), *Computers and Devices for Communication*,
Lecture Notes in Networks and Systems 147,
https://doi.org/10.1007/978-981-15-8366-7_46

2 Device Structure

Figure 1 shows the schematic of the considered waveguide structure, made up of an Si core, sitting on an SiO_2 substrate and covered by SiO_2 as its cladding. The respective dimensions are as indicated in the figure. The core is trenched from its top surface towards the bottom. The width (w_t) and depth (d_t) of the trench are two variable parameters for this study and vary from 50 to 200 nm and 0 (no trench) to 220 nm (slot waveguide), respectively, to get its appropriate dimension to avail maximum optical power in the cladding.

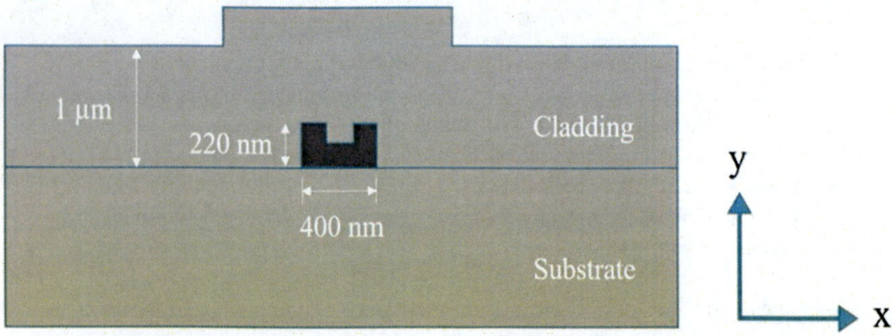

Fig. 1 Schematic of the cross-sectional view of waveguide structure

3 Results and Discussion

Simulation has been carried out using finite element simulator (COMSOL) using the dimensions of the structure mentioned in Sect. 2. Figure 2 shows the plots for the TE mode for different trench depths at a fixed trench width. Figure 2a shows the plot without any trench in the core. The mode is mostly confined within the core with a very little overlapping with the cladding. However, as the trench depth increases (as shown in Fig. 2b–d), leakage of the mode to the cladding increases through the trench sidewalls by reducing the confinement within the core increasing the density of mode in cladding.

The normalized power (P_{norm}) available in the cladding as a function of trench depth is shown in Fig. 3 for four different values of trench widths. As seen from Fig. 3, as d_t increases, the power in the cladding increases. However, for some values of trench width, the power in the cladding decreases as the depth is increased. According to the figure, it reaches maximum at $d_t = 200$ nm for $w_t = 150$ nm and at $d_t = 180$ nm for $w_t = 200$ nm. The reason for this can be explained using the next figure.

In Fig. 4, we plot the TE mode of the waveguide structure for the last three values of d_t, for a trench width of 150 nm. It can be observed from the figure that as d_t increases, the spreading of mode power also increases. Looking at the enlarged view of the marked portion of Fig. 4a, it is seen that the core dimension at the bottom after trench is thick enough (40 nm) to restrict the delivery of mode power into the substrate by a significant amount. Hence, up to this trench depth ($d_t = 180$ nm), whatever power is being released

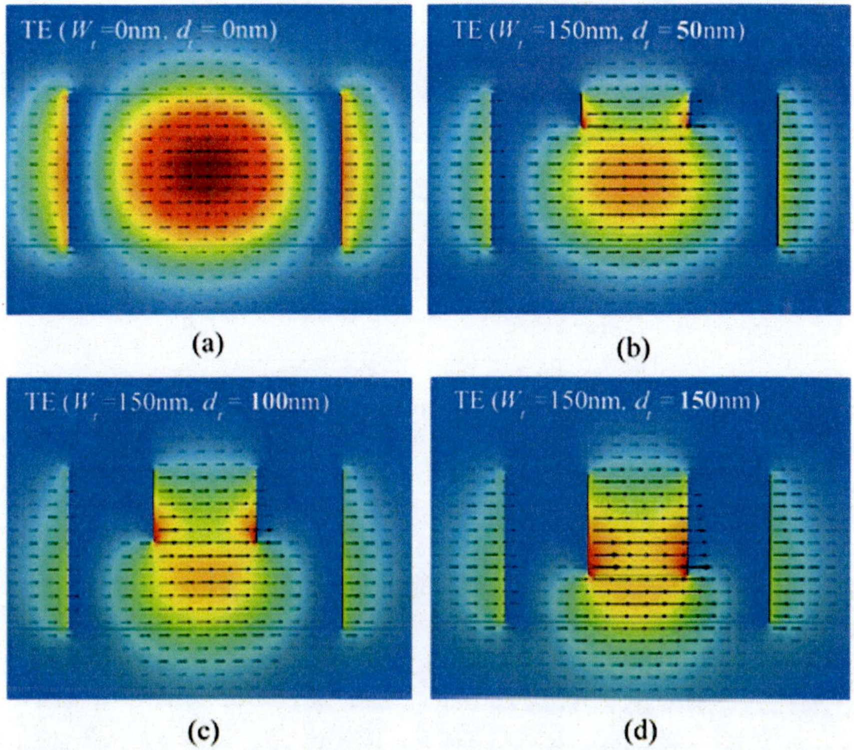

Fig. 2 Figure shows the TE mode shape for $d_t = 0$ nm (**a**), 50 nm (**b**), 100 nm (**c**), and 150 nm (**d**), respectively, with a trench width of 150 nm. The arrows are proposal to the field strength (different for each mode plot)

through the trench remains within the cladding (trench) region. However, with further increment in d_t ($d_\mathrm{t} = 200$ nm; Fig. 4b), the thickness of the core below the trench becomes very small and the mode power started leaking into the substrate from the cladding (trench). This results in decay of the power in cladding region. Further, when the trench is made up to its full depth, i.e., $d_\mathrm{t} = 220$ nm (Fig. 4c), there is no core left below the trench by allowing quite a good amount of power released into the substrate. This results in further reduction of power in the cladding region. Therefore, to get more power in the cladding, trench depth should be more which is also a function of the trench width.

4 Conclusion

The present study shows that introducing a trench in the conventional rectangular core increases the mode power available in cladding. However, more power can be obtained by increasing the depth and decreasing the width. This can be helpful for design of efficient photonic phase actuators and sensors.

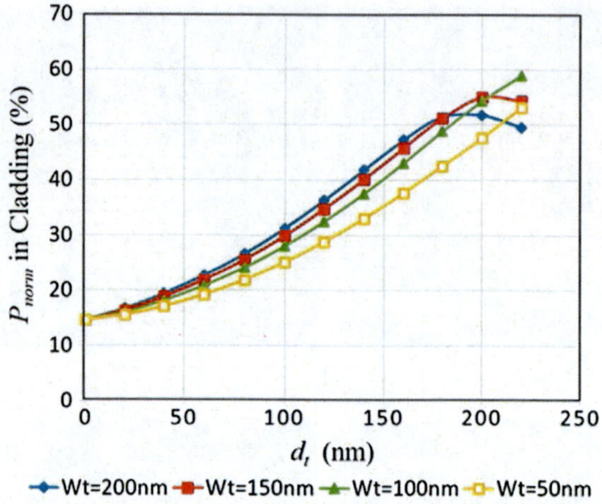

Fig. 3 Percentage of normalized power available in cladding with respect to various trench depths for four different trench widths

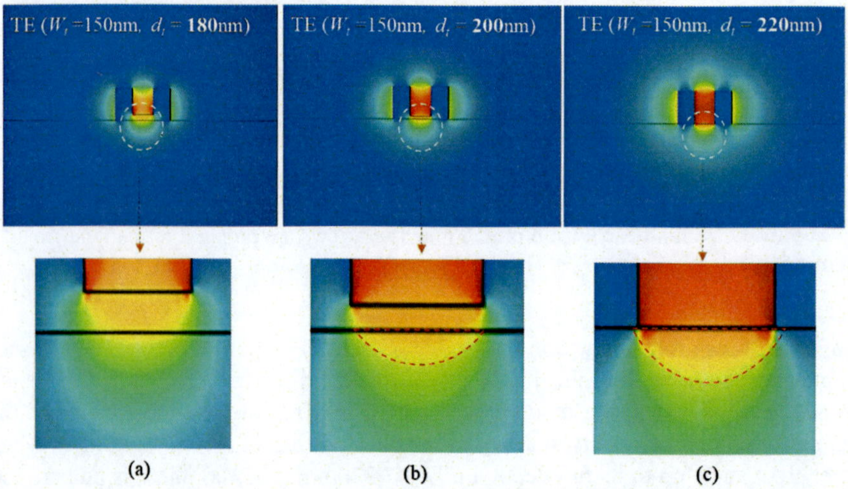

Fig. 4 Figure shows the TE mode plot for the structure for d_t = 180 nm (**a**), 200 nm (**b**), and 220 nm (**c**), respectively, with a trench width of 150 nm. The bottom row shows the enlarged view of the marked region (by circle) of the top row figures

Acknowledgements. This work is financially supported by INSPIRE Fellowship program [IF 150280] of Department of Science and Technology (DST), India.

References

1. Roelkens, G., Dave, U.D., Gassenq, A., Hattasan, N., Hu, C., Kuyken, B., Leo, F., Malik, A., Muneeb, M., Ryckeboer, E., Sanchez, D., Uvin, S., Wang, R., Hens, Z., Baets, R., Shimura, Y.,

Gencarelli, F., Vincent, B., Loo, R., Van Campenhout, J., Cerutti, L., Rodriguez, J.-B., Tournié, E., Chen, X., Nedeljkovic, M., Mashanovich, G., Shen, L., Healy, N., Peacock, A.C., Liu, X., Osgood, R.: Silicon-based photonic integration beyond the telecommunication wavelength range. IEEE J. Sel. Top. Quant. Electron. **20**, 394 (2014)
2. Li, M., Member, IEEE, Chen, X., Member, IEEE, Su, Y., Senior Member, IEEE, Wang, X., Chen, M., Dai, D., Member, IEEE, Liu, J., Zhu, N.H.: Photonic integration circuits in China. IEEE J. Quantum Electron. **52**, 0601017 (17 p) (2016).
3. Kopp, C., Bernabé, S., Bakir, B.B., Fedeli, J.M., Orobtchouk, R., Schrank, F., Porte, H., Zimmermann, L., Tekin, T.: Silicon photonic circuits: On-CMOS integration, fiber optical coupling, and packaging. IEEE J. Sel. Top. Quantum Electron. **17**, 498 (2011)
4. Kaminow, I.P.: Optical integrated circuits: a personal perspective. J. Lightwave Technol. **26**, 994 (2008)
5. Hasek, T., Kurt, H., Citrin, D.S., Koch, M.: Photonic crystals for fluid sensing in the subterahertz range. Appl. Phys. Lett. **89**, 173508 (3 p) (2006)
6. Luan, E., Shoman, H., Ratner, D.M., Cheung, K.C., Chrostowski, L.: Silicon photonic biosensors using label-free detection. Sensors **18**, 3519 (2018)
7. Orlandi, P., Morichetti, F., Strain, M.J., Sorel, M., Melloni, A., Bassi, P.: Tunable silicon photonics directional coupler driven by a transverse temperature gradient. Opt. Lett. **38**, 663 (2013)
8. Barrios, C.A.: Optical slot-waveguide based biochemical sensors. Sensors **9**, 4751–4765 (2009)

Circular Patch Antenna with Ring Structures for Dual X band and 5G Applications

Vivek Parimi[1], Suraj Polamaina[1], Ku Chia Hao[2], Abhirup Datta[3], and Somaditya Sen[4(✉)]

[1] Department of Electrical Engineering, Indian Institute of Technology Indore, Indore 453552, India
[2] Department of Electrical Engineering, Ming Chi University of Technology, New Taipei 24301, Taiwan
[3] Discipline of Astronomy, Astrophysics and Space Engineering, Indian Institute of Technology Indore, Indore 453552, India
[4] Department of Physics, Indian Institute of Technology Indore, Indore 453552, India
sens@iiti.ac.in

Abstract. In this paper, a novel circular patch antenna model with added ring structures has been proposed and numerically simulated using CST Microwave Studio Suite. Various progressive stages of the antenna model have been shown with corresponding improvements in response. The final antenna model shows a good simulated response in the X band (10 GHz) and 5G communication band (27 GHz) with a decent gain value ranging from 3.25 to 4.77 dB.

Keywords: Microstrip patch antenna · X band · 5G communication · Dual band

1 Introduction

In recent years, microstrip antennas have attracted attention in wireless communication system design because of their compactness and narrow impedance bandwidth. Multiple band antennas with good bandwidth and high gains are extremely desirable for communication applications. Since the advent of 5G technology, efforts are being made to create models of antennas that are compact, cheap and provide good transmission and gain [1–3]. X band is attractive for various communication applications. The band is also allocated for military purposes such as airborne radars for security and monitoring [4, 5]. Having antennas that can be used for both these applications would be very useful, and hence, in this paper, we present a model for a simple planar microstrip patch antenna for X band and 5G communication applications with narrow bandwidth and appreciable gain. This report is an initiative of creating awareness of the possibility of such a simple and cheap working model for today's communication science and industry.

© Springer Nature Singapore Pte Ltd. 2021
N. R. Das and S. Sarkar (eds.), *Computers and Devices for Communication*,
Lecture Notes in Networks and Systems 147,
https://doi.org/10.1007/978-981-15-8366-7_47

2 Antenna Design

Four antenna models are studied in this work. The geometry of the four antennas is represented in Fig. 1. A square-shaped substrate made of Rogers TMM4 of side length 15 mm and thickness 1 mm was chosen. The substrate material had a dielectric constant of 4.77 and a loss tangent of 0.02. A Cu microstrip transmission line with dimensions width 1.81 mm, length 5.9 mm and conductivity 5.8 × 107 S/m ensures an impedance of 50 Ω. A Cu ground plane was made of length 5.5 mm and width 15 mm which was formed at the opposite face of the substrate.

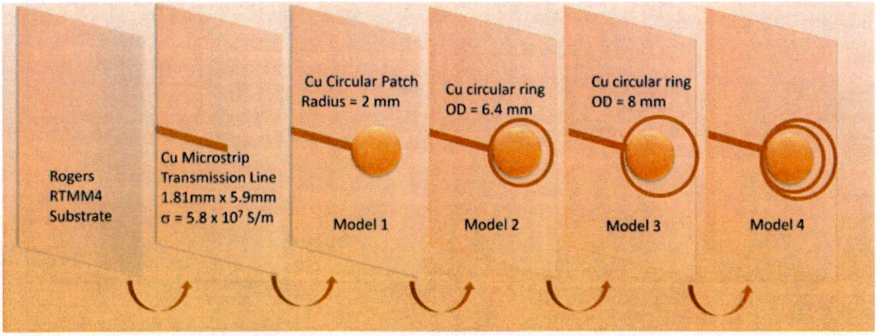

Fig. 1 Progressive development of the antenna model from a single patch to patch with two rings

The antennae consisted of circular "patch and ring" set-ups made of Cu. Four different models have been studied with different radii of the Cu circles and different positions (Fig. 1). The first model consisted of a circular patch of radius 2 mm initiating at the end of the transmission line. The second model had the same circular patch along with an additional ring of outer radius 3.2 mm and inner radius 3 mm tangential to the patch at the point of contact with the transmission line. The third model was similar to the second one with a larger ring of outer radius 4 mm and inner diameter 3.8 mm placed similarly as the second one. The fourth model consisted of the circular patch and the two rings together.

3 Simulation Technique

Microwave Studio Suite (MWS) of Computer Simulation Technology (CST) software was used to analyse the performance of the designed antennae. *S* parameter, directivity, gain, etc., were determined using this software. The simulations were carried out using a time-domain analysis in CST MWS. The model described above is created using the CST modelling and design tools. The DRA model in the software is excited using a discrete port at the transmission line. Materials with losses are used in the simulations so as to get maximum accuracy in the simulation results. Each simulation is carried out over a frequency range of 0–30 GHz. Within the time-domain analysis, the transmission line matrix (TLM) which provides an accurate broadband calculation has been used.

4 Simulation Results and Analysis

The simulated S_{11} response of the four antenna models is shown in Fig. 2. The impedance bandwidth of the antenna is defined as the frequency range over which $|S_{11}| < -10$ dB. The two frequency ranges of interest are in the X band (~10 GHz) and in the 5G range (~27 GHz).

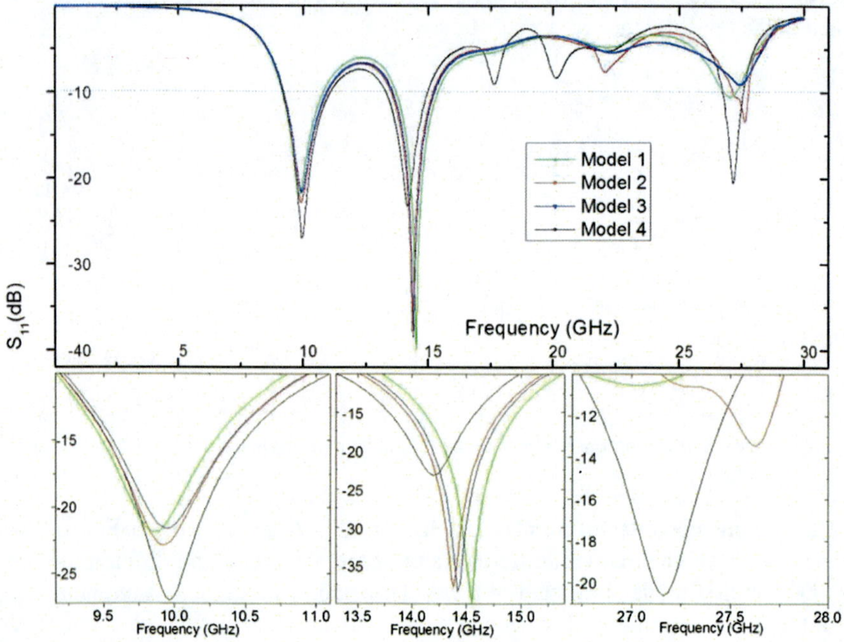

Fig. 2 Simulated S_{11} response of the antenna from 0–30 GHz with focus on 10, 15 and 27 GHz communication bands

For the X band, Model 1 shows a response at 9.84 GHz with a $|S_{11}|$ of 21.86 dB with a bandwidth ranging from 9.17–10.79 GHz. The response frequency shifts slightly to higher frequencies with the other models. The frequency ranges for Model 2, 3 and 4 were found to be at 9.17–10.96 GHz, 9.21–10.95 GHz and 9.19–11.10 GHz, respectively, with maximum $|S_{11}|$ of 22.82 dB at 9.91 GHz, 21.57 dB at 9.95 GHz and 26.96 dB at 9.96 GHz, respectively.

For the 5G band, Model 1 and 2 had responses at the frequency ranges 26.76–27.29 GHz and 27.14–27.76 GHz, respectively, with maximum $|S_{11}|$ of 10.53 dB at 27.05 GHz and 13.41 dB at 27.63 GHz, respectively. Model 3 shows no response at the 5G range. Model 4 has the best response with a bandwidth ranging from 26.74–27.56 GHz with maximum $|S_{11}|$ of 20.60 dB at 27.16 GHz. Note that, Model 4 shows a remarkably high $|S_{11}|$ at the 5G range without much depreciation in the response at the X band which makes it highly suitable for dual X band and 5G applications.

In addition to the responses at the X band and 5G band, these models also show an additional response around 15 GHz which can be used for some 5G radio applications. The following are the ranges and maximum response properties of the four models:

Model 1 Range 13.72–15.38 GHz; Maximum of 39.85 dB at 14.55 GHz.
Model 2 Range 13.52–15.19 GHz; Maximum of 37.80 dB at 14.38 GHz.
Model 3 Range 13.57–15.24 GHz; Maximum of 38.38 dB at 14.42 GHz.
Model 4 Range 13.31–14.96 GHz; Maximum of 23.22 dB at 14.20 GHz.

The gain and far-field patterns of Model 1 and Model 4 are shown in Fig. 3. The gain at 10 GHz for both antennas is 3.25 dB. At 27 GHz, the gain for Model 1 is 6.29 dB and the gain for Model 4 is 4.77 dB. While the gain for Model 1 at 27 GHz is higher than the gain of Model 4, it is important to note that the Model 1 has a much lower transmission and bandwidth at the 5G range than Model 4.

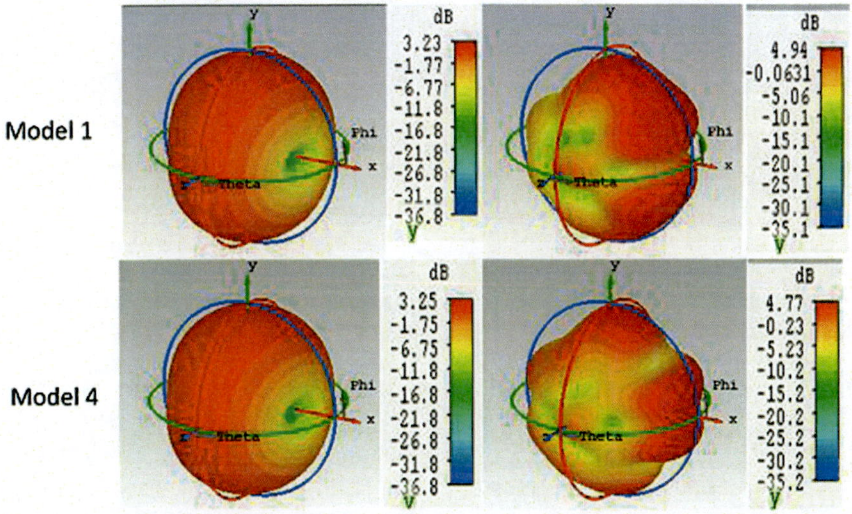

Fig. 3 Simulated values of gain and far-field pattern of Model 1 and Model 4 antennas at 10 GHz (left column) and 27 GHz (right column)

One of the possible reasons for the improvement in response of the circular patch antenna by adding rings is due to a capacitive coupling among the rings that resonates and enhances the response at 27 GHz.

5 Conclusion

A simple and cheap circular patch antenna model for X band and 5G applications is proposed. The antenna requires a Rogers TMM4 square substrate, a strip, disc, and rings made of copper. Improvements in performance by adding the ring structures have been detailed to enhance the transmission in the 5G range. The final model of the antenna displays a healthy bandwidth of 9.19–11.10 GHz and 26.74–27.56 GHz with gains of 3.25 dB and 4.97 dB, respectively.

References

1. Manekar, A.P., Dr. Varade, S.W.: Design and simulation of directional antenna for millimeter wave mobile communication. Int. J. Adv. Res. Comput. Commun. Eng. **5**(6) (2016)
2. Chauhan, B., Vijay, S., Gupta, S.C.: Millimeter wave mobile communications microstrip antenna for 5G-A future antenna. Int. J. Comput. Appl. **99**(19), 15–18 (2014)
3. Mamunur Rashid, M.D., Hossain, S.: Antenna solution for millimeter wave mobile communication (MWMC): 5G. Int. J. Sci. Res. Eng. Technol. (IJSRET) **3**(8) (2014). ISSN 2278-0882
4. Sidhu, A.K., et al.: Microstrip rectangular patch antenna for S and X band applications. In: IEEE WiSPNET 2016 Conference, pp. 248–251 (2016)
5. Lakrit, S.: Design of dual and wideband rectangular patch antenna for C and X band applications. Adv. Electromagn. **7**, 145–150 (2018)

Photon Density Distribution in Quantum Dot-Based Light-Emitting Diode

Shampa Guin(✉) and Nikhil Ranjan Das

Institute of Radio Physics and Electronics, University of Calcutta, 92, Acharya Prafulla Chandra Road, Kolkata 700009, W.B, India
shampaphysics@gmail.com, nrd@ieee.org

Abstract. In this study, the distribution of photons within a quantum dot (QD) light-emitting diode (LED) structure is presented by solving steady-state rate equations. The results show that the total photon density within the structure is distributed in a nonlinear manner along the direction of light propagation. Then the variation of photon density, contributing to LED power output, with facet reflectivity is shown for different injection current densities.

Keywords: Quantum dot · Facet reflectivity · Light-emitting diode · Photon distribution · Optical power

1 Introduction

Light emitters with high power and broad linewidth have attracted the researchers' interest because of its need for certain applications requiring high resolution and large probing depth [1, 2]. High power provides high detection sensitivity and wide dynamic range of operation, while large bandwidth provides high resolution and contrast. Superluminescent LED is such a source satisfying these requirements and finds application in optical coherence tomography, ghost imaging, compact random number generation, structural health monitoring, optical fiber sensors, etc. [2, 3]. Quantum Dot (QD) in the active layer may lead to significant improvement in the performance of the SLED due to its complete carrier confinement in all the spatial directions [4].

2 Theoretical Background

The schematic cross section of a QD SLED is shown in Fig. 1. The bias is applied between the top and bottom contacts, so the input current flows along the x-direction. The carriers will recombine within the active layer to generate photons, which are assumed to propagate along the z-direction. This gives rise to a relationship between the carriers and the photons. So, the carrier and photon rate equations are coupled to each other and have to solve simultaneously to find out the output optical power. Since the generated photons are spontaneously emitted, so they can propagate in all the directions. To include this effect, both the forward and backward propagating photons are assumed for the analysis. Under steady-state condition, the photon rate equation for forward propagating photons that travel along $+z$-direction can be written as follows:

© Springer Nature Singapore Pte Ltd. 2021
N. R. Das and S. Sarkar (eds.), *Computers and Devices for Communication*,
Lecture Notes in Networks and Systems 147,
https://doi.org/10.1007/978-981-15-8366-7_48

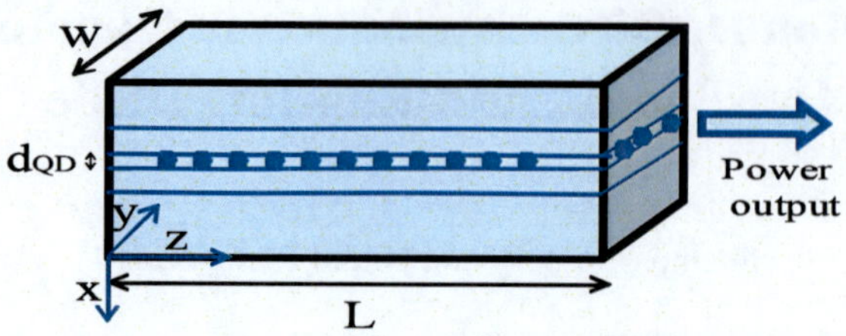

Fig. 1 Schematic structure of the LED structure with QD active layer

$$\frac{dP}{dz} = g(\langle n\rangle, \lambda)P(\langle n\rangle, \lambda, z) + \delta_{f\Omega}\delta_{f\lambda}\frac{\langle n\rangle N_{QD}}{d_{QD}\tau_{sp}v} \quad (1)$$

Similarly, the steady-state rate equation for the backward propagating photons,

$$\frac{dQ}{dz} = -g(\langle n\rangle, \lambda)Q(\langle n\rangle, \lambda, z) - \delta_{r\Omega}\delta_{r\lambda}\frac{\langle n\rangle N_{QD}}{d_{QD}\tau_{sp}v} \quad (2)$$

The steady-state carrier rate equation that relates both the carrier and photon is given by,

$$\frac{J}{vqd_{QD}L_{QD}} = \sum_{\lambda}\left[g(\langle n\rangle, \lambda)\{P(\langle n\rangle, \lambda) + Q(\langle n\rangle, \lambda)\}\right] + \frac{\langle n\rangle N_{QD}}{d_{QD}\tau_{sp}v} \quad (3)$$

where J is the injected current density, P and Q are forward and backward propagating photon densities, respectively, g is the material gain, $\delta_{f\Omega}\delta_{f\lambda}(\delta_{r\Omega}\delta_{r\lambda})$ is the fraction of produced emission contributing along the propagation line, v is the photon velocity within the active region, $<n>$ is the average number of carriers per QD, N_{QD} is the QD areal density, i.e., number of QD per unit area, d_{QD} is the thickness of one layer, τ_{sp} is the spontaneous recombination lifetime.

Assuming both fractions of forward and backward photons contributing at the output power are equal, δ_{Ω} can be expressed as [5],

$$\delta_{f\Omega} = \delta_{r\Omega} = \frac{\cos^{-1}\left(\frac{\eta_2}{\eta_1}\right).\frac{w}{L}}{2\pi} \quad (4)$$

where $\eta_1(\eta_2)$ are the refractive index of the dot (barrier) layer, L and w are the length and width of the active layer or the stripe contact, respectively.

If it is assumed that $<n>$ is constant, then the solution of (1) and (2) is found to be as follows:

$$P(\langle n\rangle, z, \lambda) = P_0e^{g(\langle n\rangle,\lambda)z} + \left(e^{g(\langle n\rangle,\lambda)z} - 1\right)P_{sp} \quad (5)$$

$$Q(\langle n\rangle, z, \lambda) = Q_0e^{-g(\langle n\rangle,\lambda)z} + \left(e^{-g(\langle n\rangle,\lambda)z} - 1\right)Q_{sp} \quad (6)$$

where P_0, Q_0 are the forward and backward traveling photon densities at $z = 0$ and

$$P_{sp} = \delta_{f\Omega}\delta_{f\lambda}\frac{\langle n\rangle N_{QD}}{d_{QD}\tau_{sp}v.g(\lambda)} \quad \text{and} \quad Q_{sp} = \delta_{r\Omega}\delta_{r\lambda}\frac{\langle n\rangle N_{QD}}{d_{QD}\tau_{sp}v.g(\lambda)} \quad (7)$$

From (5) and (6), it can be noted that photon densities are z-dependent. So, it is possible to substitute (5) and (6) into (3) to obtain a z-dependent carrier density. Using boundary conditions, the optical power at the output facet is calculated.

3 Results and Discussion

For the study, InAs QD having height 10 nm with GaAs barrier of height 30 nm is taken for example. Figure 1 shows the distribution of total photon density, which is the sum of the forward and backward propagating photon densities, for different injected current densities. It was seen that both the forward and backward traveling photon densities increase along their respective propagation directions. This is due to the fact that the emitted photons are getting amplified as they propagate through the active gain medium. It may be noted from the inset of Fig. 2 that there exists a minimum at which average number of carriers is maximum. This is because the photons are generated by the recombination of carriers. It may also be noted that the minimum photon density occurs at the middle of the device length because of equal facet reflectivities. The device with unequal facet reflectivities shows that the minimum photon density position does not occur at the midpoint of the device length.

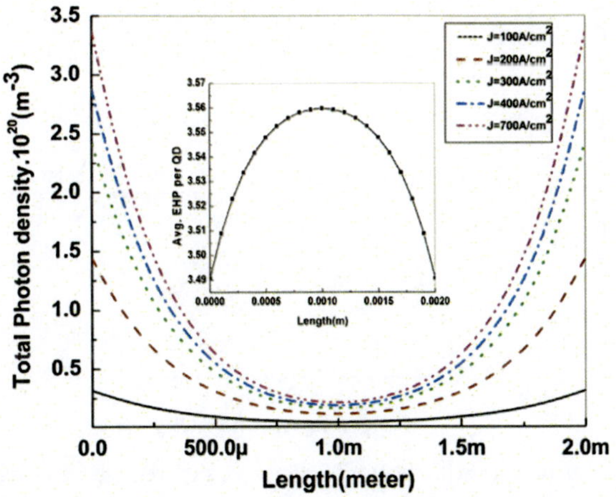

Fig. 2 Distribution of photon density along the length for different injection current densities. Inset shows the distribution of carriers for 100 A cm^{-2}

Figure 3 shows the variation of spontaneous recombination lifetime with energy and it is around ns, which suggests that fast spontaneous emission is responsible for high output power. The variation of output photon density with facet reflectivity for different injection current densities is shown in Fig. 4. Output photon density decreases as the facet reflectivity increases. This is because the backward reflection of photons increases the backward photon density which decreases the carries within the device and results in gain saturation. As a result, the forward photon density decreases which in turn decreases the output power.

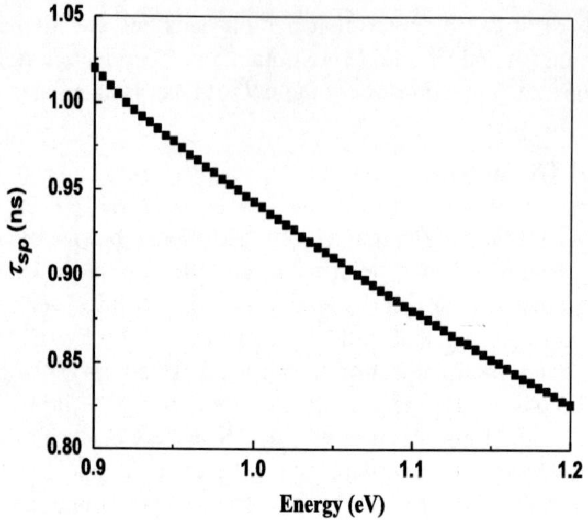

Fig. 3 Recombination lifetime (spontaneous emission) versus energy

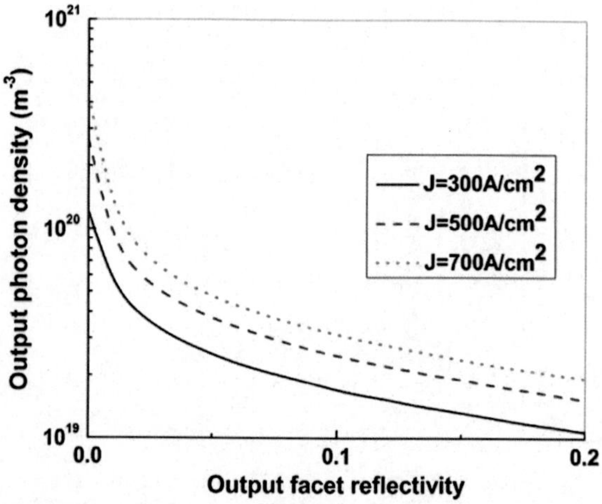

Fig. 4 Output photon density with facet reflectivity for different injection current densities

4 Conclusion

To obtain the device characteristics of a quantum dot light-emitting diode, the steady-state rate equations for carriers and both forward and backward traveling photon densities are solved. The forward and backward traveling photon densities increase along their respective propagation directions. As a result, there exists a minimum point within the device where number of carriers is maximum. The output photon density decreases with increase in facet reflectivity. To obtain high output power, the facet reflectivity should be very small.

References

1. Heo, D.C., Song, J.D., Choi, W.J., Lee, J.I., Jung, J.C., Han, I.K.: High power broadband InGaAs/GaAs quantum dot superluminescent diodes. Electron. Lett. **39**(11), 863–865 (2003)
2. Goldberg, G.R., Ivanov, P., Ozaki, N., Childs, D.T.D., Groom, K.M., Kennedy, K.L., Hogg, R.A.: Gallium nitride light sources for optical coherence tomography. In: Proceedings, vol. 10104, Gallium Nitride Materials and Devices XII; 101041X (2017)
3. Vizbaras, K., Dvinelis, E., Simonyt, I., Trinkunas, A., Greibus, M., Songaila, R., Zukauskas, T., Kausylas, M., Vizbaras, A.: High power continuous-wave GaSb-based superluminescent diodes as gain chips for widely tunable laser spectroscopy in the 1.95–2.45 μm wavelength range. Appl. Phys. Lett. **107**(011103), 1–4 (2015)
4. Guin, S., Das, N.R.: Enhancement of optical gain in quantum dot ensemble with electric field. Superlatt. Microstruct. **125**, 151–158 (2019)
5. Park, J., Li, X.: Theoretical and numerical analysis of superluminescent diodes. IEEE J. Light. Tech. **24**, 2473–2480 (2006)

Check for updates

Parabolic Pulse Generation by Dispersion Increasing Chalcogenide Fiber (DICF) in Normal Dispersion Regime

Binoy Krishna Ghosh[1(✉)], Somen Adhikary[1], Roshmi Chatterjee[1], Debasruti Chowdhury[1], Navonil Bose[2], Dipankar Ghosh[3], and Mousumi Basu[1]

[1] Department of Physics, Indian Institute of Engineering Science and Technology, Shibpur, Howrah, West Bengal 711103, India
binoy.phys@gmail.com, adhikary.somen@gmail.com, roshmic987@gmail.com, debasruti.chowdhury@gmail.com, mousumi.basu@gmail.com

[2] Department of Physics, Supreme Knowledge Foundation Group of Institutions, 1 Khan Road, Mankundu, Chandannagar, Hooghly, West Bengal 712139, India
navonil05@gmail.com

[3] Department of Basic Science, MCKV Institute of Engineering, Liluah, Howrah, West Bengal 711204, India
ghosh.dip@gmail.com

Abstract. The parabolic pulse (PP) generation within an optical fiber occurs in the presence of gain. The required amount of gain can be supplied either physically or virtually by suitably tapering the core radius throughout the length of the fiber. Regarding PP generation, designing a dispersion increasing fiber (DIF) is undesirable as it introduces negative gain within the fiber. In this work, we have developed a special kind of a dispersion increasing chalcogenide fiber (DICF) where the nonlinearity-induced virtual gain plays the dominant function. In spite of the role of dispersion in reducing the effective gain, an efficient PP is generated in a sufficiently shorter fiber length without supplying any physical gain due to the much more substantial contribution from the nonlinearity of the fiber.

Keywords: Parabolic pulse (PP) · Virtual gain · Dispersion increasing fiber (DIF) · Normal dispersion chalcogenide fiber (NDCF)

1 Introduction

Recently advanced nonlinear fiber optics has motivated researchers in generation of a new kind of pulse, namely optical similaritons [1, 2]. Conserving definite correlations among pulsewidth, frequency chirp, and energy, this type of pulse can sustain its stability against perturbation and collisions. Once converging asymptotically to this kind of shape, a pulse can endure high value of nonlinearity and is free from the deleterious effect of wave breaking [3]. Plethora of research activities toward parabolic similariton generation have been observed in last two decades [4, 5], as they have tremendous

© Springer Nature Singapore Pte Ltd. 2021
N. R. Das and S. Sarkar (eds.), *Computers and Devices for Communication*, Lecture Notes in Networks and Systems 147,
https://doi.org/10.1007/978-981-15-8366-7_49

applications in supercontinuum generation through pulse compression [6], pulse reshaping, in high-power femtosecond lasers, all optical signal regeneration, processing, and many more [7]. Since the formation of parabolic pulse (PP) is possible only in a gain-induced Normal Dispersion Fiber (NDF) [8], the amount of gain as necessity can be supplied by an amplifier or by the use of a virtually gained dispersion decreasing fiber [9]. Off late, researchers also show considerable interest in highly nonlinear fibers such as chalcogenide-made specialty fibers [10]. Enormously wide transmission window ranging from 1.3 μm up to 20 μm within the mid-infrared [11] and large nonlinear coefficient of this kind of materials make them as one of the best choices. In the present work, our objective is to design a dispersion increasing fiber made of chalcogenide materials possessing very high nonlinear coefficients and thereby generating PP at a shorter fiber length by suitable management of the nonlinearity-induced virtual gain without any aid of the physical gain.

2 Parabolic Pulse Production Through a Tapered Fiber

The pulse propagation through a nonlinearity varying fiber can be perfectly represented by the nonlinear Schrödinger equation (NLSE) in presence of gain, as given by Ghosh et al. [11], and Ghatak and Thyagarajan, [12]

$$i\frac{\partial A}{\partial z} = \frac{1}{2}\beta_2(0)D(z)\frac{\partial^2 A}{\partial T^2} - \gamma(0)\Gamma(z)|A|^2A + \frac{i}{2}(g-\alpha)A \tag{1}$$

where the slowly varying envelope of the pulse is $A(z,T)$, $\gamma(0)$ and $\beta_2(0)$ are the initial value of nonlinearity and GVD, whereas $\Gamma(z)$ and $D(z)$ represent the respective normalized variations along the fiber length. Each of these lengthwise variations introduces some amount of virtual gain, so that the effective gain coefficient can be represented in terms of physical fiber length as

$$\begin{aligned}\delta(z) &= \frac{1}{D(z)}\left\{G_0 - \frac{1}{\beta_2(z)}\frac{d\beta_2(z)}{dz} + \frac{1}{\gamma(z)}\frac{d\gamma(z)}{dz}\right\}\\ &= \frac{G_0}{D(z)} + G_D(z) + G_N(z)\end{aligned} \tag{2}$$

where $G_0 = (g - \alpha)$ denotes the overall gain above the loss; now the dispersion-induced and nonlinearity-induced virtual gains are, respectively,

$$\begin{aligned}G_D(z) &= -\frac{1}{D(z)\beta_2(z)}\frac{d\beta_2(z)}{dz},\\ G_N(z) &= \frac{1}{D(z)\gamma(z)}\frac{d\gamma(z)}{dz}.\end{aligned} \tag{3}$$

With an intention of PP geneartion through a fiber, the symmetrized split-step Fourier method can be applied to solve the NLSE numerically, and the accuracy of the obtained pulse compared to an ideal parabolic pulse is estimated by the misfit parameter [11, 13] as described below:

$$\mu^2 = \int\left[|u|^2 - |u_P|^2\right]^2 d\tau / \int|u|^4 d\tau \tag{4}$$

where $|u(\tau)|^2$ and $|u_P(\tau)|^2$ are the normalized power profiles of the propagating pulse and the exact parabolic pulse, respectively.

3 Designing a Tapered Chalcogenide Fiber for PP Generation

It is well known that the virtual gain originated due to the GVD variation along the fiber length is much more prominent than the same due to the variation of the nonlinear coefficient in case of the silica-based fibers. Hence, in order to realize the effect of virtual gain due to nonlinearity in case of generation of PP, one must design a fiber which is able to produce a higher lengthwise variation of nonlinearity. To achieve this goal, a dispersion increasing chalcogenide fiber (DICF) with step index profile is designed with As_2S_5 as core [14] and $Ge_{17}Ga_4Sb_{10}S_{69}$ glass [15] as cladding material. The scalar wave equation [12] is solved to estimate β_2 and γ as a function of core radius of the DICF at the operating wavelength of 1550 nm as shown in Fig. 1. From this figure, two specific core radii of the fiber, viz $a \sim 2.24$ µm and $a \sim 3.44$ µm can be identified where the GVD and nonlinearity attain their maximum values, respectively ($\beta_2 \sim 368.28$ ps^2/km and $\gamma \sim 170.69$ W^{-1} km^{-1}). Since our intention is to obtain PP through dispersion increasing profile, we have selected the operating core radius region from $a \sim 1.73$ µm to $a \sim 2.24$ µm within which both the GVD and the nonlinearity increase. Although β_2 has a slight rise from 363.5 to 368.28 ps^2/km, γ varies sharply between 26.53 and 98.85 W^{-1} km^{-1} in this region.

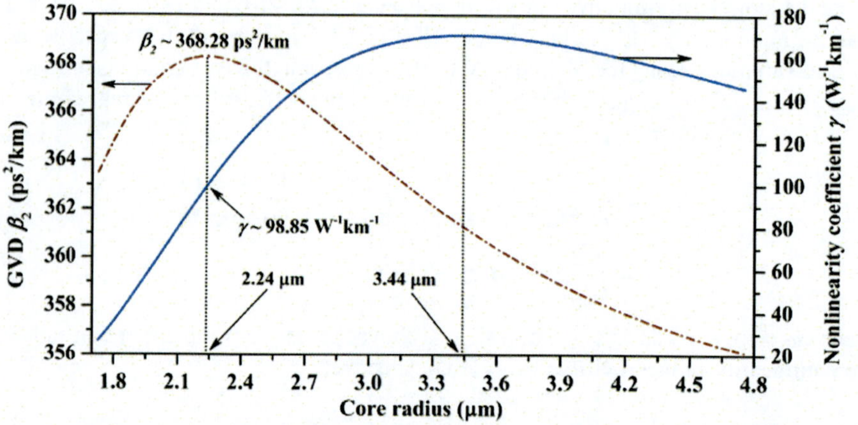

Fig. 1 Core radius dependence of the GVD and nonlinear coefficient of the proposed fiber

To study the effect of virtual gain due to nonlinearity on PP generation by the proposed DICF, one linear profile (LP) and a number of nonlinear profiles (NLPs) of core radius variations along the fiber length are considered. In each case, the core radius of the fiber has been increased for a length of 60 m, and the corresponding change in nonlinearity has been depicted in Fig. 2.

In order to observe the pulse progress through the profiles mentioned above, an initially chirped Gaussian pulse with pulse width $T_0 = 2$ ps, peak power $P_0 = 8$ W, and

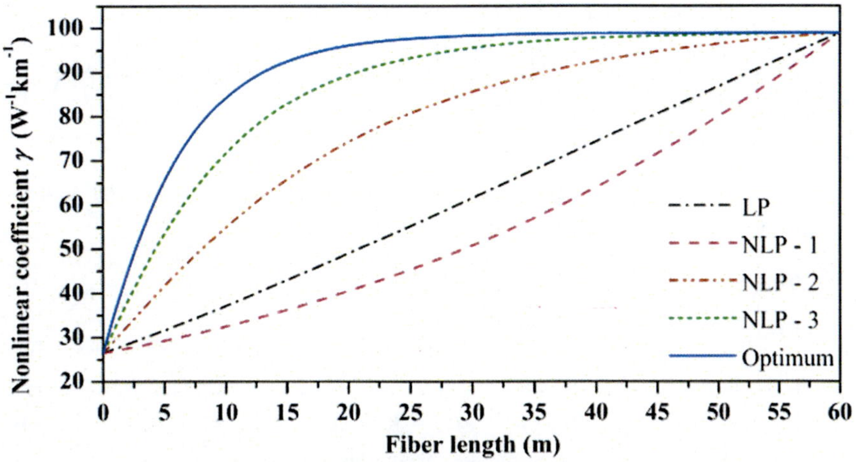

Fig. 2 Variation of the nonlinear coefficient with fiber length

initial chirp parameter $C = 2$ has been fed through each of the passive DICFs separately. In all the cases, G_0 is considered to be zero. As chalcogenide fibers generally exhibit higher amount of attenuations (~1.5 dB/m), the same amount of physical gain has been considered to be supplied in this case to make the overall gain zero.

The reshaping of the pulse into the parabolic form is verified by calculation of the misfit parameter μ [11, 13] throughout the fiber length. Here, the pulse form with $\mu \leq 0.02$ has been considered as an efficient parabolic one and the minimum fiber length at which the parabolic pulse reshaping is obtained is considered as optimum length (L_{opt}). The lengthwise evolution of μ through each fiber profile identifies the optimum length of PP generation as depicted in Fig. 3. It is observed from the figure that generation of parabolic pulses within 60 m of fiber length is not possible through all the core radius profiles. Thereby, the core radius profiles permitted for the PP generation through the DICF are found out, and the profile which is capable of generating efficient PP form at the minimum length among all types of fiber profiles is regarded as the optimum fiber profile. Amount of the dispersion as well as the nonlinearity-induced virtual gain has been calculated for each fiber profile and is presented in Table 1.

It is clearly evident from the table that the nonlinearity-induced virtual gain in each case over-compensates the negative dispersion-induced virtual gain and thus has a visibly dominant contribution in efficient PP formation within a very short optimum fiber length. The lengthwise variations of the dispersion and nonlinearity-induced virtual gain in case of optimum fiber profile are presented in Fig. 4.

Finally, the merit of our proposed DICF with optimum radius profile is examined by comparing it with a normal dispersion chalcogenide fiber (NDCF). This NDCF is formed by the same material combination, and its core radius is taken to be fixed at that particular value for which the GVD reaches its maximum value for the DICF (as depicted in Fig. 1). An input Gaussian pulse with same initial parameters has been fed to the fiber input. Here, a constant gain is maintained throughout the fiber length, and its value (~43.48 km^{-1}) is taken to be equal to the path-averaged total virtual gain obtained in

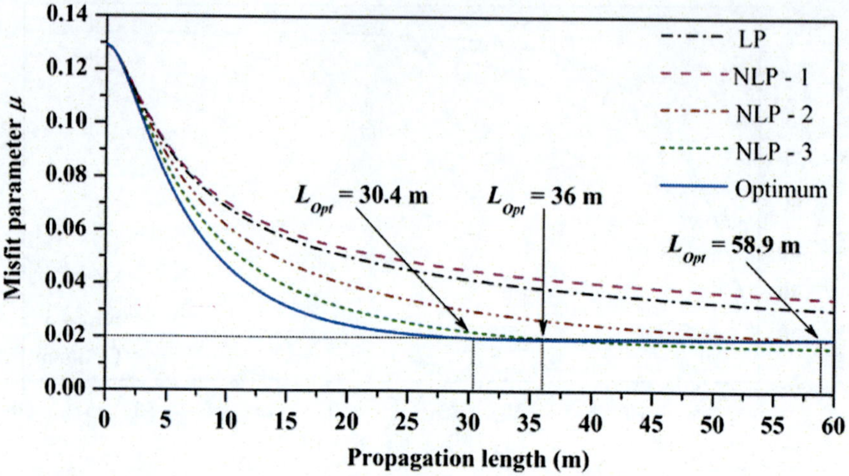

Fig. 3 Lengthwise variation of misfit parameter for different core radius profiles

Table 1 Dispersion and nonlinearity-induced virtual gain for different profiles

Profile name	L_{Opt}(m)	Path-averaged virtual gain (km^{-1})	
		G_D due to β_2	G_N due to γ
Linear (LP)	–	−0.22	22.04
NLP-1	–	−0.22	22.83
NLP-2	58.9	−0.23	23.04
NLP-3	36.0	−0.37	36.45
Optimum	30.4	−0.44	43.92

case of the DICF with the optimum radius profile. The lengthwise evolution of μ toward the PP generation through this fiber is plotted in Fig. 5. This figure substantiates the fact that in spite of the equal amount of gain in both the cases, our proposed fiber is efficient enough to generate PP in sufficiently shorter length (L_{Opt} ~ 30.4 m) in comparison with the NDCF (L_{Opt} ~ 66 m). The output pulses from both the fibers as depicted in Fig. 6 show the linear frequency chirp throughout the pulse profiles. The lengthwise change in pulse width and power of an initial Gaussian pulse into a parabolic shape through our proposed fiber has been illustrated by the contour plot in Fig. 7.

4 Conclusion

The essential amount of gain for PP generation can be provided in a passive way in dispersion decreasing fibers by the proper tapering of the core radius where the dispersion-induced virtual gain usually acts as the chief contributor. Therefore, the dispersion increasing nature in silica fibers is commonly avoided as it supplies the negative

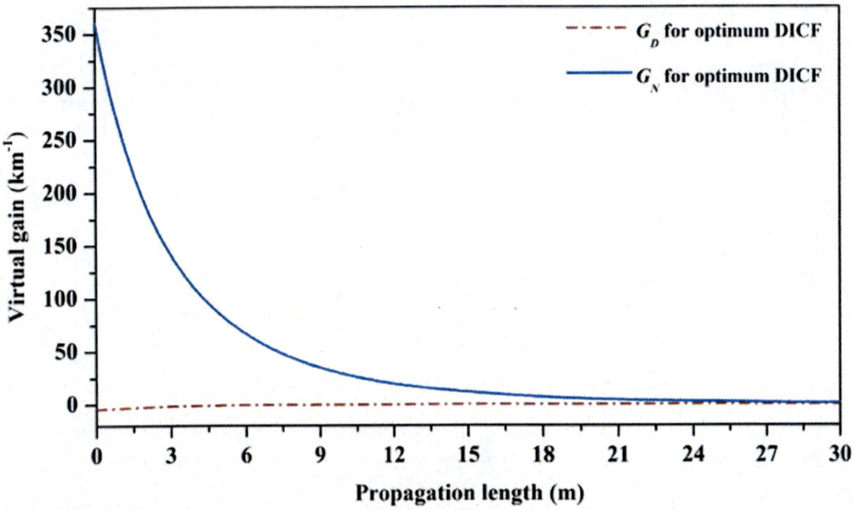

Fig. 4 Lengthwise variation of the nonlinearity and GVD induced virtual gain for optimum nonlinear fiber profile

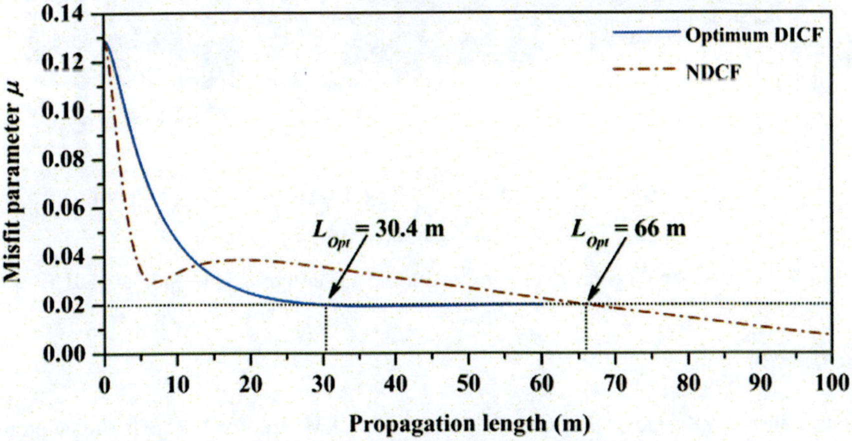

Fig. 5 Lengthwise variation of misfit parameter for optimum DICF and the NDCF

virtual gain which is harmful for PP formation. In this work, a dispersion increasing chalcogenide fiber (DICF) is designed and optimized with the optimum core radius profile in such a way that the nonlinearity-induced virtual gain overshadows the negative dispersion-induced virtual gain. As a result, when compared to a standard normally dispersive chalcogenide fiber with the external gain equal to the path averaged value of virtual gain of the DICF, our proposed fiber exhibits superior performance by reducing almost 54% of optimum length for PP generation.

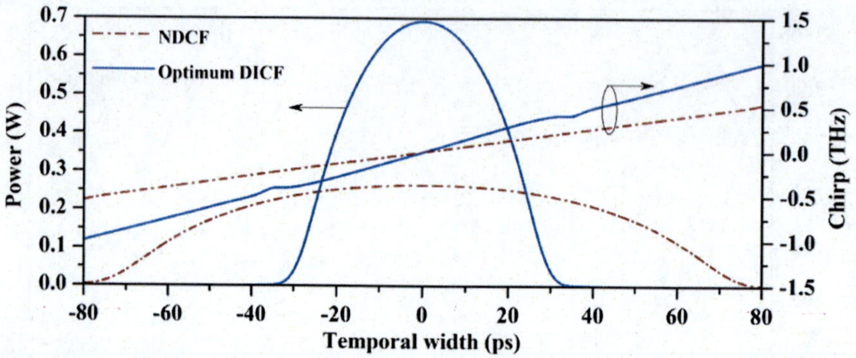

Fig. 6 Output temporal pulse form and chirp value for the NDCF and the optimum DICF

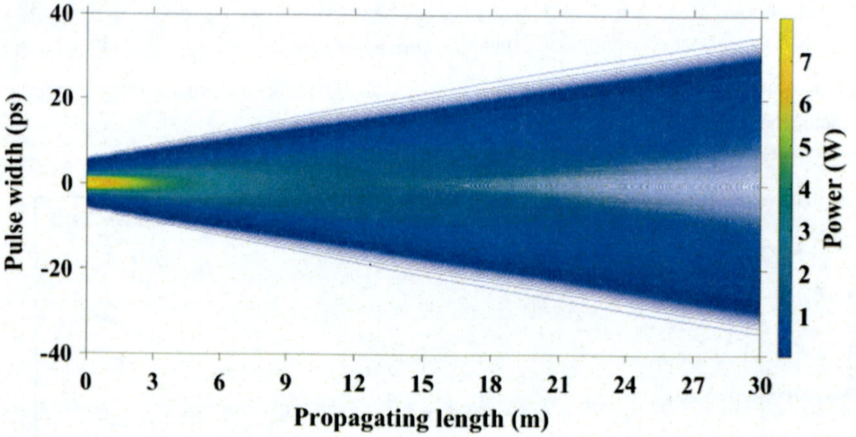

Fig. 7 Pulse reshaping in terms of pulse width and power through the optimum DICF

References

1. Fermann, M.E., Kruglov, V.I., Thomsen, B.C., Dudley, J.M., Harvey, J.D.: Self-similar propagation and amplification of parabolic pulses in optical fibers. Phys. Rev. Lett. **84**, 6010–3 (2000)
2. Yu, W., Zhou, Q., Mirzazadeh, M., Liu, W., Biswas, A.: Phase shift, amplifcation, oscillation and attenuation of solitons in nonlinear optics. J. Adv. Res. **15**, 69–76 (2019)
3. Anderson, D., Desaix, M., Karlsson, M., Lisak, M., Quiroga Teixeiro, M.L.: Wave-breaking-free pulses in nonlinear-optical fibers. J.Opt. Soc. Am. B **10**, 1185–1190 (1993)
4. Mei, C., Yuan, J., Li, Feng., Yan, B., Sang, X., Zhou, X., Wu, Q., Wang, K., Long, K., Yu, C.: Generation of parabolic pulse in a dispersion and nonlinearity jointly engineered silicon waveguide taper. Opt. Comm. **448**, 48–54 (2019)
5. Finot, C., Dudley, J.M., Kibler, B., Richardson, D.J., Millot, G.: Optical parabolic pulse generation and applications. IEEE J. Quant. Elec. **45**, 1482–1489 (2009)
6. Li, Y., Sang, X.: Mid-infrared supercontinuum generation and its application on all-optical quantization with different input pulses. Chin. Phys. B **28**, 054206 (2019)

7. Hirooka, T., Nakazawa, M.: All-optical 40-GHz time-domain Fourier transformation using XPM with a dark parabolic pulse. IEEE Photon. Technol. Lett. **20**, 1869–1871 (2008)
8. Finot, C., Millot, G. and Dudley, J.M.: Asymptotic characteristics of parabolic similar ton pulses in optical fiber amplifiers. Opt. Lett. **29**, 2533–2535 (2004)
9. Wabnitz, S., Finot, C.: Theory of parabolic pulse propagation in nonlinear dispersion decreasing optical fiber amplifiers. J. Opt. Soc. Am. B **25**, 614–21 (2008)
10. Cheng, T., Nagasaka, K., Tuan, T.H., Xue, X., Matsumoto, M., Tezuka, H., Suzuki, T., Ohishi, Y.: Mid-infrared supercontinuum generation spanning 2.0 to 15.1 μm in a chalcogenide step-index fiber. Opt. Lett. **41**, 2117–2120 (2016)
11. Ghosh, D., Basu, M., Sarkar, S.: Generation of self-similar parabolic pulses by designing normal dispersion decreasing fiber amplifier as well as its staircase substitutes. J. Lightwave Technol. **27**, 3880–7 (2009)
12. Ghatak, A., Thyagarajan, K.: Introduction to Fiber Optics. Cambridge University Press (1999).
13. Ghosh, B.K., Ghosh, D., Basu, M.: Potential use of nonlinearity-induced virtual gain on parabolic pulse formation in highly nonlinear tapered fiber system. J. Opt. **21** 045503 (2019)
14. Salem, A.B., Diouf, M., Cherif, R., Wague, A.d., Zghal, M.: Ultraflat-top mid infrared coherent broadband super continuum using all normal As2S5-borosilicate hybrid photonic crystal fiber. Opt. Eng. **55**(6), 066109 (2016)
15. Kohoutek, T., Mizuno, S., Suzuki, T., Ohishi, Y., Matsumoto, M., Misumi, T.: Third-harmonic generation measurement of nonlinear optical susceptibility $\chi 3$ of Ge-Ga-Sb-S chalcogenide glasses proposed for highly nonlinear photonic fibers. J. Opt. Soc. Am. B **28**(2), 298–305 (2011)

Check for updates

Modes and Coupling in Seven-Core Optical Fiber

Sonali Basak(✉), Santu Sarkar, and Nikhil Ranjan Das

Institute of Radio Physics and Electronics, University of Calcutta, 92, Acharya Prafulla Chandra Road, Kolkata, West Bengal 700009, India
basaksonali1@gmail.com, santu.irpe@gmail.com, nrd@ieee.org

Abstract. This paper presents a study on the propagation of modes of electromagnetic wave through a homogeneous multicore fiber. Complete view for fundamental modes for each core and other linearly polarized modes are obtaining here. Coupling between fundamental modes and other different cladding modes are presented in this paper. This study includes the coupling coefficient between multiple modes under periodic perturbation conditions for hexagonal seven core configuration.

Keywords: Multicore fiber · Space division multiplexing · Mode · Coupling coefficient

1 Introduction

The continuous demand for large transmission capacity leads to the research of high-capacity optical transmission systems and networks. The theoretical limit of single-mode single-core fiber is reaching up to 100 Tb/s due to nonlinear effect and limit of power transmission through single mode [1]. To further increase the capacity, space division multiplexing (SDM) has been proposed [2]. Two possible ways of SDM can be realized. First one is to accommodate multiple spatial channels in a single fiber cladding which is known as multicore fiber transmission and the another is to realized multiple modes in a fiber, which is few-mode fiber or multimode fiber [3]. Recently, space division multiplexing using multicore fibers (MCFs) is an emerging technology to overcome the limit of transmission capacity by supporting single mode or number of modes within a single core, and it does not require complex multiple-input multiple-output signal processing at the receiver side. Multicore fibers are specially two types: one is solid and another is holey fiber. Holey fibers are completely made from photonic crystal (PCF). Further MCFs can be categorized into homogeneous and heterogeneous MCFs. In homogeneous MCFs, the cores are identical (core size and/or refractive indices) and have been fabricated to realize long haul transmission with low crosstalk [4, 5]. In case of heterogeneous MCFs, all non-identical cores are arranged to limit the core-to-core coupling than the homogeneous structure [6]. In addition to this, a special kind of structure such as trench-assisted MCF (TA-MCF) and hole-assisted MCF (HA-MCF) that realizes much smaller crosstalk

© Springer Nature Singapore Pte Ltd. 2021
N. R. Das and S. Sarkar (eds.), *Computers and Devices for Communication*,
Lecture Notes in Networks and Systems 147,
https://doi.org/10.1007/978-981-15-8366-7_50

and larger effective area (A_{eff}) comparing to normal MCF with step-index profile also has been proposed [7, 8]. The analyses of coupling coefficient in most works utilize the two-core model, i.e., directional coupler, as a guideline to investigate crosstalk. Intercore crosstalk in MCFs is an issue in expanding the transmission capacity of MCFs. Recently, the crosstalk mechanism in MCF has been uncovered by analytically and experimentally. Some implementations of various types of MCFs already started in few years ago. Taylor et al. (2006) implemented multicore fiber as an interconnect operating at 850 nm has low crosstalk and can support 1150 channels/mm [9]. Ishida et al. (2013) investigated the longitudinal power decay of weakly coupled multicore fiber (MCF) crosstalk estimation [10].

In this paper, mode-coupling dynamics for homogeneous seven-core MCFs with hexagonal geometry and specific index profiles are chosen that are compatible with commercially available multicore fiber in the market. The different modes for core and cladding are derived by finite element method. The coupling coefficients for between core and cladding modes for seven-core hexagonal MCF with various perturbations are calculated and that are not studied in previous work.

2 Theory

2.1 Analysis of Refractive-Index Distribution in the Coupled Region

The refractive-index profile of the step-index multicore fiber is shown Fig. 1. The refractive indices for the core and cladding are n_{co} and n_{cl}, respectively. The relative refractive-index difference between core and cladding is Δ_1. The radius of each is a_{co}, and the diameter of cladding is a_{cl}.

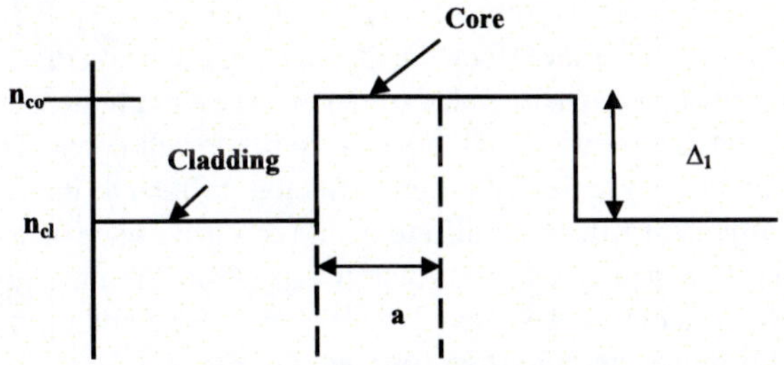

Fig. 1 Refractive-index profiles and cross-sectional dimensions for normal step-index structure

2.2 Mode Coupling Coefficient in the Coupled Region

The expression for the coupling coefficient considering the acoustic wave perturbation form first core (m) to second core (n) of MCF can be written as

$$C_{12} = \frac{\omega \varepsilon_0}{4P_{\text{Total}}} \iint E_1^*(x, y) \Delta \varepsilon_r(x, y) E_2(x, y) \text{d}x\text{d}y \tag{1}$$

where P_{total} is the addition of the optical mode powers that are involved in the coupling process, and it is given as following

$$P_{Total} = \frac{1}{2} \iint \mathrm{Re}\left[E_x H_y^* - E_y H_x^*\right] dxdy \tag{2}$$

The dielectric perturbation $\Delta\varepsilon_r(x, y)$ can be expressed as [11]

$$\Delta\varepsilon_r(x, y) = 2n_0 n_0 (1 + \chi) K^2 y u_0 \tag{3}$$

where n_0 is the refractive index of the effective region, $\Delta\varepsilon_r(x, y)$ for core region, $n_0 = n_{co}$ and for cladding region, $n_0 = n_{cl}$. The constant χ is the manifestation of the photoelastic effect, and u_0 is the amplitude of the fiber displacement, or equivalently the depth of the modulation.

3 Results and Discussion

The structure given in Fig. 2 is two dimensional, and consequently two-dimensional simulations were carried out to obtain mode field distributions. The MCF is a new structure that are recently under experimental investigation. The propagation dynamics of homogeneous seven-core MCFs consisting of a center core (namely, core 1) surrounded by six cores (core 2, 3 … 7) arranged in a hexagonal configurations. Diameters of cores and claddings are 8.3 μm ($2a_{co}$) and, 125 μm ($2a_{cl}$). The parameters of the structure taken in this study are standard and found from literature. The cladding was surrounded with an air boundary with dimensions of 200 μm × 200 μm square region as shown in Fig. 2. The configuration with air boundary realized to guide the cladding modes along the longitudinal direction. Refractive indices for core, cladding, and air regions were adjusted as 1.4507, 1.4367, and 1, respectively, as an example [11]. Figure 3 shows the fundamental core modes in the seven-core hexagonal configuration that are simulated in COMSOL. The normalized intensity value of the electric field indicated with the color bar as shown in figure. The mode amplitudes are same for all cores. The fundamental core modes in given core systems have same effective refractive indices for 1550 nm wavelength. In reality, their values are different, and this difference occurs at seventh or eighth digit, and it can be neglected for further investigation. We have given all the core modes of seven-core systems, but some of the higher-order cladding modes (different effective indices) for those configurations presented in Fig. 4.

When the effective indices are less than the effective indices of the fundamental modes (1.44666), this creates the electric field totally concentrated at the cladding region, and some portion of field are in cores. Electric field is concentrated at the core when effective indices go near the refractive index of the core. The field intensity reaches maximum up to 434(v/m) when it confines into the core region. The expression for coupling coefficient was given in Eq. (1) which is used for finding coupling coefficient between core and cladding region. $E_1(x, y)$ and $E_2(x, y)$ are the electric field for core and cladding modes that are part of the coupling process. The total power is addition of cladding and core power. Figure 4 shows coupling coefficient variations with respect to different u_0 for coupling between the center core to a first cladding mode (having

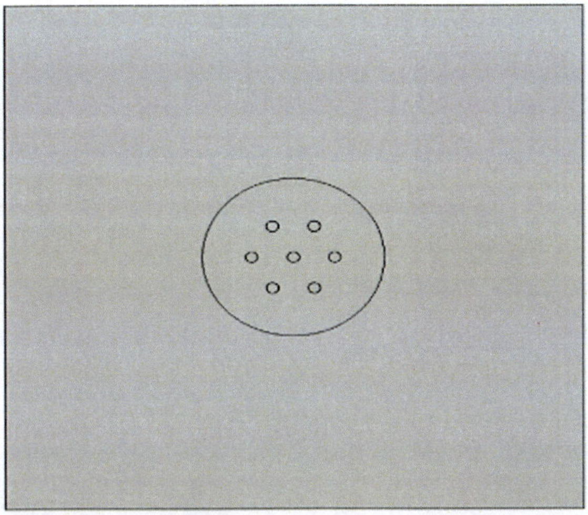

Fig. 2 Two-dimensional seven-core hexagonal multicore fiber configurations simulated in COMSOL

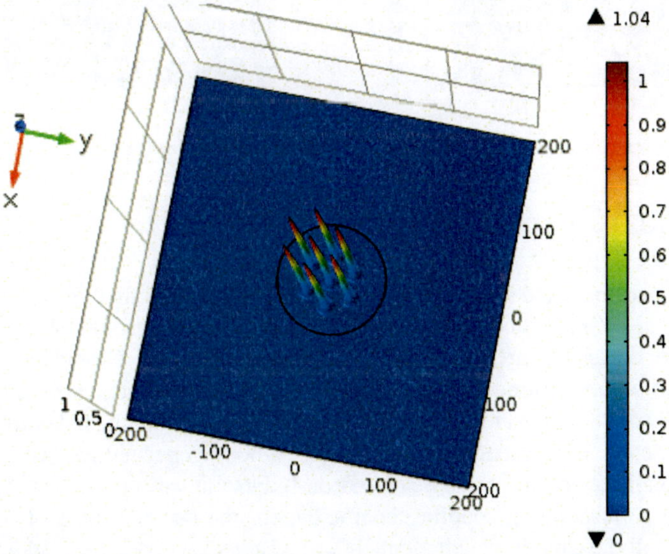

Fig. 3 Distributions of fundamental core modes for seven-core hexagonal configuration

effective refractive index 1.4376 (1)). The analysis is done by launching the light into the center core. This analysis is helpful to understand the coupling dynamics in a better way and provide us a tool to investigate the effects of different higher order cladding modes. The solution for coupling coefficient will be used to find the coupling length with different depth of index modulation.

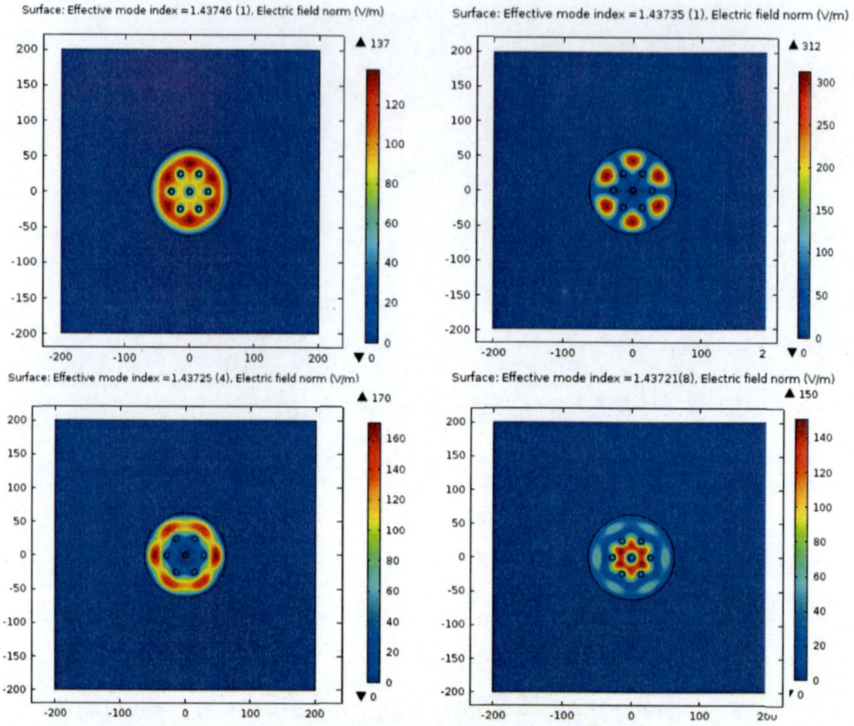

Fig. 4 Some of the supported higher-order cladding modes for the seven-core system

4 Conclusions

In this paper, we presented a detailed simulation of mode-coupling dynamics for seven-core hexagonal MCFs with identical cores. And the complete solutions for the mode amplitudes at distance z are obtained for all launching cores. Refractive-index profiles and cross-sectional dimensions for normal step-index structure and coupling theory under periodic perturbation are covered here. The core and cladding modes are classified by the numerically. The propagation dynamics of a homogeneous seven-core MCF can be applied to all the homogeneous seven-core MCFs with various geometry and index profiles. The corresponding coupling coefficient for the center core is also obtained. Furthermore, all the coupling coefficients $[C_{\text{mn}}]$ of adjacent cores and cladding or non-adjacent cores as a function of the depth of index modulation will be discussed in this process. It can be concluded that the coupling coefficients C_{12} can be adjusted by changing the depth of modulation. The mode-coupling coefficients of the cores will be helpful for evaluation of the crosstalk of the entire optical fiber. This process can be applicable for the further study of coupling coefficient of different types of MCFs.

References

1. Cai, J.-X., et al.: 64QAM based coded modulation transmission over transoceanic distance with > 60 Tb/s capacity. Presented at the Optics Fiber Communication Conference, Los Angeles, USA (2015)
2. Morioka, T.: New generation optical infrastructure technologies: "EXACT initiative" towards 2020 and beyond. In: Proceedings of 14th Opto Electronics and Communications Conference (Institute of Electrical and Electronics Engineers) (2009)
3. Mizuno, T., et al.: Dense space division multiplexed transmission over multicore and multimode fiber for long-haul transport systems. IEEE J. Lightwave Technol. **34**, 1484–1493 (2016)
4. Takenaga, K., et al.: An investigation on crosstalk in multi-core fibers by introducing random fluctuation along longitudinal direction. IEICE Trans. Commun. E **94**-B, 409–416 (2011)
5. Sakaguchi, J., et al.: 305 Tb/s space division multiplexed transmission using homogeneous 19-core fiber. IEEE J. Lightwave Technol. **31**, 554–562 (2013)
6. Koshiba, M., et al.: Heterogeneous multi-core fibers: proposal and design principle. IEICE Electron. Express **6**(2), 98–103 (2009)
7. Ye, F., et al.: Simple analytical expression for crosstalk estimation in homogeneous Trench-assisted multi-core fibers. Opt. Soc. Am. **22**, 23007–23018 (2014)
8. Ziolowicz, A., et al.: Hole-assisted multicore optical fiber for next generation telecom transmission systems. Appl. Phys. Lett. **105**, 081106-081106-4 (2014)
9. Taylor, D.M., et al.: Demonstration of multi-core photonic crystal fiber in an optical interconnect. Electron. Lett. **42**(6), 331–332 (2006)
10. Ishida, I., et al.: Longitudinal power decay of weakly coupled multi-core fiber. IEEE Photon. Technol. Lett. **25**(13), 1270–1273 (2013)
11. Atasever, T.C.: Periodic Microbending Induced Coherent Mode Coupling in Multicore optical Fibers Thesis. University of California, Irvine (2016)

Design of an Ultra-Wideband Polarization-Insensitive Frequency-Selective Absorber

Ankita Indu[1](✉), S. Mondal[1], and P. P. Sarkar[2]

[1] Institute of Radio Physics and Electronics, University of Calcutta, 92, Acharya Prafulla Chandra Rd,, Kolkata, West Bengal 700009 , India
nkitaindu@gmail.com, santanumondal2008@rediffmail.com
[2] Department of Engineering and Technological Studies, University of Kalyani, Kalyani, West Bengal 741235, India
parthabe91@yahoo.co.in

Abstract. This paper presents a miniaturized novel frequency-selective surface (FSS) absorber having ultra-wide absorption band. The proposed FSS absorber consists of a single-layer FSS printed on a grounded lossy dielectric with an air spacer and lumped resistors. The proposed design shows almost more than 90% absorptivity from 3.61 to 10.63 GHz, i.e., fractional bandwidth is 98.6%. A wide absorption band has been achieved by combining the effect of dielectric loss and ohmic loss of lumped resistors. The proposed structure is polarization insensitive being fourfold symmetric in nature. The structure shows more than 80% absorptivity upto 45° under oblique incidence at TE polarization.

Keywords: Frequency-selective absorber · Absorptivity · Polarization insensitive · Ultra-wideband

1 Introduction

Microwave absorbers have been used in wide range of applications such as radar cross section reduction of objects, electromagnetic interference (EMI) reduction, and radar-absorbing materials (RAM) in stealth technology [1–3]. Jaumann absorber or wedge-tapered absorber, and plasma absorber are used as conventional EM absorbers, but these structures are bulky in nature [4–6]. This drawback can be overcome by using absorptive frequency-selective surface (AFSS), which are ultra-thin, compact, and ease of fabrication. A triple-band metamaterial absorber was proposed in [7], but this kind of absorber suffers from narrow bandwidth problem due to the resonance-based absorption mechanism. The absorption bandwidth of these absorbers can be enhanced either by introducing multiple resonating structure within a single-layer unit cell [8] or using multilayer structure [9]. A single-layer ultra-thin ultra-wideband metamaterial absorber has been reported in [10], but this structure is sensitive to polarization. An absorptive

© Springer Nature Singapore Pte Ltd. 2021
N. R. Das and S. Sarkar (eds.), *Computers and Devices for Communication*,
Lecture Notes in Networks and Systems 147,
https://doi.org/10.1007/978-981-15-8366-7_51

FSS structure is reported with absorbption band at lower frequency and a transparent window above this band introducing a bandpass FSS [11]. A triband absorber is designed using Sierpinski–Minkowski fractal structure, and lumped resistors are used to increase the absorption ratio by controlling the induced surface current [12]. In [13], a single-layer Circuit analog absorber is reported. The absorber in [14] is composed of artificial impedance surfaces and resistor–capacitor layer, but bandwidth is not sufficiently increased. A wideband absorption bandwidth is obtained using a random metasurface nonlinear pattern on resistive sheet [15], which leads to design complexity. A single-layer microwave absorber is designed using cross-dipole array and lumped resistors to reduce the RCS [16], still bandwidth is below 75%.

In this paper, a novel FSS absorber is proposed, and its bandwidth is enhanced using lumped resistors and an air spacer with height of $0.084\lambda_L$, where λ_L corresponds to lowest frequency of absorption. The proposed design is polarization insensitive and has almost 90% absorptivity from 3.61 to 10.63 GHz. A parametric study is carried out to investigate the effects of different thickness of air spacer and lumped resistors on absorption band.

2 Design of the Microwave Absorber

The FSS unit cell is printed on a lossy FR4 dielectric substrate (relative permittivity $\varepsilon_r = 4.3$ and dielectric losstangent = 0.025) with thickness $h_1 = 0.5$ mm. An air spacer of thickness h_2 has been added between the FSS and metal ground. The top FSS is made of copper (conductivity $(\sigma) = 5.8 \times 10^7$ S/m) having thickness of 0.035 mm. The air spacer has a height of 7 mm ($0.084\lambda_L$). Overall dimension of the designed unit cell is 11.8 mm × 11.8 mm × 7.5 mm ($0.142\lambda_L \times 0.142\lambda_L \times 0.09\lambda_L$) as shown in Fig. 1. The unit cell is simulated under Floquet port boundary conditions in CST Studio 2014. The absorptivity (A) can be expressed as $A = 1 - |S_{11}|^2 - |S_{21}|^2$, where $|S_{11}|^2$ is the reflected power and $|S_{21}|^2$ is the transmitted power. Since the FSS is metal backed, it will act as a perfect reflector resulting zero transmission, i.e., $S_{21} = 0$. Therefore, the high absorptivity can be achieved by reducing the reflected power ($|S_{11}|^2$). The unit cell consists of eight lumped resistors each of 200 Ω to enhance the absorption bandwidth.

3 Simulated Results

The proposed absorber is wideband in nature. The simulated absorptivity characteristic is shown in Fig. 2. It is observed from Fig. 2 that the absorption bandwidth above 90% ranges from 3.61 to 10.63 GHz, i.e., fractional bandwidth is 98.6% and full width half maximum (FWHM) is 14.22 GHz (2.836–17.056 GHz). In Figs. 3 and 4, the surface current distributions at 4.438 and 8.992 GHz are shown. It is observed that most of the surface currents are confined within the outer ring and inner ring at 4.438 GHz and 8.992 GHz, respectively. So the lower and higher absorption bandwidth is controlled by the dimensions of the outer and inner ring of the design, respectively. It is also found that the surface current at top and bottom are opposite in direction, and hence, create

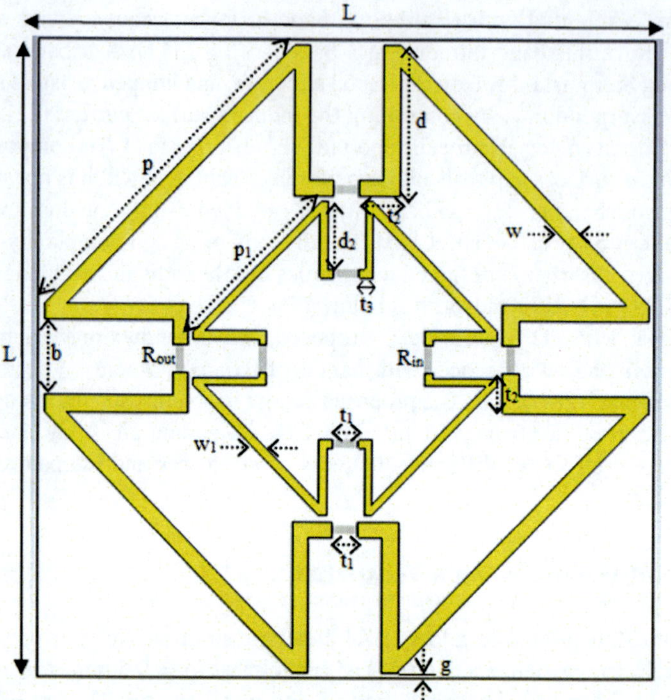

Fig. 1 Top view of the proposed design: $L = 11.8$ mm, $g = 0.102$ mm, $t_1 = 0.5$ mm, $p = 6.78$, $b = 1.4$ mm, $d = 2.8$ mm, $t_2 = 0.75$ mm, $t_3 = 0.25$ mm, $w = 0.3$ mm, $p = 6.785$ mm, $p_1 = 3.386$ mm, $w_1 = 0.15$ mm, $d_2 = 1.25$ mm, $R_{out} = 200\ \Omega$, $R_{in} = 200\ \Omega$

circulating loop perpendicular to the direction of incident magnetic field, thus creates magnetic excitation. Induced electric field is along the direction of incident electric field, leading electric excitation. A strong absorption is occurred at these frequencies due to presence of both electric and magnetic excitation.

The proposed design is analyzed under various angle of polarization, and it is observed from Fig. 5 that it maintains almost 90% absorptivity under all the polarization angles due to its fourfold symmetry. The proposed structure is studied under oblique angle of incidence under TE polarization, and it is found that the proposed design maintains 80% absorptivity upto 45° angle of incidence under TE polarization as shown in Fig. 6.

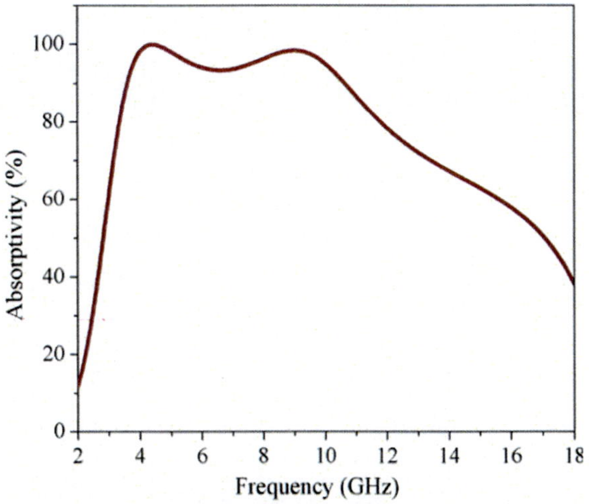

Fig. 2 Plot of absorption characteristics under normal incidence of the proposed structure

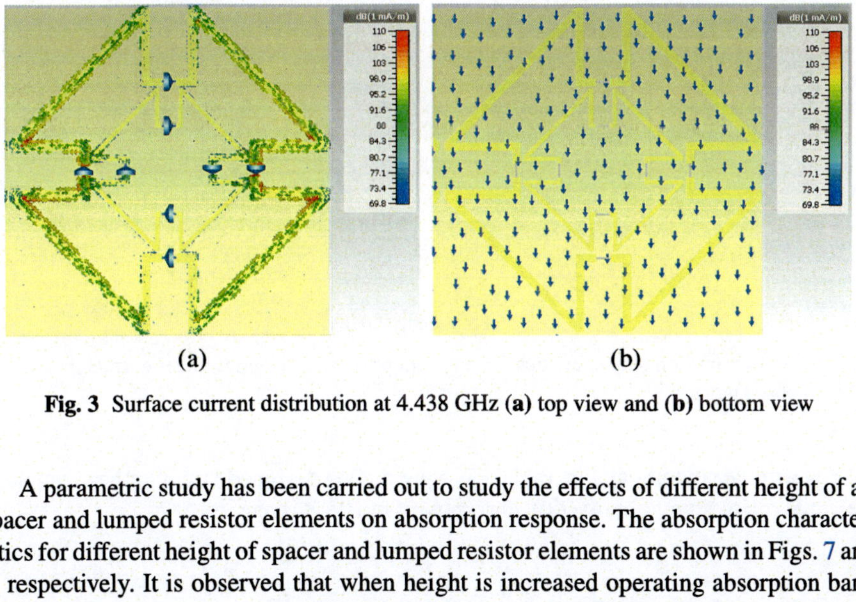

Fig. 3 Surface current distribution at 4.438 GHz (**a**) top view and (**b**) bottom view

A parametric study has been carried out to study the effects of different height of air spacer and lumped resistor elements on absorption response. The absorption characteristics for different height of spacer and lumped resistor elements are shown in Figs. 7 and 8, respectively. It is observed that when height is increased operating absorption band shifts to lower frequency, but bandwidth little decreases. In Fig. 8, significant variations of absorptivity are observed for different values of lumped resistance. In this design, the highest absorptivity is obtained at $h_2 = 7$ mm and $R_{in} = R_{out} = 200\ \Omega$.

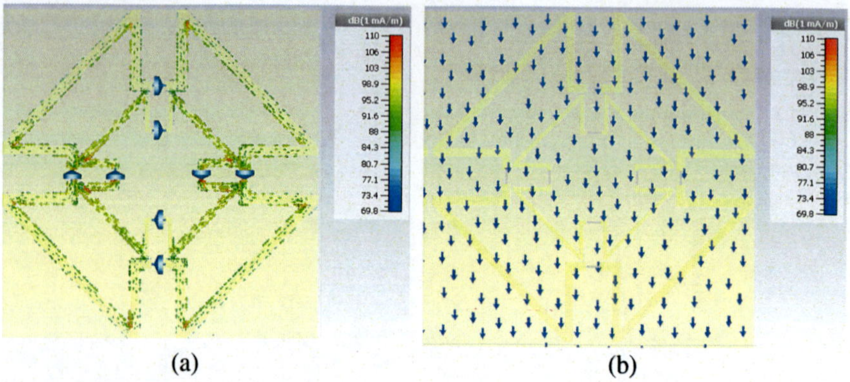

Fig. 4 Surface current distribution at 8.992 GHz (**a**) top view and (**b**) bottom view

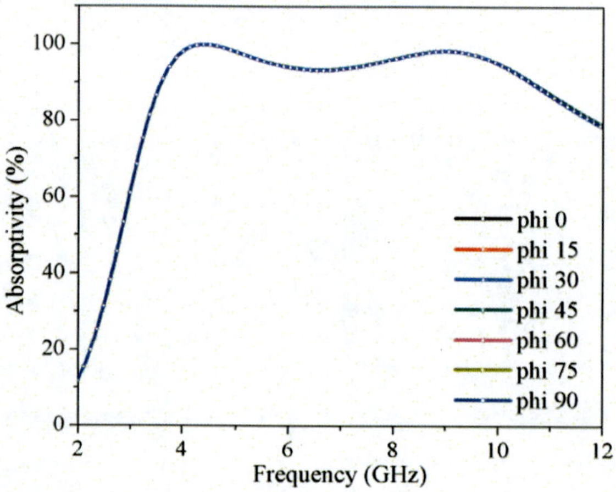

Fig. 5 Plot of absorption characteristics for different polarization angles under normal incidence of the proposed structure

A performance comparison Table 1 is shown. Our designed structure has compact size while maintaining almost 90% absorptivity in UWB range.

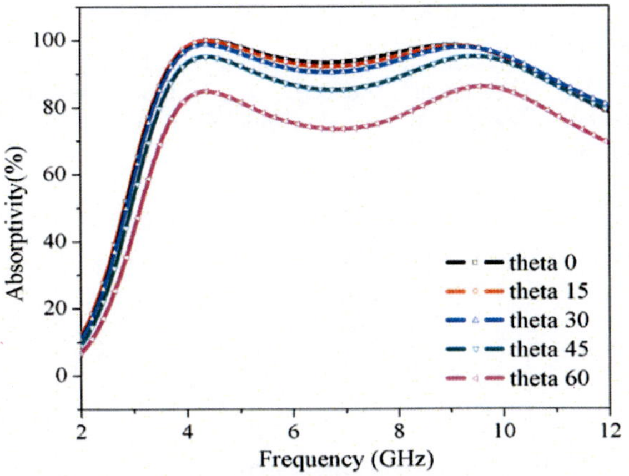

Fig. 6 Plot of absorption characteristics for different incidence angles under TE polarization of the proposed structure

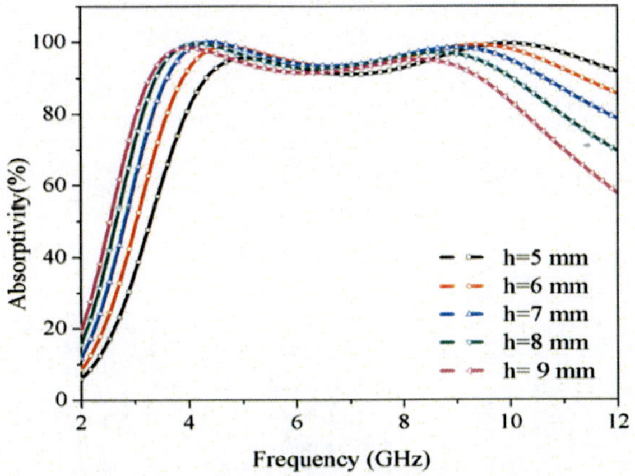

Fig. 7 Plot of absorption characteristics for different height of spacer

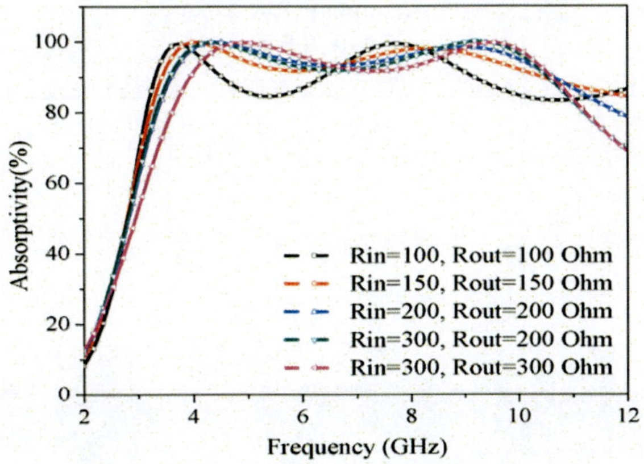

Fig. 8 Plot of absorption characteristics for different lumped resistance values

Table 1 Performance comparison table

References	90% absorption bandwidth	Fractional bandwidth (%)	Unit cell characteristics	Relative thickness	Polarization insensitive	Oblique incidence (TE)
[12]	1.7–2.6 4.56–5.59 12.37–13.29	99.8 99.1 99.9	Single layer + air spacer + 16 resistor (20 × 20 × 14.1) mm^3	$0.08\lambda_L$	Yes	30°
[13]	2–10	128.7	Single layer + spacer + 16 resistor (24.9 × 24.9x13.2) mm^3	$0.088\lambda_L$	NA	30°
[14]	6.79–14.96	75.13	Double layer + air spacer (13.8 × 13.8 × 5) mm^3	$0.146\lambda_L$	Yes	50°
[15]	5.10–14.07	93.6	Resistive film with sheet resistivity $70\Omega/m^2$ + spacer (8.12 × 8.12 × 6.6) mm^3	$0.11\lambda_L$	NO	60°
[16]	5.3–11.2	70.7	Single layer, 4 resistors (13.6 × 13.6 × 4.47) mm^3	$0.077\lambda_L$	NA	NA
This work	3.61–10.63	98.6	Single layer + air spacer + 8 resistor (11.8 × 11.8 × 7.5) mm^3	$0.09\lambda_L$	Yes	45°

4 Conclusion

In this paper, a miniaturized ultra-wideband polarization insensitive FSS absorber is designed. An air gap spacer and lumped resistor parameters are introduced to enhance the absorption bandwidth. The proposed absorber offers the 10 dB reflection bandwidth of 98.6% (3.61–10.63 GHz), whereas it has the thickness at the lowest cut off frequency of 0.09 λ_L, and a full width half maximum (FWHM) is 14.22 GHz (2.836–17.056 GHz). This structure shows 90% absorptivity under all the polarization angles, and also it maintains 80% absorptivity upto 45° angle of incidence under TE polarization. A parametric study is carried out to study the effects of spacer height and lumped parameters on absorptivity, and highest absorbtivity is obtained at spacer height $h = 7$ mm and $R_{\text{in}} = R_{\text{out}} = 200\ \Omega$.

References

1. Munk, B.A.: Frequency Selective Surfaces: Theory and Design. Wiley, New York, USA (2000)
2. Panwar, R., Puthucheri, S., Singh, D., Agarwala, V.: Design of ferrite—graphene-based thin broadband radar wave absorber for stealth application. IEEE Trans. Magn. **51**(11), 1–4 (2015)
3. Knott, E.F., Lunden, C.D.: Two-sheet capacitive Jaumann absorber. IEEE Trans. Antennas Propag. **43**(11), 1339–1343 (1995)
4. Bucci, O.M., Franceschetti, G.: Scattering from wedge-tapered absorbers. IEEE Trans. Antennas Propag. **19**(1), 96–104 (1971)
5. Ha, J., Shin, W., Lee, J.H., Kim, Y., Kim, D., Lee, Y., Yook, J.G.: Effect of plasma area on frequency of monostatic radar cross section reduction. J. Electromagn. Eng. Sci. **17**(3), 153–158 (2017)
6. Wang, Z., Fu, J., Zeng, Q., Son, M., Denidni, T.A.: Wideband transmissive frequency-selective absorber. IEEE Antennas Wirel. Propag. Lett. **18**(7), 1443–1447 (2019)
7. Zhai, H., Zha, C., Li, Z., Liang, C.: A triple-band ultrathin metamaterial absorber with wide-angle and polarization stability. IEEE Antennas Wirel. Propag. Lett. **14**, 241–244 (2015)
8. Kundu, D., Mohan, A., Chakraborty, A.: Ultrathin polarization independent absorber with enhanced bandwidth by incorporating Giusepe Peano fractal in square ring. Microw. Opt. Technol. Lett. **57**(5), 1072–1078 (2015)
9. Sun, J., Liu, L., Dong, G., Zhou, J.: An extremely broad band metamaterial absorber based on destructive interference. Opt. Express **19**(22), 21155–21162 (2011)
10. Ghosh, S., Bhattacharyya, S., Chaurasiya, D., Srivastava, K.V.: An ultra-wideband ultra-thin metamaterial absorber based on circular split rings. IEEE Antennas Wirel. Propag. Lett. **14**, 1172–1175 (2015)
11. Chen, Q., Yang, S., Bai, J., Fu, Y.: Design of absorptive/transmissive frequency-selective surface based on parallel resonance. IEEE Trans Antennas Propag. **65**(9), 4897–4902 (2017)
12. Amiri, M., Tofigh, F., Shariati, N., Lipman, J., Abolhasan, M.: Miniature tri-wideband Sierpinski-Minkowski fractals metamaterial perfect absorber. IET Microwaves Antennas Propag. **13**(7), 991–996 (2019)
13. Shang, Y., Shen, Z., Xiao, S.: On the design of single-layer circuit analog absorber using double-square-loop array. IEEE Trans. Antennas and Propag **61**(12), 6022–6029 (2013)
14. Yoo, M., Sungjoon Lim, S.: Polarization-independent and ultra-wideband metamaterial absorber using a hexagonal artificial impedance surface and a resistor-capacitor layer. IEEE Trans. Antennas Propag. **62**(5), 2652–2658 (2014)

15. Yuan, X., Zhang, C., Chen, M., Cheng, Q., Cheng, X., Huang, Y., Fang, D.: Wideband high-absorption electromagnetic absorber with chaos patterned surface. IEEE Antennas Wirel. Propag. Lett. **18**(1), 197–201 (2019)
16. Kundu, D., Mohan, A., Chakrabarty, A.: Single layer wideband microwave absorber using array of crossed dipoles. IEEE Antennas Wirel. Propag. Lett. **15**, 1589–1592 (2016)

Fizeau Interferometers: Extracting Sub-band Information

Siddharth Savyasachi Malu[1](✉), Abhirup Datta[1], and Peter Timbie[2]

[1] Discipline of Astronomy, Astrophysics and Space Engineering, Indian Institute of Technology Indore, Simrol, Khandwa Road, Indore, Madhya Pradesh 453552, India
siddharth@iiti.ac.in

[2] Department of Physics, University of Wisconsin-Madison, Chamberllin Hall, 1150 University Avenue, Madison, WI 53705, USA

Abstract. Measurements of the 2.7 K CMB radiation provide the most stringent constraints on cosmological models. The power spectra of the temperature anisotropies and the E-mode polarization of the CMB are explained well by the inflationary paradigm. The next generation of CMB experiments aim at providing the most direct evidence for inflation through the detection of B-modes in the CMB polarization, presumed to have been caused by gravitational waves generated during inflation. B-mode polarization signals are very small (~10^{-8} K) compared with the temperature anisotropies (~10^{-4} K). Systematic effects in CMB telescopes can cause leakage from temperature anisotropy into polarization. Bolometric interferometry is a novel approach to measuring this small signal with lower leakage. Subdividing the frequency passband of a Fizeau interferometer would mitigate the problem of "fringe smearing." The sub-band splitting method described here is general and can be applied to broadband Fizeau interferometers across the electromagnetic spectrum.

1 Introduction

Current [1, 2] and planned CMB instruments [3] all use some type of imaging technique. With focal-plane arrays of hundreds of background-limited detectors, they are capable of detecting the B-mode signals predicted in the most optimistic models, at the level of ~10^{-8} K. Systematic effects have been extensively studied for imaging polarimeters in the context of CMB measurements and appear to be controllable at this level as well [4, 5]. However, at some level, all instruments can "mix" the relatively large temperature anisotropy and E-mode polarization signals into B-modes.

Adding interferometers have the ability to correlate large numbers of inputs over wide bands. Here, we present an adding interferometer based on a Fizeau beam combiner. Combined with the optimal phase-shifting scheme described in [6], this is a promising approach for measuring the B-mode polarization. The technique is compatible with either coherent receivers (amplifiers) or incoherent detectors (bolometers).

© Springer Nature Singapore Pte Ltd. 2021
N. R. Das and S. Sarkar (eds.), *Computers and Devices for Communication*,
Lecture Notes in Networks and Systems 147,
https://doi.org/10.1007/978-981-15-8366-7_52

2 The Fizeau Beam Combination Technique to Extract Sub-bands

For the correct BibTeX to combine signals from $N > 2$ antennas, we use a Fizeau beam combiner, a type of "image plane" combiner [7]. This technique is analogous to the simplest interferometer in 1D: the Young's double (or multiple)-slit interferometer. The beams from the apertures can directly illuminate an array of detectors, or, more typically, they pass through a lens or telescope first to reduce the size of the instrument. While Fizeau combining is well known, we stress here the fact that there are two types of path differences (and therefore phase differences) for rays traveling from a source to a detector: one path difference occurs outside the instrument, and the other inside the instrument.

Compare this to a conventional interferometer, where rays only undergo a phase difference before they enter the antennas. (Note that long-baseline optical interferometers usually include a "delay line" between the apertures and the beam combiner. This element introduces an equal path length for all rays entering an aperture. In contrast, the internal path differences that we are concerned with are different for each pixel in the focal plane.)

Let us explore what this combination of path differences achieves. We start by noting that the "external" phase differences, which are present in any interferometer, are the reason that the visibility function is a Fourier transform of the image on the sky. The visibility measured by a single baseline essentially selects one Fourier mode from the image. In the Fizeau system, we have an additional set of phase differences. Without loss of generality, we may assign a negative sign to the phases introduced inside.

With these, the output at every detector on the focal plane can be written as

$$O_j = \sum_{\alpha=1}^{m} \Re\left[V_\alpha \exp i\phi_{j\alpha}\right]$$

which gives us a system of equations. Here,
O_j = output at the jth detector
V_α = visibility due to one baseline in the αth sub-band
$\phi_{j\alpha}$ = phase at the jth detector for the αth sub-band.
The quantity that we wish to infer is V_α, the visibility in the αth sub-band—this is a complex quantity, so that for m visibilities, we really need to infer 2 m quantities. However, there are only m equations above with 2 m unknowns.

In order to be able to write another unique set of m equations, we can alter the phase for this baseline by an amount $\Delta\phi$. Further, we can approximate the fringes over each detector, and assume that no fringe varies across the area of the detector, provided that the size of the detector is small compared with the fringe.

Representing the average fringe over each detector area as $F(x, \nu\alpha)$, we can write the outputs from the detectors without the extra applied phase shift, and with this phase shift as

$$O_j = A \sum_{\alpha=1}^{m} \Re\left[V_\alpha \exp i\phi_{j\alpha}(\mathbf{x}) F(\mathbf{x}, \nu_\alpha)\right]$$

$$O_j' = A \sum_{\alpha=1}^{m} \Re\left[V_\alpha \exp i(\phi_{j\alpha}(\mathbf{x}) + \Delta\varphi)F(\mathbf{x}, \nu_\alpha)\right]$$

where

A = area of a single detector.

The above equations have 2 m unknown quantities, and there are 2 m equations—these can therefore be solved to get the visibilities for the m sub-bands.

We stress here that the number of distinct phases needed in order to solve for 2 m quantities is 2 m. The difference between any two successive phases is therefore ~π/m—this is the difference in phases required for this technique to work. Since B-modes in CMB polarization are ~4 orders of magnitude lower than temperature anisotropies, therefore, it is important to maintain the errors in phase determination low.

In order to separate galactic foregrounds, spectral information is needed within each wide band—this implies that the Fizeau system which we have proposed here can be used to provide this spectral information, without the need for any specialized spectrum analysis hardware.

We have shown that it is possible to treat the division of fringes over a focal plane as division into sub-bands, for a single baseline, using phase shifting.

In addition, by introducing phase modulators discussed in [6, 8], we can measure visibilities for all baselines in a Fizeau system. While it is possible to divide the bandwidth into many different sub-bandwidths, it is not possible to do this indefinitely. The beam for a single antenna determines the FOV of the instrument and limits the resolution in the *u*–*v* plane.

In conclusion, the Fizeau system introduced here is potentially powerful tool for astrophysics: It could allow the recovery of more information than is possible with traditional interferometers or imagers and does not need significantly more resources to build. Its application in CMB cosmology is straightforward and can be demonstrated in future versions of B-mode experiments (Fig. 1).

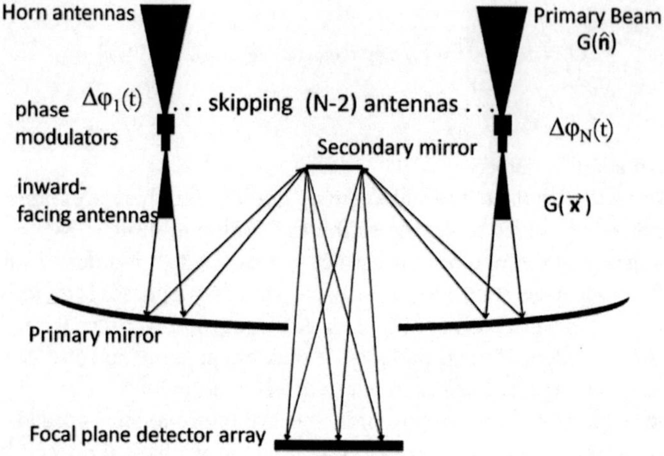

Fig. 1 Block diagram of an adding interferometer with $N > 2$. Each phase shifter is modulated in a sequence that allows recovery of the interference terms (visibilities) by phase-sensitive detection at the detectors. The signals are mixed in the beam combiner and detected. The beam combiner can be implemented either using guided waves (e.g., in a Butler combiner) or quasioptically (Fizeau combiner), as above. The top triangles represent corrugated conical horn antennas. For the case of an interferometer using coherent receivers, amplifiers and/or mixers could be placed before the beam combiner

Acknowledgements. SSM and AD acknowledge a generous grant for Astronomy from IIT Indore.

References

1. Takahashi, Y.D., Ade, P.A.R., Barkats, D., et al.: Characterization of the BICEP telescope for high-precision cosmic microwave background polarimetry, ApJ **711**, 1141 (2010)
2. Hinderks, J.R., Ade, P., Bock, J., et al.: QUaD: a high-resolution cosmic microwave background polarimeter, ApJ **692**, 1221 (2009)
3. Bock, J., Aljabri, A., Amblard, A., et al.: Study of the Experimental Probe of Inflationary Cosmology (EPIC)-Intemediate Mission for NASA's Einstein Inflation Probe, ArXiv e-prints (2009)
4. Bock, J., Church, S., Devlin, M., et al.: Task Force on Cosmic Microwave Background Research, ArXiv Astrophysics e-prints (2006)
5. Hu, W., Hedman, M.M., Zaldarriaga, M.: Benchmark parameters for CMB polarization experiments, Phys. Rev. D **67**, 043004 (2003)
6. Charlassier, R., Hamilton, J., Br´eelle, E., et al.: An efficient phase-shifting scheme for bolometric additive interferometry, A&A, **497**, 963 (2009)
7. Traub, W.A.: Beam Combination and Fringe Measurement, In: Lawson, P.R. (ed.) Principles of Long Baseline Stellar Interferometry, chapter 3, vol. 31 (2000)
8. Hyland, P., Follin, B., Bunn, E.F.: Phase shift sequences for an adding interferometer, MNRAS **393**, 531 (2009)

A Comparative Study on Determination of Optimum Detection Threshold for Minimum BER in a WDM Receiver with SRS and FWM Crosstalk

Santu Sarkar[1], Pinakpani Mukherjee[2](✉), and Nikhil Ranjan Das[1]

[1] Institute of Radio Physics and Electronics, University of Calcutta, 92 A. P. C. Road, Kolkata, India

[2] Department of Electronics and Communication Engineering, Academy of Technology, Adisaptagram, Hooghly, India

ppmukherjee@yahoo.co.in

Abstract. The study presented in this paper has been focused on the investigation of the WDM receiver performance in presence of the nonlinear crosstalks—namely, stimulated Raman scattering and four-wave mixing—and, hence, suggesting optimum detection thresholds to minimize bit error rate. To calculate the bit error rate, good representative models which are valid for any finite number of interferers are followed and the models are non-Gaussian in nature. The optimum detection thresholds calculated for different number of interfering channels keeping signal-to-noise ratio fixed are also summarized in tabular form.

Keywords: WDM · Crosstalk · FWM · SRS · Bit error rate and optimum detection threshold

1 Introduction

In this paper, the performance of an optical WDM system is studied in presence of nonlinear crosstalks—namely stimulated Raman scattering (SRS) and four-wave mixing (FWM) crosstalks [1, 2]. Stimulated Raman scattering (SRS) and four-wave mixing (FWM) crosstalks occur due to the nonlinearity of optical fiber for large power. Nonlinearity in the form of nonlinear inelastic scattering processes gives rise to SRS crosstalk while the nonlinear variations of the refractive index promote FWM crosstalk [3–11]. As these crosstalks have different origin and visible in different conditions, for comparing the performance, some common parameters are chosen. Assuming the receiver noise as of thermal nature, the overall noise in the receiver output is the combination of crosstalk and thermal noise. Assumption is also made that the power fed to each channel is identical. The frequencies (wavelengths) of the channels cannot be chosen similar in all the cases mainly because of the nature of different types of crosstalk. Thus, the frequencies are chosen to be different for the two kinds of crosstalks. Based on the models

© Springer Nature Singapore Pte Ltd. 2021
N. R. Das and S. Sarkar (eds.), *Computers and Devices for Communication*, Lecture Notes in Networks and Systems 147,
https://doi.org/10.1007/978-981-15-8366-7_53

discussed in [8] and [10], bit error rate (BER) and optimum detection thresholds [8, 9] are compared for SRS and FWM crosstalk, respectively. In Sect. 2, the theory of SRS and FWM crosstalks is discussed in brief, and in afterward sections, the comparative study on BER and optimum detection thresholds for minimum BER is presented.

2 Theory of SRS and FWM Crosstalk and Formulation of BER

In the previous section, nonlinear crosstalk is introduced. The crosstalk is said to be nonlinear, when the transmission medium behaves nonlinearly [2] when the transmitted power is high. Stimulated Raman scattering [3–8] results in the generation of stokes wave of wavelength higher than that of the pump wave. The stokes wave continues to propagate along with the pump wave in the fiber. In a WDM communication system, if the wavelength of the stokes wave coincides with that of an existing channel, it induces crosstalk. It decreases the information-carrying capacity and increases the probability of bit error of the system [3]. Several literatures report the implications of SRS crosstalk on the system performance. In most of the cases, the probability density function (pdf) of the power depletion is assumed Gaussian to calculate the bit error rate [6]. Though the Gaussian model is simple, it cannot accurately describe the crosstalk noise particularly when the number of interfering channels is not very large. Here, we followed a non-Gaussian model which presents the probability of power depletion due to stimulated Raman scattering in a non-Gaussian form and is applicable for any number of interferers. The calculation of probability of power depletion due to stimulated Raman scattering is given in section II of [8], and bit error rate at the receiver output is calculated.

In FWM process, two or three frequency components interact with each other and generate new frequency components [1, 9]. In a WDM system with equally spaced channels, these newly generated frequencies may overlap with existing channel frequencies giving rise to crosstalk [2]. When the channel frequencies and the FWM generated frequencies are same or close to each other, phase-matching condition is satisfied. As a result, crosstalk accumulates over a long distance in the fiber. The system degradation due to four-wave mixing can be minimized by using non-uniform frequency spacing between the channels [10]. In that case, only a few FWM components may overlap with the existing channel frequencies, and also the phase-matching condition may not be satisfied. Nevertheless, the FWM frequencies are still generated at the expense of the transmitted power, giving rise to depletion of the channel power. The channel power depletion is a random variable and is assumed Gaussian in many cases [9–11]. As mentioned earlier, simple Gaussian models cannot accurately describe the signal-crosstalk noise, especially when the number of interfering channels is not very large. In this paper, we followed a non-Gaussian model for channel power depletion and study the performance of a WDM receiver in presence of FWM crosstalk as reported in [10, 12].

3 Results and Discussion

Using the expressions as obtained in [8] and [10], the BER is calculated, and the corresponding minimum BER is found for both the SRS and FWM crosstalk. The value of the particular detection threshold which leads to minimum BER is called optimum detection threshold [13, 14]. Figure 1a, b shows the variation of detection threshold (d/σ_{th}) against photocurrent (I_s/σ_{th}) for SRS and FWM crosstalks, respectively, keeping the number of channels fixed. The ideal optimum detection threshold when crosstalk is absent (i.e., $N = 0$) is shown by dashed line. It is evident from the figure that the deviation from ideal detection threshold is more for higher N because of higher crosstalk interference. The figures show that there exists a minimum BER for a particular value of detection threshold distinct for each type of crosstalk. This optimum detection threshold d_{opt}/σ_{th} for minimum BER is less than the ideal value $(I_s/\sigma_{th})/2$ because of crosstalk. In case of SRS crosstalk, the optimum detection threshold is very close to the ideal value, and for FWM, it is far from the ideal value. It is observed that for small number of channels, the effect of SRS crosstalk is low. Therefore, optimum detection threshold is very close to the ideal value. But if the number of channels is increased, the effect of SRS crosstalk also increases, and the optimum detection threshold is shifted from its ideal value. This study shows that minimum BER due to FWM crosstalk is very high. We can conclude that FWM crosstalk affects the system mostly even when the number of channel is three. It can be seen that as the number of channels decreases, the optimum detection threshold d_{opt}/σ_{th} increases and approaches toward the ideal value of $(I_s/\sigma_{th})/2$. For SRS crosstalk, the minimum channel number is two, and for FWM crosstalk, it is three. In case of SRS crosstalk, power of the shortest wavelength channel is depleted, and this is distributed among all higher wavelength channels. In case of FWM crosstalk, non-degenerate case is assumed where at least three unique channel frequencies are required to generate different frequency components. For FWM, the rate of decrease of d_{opt}/σ_{th} with number of interfering channels is very high indicating enormous increase in depletion, and the optimum detection threshold is very far from its ideal value even when the channel number is three.

In Table 1, d_{opt}/σ_{th} and the corresponding values of BER are shown for ready reference. The data are shown for two different values of I_s/σ_{th}. It can be seen from the table that as the number of interfering channels decreases, the minimum BER also decreases for all the cases and the optimum threshold of detection increases toward the ideal value $(I_s/\sigma_{th})/2$.

4 Conclusion

This study shows that minimum BER due to FWM crosstalk is very high. In case of SRS crosstalk, the optimum detection threshold is very close to the ideal value and for FWM, it is far. As the numbers of interfering channels are increased, the minimum BER also increases for all the cases and the optimum threshold of detection shifts away from the ideal value. For SRS crosstalk, the minimum channel number is two, and for FWM crosstalk, it is three.

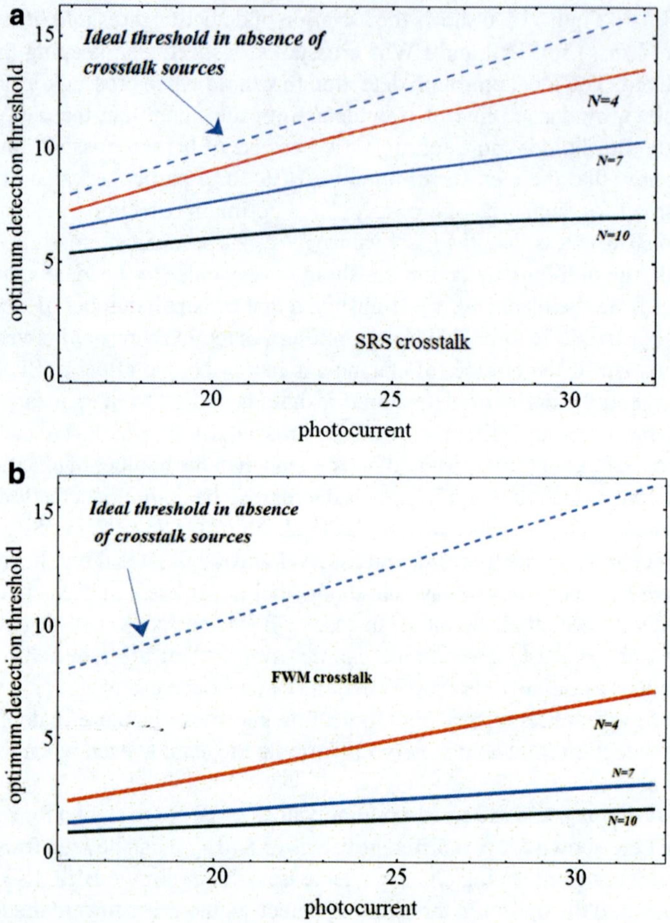

Fig. 1 **a, b** Optimum detection threshold leading to minimum BER, against photocurrent (I_s) for various interferers (N), where crosstalk level (ε_i) and thermal noise (σ_{th}) are fixed for SRS and FWM crosstalk, respectively

Table 1 Comparison of optimum detection threshold for SRS and FWM crosstalk

I_s/σ_{th}	N	d_{opt}/σ_{th}	BER	d_{opt}/σ_{th}	BER
		SRS		FWM	
16	1	–	–	–	–
	2	7.7	5.377E−15	–	–
	3	7.6	2.245E−14	4	3.207E−5
	4	7.3	1.147E−13	2.3	1.213E−2
	5	7.1	7.608E−13	1.86	3.663E−2
	6	6.8	6.118E−12	1.55	6.5571E−2
	7	6.5	5.791E−11	1.32	9.776E−2
	8	6.1	5.269E−10	1.14	1.295E−1
	9	5.7	5.045E−9	1.01	1.589E−1
	10	5.4	4.499E−8	0.9	1.854E−1
20	1	–	–	–	–
	2	9.7	1.507E−22	–	–
	3	9.4	2.216E−21	6.6	9.577E−011
	4	9.1	5.559E−20	3.4	3.458E−4
	5	8.7	2.224E−18	2.6	5.7204E−3
	6	8.2	1.128E−16	2.1	1.7514E−2
	7	7.7	6.6305E−15	1.8	3.668E−2
	8	7.2	3.8411E−13	1.5	6.188E−2
	9	6.6	1.816E−11	1.4	8.918E−2
	10	6.1	6.684E−10	1.2	1.1609E−1

References

1. Monroy, I.T., Tangdiongga, E.: Crosstalk in WDM Communication Network. Kluwer Academic Publishers, Massachusetts (2002)
2. Singh, S.P., Singh, N.: Nonlinear effects in optical fibers: origin, management and applications. Progr. Electromagnet. Res. PIER **73**, 249–275 (2007)
3. Cantono, M., et al.: On the interplay of nonlinear interference generation with stimulated Raman scattering for QoT estimation. J. Lightwave Technol. **36**(15), 3131–3141 (2018). https://doi.org/10.1109/JLT.2018.2814840
4. Ceballos-Herrera, D.E., Gutierrez-Castrejon, R., Alvarez-Chavez, J.A.: Stimulated Raman scattering and four-wave mixing effects on crosstalk of multicore fibers. IEEE Photon. Technol. Lett. **30**(1), 63–66 (2018). https://doi.org/10.1109/LPT.2017.2774501
5. Cantono, M., Curri, V., Mecozzi, A., Gaudino, R.: Polarization-related statistics of Raman crosstalk in single-mode optical fibers. J. Lightwave Technol. **34**(4), 1191–1205, 15 (2016). https://doi.org/10.1109/JLT.2015.2506481
6. Ho, K.P.: Statistical properties of stimulated Raman crosstalk in WDM systems. IEEE J. Lightwave Technol. **18**, 915–921 (2000)

7. Forghieri, F., Tkach, R.W., Chraplyvy, A.R.: Effect of modulation statistics on Raman crosstalk in WDM system. IEEE Photon. Technol. Lett. **7**, 101–103 (1995)
8. Das, N.R., Sarkar, S.: Probability of power depletion due to SRS cross-talk and optimum detection threshold in a WDM receiver. IEEE J. Quantum Electron. **47**, 424–430 (2011)
9. Liang, J., Iwashita, K.: FWM compensation in DPSK transmission by reducing detectors with digital coherent detection using backward propagation. Int. J. Inf. Electron. Eng. **1**, 99–104 (2011)
10. Das, N.R., Sarkar, S.: On the optimum detection threshold for minimum bit error rate due to four-wave mixing in a WDM system. Opt. Commun. Network. IEEE/OSA J. **5**(4), 370–377 (2013)
11. McKinstrie, C.J., Raymer, M.G.: Four-wave-mixing cascades near the zero-dispersion frequency. Opt. Exp. **14**, 9600–9610 (2006)
12. Papoulis, A.: Probability, Random Variable, and Stochastic Processes. McGraw-Hill, New York (1984)
13. Sarkar, S., Das, N.R.: Study of component crosstalk and obtaining optimum detection threshold for minimum bit-error-rate in a WDM Receiver. J. Lightwave Tech. **27**, 4366–4373 (2009)
14. Mukherjee, P., Sarkar, S., Das, N.R.: An approach for realistic estimation of BER due to signal-component crosstalk in a WDM receiver. Optik **146**, 1–7 (2017)

Enhancement in Electrical Characteristics of AlGaN/GaN HEMT Using Gate Engineered Dielectric Pocket Dual-Metal Gate

Ajay Kumar Visvkarma[1,2](✉), Khushwant Sehra[1,3], Robert Laishram[2], D. S. Rawal[2], and Manoj Saxena[3]

[1] Department of Electronic Science, Semiconductor Device Research Laboratory, University of Delhi, New Delhi, India
ajayvisvkarma13@gmail.com, sehrakhushwant@gmail.com

[2] Solid State Physics Laboratory, MMIC Fabrication Division, New Delhi, India
robertlaishramspldrdo@gmail.com, dsrawal15@gmail.com

[3] Department of Electronics, Deen Dayal Upadhyaya College, University of Delhi, New Delhi, India
saxenamanoj77@gmail.com

Abstract. This study presents enhancement of AlGaN/GaN HEMT device electrical characteristics by employing different gate engineered architectures. The dual-metal gate (DMG) structure is combined in different forms with recessed AlGaN and gate dielectric ($HfSiO_4$) in order to extract the advantages offered by the individuals. A remarkable improvement in transconductance (8%) and drain current (~13 & 7%) is achieved with dual-metal-gated HEMT as well as with the proposed dielectric pocket (DP) dual-metal-gated HEMT device. The increased OFF-state leakage with the incorporation of dual-metal gate is suppressed successfully with the implication of dielectric pocket dual-metal gate structure. Apart from this, DP-DMG HEMTs has a 0.15 V positive shift in the threshold voltage in comparison with conventional SMG-HEMT, and therefore, this dielectric pocket HEMT can be seen as an upgradation to the next generation of HEMT devices.

Keywords: AlGaN/GaN HEMT · Gate Engineering · Hafnium Silicate · Dual-Metal Gate · Silvaco's ATLAS

1 Introduction

Devices based on wide bandgap III-nitride semiconductor materials have become an integral part of defense applications, satellite communication systems, lightening applications, and many new sensing applications [1, 2]. Higher bandgap, ability to form hetero-structures with high carrier density and mobility, and presence of spontaneous and piezoelectric polarization make III-nitride materials superior over other conventionally used semiconductor materials [3, 4]. Hence, devices based on these materials like gallium nitride HEMTs are much more reliable for high temperature and power applications. GaN based electronics is being adopted for various existing space and military

© Springer Nature Singapore Pte Ltd. 2021
N. R. Das and S. Sarkar (eds.), *Computers and Devices for Communication*,
Lecture Notes in Networks and Systems 147,
https://doi.org/10.1007/978-981-15-8366-7_54

applications where high-power and high-frequency devices are required. Apart from various advantages offered by these materials, the device performance is limited due to degradation of metals contacts at high temperature or at high bias or electrical stress condition. Many research groups across the globe are working in this area. Employment of different gate geometries, dual- or tri-metal gate structures, recess gate structures, and use of various gate dielectrics has been extensively searched for device characteristics improvement [5–7]. Improvement in drain current and transconductance is seen with dual- or tri-metal gate structures [8]. Recessing of AlGaN barrier and incorporation of gate dielectric can be used to tune the threshold of the HEMT device.

In this paper, we are presenting various gate engineered AlGaN/GaN HEMT structures along with their electrical behavior. Dual-metal gate (DMG) is employed in different forms in order to utilize its great potential. With the best of knowledge, a combination of metal/insulator gate with recessed AlGaN structure is being presented for the first time. Hafnium silicate ($HfSiO_4$) is chosen as gate dielectric because of its high dielectric constant, and also it possess lower hysteresis in comparison with conventionally used gate dielectrics [9].

2 Device Structures and Models

The cross-sectional view of different device structures analyzed in this paper are shown in Fig. 1, and related dimensions are summarized in Table 1. Device width is taken as 1 μm for normalization purposes. The conventional single-metal gate (SMG) HEMT architecture and its dual-metal gate (DMG) realization are depicted in Fig. 1a, b. Application of gate recess technique and our proposed dielectric pocket (DP) beneath the gate region using DMG is presented in Fig. 1c, d. The recessed area of AlGaN and the dielectric pocket (DP) is 5 nm as labeled in Fig. 1c, d.

All performance studies have been carried out using Silvaco's TCAD device simulator tool, ATLAS [10]. For device simulation, Shockley–Read–Hall (SRH) recombination model with fixed lifetimes at 300 K temperature for both electron and holes along with Boltzmann statistics for describing the distribution of particles over energy states has been used.

For low-field mobility, analytic low-field mobility model has been invoked which is essentially based upon Caughey—Thomas mobility model. For incorporating the effect of velocity saturation, GANSAT high-field mobility was invoked [10].

3 Results and Discussions

Comparison of output characteristics of different architectures under study is depicted in Fig. 2. It is observed from Fig. 2a that DMG HEMT results in enhancement of drain current (~13%) which is due to the improvement in the average velocity of carriers under the channel region beneath the gate [11]. The enhancement of drain current is traded off with increased OFF—state leakage current as shown in Fig. 3a. This increase in leakage is due to reduction in effective work function of gate metal [8]. From Fig. 2b, it is noted that drain current is reduced when gate is recessed into AlGaN barrier layer owing to reduction of 2DEG density under the gate region. The reduction in maximum drain

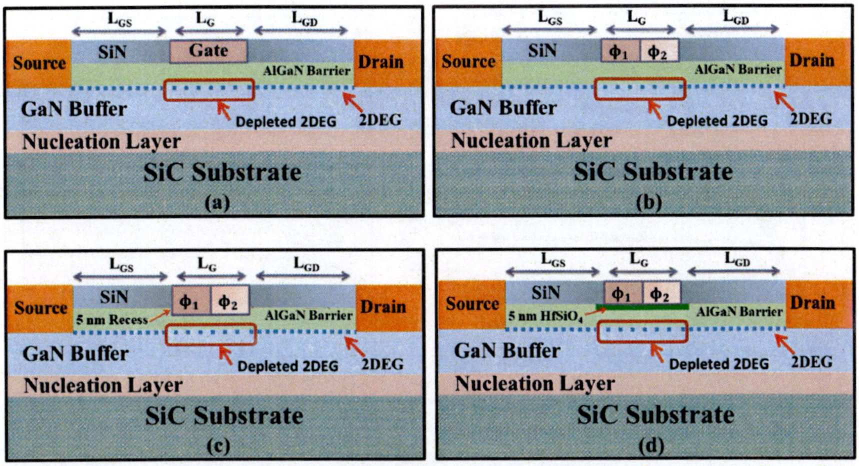

Fig. 1 Cross-sectional view of **a** conventional single-metal gate (SMG) HEMT, **b** dual-metal gate (DMG) HEMT, **c** recessed AlGaN DMG HEMT, and **d** hafnium silicate-filled dielectric pocket (DP) AlGaN DMG HEMT. [*Note* For TCAD simulations $\phi_1 > \phi_2$]

Table 1 Device dimensions of various architectures as in Fig. 1

Dimensions	Description	Value (μm)
L_G	Gate length	0.700
L_{SD}	Source—drain spacing	6.000
L_{SG}	Source—gate spacing	2.650
L_{GD}	Gate—drain spacing	2.650
$t_{Passivation}$	Thickness of passivation layer (SiN)	0.100
$t_{Barrier}$	Thickness of barrier layer ($Al_{0.25}Ga_{0.75}N$)	0.027
t_{Buffer}	Thickness of buffer layer (GaN)	2.200
t_{NL}	Thickness of nucleation layer (AlN)	0.100
$t_{Substrate}$	Thickness of substrate layer (Substrate)	5.000
ϕ_1/ϕ_2	Work function of gate	5.10 eV/4.30 eV

current can be explained on the basis of transfer characteristics which has a noticeable positive shift in the threshold voltage. It is further observed from transconductance curves given in Fig. 3b that this results in enhancement of peak transconductance (8%) which is due to the improved gate action as a result of reduction in barrier layer thickness. Moreover, the dependence of drain current and transconductance on electric field beneath the gate region and between the gate-drain region along with the carrier velocity is of great importance. An improvement in carrier velocity and gate electric field has already been reported by various groups [5, 11].

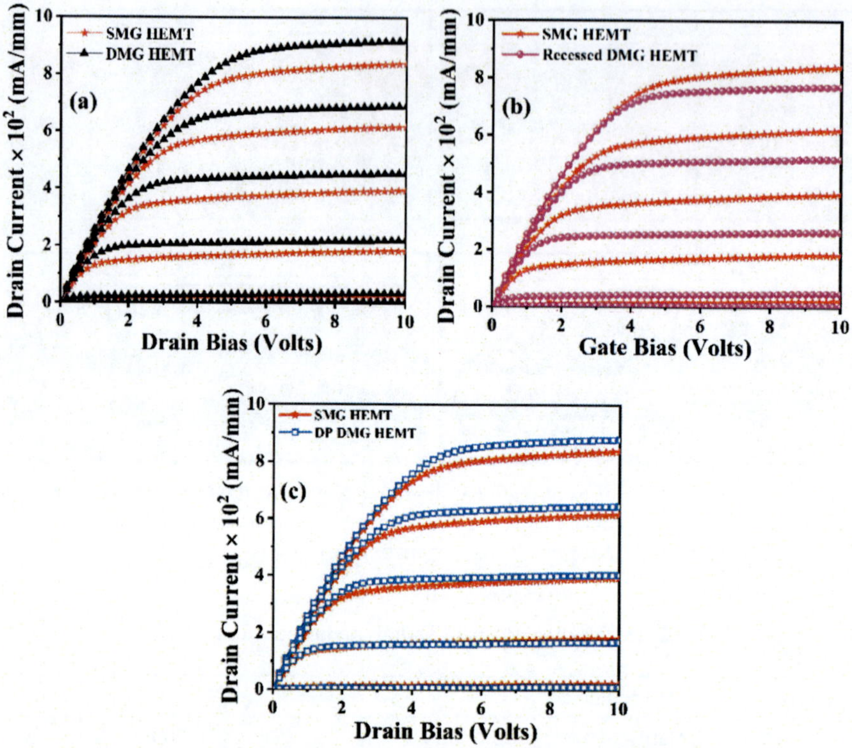

Fig. 2 Comparison of output characteristics of conventional single-metal gate (SMG) HEMT with **a** dual-metal gate (DMG) HEMT, **b** recessed AlGaN DMG HEMT, and **c** dielectric pocket (DP) AlGaN DMG HEMT. Gate bias is varied from 0 to −5 V in steps of 1 V

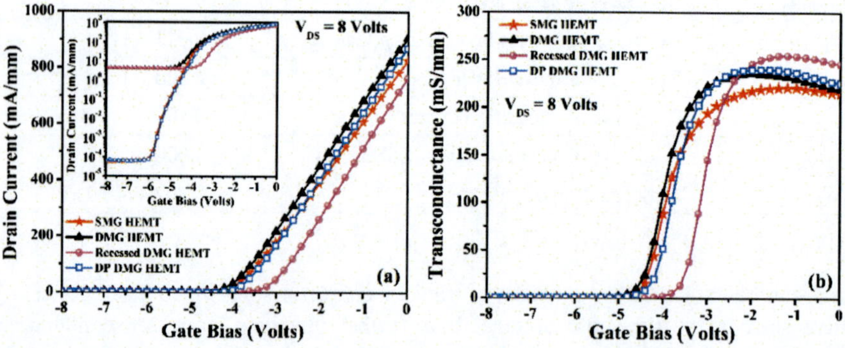

Fig. 3 Comparison of **a** transfer characteristics and **b** transconductance curves of different architectures under study

It is evident from literature that application of metal insulator semiconductor (MIS) structure results in the enhancement of drain current due to reduced gate action [12]. Further, it has been established [9] that hafnium silicate ($HfSiO_4$) so employed for

DP exhibits lower hysteresis when employed as a gate dielectric in comparison with conventionally used dielectrics. To incorporate the advantages offered by various gate techniques as discussed above, a dielectric pocket-based DMG HEMT structure is being proposed to get rid of the negative aspects of the above mentioned structures. A comparison of conventional SMG-HEMT and the proposed DP-DMG HEMT presented in Fig. 2c and Fig. 3 shows that the improvement in drain current and peak transconductance recorded with DMG structure is achieved with DP-DMG HEMT structure without sacrificing the OFF-state leakage current. Hence, the structure justifies its use for high-power and high-frequency applications.

4 Conclusions

A new dielectric pocket-based dual-metal gate architecture is being presented in this study for AlGaN/GaN HEMT devices in order to improve its electrical performance. Extensive simulations were performed using Silvaco ATLAS software. An enhancement in the drain current and transconductance is achieved with suppressed OFF-state leakage with the proposed DP-DMG HEMT structure.

Acknowledgements. The authors would like to thank DBT Star College Laboratory at Deen Dayal Upadhyaya College, University of Delhi; CARS Project No.: 1115/CARS-73/TS/SPL/18 funded by Solid State Physics Laboratory (SSPL) and MMIC Division, Solid State Physics Laboratory (SSPL), DRDO, New Delhi, India, for providing necessary tools for completion of this work.

References

1. Umesh, K., Mishra, Parikh, P., Wu, Y.-F.: AlGaN/GaN HEMTs-an overview of device operation and applications. Proc. IEEE **90**(6), 1022–1031 (2002)
2. Pengelly, R.S., Wood, S.M., Milligan, J.W., Sheppard, S.T., Pribble, W.L.: A review of GaN on SiC high electron-mobility power transistor and MMICs. IEEE Trans. Microwave Theory Tech. **60**(6), 1764–1783 (2012)
3. Ibbetson, J.P., Fini, P.T., Ness, K.D., Den Baars, S.P., Speck, J.S., Mishra, U.K.: Polarization effects, surface states, and the source of electrons in AlGaN/GaN heterostructure field effect transistor. Appl. Phys. Lett. **77**(2), 250–252 (2000)
4. Ambacher, O., Smart, J., Shealy, J.R., Weimann, N.G., Chu, K., Murphy, M., Schaff, W.J., Eastman, L.F., Dimitrov, R., Wittmer, L., Stutzmann, M., Rieger, W., Hilsenbeck, J.:Two-dimensional electron gases induced by spontaneous and piezoelectric polarization charges in n- and Ga-face AlGaN/GaN heterostructures. J. Appl. Phys. **85**(6), 3222–3233 (1999)
5. Jang, Y.I., Lee, S.H., Jae, H.S., Young, J.Y., Ra, H.K., Min, S.C., Bo, G.K., Min, Y., Lee, J.-H., Man Kang, I.: Design and analysis of AlGaN/GaN MIS HEMTs with a dual-metal-gate structure. J. Semi. Techn. Sci. **17**(2), 223–228 (2017)
6. Kang, Y., Sung, H.-k., Kim, H.: Investigation of kink effect in normally-off AlGaN/GaN recessed-gate MOS-heterostructure FETs. J. Vacuum Sci. Technol. B Nanotechnol. Microelectron. Mater. Process. Measure. Phenomena **34**, 052202 (2016)

7. Green, R.T., Luxmoore, I.J., Lee, K.B., Houston, P.A., Ranalli, F., Wang, T., Parbrook, P.J., Uren, M.J., Wallis, D.J., Martin, T.: Characterization of gate recessed GaN/AlGaN/GaN high electron mobility transistors fabricated using a $SiCl_4/SF_6$ dry etch recipe. J. Appl. Phys. **108**, 013711 (2010)
8. Visvkarma, A.K., Laishram, R., Kapoor, S., Rawal, D.S., Vinayak, S., Saxena, M.: Improvement in DC and pulse characteristics of AlGaN/GaN HEMT by employing dual metal gate structure. Semicond. Sci. and Technol (2019). https://doi.org/10.1088/1361-6641/ab3ce4
9. Nabatame, T., Maeda, E., Inoue, M., Yuge, K., Hirose, M., Shiozaki, K., Ikeda, N., Ohishi, T., Ohi, A.: Hafnium silicate dielectrics in GaN metal oxide semiconductor capacitors. Appl. Phys. Express **12**, 0110091–0110095 (2019)
10. Atlas Device User Guide Version, 5.16
11. Long, W., Chin, K.K..: Dual Material Gate Field Effect Transistor (DMGFET). IEDM Tech. Dig., pp. 549–552 (1997)
12. Lin, Y.C., Lin, T.W., Wu, C.H., Yao, J.N., Hsu, H.T., Shih, W.C., Kakushima, K., Tsutsui, K., Iwai, H., Chang, E.Y..: Optimization of gate insulator material for GaN MIS-HEMT. In: 28th International Symposium on Power Semiconductor Devices and ICs (2016)

Nanoscale Materials and Devices (NMD)

Non-Ohmic Characteristics of a Quantum Confined Degenerate Ensemble of Carriers in a Well of GaAs at Low Lattice Temperature

Bittu Roy, Sulava Bhattacharyya, and Debi Prosad Bhattacharya(✉)

Department of Physics, Jadavpur University, Kolkata, West Bengal 700032, India
dbhattacharya140@gmail.com

Abstract. High-field mobility characteristics of a degenerate two-dimensional gas (2DEG) are obtained under the condition of low lattice temperature. The characteristics are obtained by adopting two methods: (1) by solving the energy balance equation of the electron–phonon system and (2) from the current density of the non-equilibrium carriers. The quasi-elastic interactions with the deformation potential acoustic and the piezoelectric phonons have been considered. Some qualitative agreement of the numerical results thus obtained for wells of GaAs with the available experimental data has been observed.

Keywords: 2DEG · High-field mobility · Degeneracy · Low lattice temperature

1 Introduction

The high-field transport characteristics of a degenerate ensemble of electrons confined in a semiconductor quantum well is useful from the device point of view. The electric field $\vec{\varepsilon}$ dependence of the effective temperature T_e of such an ensemble in a quantum well of GaAs at low lattice temperature has been worked out recently by the present authors [1]. The results thus obtained have been interesting.

If the lattice temperature is low ($T_{\mathrm{L}} \leq 20\,K$), an electric field of the order of a fraction of a volt/cm may seem to be relatively high. Thus, the carrier ensemble gets significantly perturbed from the state of the thermodynamic equilibrium and attains a temperature T_e higher than the lattice temperature T_{L} [2].

In an ensemble of 2DEG on the surface of a compound semiconductor, which lacks inversion symmetry, the electrons interact with piezoelectric phonons besides the deformation potential acoustic phonons. The interaction with the polar optical phonons may be neglected as the number of such phonons is quite small.

Considering a degenerate ensemble of electrons in a quantum well, the present communication deals with a theoretical analysis of the high-field mobility characteristics at low lattice temperatures. Two methods have been followed: first, by solving the energy balance equation of the electron–phonon system in the presence of a relatively high electric field; second, from the current density of the non-equilibrium carriers.

© Springer Nature Singapore Pte Ltd. 2021
N. R. Das and S. Sarkar (eds.), *Computers and Devices for Communication*, Lecture Notes in Networks and Systems 147,
https://doi.org/10.1007/978-981-15-8366-7_55

2 Theory

Method 1: The energy balance condition for the electron–phonon system due to the combined interaction with the deformation potential acoustic (ac) and the piezoelectric (pz) phonons may be expressed as [2]

$$e\mu(\varepsilon)\varepsilon^2 = \left\langle \frac{\partial E_{\vec{k}}}{\partial t} \right\rangle_{\text{ac.}} + \left\langle \frac{\partial E_{\vec{k}}}{\partial t} \right\rangle_{\text{pz.}}, \tag{1}$$

where e is the electronic charge, $\mu(\varepsilon)$ is the high-field mobility, $\left\langle \frac{\partial E_{\vec{k}}}{\partial t} \right\rangle_{\text{ac.(pz)}}$ are the average energy loss rate of the electrons due to the corresponding ac. and pz. interactions, respectively. The energy loss rate for any interaction may be calculated from [2]

$$\left\langle \frac{\partial E_{\vec{k}}}{\partial t} \right\rangle = \int_0^\infty E_{\vec{k}} \left(\frac{\partial f_0\left(\vec{k}\right)}{\partial t} \right) d\vec{k} \Big/ \int_0^\infty f_0\left(\vec{k}\right) d\vec{k}, \tag{2}$$

here $f_0\left(\vec{k}\right)$ is the spherical part of the high-field distribution function of the degenerate ensemble, $\vec{k}$ is the wave vector of an electron. In order to carry out the integration in (2), one can in principle use the Fermi–Dirac (F.D.) function at the effective temperature T_e of the electrons. But this would lead to much difficulty to carry out the integration analytically. As such, we make use of the well tested alternative form for the F.D. distribution as described in details in [3]. The whole energy range is divided into three regions: $0 < E_{\vec{k}} \le \beta_1 E_F$, $\beta_1 E_F < E_{\vec{k}} \le \beta_2 E_F$ and $\beta_2 E_F < E_{\vec{k}} < \infty$, where E_F is Fermi energy, $\beta_1 = 1 - (K_B T_L / E_F)$ and $\beta_2 = 1 + (K_B T_L / E_F)$. For each region, the distribution is approximated by different simple functions. It may be mentioned here that the energy loss rate for any interaction turns out to be a function of T_e. The field corresponding to T_e may be obtained from [1]. Hence, $\mu(\varepsilon)$ may be obtained from (1) for any value of ε.

Method 2: The high-field mobility of a degenerate ensemble of 2DEG may also be obtained from current density of the non-equilibrium carriers [4] as

$$\mu(\varepsilon) = -\frac{e}{m_\parallel^*} \int_{E_0}^\infty \left(E_{\vec{k}} - E_0\right) \tau\left(E_{\vec{k}}\right) \frac{\partial f_0\left(\vec{k}\right)}{\partial E_{\vec{k}}} d\vec{k} \Big/ \int_{E_0}^\infty f_0\left(\vec{k}\right) d\vec{k}, \tag{3}$$

where $m_\parallel^*$ the effective mass of the carriers parallel to the surface, $\tau\left(E_{\vec{k}}\right)$ is the momentum relaxation time of various collision processes. Under the prevalent conditions, the electrons are assumed to be occupy only the lowest subband with energy E_0. The integration in (3) may now be carried out using the expression for $\tau_{ac,pz}\left(E_{\vec{k}}\right)$ from [1]. The combined mobility can then be found by Mattheisen's rule.

3 Results

The high-field mobility of electrons in GaAs is calculated by using same parameter values as that used in [1].

Figure 1 shows how does a relatively high field significantly effect the mobility characteristics. The mobility assumes a maximum value at lower fields and falls off with the increase of the field. Curves marked 2a and 2b are compared with the experimental [5] curve 2c. The qualitative agreement of our theoretical results with the experimental data is observed. The $T_e/T_L - \varepsilon$ characteristics has also been given in Fig. 1 from [1] for ready reference. Figure 2 shows that the average energy loss rate of carriers increases with the rise of the carrier temperature and rate of average energy loss of electrons is more for higher concentration.

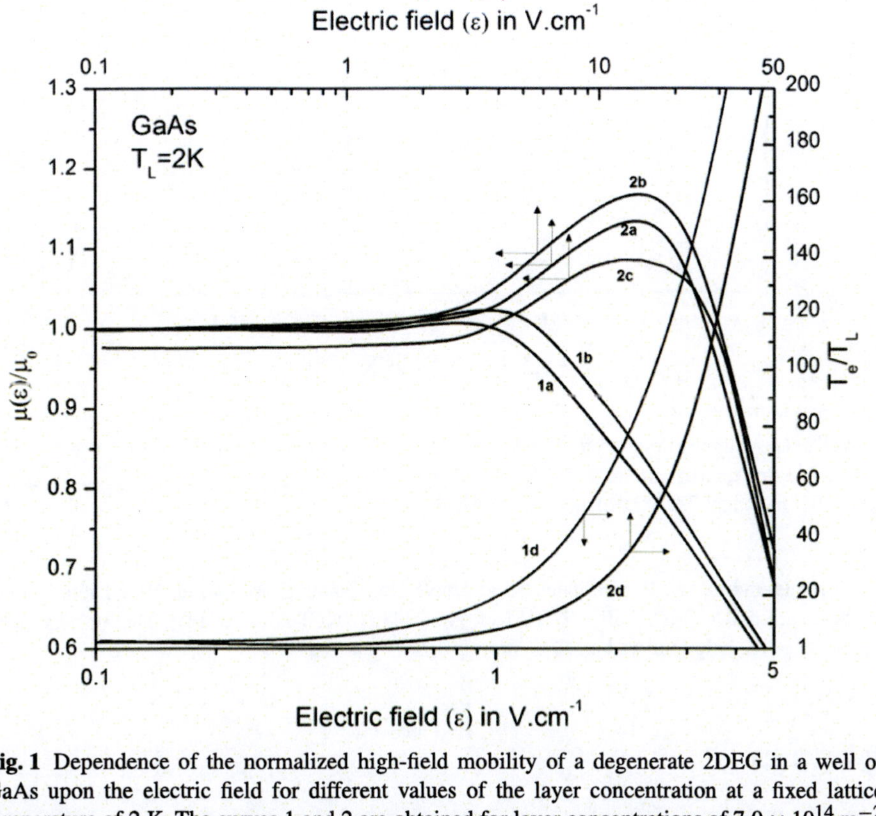

Fig. 1 Dependence of the normalized high-field mobility of a degenerate 2DEG in a well of GaAs upon the electric field for different values of the layer concentration at a fixed lattice temperature of 2 K. The curves 1 and 2 are obtained for layer concentrations of $7.0 \times 10^{14}\,m^{-2}$ and $7.0 \times 10^{15}\,m^{-2}$, respectively. The curves a and b represent the normalized mobility from method 1 and 2, respectively. The curve marked 2c is the experimental curve [5]. The curves marked d represent the field dependence of the effective electron temperature obtained from [1]

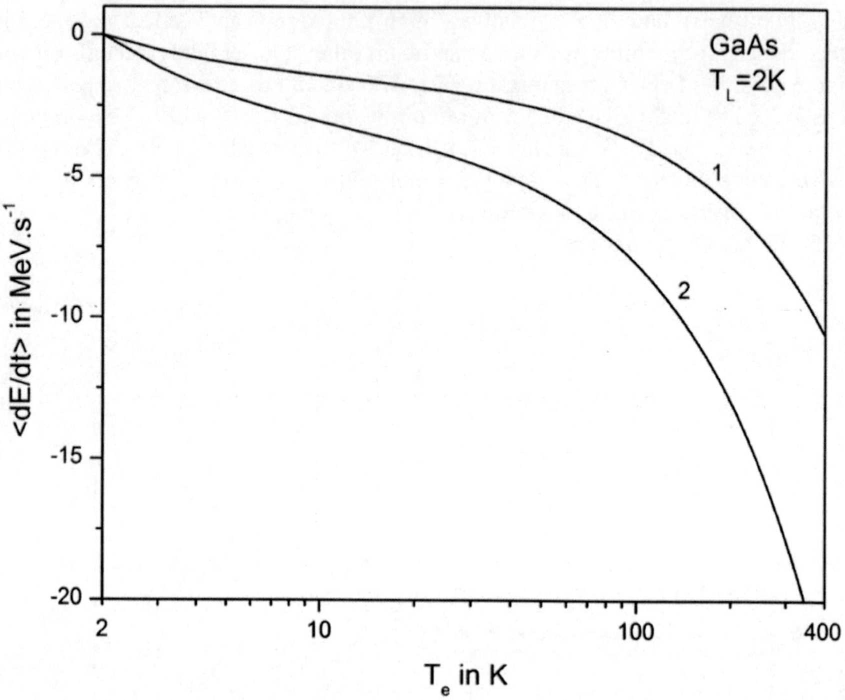

Fig. 2 Dependence of the average energy loss rate of electron in a degenerate ensemble of 2DEG in a well of GaAs upon effective electron temperature for different values of the layer concentration at a fixed lattice temperature of 2 K. The curves 1 and 2 are obtained for layer concentrations of $7.0 \times 10^{14}\ m^{-2}$ and $7.0 \times 10^{15}\ m^{-2}$, respectively

Acknowledgements. B. Roy is indebted to University Grants Commission, New Delhi, India for the award of UGC-NET JRF [UGC-Ref. No.: 34121/(NET-DEC. 2014)]. The authors also acknowledge the help rendered by S. Saha.

References

1. Roy, B., Bhattacharyya, S., Bhattacharya, D.P.: Heating of a degenerate electron ensemble in a well of compound semiconductors at low lattice temperatures. Appl. Phys. A. **125**, 223 (2019)
2. Conwell, E.M.: High Field Transport in Semiconductors. Academic Press, New York (1967)
3. Das, B., Basu, A., Das, J., Bhattacharya, D.P.: Effective temperature of the non-equilibrium electrons in a degenerate semiconductor at low lattice temperature. Phys. B. **474**, 21–26 (2015)
4. Nag, B.R.: Physics of Quantum Well Devices. Kluwer Academic Publishers, Dordrecht (2000)
5. Shah, J., Pinczuk, A., Störmer, H.L., Gossard, A.C., Wiegmann, W.: Electric field induced heating of high mobility electrons in modulation-doped GaAs-AlGaAs heterostructures. Appl. Phys. Lett. **42**(1), 55–57 (1983)

Resistorless Electronically Tunable Quadrature Oscillator Using Single CDTA

Rupam Das[1] and Sajal K. Paul[2(✉)]

[1] Department of ECE, Asansol Engineering College, Asansol, India
das_rupam@rediffmail.com
[2] Department of EE, IIT ISM, Dhanbad, India
sajalkpaul@rediffmail.com

Abstract. A new current-mode sinusoidal quadrature oscillator is presented using current differencing transconductance amplifier (CDTA). This oscillator circuit is designed using one CDTA and two grounded capacitors. By adjusting the bias of CDTA, the oscillation condition and the oscillation frequency can be controlled. The output of the oscillator is current and has high output impedance. It is resistorless and uses grounded capacitors, which ease the IC implementation. PSPICE simulation satisfies the theoretical results.

Keywords: Quadrature oscillator · CDTA · Low component count

1 Introduction

Recently for signal generating and signal processing, CDTA is one of the most widely used analog building blocks. This block was first invented by Bioleck in 2003 [1, 2]. CDTA has low input and high output impedance and also useful for electronic adjustment of various parameters. Two sinusoidal signals having 90° phase difference provided from an oscillator are called quadrature oscillator. Several sinusoidal quadrature oscillators using CDTA are reported in the technical literature [3–9]. They have the following shortcomings:

1. use more than one CDTA in the count [3–5].
2. use more than two passive components in the count [5–8].
3. use floating passive components [7, 8].
4. unavailability of electronic control of the oscillation frequency [6–8].
5. use of passive resistor [5–8].

A new current-mode (CM) quadrature oscillator is proposed using one CDTA building block. It uses two capacitors and no resistor. The frequency of oscillation (FO) is tunable by the bias currents of CDTA. It consumes low power and enjoys low total harmonic distortion. The output impedance of the oscillator is high and hence does not require any buffer to connect to the application circuits.

© Springer Nature Singapore Pte Ltd. 2021
N. R. Das and S. Sarkar (eds.), *Computers and Devices for Communication*,
Lecture Notes in Networks and Systems 147,
https://doi.org/10.1007/978-981-15-8366-7_56

2 Description of CDTA

The CDTA symbol is shown in Fig. 1 [9].

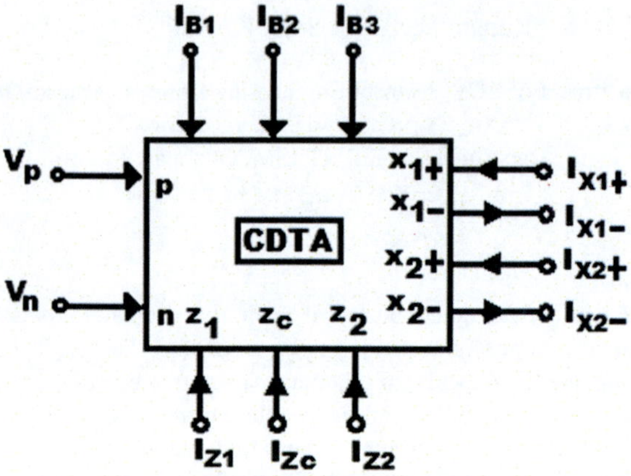

Fig. 1 Symbol of CDTA

The port relationships of the CDTA is as below:

$$V_p = V_n = 0,\ I_z = I_p - I_n,\ I_{x1}\pm = \pm g_{m1} V_{z1},\ I_{x2}\pm = \pm g_{m2} Vz2 \tag{1}$$

where

$$g_{m1} = \sqrt{2k_1 I_{B2}},\ g_{m2} = \sqrt{2k_2 I_{B3}} \tag{2}$$

$k_1 = \mu_n C_{ox}(W/L)_{47,48}$, and $k_2 = \mu_n C_{ox}(W/L)_{53,54}$.

Here k_1 and k_2 are the physical parameters of CMOS transistors, I_{B1}, I_{B2}, and I_{B3} are biasing current of CDTA, μ_n is the mobility of the *n*MOS transistor, C_{ox}, W and L are unit area oxide capacitance, channel width, and length of the MOSFETs.

3 Proposed Circuit

The CM quadrature oscillator (QO) circuit is given in Fig. 2:

Analysis results in the characteristic equation of the QO as:

$$s^2 C_1 C_2 + s(g_{m1} C_2 - g_{m2} C_1) + g_{m1} g_{m2} = 0 \tag{3}$$

The condition of oscillation (CO) and the frequency of oscillation (FO) can be written as:

$$\text{CO: } g_{m2} C_1 = g_{m1} C_2 \tag{4}$$

$$\text{FO:} \omega_0 = \sqrt{\frac{g_{m1} g_{m2}}{C_1 C_2}} \tag{5}$$

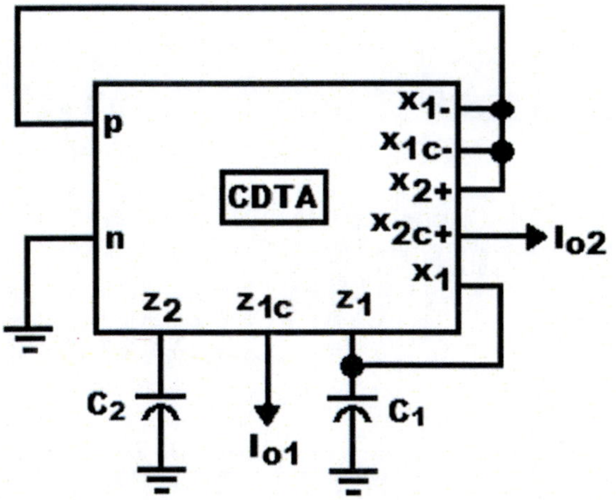

Fig. 2 Quadrature oscillator circuit using CDTA

The quadrature relationship between the outputs $I_{01}(s)$ and $I_{02}(s)$ is obtained as

$$\frac{I_{02}(s)}{I_{01}(s)} = \frac{g_m}{sC_2} = \frac{g_m}{\omega C_2} \mathrm{e}^{-j90^0} \tag{6}$$

Hence, the phase difference φ between I_{01} and I_{02} is 90°.

The sensitivity of the oscillator frequency (ω_0) for various active and passive components are found as

$$S^{\omega_0}_{C_1,C_2} = -\frac{1}{2}, S^{\omega_0}_{g_{m1},g_{m2}} = \frac{1}{2} \tag{7}$$

4 Simulation Results

It is simulated with Cadence PSPICE simulator using a 0.18 μm CMOS (TSMC) model parameter to confirm the theoretical analysis. The simulation is done by using supply voltage of ±1.5 V, bias currents of $I_{B2} = I_{B3} = 25$ μA and $I_{B1} = 750$ μA. The W/L ratio of various CMOS transistors are taken from [9]. Figures 3 and 4 show the output of the QO during transient and steady state, respectively. The simulated frequency of oscillation of 30.1 kHz is found to be very close to the theoretical value of 29.4 kHz. Quadrature relationships between the generated waveforms are verified by the Lissajous Figure as shown in Fig. 5. It is elliptical due to the slightly unequal magnitude of I_{01} and I_{02}. The phase difference is found to be 89°. Total harmonic distortion (%THD) of I_{01} and I_{02} for this oscillator is 1.95% and 1.01%, respectively. Moreover, the power dissipation of the oscillator is found to be 0.24 mW.

5 Comparison

The comparison of the proposed oscillator is given in Table 1. It is found that the proposed work uses the least number of CDTA and passive components in the count. Although in

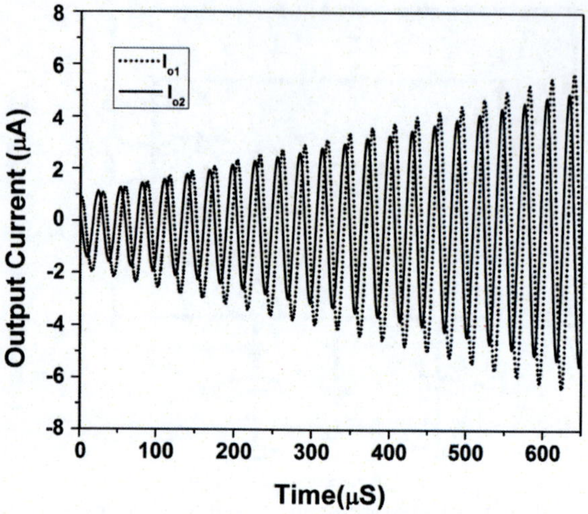

Fig. 3 Oscillator output in transient state

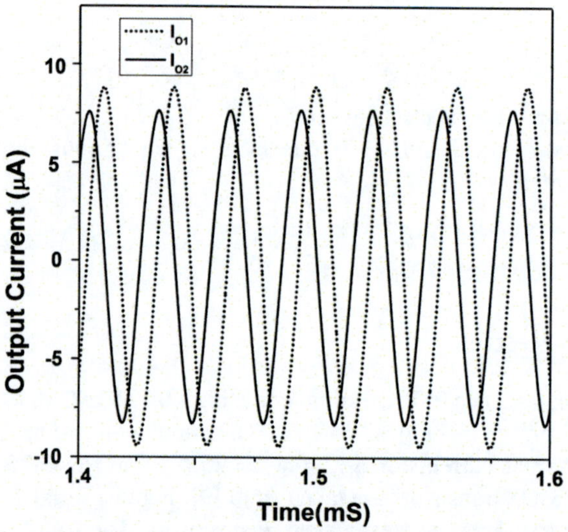

Fig. 4 Quadrature output of oscillator in steady state

[6–8] use of CDTA is one as that of our work, the number of passive components in the count is more and also the capacitors are not grounded.

6 Conclusion

It uses only one CDTA and two grounded capacitors. The CO and FO can be electronically controlled. The proposed oscillator does not use any external resistor which

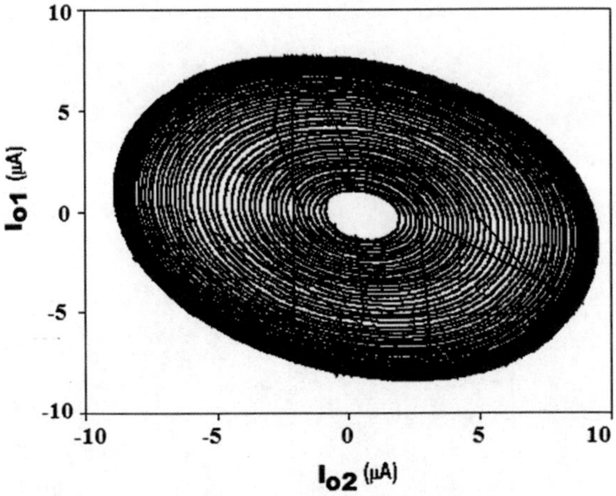

Fig. 5 Plot of I_{01} and I_{02} shows quadrature nature output

eases the IC fabrication. The sensitivity is found to be low. Simulated results verify the theoretical results well.

Table 1 Comparison with the previously published CDTA-based QOs

References	No. of ABB	Supply	Electroniccontrol of CO & FO	Grounded C	No of R + C	CM QO output	Power dessipassion (mW)	THD (%)	Phase difference	Sensitivity deviation from frequency
[3]	3	± 3 V	Yes	Yes	0 + 2	Yes	NA	2.5	NA	Low
[4]	3	± 3 V	Yes	Yes	0 + 2	Yes	NA	2.5	90^0	Low
[5]	2	± 2.5 V	Yes	Yes	1 + 2	Yes	NA	6	90^0	Low
[6]	1	± 6 V	No	No	2 + 2	No	NA	NA	NA	Low
[7]	1	± 3 V	No	No	1 + 2	Yes	NA	3.01	NA	Low
[8]	1	NA	No	No	1 + 2	Yes	NA	3	NA	NA
Our work	1	± 1.5 V	Yes	Yes	0 + 2	Yes	0.24	1.01	90^0	Low

References

1. Biolek, D.: CDTA—a building block for current-mode analog signal processing. In: Proceedings of the European Conference on Circuit Theory and Design, pp. 397–400. Krakow, Poland (2003)
2. Biolek, D., Biokova, V.: Universal biquard using CDTA elements for cascading filter design. In: Proceedings of the 13th International Multi-Conference on Circuits Systems Communications and Computers, (CSCC-2003), Greece, pp. 8–12 (2003)
3. Tangsrirat, W., Tanjareon, W.: Current-mode sinusoidal quadrature oscillator with independent control of oscillation frequency and condition using CDTAs. Indian J. Pure Appl. Phys. **48**, 363–366 (2010)
4. Tanjaroen, W., Tangsrirat, W.: Resistorless current-mode quadrature sinusoidal oscillator using CDTAs. In: Proceedings of the Asia-Pacific Signal and Information Processing Association Annual Summit and Conference, pp. 307–310 (2009)
5. Kumngern, M., Dejhan, K.: Electronically Tunable Current-Mode Quadrature Oscillator Using Current Differencing Transconductance Amplifiers. TENCON, pp. 1–4 (2009). 978-1-4244-4547-9/09/2009
6. Prasad, D., Bhaskar, R.D., Singh, K.A.: Realisation of single-resistance-controlled sinusoidal oscillator: a new application of the CDTA. WSEAS Trans. Electron. **5**, 257–259 (2008)
7. Lawanwisut, S., Biolek, D., Siripruchyanun, M.: A simple current-mode quadrature oscillator using only single CDTA. In: Proceedings of the 1st International Conference on Technical Education (ICTE2009), January 21–22, Bangkok, Thailand, pp. 119–122 (2009)
8. Jin, J., Wang, H.C.: Single CDTA based current-mode quadrature oscillator. Int. J. Electron. Commun. (AEU) **66**, 933–936 (2012)
9. Jin, J., Wang, H.C.: Current-mode universal filter and quadrature oscillator using CDTAs. Turkish J. Electr. Eng. Comput. Sci. **22**, 276–286 (2014)

Effect of Energy Loss Due to 1$s \rightarrow$ 2p Excitation and Ionization of Neutral Impurities on the Non-Ohmic Characteristics of a Compound Semiconductor at Low Lattice Temperature

Souma Saha, Subhadipta Mukhopadhyay, and Debi Prosad Bhattacharya(✉)

Department of Physics, Jadavpur University, Kolkata, West Bengal 700032, India
dbhattacharya140@gmail.com

Abstract. At low lattice temperatures ($T_L \leq 20\,K$), an apparently low electric field may effectively serve as high enough to significantly perturb an electron ensemble in a semiconductor from the state of thermodynamic equilibrium with the lattice atoms. The energy loss rate by an electron of the ensemble through impact ionization and excitation of neutral impurities may turn out to be comparable with the loss rate through interactions with the prevalent phonons and this takes part in controlling the non-Ohmic characteristics of the material. The present analysis deals with the calculation of the net energy loss rate of an electron and the subsequent effective electron temperature characteristics. The results obtained for InSb are compared with other theoretical and available experimental data. The agreement with the experiments is quite satisfactory. Moreover, the effects of impact ionization and neutral impurities at low temperatures are indeed not always negligible.

Keywords: Neutral impurities · Impact ionization and excitation · Non-equilibrium carriers · Effective electron temperature · Low lattice temperature

1 Introduction

It is well known that in the presence of a high electric field (ε), the electron ensemble in a semiconductor may be significantly perturbed from the state of thermodynamic equilibrium with the lattice atoms. The average energy of an electron then exceeds its thermal equilibrium value. Such an ensemble is sometimes picturized as one with a field dependent effective temperature $T_e(\varepsilon)$ that also exceeds the lattice temperature T_L. Under these conditions, the material exhibits some unexpected, novel properties which are technically of much importance from the device point of view. However, the electric field at which the high field effects are exhibited in a particular material depends upon the lattice temperature. For low lattice temperatures ($T_L \leq 20\,K$), even a field of a fraction

© Springer Nature Singapore Pte Ltd. 2021
N. R. Das and S. Sarkar (eds.), *Computers and Devices for Communication*,
Lecture Notes in Networks and Systems 147,
https://doi.org/10.1007/978-981-15-8366-7_57

of volt/cm may seem to be effectively high [1, 2]. The low lattice temperature has a number of specific features, which are hardly consistent with higher lattice temperatures. In developing the theory of electron transport for the higher temperatures, such low temperature features are usually ignored even when they occur in order to make the theoretical analysis easily tractable [3].

Recently, from a solution of the energy balance equation of the electron–phonon system, one of the present authors with some other analyzed the effective electron temperature characteristics $T_e(\varepsilon)$ in a compound semiconductor taking into account some of the low temperature features [4]. The results reported there for InSb has been obtained for $T_L = 1.35K$ considering electronic interaction only with deformation acoustic phonons. Some cursory agreement of the theoretical results with the experimental data has been observed. At such low temperatures, the electrons also interact with piezoelectric phonons in a compound semiconductor and thus also contribute to the process of transport. Apart from that, the energy loss rate due to $1s \to 2p$ excitation and ionization of the neutral impurities should also come into play [5].

The aim of this short communication is to study the effects of (i) interaction of the electrons with the piezoelectric phonons and (ii) the excitation and ionization of the neutral impurities by the energetic electrons in an effectively high field, on the $T_e(\varepsilon)$ characteristics of a compound semiconductor. The numerical results obtained for InSb at $T_L = 1.35K$ clearly exhibit how important it is to consider (i) the energy loss due to excitation and ionization of the neutral impurities and (ii) the electronic interaction with the piezoelectric phonons. The agreement of our theoretical results with the available experimental data seems to be quite satisfactory.

2 Theory

Let us consider a non-degenerate ensemble of electrons in an extrinsic compound semiconductor, which is doped with both donor and acceptor impurities at a low lattice temperature such that the impurities are not completely ionized. If N_D and N_A are the donor and acceptor concentrations, respectively, the concentration of the neutral impurities $N = N_D - N_A$ and the compensation ratio $C = N_A/N_D$ [5]. In the presence of an effectively high electric field, the energy distribution of the non-equilibrium ensemble of electrons is assumed to be the Maxwellian distribution function with an effective electron temperature $T_e(\varepsilon)$ [1, 2, 6].

The effective electron temperature characteristics may be determined from the energy balance equation [1].

$$\sum \left\langle \frac{\partial E_{\vec{k}}}{\partial t} \right\rangle = e\mu(\varepsilon)\varepsilon^2 \tag{1}$$

where $E_{\vec{k}}$ is the energy of an electron with wave vector $\vec{k}$. $\sum \left\langle \frac{\partial E_{\vec{k}}}{\partial t} \right\rangle = \left\langle \frac{\partial E_{\vec{k}}}{\partial t} \right\rangle_{ac.} + \left\langle \frac{\partial E_{\vec{k}}}{\partial t} \right\rangle_{pz.} + \left\langle \frac{\partial E_{\vec{k}}}{\partial t} \right\rangle_{ex} + \left\langle \frac{\partial E_{\vec{k}}}{\partial t} \right\rangle_{ion}$ are the average energy loss rate of electrons due to interaction with acoustic (ac.) and piezoelectric (pz.) phonons and due to $1s \to 2p$ excitation (ex.) and ionization (ions.) of the neutral impurities, respectively, e is the electronic charge, $\mu(\varepsilon)$

is the high-field mobility. For interaction with acoustic or piezoelectric phonons, the loss rate is calculated in the standard way using perturbation theory.

The field-dependent mobility due to different interactions is obtained as: $\mu_{ac} = \mu_{ac}^0 (T_L/T_e)^{1/2}$, $\mu_{pz} = \mu_{pz}^0 (T_e/T_L)^{1/2}$. Apart from that, there will be elastic interactions with the ionized (I) and the neutral (N) impurities and one can obtain: $\mu_I = \mu_I^0 (T_e/T_L)^{3/2}$, $\mu_N = \mu_N^0$, here $\mu_{ac}^0, \mu_{pz}^0, \mu_I^0, \mu_N^0$ are the zero field motilities. The combined mobility can then be found by Mattheisen's rule.

The loss due to excitation and ionization (non phonon processes) may be obtained from $A_I(T_e)NI$ [5] where A_I is the excitation or the ionization coefficient, I is the excitation or ionization energy. A_I may be calculated following the standard method [7].

3 Results and Discussions

The average energy loss rate of electrons and the effective electron temperature characteristics in InSb are calculated by using the following parameter values: effective mass $m^* = 0.0145 m_0$; deformation potential constant $E_1 = 30\,\text{eV}$; piezoelectric coupling constant $k_m = 0.027$; mass density $\rho_v = 5.7 \times 10^3\ \text{kg.m}^{-3}$; acoustic velocity $u_l = 3.7 \times 10^3\ \text{m s}^{-1}$; static dielectric constant $\in_{sc} = 17.51\ \in_0$, $\in_0$ being the free space permittivity, neutral impurity concentration $N = 10^{14}\ \text{cc}^{-1}$; compensation ratio $C = 0.05$; excitation energy $E_{ex} = 7.2 \times 10^{-23}$ J; and ionization energy $E_{ion} = 9.6 \times 10^{-23}$ J; Bohr radius [8] $a_0 = 1207 r_0$, where $r_0 = \hbar^2/m_0 e^2$,$\hbar = h/2\pi$, h being Planck's constant.

Here, we make use of the widely adopted simplest assumption that the cross section for the impact of the energetic electrons with the neutral impurities σ_i as constant, independent of energy $E_{\vec{k}}$. Since the cross section hardly corresponds to any physical quantity in the semiconductor, its choices by many others seem to be somewhat arbitrary and usually the value is adjusted to the experimental data, thus introduces an adjustable parameter of the analysis. The value chosen here is $\sigma_i = 10^{-14}\ \text{cm}^2$ [9]. The numerical results obtained from the present analysis are described in Figs. 1 and 2.

Figure 1 shows that the energy loss rate due to ionization and excitation of neutral impurities by the energetic electrons is not negligibly small compared to the same rate due to electronic interaction with the acoustic and the piezoelectric phonons. Moreover, the interaction with the piezoelectric phonons makes comparable contribution in the process of energy loss of the non-equilibrium electrons in a compound semiconductor. The process of energy loss due to the excitation and the ionization of the neutral impurities makes the major contribution in determining the effective electron temperature characteristics of a semiconductor at low lattice temperatures. Moreover, the agreement of our theoretical results of $T_e(\varepsilon)$ characteristics now agrees much more that what has been obtained in [4].

Figure 2 exhibits that the electronic interaction with the piezoelectric phonon makes comparable contribution in determining the effective electron temperature characteristics of a compound semiconductor at low lattice temperature. But above all, the contribution of the process of excitation and ionization of the neutral impurities seem to be much more under the prevalent conditions of interest here. The values of the parameters of

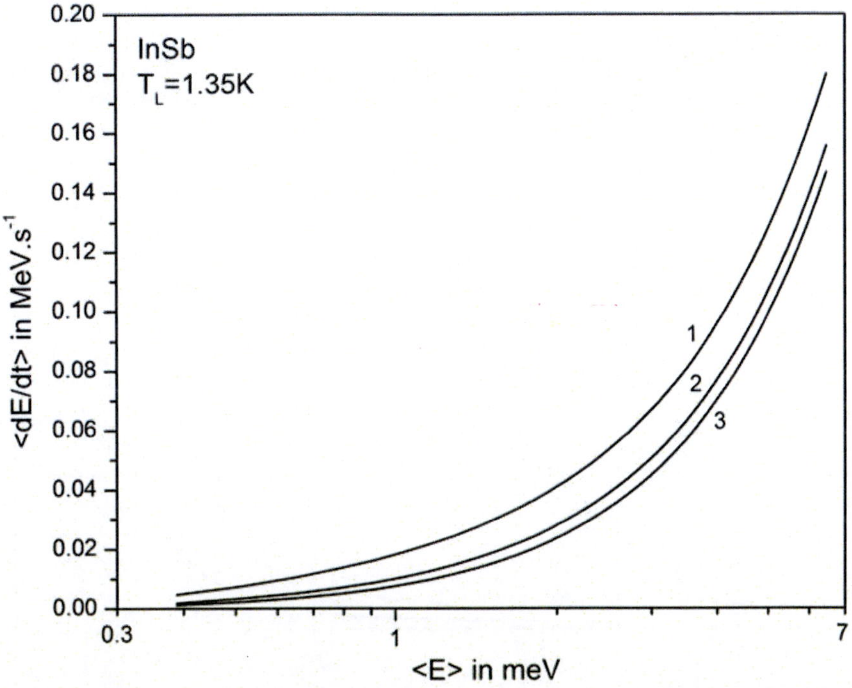

Fig. 1 Dependence of the average energy loss rate of an electron in a non-degenerate ensemble of electrons in InSb upon average electron energy $(3K_B T_e/2)$ at a lattice temperature of 1.35 K. The curve 1 is obtained considering the phonon as well as the non-phonon processes (interaction with acoustic and piezoelectric phonons and the $1s \rightarrow 2p$ excitation and ionization of the neutral impurities). The curve 2 is obtained for the combined interaction with the acoustic and the piezoelectric phonons and the curve 3 is obtained for interaction only with acoustic phonons

the experimental samples are not quoted in [6]. As such, the little discrepancy observed between the net theoretical results which consider the electronic interaction with the acoustic and the piezoelectric phonon as well as the energy loss due to excitation and ionization of neutral impurities, and the experimental characteristics may be ascribed due to the unknown parameter values of the experimental sample. The results which have been obtained here being interesting inspires further work taking due account of such low temperature features.

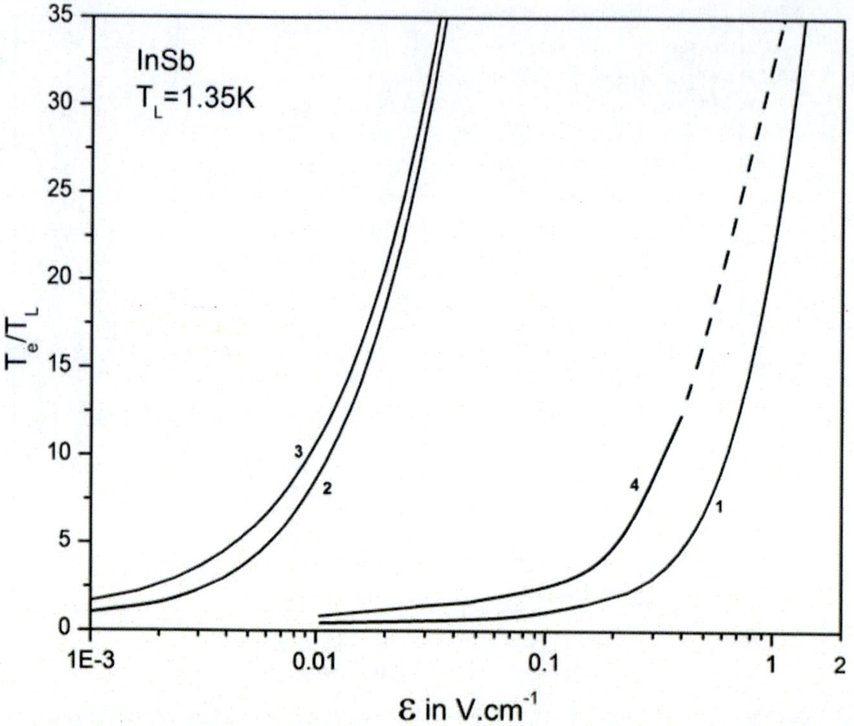

Fig. 2 Dependence of the normalized electron temperature upon the electric field in InSb at a lattice temperature 1.35 K. The curve 1 is obtained considering the energy loss due to the phonon and the non-phonon processes (interaction with acoustic and piezoelectric phonons and the $1s \rightarrow 2p$ excitation and ionization of the neutral impurities). The curve 2 is obtained for the combined interaction with the acoustic and the piezoelectric phonon. The curve 3 is obtained for interaction only with acoustic phonons. The curve marked 4 is the experimental curve [6]

Acknowledgements. S. Saha is indebted to Higher Education, Science and Technology and Biotechnology Department, Government of West Bengal, India for providing Swami Vivekananda Merit Cum Means Scholarship Scheme. The authors also acknowledge the help rendered by B. Roy.

References

1. Conwell, E.M.: High Field Transport in Semiconductors. Academic Press, New York (1967)
2. Nag, B.R.: Electron Transport in Compound Semiconductor. Springer, Berlin Heidelberg (1980)
3. Canali, C., Jacoboni, C., Nava, F., Ottaviani, G., Alberigi-Quaranta, A.: Electron drift velocity in silicon. Phys. Rev. B. **12**, 2265–2284 (1975)
4. Bhattacharya, D.P., Pramanik, T.K.: Effect of finite energy of deformation-potential acoustic phonons on the temperature of non-equilibrium carriers. J. Phys. Chem. Sol. **52**, 735–744 (1991)

5. Kachlishvili, Z.S.: Inelastic impurity scattering of hot electrons in semiconductors. Phys. Stat. Sol. **48**, 65 (1971)
6. Bauer, G.: Springer Tracts in Modern Physics, vol. 74. Springer, New York (1974)
7. Bonch-Bruevich, V.L., Landsberg, E.G.: Recombination mechanisms. Phys. Stat. Sol. **29**(1), 9–43 (1968)
8. Kittel, C.: Introduction to Solid State Physics. Wiley, India (2013)
9. Zylbersztejn, A.: Theory of low-temperature impact ionization in the high-purity germanium. Phys. Rev. **127**(3), 744 (1962)

Check for updates

Optimization of a Dual-Material Double-Gate TFET for Low Power Digital Application

Jayabrata Goswami[1](✉), Anuva Ganguly[2], Aniruddha Ghosal[3], and J. P. Banerjee[3]

[1] Netaji Subhas Open University, Kolkata, India
goswamijayabrata@gmail.com
[2] Dr. Sudhir Chandra Sur Degree Engineering College, Dumdum, Kolkata, India
gangulyanuva@gmail.com
[3] Institute of Radio Physics and Electronics, University of Calcutta, 92, Acharya Prafulla Chandra Road, Kolkata, West Bengal 700009, India
aghosal2008@gmail.com, scope.jcb@gmail.com

Abstract. Studies are carried out to design and optimize the performance of next-generation dual-material double-gate (DMDG) P-channel TFET for low power digital application using Si as the channel material and SiO_2 as gate dielectric. The channel length of the device has been varied in the range of 20–30 nm. The DC properties and performance of DMDG TFET have been compared with that of a dual-material single-gate (DMSG) TFET with respect to low power application of the device. 2D Poisson's equation is solved to obtain the band diagrams for both on and off states, the electric field profile and surface potential of the device. The performance parameters of the optimized DMDGTFET such as on–off current ratio and subthreshold swing (SS) are compared with those of DMSG TFET. The results show that the DC properties of next-generation DMDG TFET excel those of DMSG TFET for a channel length of 20 nm. The on–off current ratio in DMDG TFET is 2.18×10^4 while the same in its DMSG counterpart is one order of magnitude lower. The values of SS in DMDG and DMSG TFET are found to be 20.1 and 36 mV/decade, respectively.

Keywords: TFET · DMDG · DMSG · Sub-threshold swing · On–off current ratio

1 Introduction

In recent years, tunnel field-effect transistors (TFETs) are emerging as extremely low power devices which may be used for future replacement of complementary metal–oxide–semiconductor (CMOS) devices in ultra-large-scale integration (ULSI) chips. The main advantage of TFETs is that the subthreshold swing of these devices can be reduced below the thermionic limit of 60 mV/decade at room temperature exhibited by CMOS devices [1]. Moreover the leakage current is lower and on–off current ratio is higher in TFETs than in CMOS devices. In tunnel field-effect transistor (TFET), current

© Springer Nature Singapore Pte Ltd. 2021
N. R. Das and S. Sarkar (eds.), *Computers and Devices for Communication*,
Lecture Notes in Networks and Systems 147,
https://doi.org/10.1007/978-981-15-8366-7_58

is produced by band to band tunneling between the source and channel due to which on–off current ratio is enhanced with lower off state leakage current. The other advantages of tunnel FETs over CMOS are less intrinsic delay, improved electrical noise performance and nearly stable subthreshold swing. Two types of device structure of TFETs considered in the present study are DMDG and DMSG. In this work, we present analytical and numerical studies on the design of channel length of the device with respect to optimum DC performance for two types of device structure of TFETs. The channel length is varied in the range of 20–30 nm appropriate to the latest technology benchmark set by ITRS.

2 Structure of Device

The cross-sectional view of DMDG TFET is shown in Fig. 1. The TFET has highly doped n^+ source, p^+ channel and p^+ drain regions with doping concentrations of 3.4 $\times$ 10^{20} cm^{-3}, 10^{17} cm^{-3} and 1.1 $\times$ 10^{20} cm^{-3}, respectively. The oxide and substrate thicknesses are taken to be $t_{ox} = 2$ nm and $t_{si} = 4$ nm, respectively. The SiO_2 dielectric is deposited on both top and bottom sides between the metal gate and channel regions. The top and bottom gates are made of two different materials having different work functions. Two-dimensional Poisson's equation is analytically and numerically solved subject to appropriate boundary conditions to obtain the energy band diagram in both on and off states, the electric field profile and the surface potential of the device.

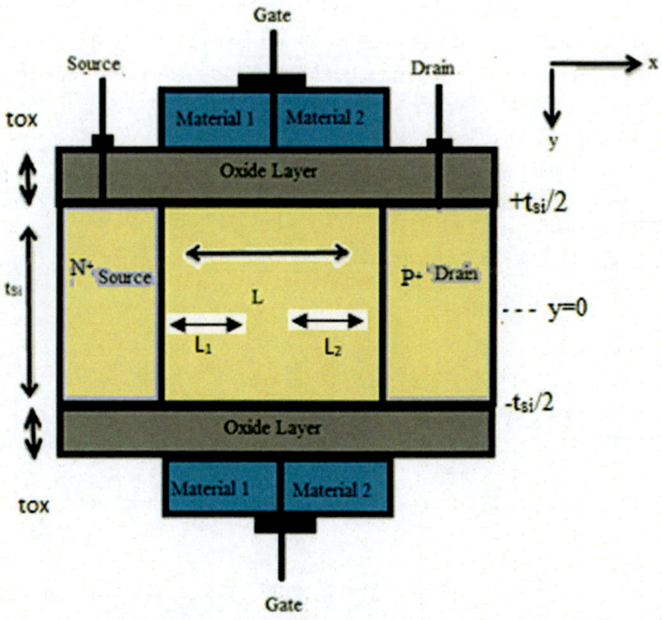

Fig. 1 Cross-sectional view of dual-material double-gate TFET

3 Analysis and Simulation

The 2D Poisson's equation in the gate–oxide region is given by

$$\frac{\partial^2 \varphi_i(x,y)}{\partial x^2} + \frac{\partial^2 \varphi_i(x,y)}{\partial y^2} = 0 \tag{1}$$

In Eq. (1), $\varphi_i(x, y)$ is the 2D potential function.

The 2D potential function can be expressed as a second-order polynomial given below

$$\varphi_i(x,y) = a_0(x) + a_1(x)y + a_2(x)y^2 \tag{2}$$

Equation (1) has been analytically solved using Eq. (2) subject to the following boundary conditions.

(i) Taking $y = +t_{Si}/2$ for the top surface, the electric field equations at the top gate–oxide interface are

$$\frac{\partial \varphi_{i1}(x,y)}{\partial y} = \frac{\varepsilon_{ox}}{\varepsilon_{Si}} \frac{\varphi_{Surf1}(x) - \psi_{g1}}{t_{ox}} \quad \text{at } 0 < x < L_1 \text{ and } y = +\text{tsi}/2 \text{ for material 1} \tag{3}$$

$$\frac{\partial \varphi_{i2}(x,y)}{\partial y} = \frac{\varepsilon_{ox}}{\varepsilon_{Si}} \frac{\varphi_{Surf2}(x) - \psi_{g2}}{t_{ox}} \quad \text{at } y = +\text{ tsi}/2 \text{ and } L_1 < x < L_2 \text{ For material 2} \tag{4}$$

In Eqs. (3) and (4), $\psi_{g1} = V_{GS} - \phi_{M1} + \chi + E_g/2$.
and $\psi_{g2} = V_{GS} - \phi_{M2} + \chi + E_g/2$.

ϕ_{M1} and ϕ_{M2} are the metal gate work functions of gate materials 1 and 2, respectively, and χ is the electron affinity.

(ii) Taking $y = y = -t_{Si}/2$ for the bottom surface, the electric field equations at the bottom gate- oxide interface are

$$\frac{\partial \varphi_{i1}(x,y)}{\partial y} = \frac{\varepsilon_{ox}}{\varepsilon_{Si}} \frac{\psi_{g1} - \varphi_{Surf1}(x)}{t_{ox}} \quad \text{at } 0 < x < L_1 \text{ and } 0 < y < \text{-tsi}/2 \text{ for material 1} \tag{5}$$

$$\frac{\partial \varphi_{i2}(x,y)}{\partial y} = \frac{\varepsilon_{ox}}{\varepsilon_{Si}} \frac{\psi_{g2} - \varphi_{Surf2}(x)}{t_{ox}} \quad \text{at } L_1 < x < L_2 \text{and } 0 < y < \text{-tsi}/2 \text{ for material 2} \tag{6}$$

(iii) At the source and drain ends of the channel, the potential functions for materials 1 and 2 are

$$\varphi_{i1}(x,y)$$

$$= \varphi_{\text{Surf1}}\left(0, \left|\frac{t_{\text{Si}}}{2}\right|\right) = V_{\text{BI}} \quad \text{at}\, x = 0 \,\text{and}\, y = \text{ tsi/2 at source end} \tag{7}$$

$$\varphi_{i2}(L, 0) = \varphi_{\text{Surf2}}\left(L, \left|\frac{t_{\text{Si}}}{2}\right|\right)$$
$$= V_{\text{BI}} + V_{\text{DS}} \quad \text{at}\, x = L \,\text{and}\, y = \text{ tsi/2 where } (L = L_1 + L_2) \text{ at drain end.} \tag{8}$$

In Eqs. (7) and (8), ϕ_{surf1} and ϕ_{surf2} are the surface potentials on materials 1 and 2, respectively, ε_{Si} and ε_{ox} are the relative permittivity of silicon and relative permittivity of SiO_2, V_{BI} is the built-in potential, E_g is the band gap energy and V_{GS} and V_{DS} are the gate to source and drain to source voltages, respectively.

Now using the boundary condition (3) and (4) and substituting Eq. (2) in Eq. (1), we obtain the surface potentials at material 1 and material 2 as

$$\varphi_{\text{Surf1}}(x) = M\mathrm{e}^{\lambda x} + N\mathrm{e}^{-\lambda x} - \psi_{g1} \tag{9}$$

$$\varphi_{\text{Surf2}}(x) = Q\mathrm{e}^{\lambda x} + R\mathrm{e}^{-\lambda x} - \psi_{g2} \tag{10}$$

where the parameters M, N, Q, and R can be expressed as

$$M = \frac{[(V_{\text{BI}} - \psi_{g1})\exp(\lambda(L))] - [V_{\text{BI}} + V_{\text{DS}} - \psi_{g2}] + [(\psi_{g1} - \psi_{g2})\cosh(\lambda L_2)]}{\exp(\lambda(L_1 - L_2)) - \exp(\lambda(L_1 + L_2))} \tag{11}$$

$$N = \frac{[V_{\text{BI}} + V_{\text{DS}} - \psi_{g2}] - [(V_{\text{BI}} - \psi_{g1})\exp(\lambda(L))] - [(\psi_{g1} - \psi_{g2})\cosh(\lambda L_2)]}{\exp(\lambda(L_1 - L_2)) - \exp(\lambda(L_1 + L_2))} \tag{12}$$

$$Q = M\exp(\lambda L_1) + (\psi_{g1} - \psi_{g2})/2 \tag{13}$$

$$R = N\exp(-\lambda L_1) + (\psi_{g1} - \psi_{g2})/2 \tag{14}$$

where λ is a parameter given by

$$\lambda = \sqrt{\frac{2\varepsilon_{\text{ox}}}{\varepsilon_{\text{Si}} t_{\text{ox}} t_{\text{Si}}}}$$

The electric field in the channel of the device is obtained by

$$\xi = \sqrt{\xi_x^2 + \xi_y^2} \tag{15}$$

The interband tunneling is the fundamental phenomenon for the generation of tunneling current or drain current of the device. The tunneling generation rate (G) of the device is calculated by using Kane model. Therefore, the drain current is given by [2]

$$I_{\text{D}} = q\iint G\mathrm{d}x\mathrm{d}y \tag{16}$$

where the tunneling generation rate G can be expressed as

$$G(\xi) = A_1 \frac{|\xi|^2}{\sqrt{E_g}} \exp\left[-B_1 \frac{E_g^{3/2}}{|\xi|}\right] \tag{17}$$

where the parameters A_1 and B_1 are taken as 4×10^{14} cm$^{-1/2}$ V$^{-5/2}$ s^{-1} and 1.9×10^7 V/cm [3], respectively.

Similar analytical solution for dual-material single-gate TFET has been obtained carried out with the same parameters as given above.

4 Results

The simulated energy band diagrams for the optimized dual-material double-gate TFET in the off state ($V_{GS} = 0$ V) and on state ($V_{GS} = 0.5$ V, $V_{DS} = 0.5$ V) are shown in Fig. 2a, b, respectively. The optimization process is done by varying the channel length from 20 to 30 nm. The metal work functions for material 1 and 2 are taken to be $\varphi_{M1} =$ 3.8 eV and $\varphi_{M2} = 4.4$ eV.

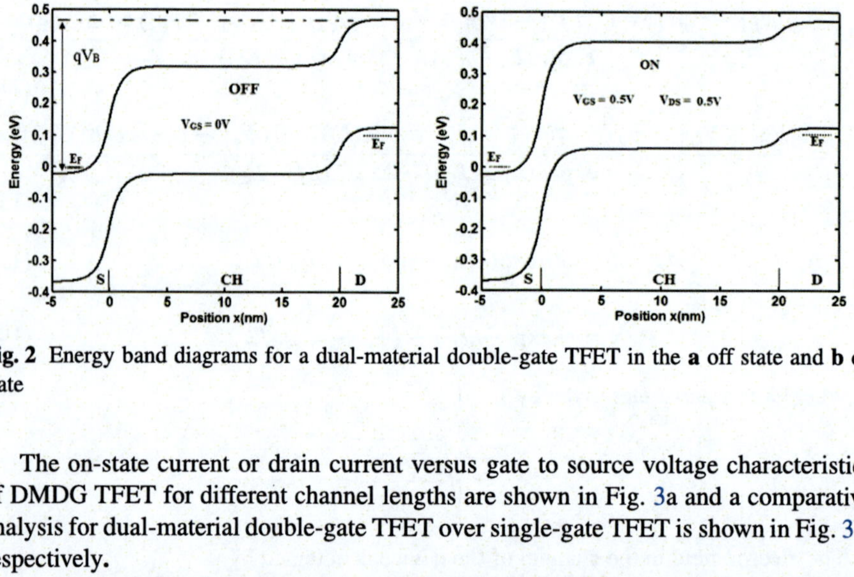

Fig. 2 Energy band diagrams for a dual-material double-gate TFET in the **a** off state and **b** on state

The on-state current or drain current versus gate to source voltage characteristics of DMDG TFET for different channel lengths are shown in Fig. 3a and a comparative analysis for dual-material double-gate TFET over single-gate TFET is shown in Fig. 3b, respectively.

Figure 3a shows the drain characteristics of DMDGTFET for three channel lengths viz. $L = 20$, 25 and 30 nm, respectively. As the channel length increases, tunneling probability decreases, and therefore, the on-state drain current decreases. Figure 3b shows the drain characteristics for both DMSGTFET and DMDGTFET for the optimized channel length of 20 nm. It is observed from the figure that the drain current is higher in DMDG TFET than in DMSG TFET at any particular value of gate to source voltage for the channel length of 20 nm.

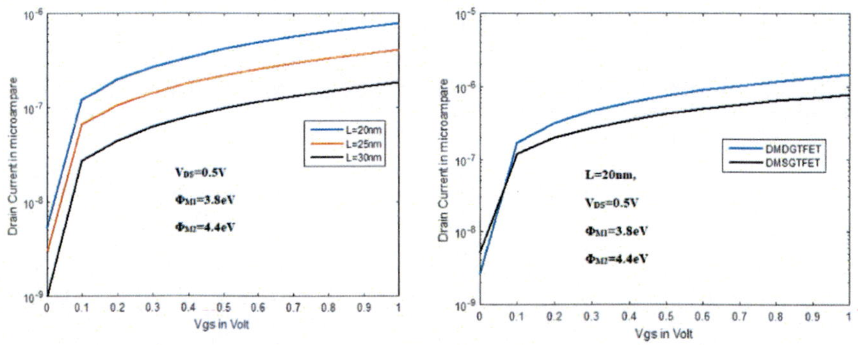

Fig. 3 **a** Drain current versus gate to source voltage for different channel lengths dual-material DGTFET and **b** drain current analysis for dual-material double-gate TFET over single-gate TFET

The subtherhold swing (SS) of the device is calculated by the following equation.

$$SS = \frac{\partial V_{GS}}{\partial(\log I_D)} \text{ mV/decade} \tag{18}$$

Table 1 shows the performance parameters of both DMDGTFET and DMSGTFET for different channel length in the range of 20–30 nm. It is observed from the table that the on–off current ratio is maximum and subthreshold swing is minimum for the channel length, $L = 20$ nm ($L_1 = 10$ nm, $L_2 = 10$ nm). With the increase of channel length, both the performance parameters of the device degrade with respect to on–off current ratio and subthreshold swing. It is observed that DMDGTFET excels with respect to higher on–off current ratio, lower subthreshold slope for the designed value of the channel length as 20 nm.

5 Conclusion

The design and optimization study of next-generation TFET are carried out for low power digital application using Si as channel material and SiO_2 as gate dielectric. It is observed that the optimized channel length for which DMDGTFET exhibits better performance with respect to higher on–off current ratio and lower subthreshold slope is 20 nm, which is appropriate to the latest technology demand specified by ITRS. The comparative study on the performance of DMDGTFET and DMSGTFET shows conclusively that the former excels the later with respect to both on–off current ratio and subthreshold swing.

Table 1 Comparative study of DMDGTFET and DMSMTFET

Device structure	Metal gate work function	Channel length $L = (L_1 + L_2)$ (nm)	I_{on} (μA)	I_{off} (nA)	I_{on}/I_{off}	Subtherhold swing(SS) mV/decade
Dual-material double-gate TFET	$\varphi_{M1} =$ 3.8 eV $\varphi_{M2} =$ 4.4 eV	$L = 20$ nm $L_1 = 10$ nm, $L_2 = 10$ nm	12.7	0.58	2.18×10^4	20.1
		$L = 25$ nm $L_1 =$ 12.5 nm, L_2 = 12.5 nm	7.7	0.54	1.42×10^4	23.5
		$L = 30$ nm $L_1 = 15$ nm, $L_2 = 15$ nm	5.11	0.60	8.51×10^3	31
Dual-material single-gate TFET	$\varphi_{M1} =$ 3.8 eV $\varphi_{M2} =$ 4.4 eV	$L = 20$ nm $L_1 = 10$ nm, $L_2 = 10$ nm	11.9	0.00911	1.3×10^3	36
		$L = 25$ nm $L_1 =$ 12.5 nm, L_2 = 12.5 nm	7.3	0.012	6.0×10^2	43
		$L = 30$ nm $L_1 = 15$ nm, $L_2 = 15$ nm	4.6	0.028	1.6×10^2	48

Reference

1. Zhang, Q. , Zhao, W., Seabaugh, A.: Low subthreshold- swing tunnel transistors. IEEE Electron Device Lett. **27**(4), 297–300 (2006)
2. Kane, E.O.: Zener tunneling in semiconductors. J. Phys. Chem. Solides **12**(2), 181–188 (1960)
3. ATLAS User's Manual 2010 Silvaco Inc. Santa Clara

Nonmonotonic Electron Mobility in Asymmetrically Doped V-shaped Coupled Quantum Well Field-Effect Transistor Structure

A. K. Panda(✉), Devika Jena, Sangeeta K. Palo, and Trinath Sahu

Department of Electronics and Communication Engineering, National Institute of Science and Technology, Palur Hills, Berhampur, Odisha 761008, India
akpanda62@hotmail.com

Abstract. In this paper, we study the nonlinear channel electron mobility μ in an asymmetrically doped double quantum well field-effect transistor (QWFET) structure. The double quantum well consists of V-shaped channels by tailoring the conduction band edge of $Al_xGa_{1-x}As$ alloy through suitable variation of the alloy concentration x. We vary the widths of the wells asymmetrically and analyze their effect on the potential profile which causes drastic changes in the subband electron wave functions (ψ_n) and energy levels (E_n). In V-shaped potential, ψ_n are more localized than that of a square well. The change in the subband electronic structure induces change in occupation of subbands leading to intersubband interactions. We show that oscillatory enhancement in mobility under double subband can be obtained through the ionized impurity scattering through intersubband effects. The typical change of alloy concentration affects the alloy scattering which influences the overall total mobility μ.

Keywords: QWFET · Wave functions · Energy levels

1 Introduction

In past decades, the study of non-square quantum well (QW) structures is getting more attention due to their ability of manipulating the subband electron states 1–4. In fact, their specific shape of structure potentials can be employed for optoelectronic devices [1, 2]. By altering the structure parameters like well and barrier widths and doping concentrations, the two-dimensional electron gas (2DEG) transport properties can be changed [3–10]. The tunneling effect of double quantum well (DQW) structure is used in the development of electro-optical devices. The energy levels of a DQW split due to coupling of subband wave functions between the wells. As the structure parameters change, the coupling effect also changes. The electron mobility is studied in DQW of different shapes, like square double quantum well (SDQW), cubic DQW (CDQW), parabolic DQW (PDQW) and V-shaped DQW (VDQW) [10].

We study the effect of asymmetry in doping concentrations and well widths for a VDQW grown from $Al_xGa_{1-x}As$ by linearly varying the alloy fraction $x = 0$ at the center

© Springer Nature Singapore Pte Ltd. 2021
N. R. Das and S. Sarkar (eds.), *Computers and Devices for Communication*,
Lecture Notes in Networks and Systems 147,
https://doi.org/10.1007/978-981-15-8366-7_59

to $x_v = 0.1$ near the edge of the well. In VDQW, due to the alloy layer, the alloy disorder scattering potential and potential due to confinement are higher as compared to SDQW. We study the mobility as a function of asymmetry in well widths in a VDQW field-effect transistor. We show that substantial change in the amplitudes of subband wave functions near the resonance of subband states causes oscillatory behavior of mobility in double subband occupancy through intersubband effects. The nonlinearity in μ basically occurs through the ionized impurity scattering. Our study of nonmonotonous mobility due to asymmetric structure parameter in VDQW structure can be made use to study the field-effect transistors.

2 Theory

We propose a coupled VDQW field-effect transistor structure made of $Al_xGa_{1-x}As$. The schematic structure layout is described in Fig. 1 where the two wells of widths w_1 (substrate side) and w_2 (surface side). The central barrier is of width b. To reduce the scattering effect, delta doping is done in both the extreme barriers at a length s from the edges of the well and barrier. N_{d1} and N_{d2} are the doping concentrations of width d toward the substrate and surface side, respectively.

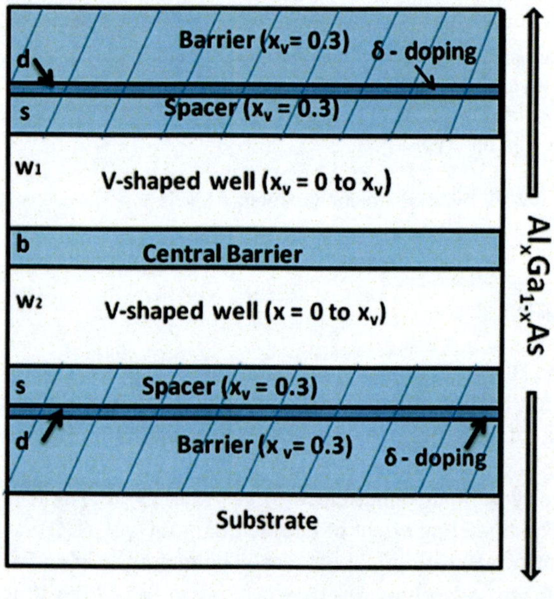

Fig. 1 Structural layout of $Al_xGa_{1-x}As$ V-shaped double quantum well-based HEMT system

The band banding is due to the diffusion of electrons from the outside barriers into the well. The wave function $\psi_n(z)$ and energy levels$E_n$can be attained with the help of transmission coefficient across the whole potential. The electrons transfer and confine in the V-shaped potential wells. Taking the growth direction along z-axis, the one-dimensional Schrödinger equation satisfies [11]:

$$\left[\frac{-\hbar^2}{2m^*}\frac{d^2}{dz^2} + V(z)\right]\psi_n(z) = E_n\psi_n(z) \tag{1}$$

m^* is the electron effective mass. $V(z) = V_s(z) + V_H(z) + V_F(z)$. $V_s(z)$ is the potential of the structure, $V_F(z)$ is the potential for the applied electric field F, i.e., $V_F\,(z) = eFz$. The Hartree potential $V_H(z)$ can be got using Poisson's equation.

$$\frac{d^2V_H}{dz^2} = \frac{4\pi e^2}{\varepsilon_0}[N(z) - n(z)] \tag{2}$$

$N(z)$ is the concentration of impurities and $n(z)$ is the distribution of electrons. The subband mobility is related to the transport life time τ_n. The multisubband electron transport life time τ_n is derived—from the Boltzmann transport equation [11]:

$$\sum_{n=0}^{M} C_{mn}\tau_n = 1\ , \tag{3}$$

where M is the different occupied levels. m and n are indices of subbands. For two occupied subbands, $m, n = 0$ and 1. The transport matrix elements C_{mn} stand for single and double subband occupancy. When lowest one is occupied, $m = n = 0$, $C_{00} = A_{00}$ $= 1/\tau_0$. Whereas when two subbands are occupied, τ_0 and τ_1 are:

$$\frac{1}{\tau_0} = \frac{(A_{00} + B_{01})(A_{11} + B_{10}) - K_{01}K_{10}}{(A_{11} + B_{10}) + (E_{F1}/E_{F0})^{1/2}K_{01}} \tag{4}$$

$$\frac{1}{\tau_1} = \frac{(A_{00} + B_{01})(A_{11} + B_{10}) - K_{01}K_{10}}{(A_{00} + B_{01}) + (E_{F0}/E_{F1})^{1/2}K_{10}} \tag{5}$$

Here A_{mm} is intrasubband while B_{mn} and K_{mn} are the intersubband scattering rate matrix elements, expressed through $V^{\text{eff}}{}_{mn}(q_{mn})$ as

$$\begin{aligned}\left|V_{nm}^{\text{Imp}}(q)\right|^2 = {} & \frac{4\pi^2N_d e^4}{\varepsilon_0^2 q^2}\left[\int_{-(d_P+s+L+b/2)}^{-(s+L+b/2)} dz_i\left|\sum_{n'm'}\varepsilon^{-1}_{nm,n'm'}(q)Z_{n'm'}(q, z_i)\right|^2\right] \\ & + \frac{4\pi^2N_d e^4}{\varepsilon_0^2 q^2}\left[\int_{(s+L+b/2)}^{(d+s+L+b/2)} dz_i\left|\sum_{n'm'}\varepsilon^{-1}_{nm,n'm'}(q)Z_{n'm'}(q, z_i)\right|^2\right]\end{aligned} \tag{6}$$

where

$$Z_{n'm'}(q, z_i) = \int_{-\infty}^{\infty} dz\,\psi_{n'}(z)\psi_{m'}(z)e^{-q|z-z_i|} \tag{7}$$

$$\left|V_{nm}^{Al}(q)\right|^2 = \left[a^3(\delta V)^2x(1-x)/4\right] \times \int dz\left|\sum_{n'm'}\psi_{n'}(z)\psi_{m'}(z)\varepsilon^{-1}_{nm,n'm'}(q)\right|^2 \tag{8}$$

$\varepsilon^{-1}_{nm,n'm'}$ is the inverse of dielectric screening matrix. The subband mobility $\mu_n{}^{\text{Imp/}Al}$ $(E_F) = [e/m]\ \tau_n{}^{\text{Imp/}Al}\ (E_F)$. For Imp/$Al$ scattering, $\mu^{\text{Imp}Al} = \sum_n n_n\mu_n{}^{\text{Imp/}Al}/\sum_n n_n$. The mobility μ is obtained with the help of Matthiessen's rule.

3 Results and Discussion

We study the impact of asymmetric well width variation on mobility μ in VDQW. The specifications of the structure are d = 20 Å, s = 60 Å, b = 40 Å, N_{d1} = 0 and N_{d2} = 3 × 10^{18} cm^{-3}. We linearly vary x within in the well, 0 at the center and x_v = 0.1 at the edges. The interface roughness Δ = 2.83 Å and Λ = 100 Å. The alloy disorder scattering potential δV = 1.56 eV [12]. The sum of the well widths is considered as fixed, i.e., $w_1 + w_2$ = 200 Å. So when we change w_1 accordingly w_2 changes, which causes asymmetry in the structure.

In Fig. 2, we show the mobility μ^{Imp}, μ^{Al} and μ in 10^4 cm^2/Vs as functions of well width w_1. The mobility μ is driven by both μ^{Imp} and μ^{Al}. For w_1 = 50 Å, single subband is occupied and it continues up to w_1 = 109 Å where μ mostly constant. The range of double subbands occupancy continues from w_1 = 109 Å to w_1 = 140 Å. At the transition point, μ^{Imp} decreases but μ^{Al} increases due to intersubband effects. Within two subbands occupied range, μ rise and fall nonlinearly. Even though μ^{Imp} dominates the fluctuation, μ^{Al} just contributes to the general magnitude of μ. The detailed nature of μ^{Imp} and μ^{Al} can be described with the help of their subband mobilities. Figure 3 presents ${\mu_0}^{\text{Imp}}$, ${\mu_1}^{\text{Imp}}$, ${\mu_0}^{\text{Al}}$, and ${\mu_1}^{\text{Al}}$ with respect to the well width w_1. In the two-subband occupied case, there is a hump at w_1 = 122 Å which is due to ${\mu_1}^{\text{Imp}}$. The central barrier width b also affects the height of this hump. In Fig. 4, we plot μ versus w_1 for b = 20, 40, 60 Å. As we increase b, the elevation of humps increases causing increase in the oscillatory character of μ.

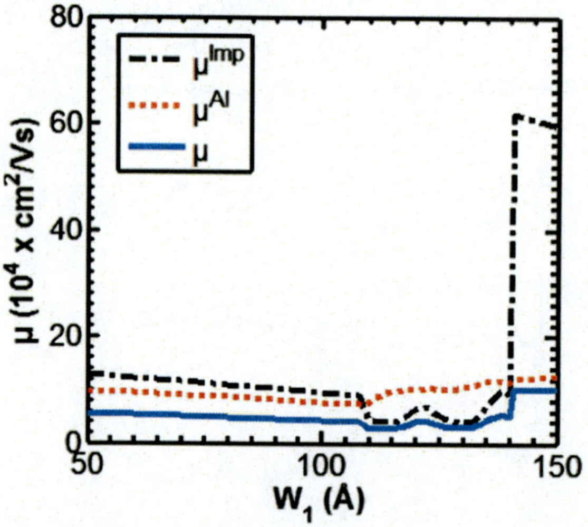

Fig. 2 μ^{Imp}, μ^{Al} and μ for d = 20 Å, b = 40 Å, s = 60 Å, $w_1 + w_2$ = 200 Å, x_v = 0.1, N_{d1} = 0, N_{d2} = 3 × 10^{18} cm^{-3}

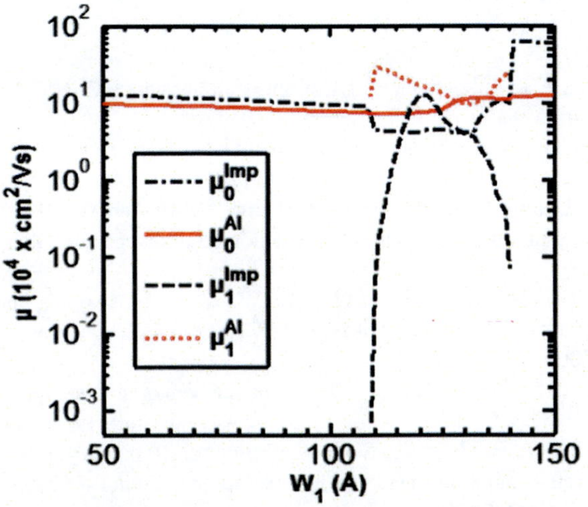

Fig. 3 $\mu_0{}^{\text{Imp}}$, $\mu_1{}^{\text{Imp}}$, $\mu_0{}^{\text{Al}}$ and $\mu_1{}^{\text{Al}}$ for $d = 20$ Å, $b = 40$ Å, $s = 60$ Å, $w_1 + w_2 = 200$ Å, $x_v = 0.1$, $N_{d1} = 0$, $N_{d2} = 3 \times 10^{18}\ \text{cm}^{-3}$

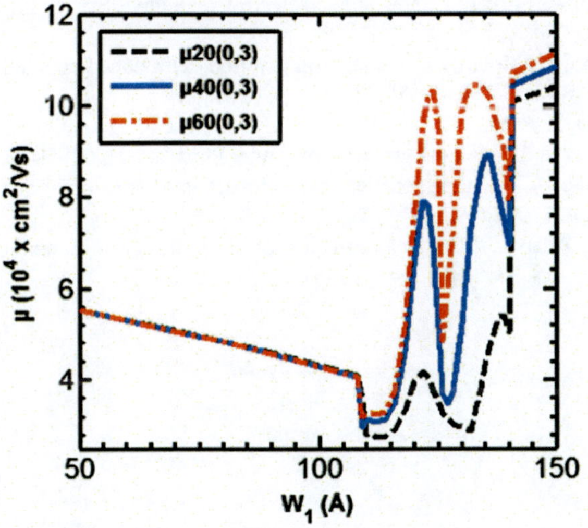

Fig. 4 Comparison of μ for $b = 20, 40, 60$ Å, $d = 20$ Å, $s = 60$ Å, $w_1 + w_2 = 200$ Å, $x_v = 0.1$ $N_{d1} = 0$ and $N_{d2} = 3 \times 10^{18}\ \text{cm}^{-3}$

4 Conclusion

The principal work here is the study of the nonmonotonic electron mobility in a V-shaped double quantum well (VDQW) as a function of asymmetry in the well widths of the DQW. Within two subbands occupancy, oscillatory increase of mobility occurs through ionized impurity scattering. The nonlinearity of μ rises with rise in barrier width. Our results can be used to study of the coupled quantum well based nano-devices.

References

1. Choi, R.J., Hahn, Y.B.: Efficient blue-light emitting diodes with InGaN/GaN triangular shaped multiple quantum wells. Appl. Phys. Lett. **82**, 2764 (2003)
2. Ozturk, E.: Nonlinear intersubband transition in different shaped quantum wells under intense laser field. Sperlattices Microstruct. **82**, 303–312 (2015)
3. Mamani N.C., Duarte, C.A., Gusev, G.M., Quivy, A.A., Lamas, T.E.: Magnetotransport inAlxGax−1As quantum wells with different potential shapes. Brazilian J. Phys. **36**(2A) (2006)
4. Gao, K.H., Yu, G., Zhou, Y.M., Zhou, W.Z., Lin, T., Chu, J.H., Dai, N., Spring Thorpe, A.J., Austing, D.G.: Transport properties of AlGaAs/GaAs parabolic quantum wells. Appl. Phys. **105**, 013712 (2009)
5. Das, S., Nayak, R.K., Sahu, T., Panda, A.K.: Enehancement of multisubband electron mobility in asymmetrically doped coupled double quantum well structure. Phys. B **476**, 91–95 (2015)
6. Nayak, R.K., Das, S., Panda, A.K., Sahu, T.: Structural asymmetry induced size quantized nonmonotus electron mobility in GaAs/AlGaAs double quantum well structure. Sperlattices Microstruct. **89**, 75–82 (2016)
7. Sahoo, N., Panda, A.K., Sahu, T.: Enhancement of electron mobility in square-parabolic asymmetric double quantum well structure. Sperlattices Microstruct. **105**, 11–21 (2017)
8. Palo, S.K., Sahoo, N., Panda, A.K., Sahu, T.: Oscillation of electron mobility in V-shaped double quantum well structure under applied electric field. Phys. Stat. Solid. (b) **256**, 1800337 (2019)
9. Sahoo, N., Sahu, T.: Mobility modulation in inverted delta doped coupled double quantum well structures. Phys. B **498**, 49–54 (2016)
10. Palo, S.K., Sahu, T., Panda, A.K.: Effect of non-square structure potential on the multisubband electron mobility in double quantum well structure. Phys. B **545**, 62–68 (2018)
11. Ando, T., Fowler, A.B., Stern, F.: Electronic properties of two dimensional systems. Rev. Mod. Phys. **54**, 437–672 (1982)
12. Saxsena, A.K., Adams, A.D.: Determination of alloy scattering potential in $Ga_{1\text{-}x}Al_xAs$ alloys. J. Appl. Phys. **58**, 2640–2645 (1985)

Comparative Study of Threshold Characteristics in Low-Dimensional TFET with Quantum Confinement

Sharmistha Shee Kanrar[1](✉), Dinesh Kumar Dash[1,2], and Subir Kumar Sarkar[1]

[1] Department of ETCE, Jadavpur University, Kolkata, West Bengal 700032, India
sharmi.shee@gmail.com, dineshdash123@gmail.com,
su_sircir@yahoo.co.in

[2] Department of Electronic and Telecommunication Engineering, Parala Maharaja Engineering College, Berhampur, India

Abstract. In this paper, we develop analytical model of threshold characteristics for a Dual Material Double Gate tunnel FET to obtain compact and useful expressions. For the first time in literature, the influence of quantum confinement effects on threshold characteristics in short-channel tunnel FET are measured. For this purpose, we deploy the novel 'surface potential-based approach' which incorporates solutions of Schrodinger equation and Poisson's equation. Using these models, a detailed quantitative comparison between classical and quantum models is plotted. The analytical results are validated with simulation results.

Keywords: Tunnel FET · Low dimensional · Analytical modeling · Quantum confinement · Tunneling barrier · Threshold voltage

1 Introduction

Owing to severe limitation of conventional MOSFET devices in very low dimensions, the tunnel field effect transistors (TFETs) have emerged as new promising candidate. TFETs conquer the limitation of '60 mV/decade subthreshold swing' as well as provide 'great immunity against short-channel effects (SCEs)' [1].

In contrast to MOSFETs, TFETs exhibit complex nature of threshold voltage in terms of 'Gate Threshold Voltage' and 'Drain Threshold Voltage' [2]. These threshold voltages had conventionally been extracted by the researchers from constant current model or transconductance model on basis of numerical simulation [3]. But these methods have subtle logical base and moreover, such methods cannot be inscribed in terms of complete analytical expressions. Our aim in this paper is to derive threshold voltage in TFET in a completely analytical manner. Novelty lies in the 'surface potential'-based approach.

Though TFETs are greatly immune to SCEs, quantum mechanical effects (QMEs) prevail at dimensions comparable to the de Broglie wavelength. Conduction band and valence band develop into discrete spectrums (i.e., subband quantization) which compel

© Springer Nature Singapore Pte Ltd. 2021
N. R. Das and S. Sarkar (eds.), *Computers and Devices for Communication*,
Lecture Notes in Networks and Systems 147,
https://doi.org/10.1007/978-981-15-8366-7_60

BTBT to occur only between the bound states [1]. In this paper, we address shift of threshold characteristics owing to QMEs and propose the corresponding analytical models. The obtained results are duly validated with Silvaco ATLAS version 5.18.3.R [4].

2 Device Structure

Here, a dual material double gate TFET structure is deployed (vide Fig. 1a) and its corresponding energy band diagram before quantization is depicted Fig. 1b. After quantization, modified band diagram is plotted in Fig. 1c. We consider M_1 and M_2 of length L_1 and L_2, respectively, and $\varphi_{M1} < \varphi_{M2}$ so that a 'gate tunable barrier' is created to enhance I_{on}/I_{off} [5]. The channel is uniformly doped and effects of fixed oxide charges are not considered.

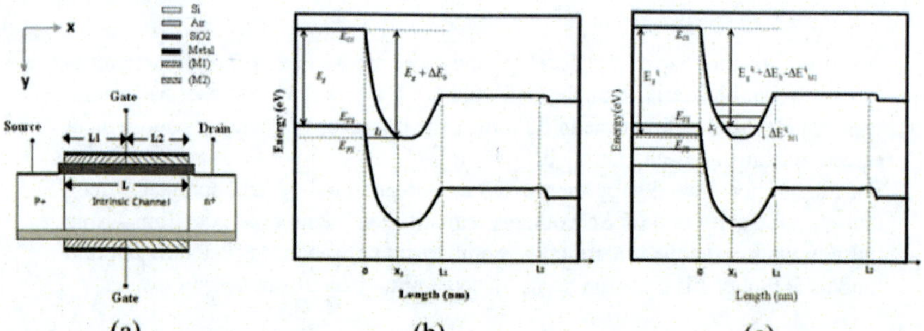

Fig. 1 **a** Device structure. **b** Energy band diagram without QME. **c** Energy band diagram with QME

3 Analytical Models

3.1 Classical Model

3.1.1 For our structure, the 2D Poisson is

$$\frac{\partial^2 \phi(y,x)}{\partial y^2} + \frac{\partial^2 \phi(y,x)}{\partial x^2} = \frac{qN_{Ch}}{\varepsilon_{Si}} \tag{1}$$

$$\phi(x,y) = \phi_S(x) + c_1(x)y + c_2(x)y^2 \tag{2}$$

$$\text{For } 0 \le x \le L_1, \quad 0 \le y \le t_{Si} \quad \phi_1(x,y) = \phi_{S1}(x) + c_{11}(x)y + c_{12}(x)y^2.$$

For

$$\text{For } L_1 \le x \le L, \quad 0 \le y \le t_{Si} \quad \phi_2(x,y) = \phi_{S2}(x) + c_{21}(x)y + c_{22}(x)y^2$$

where $L = L_1 + L_2$. $C_{11}, C_{12}, C_{21}, C_{22}$ are obtained by applying boundary conditions [4] and we get

$$\text{For } 0 \le x \le L_1,\quad 0 \le y \le t_{Si}\quad \phi_{S1}(x) = a\mathrm{e}^{\eta x} + b\mathrm{e}^{-\eta x} - \beta_1/\alpha. \tag{3}$$

$$\text{For } L_1 \le x \le L,\quad 0 \le y \le t_{Si}\quad \phi_{S2}(x) = c\mathrm{e}^{\eta(x-L_1)} + d\mathrm{e}^{-\eta(x-L_1)} - \beta_2/\alpha. \tag{4}$$

$$a = (V_{bi} + V_{ds} + \sigma_2) - \left[(V_{bi} + \sigma_1) + \frac{(\sigma_1 - \sigma_2)\cosh(\eta L_2)}{1 - \mathrm{e}^{-2\eta L}}\right]\mathrm{e}^{-\eta L},$$
$$b = \frac{(V_{bi} + \sigma_1) - (V_{bi} + V_{ds} + \sigma_2)\mathrm{e}^{-\eta L}}{1 - \mathrm{e}^{-2\eta L}} - \frac{(\sigma_1 - \sigma_2)\cosh(\eta L_2)\mathrm{e}^{-\eta L}}{1 - \mathrm{e}^{-2\eta L}}$$
$$c = a\mathrm{e}^{\eta L_1} - \frac{(\sigma_1 - \sigma_2)}{2},\quad d = b\mathrm{e}^{-\eta L_1} - \frac{(\sigma_1 - \sigma_2)}{2},$$
$$\alpha = 2\left[1 + \frac{C_{ox}}{C_{Si}} + \frac{C_{ox}}{C_{air}}\right] \Big/ t_{Si}^2\left(1 + \frac{2C_{Si}}{C_{air}}\right),$$
$$\beta_j = \frac{qN_{Ch}}{\varepsilon_{Si}} - 2V'_{Gj}\left[\frac{C_{ox}}{C_{air}} + \frac{C_{ox}}{C_{Si}}\right] - 2V'_{Gj} \Big/ t_{Si}^2\left(1 + \frac{2C_{Si}}{C_{air}}\right),$$
$$\sigma_j = \frac{\beta_j}{\alpha},\quad \eta = \sqrt{\alpha},\quad V'_{Gj} = V_{gs} - V_{FBj},\quad V_{FBj} = \varphi_{Mj} - \varphi_{Si}. \tag{5}$$

3.1.2 Now, the threshold voltage is supposed to be the gate voltage which creates the onset of electron transfer through BTBT. This also marks the foundation of tunnel path ($L_{\min} = x_1$) where the difference between the conduction band of the source and the channel reaches $Eg + \Delta E_b$ vide Fig. 1b. Where,

$$\Delta E_b = E_{VS} - E_{FS}. \tag{6}$$

E_{VS}, E_{FS} are, respectively, the valence band and Fermi level at source.

To find x_1, we equate the surface potential at $x = x_1$ with $(E_g + \Delta E_b)/q$ at (3) and we get

$$\phi_1(x, y)|_{x=x_1} = \phi_{S1}(x) = a\mathrm{e}^{\eta x_1} + b\mathrm{e}^{-\eta x_1} - \beta_1/\alpha = (E_g + \Delta E_b)/q. \tag{7}$$

$$x_1 = \left[(E_g + \Delta E_b/q) + a + b + (\beta_1/\alpha)\right]/[\eta(a - b)]. \tag{8}$$

3.1.3 Besides, it is reported that in TFETs, the tunneling barrier exhibits a transition from strong V_{gs} dependence to weak V_{gs} dependence at $V_{gs} = V_{th}$ [3] and at such point,$\phi_{S1}(x)|_{x=x_1} = V_{ds} + (kT/q)\ln(N_{drain}/N_{ch})$.

$$\phi_{S1}(x)|_{x=x_1} = a\mathrm{e}^{\eta x_1} + b\mathrm{e}^{-\eta x_1} - \beta_1/\alpha = V_{ds} + {}^{kT}\!/_{q}\ln(N_{\mathrm{drain}}/N_{ch}). \tag{9}$$

$$V_{th} = \frac{qN_{ch}t_{si}}{2C_{ox}} + \frac{\left[(V_{bi} + V_{ds} + V_{fb2}) - \ell_1(V_{bi} + V_{ds} + V_{fb1}) + \ell_2(V_{fb1} - V_{bi}) - \gamma)\right]}{1 - \ell_1 + \ell_2} \tag{10}$$

(10) is obtained by substituting x_1 from (8), where, $\ell_1 = [\sinh(\alpha L)]/[\sinh(\alpha x_1)]$,$\ell_2 = [\sinh(\alpha(L - x_1))]/[\sinh(\alpha x_1)]$, $\gamma = (V_{FB2} - V_{FB1})\cosh((\alpha(L - x_1))$.

3.2 QME Correction

3.2.1 At very low dimensions, energy states are quantized in potential well under M_1 vide Fig. 1c. We use Schrödinger equation (separation of variables method) (vide 11) to derive subband energy values under $M_1(E^q_{CM_1})$ (vide 12)

$$\frac{\partial^2\psi}{\partial y^2} + \frac{2qm^*_{l,t}}{\hbar^2}(E - E_0)\psi = 0. \tag{11}$$

$$E^q_{CM_1} = E_{CM_10} + \frac{\hbar^2\pi^2 i^2}{2qm^*_{l,t}t^2_{Si}} + \frac{C_{22}t^2_{si}}{6}\left[1 + \frac{3}{\pi^2 i^2}\right], \quad i = 1, 2\ldots \text{ (Index of subbands)}. \tag{12}$$

In (12) E_{CM_10} is the classical value of conduction band bottom, 2nd term is the separation of first subband from conduction band bottom (after quantization) and the last term defines 'first level perturbation' owing to quasi-parabolic nature of the quantum well.

3.2.2 Similarly at source, the valance band splits into subbands defined by E^q_{VS}

$$E^q_{VS} = E_{VS0} + \frac{\hbar^2\pi^2 j^2}{2qm^P_0 m^P_h t^2_{Si}}, \quad j = 1, 2\ldots \text{ (Index of subbands)}. \tag{13}$$

where $m_h = m_{hh}$ for heavy holes or m_{lh} for light holes. The spin orbit splitting is 0.044 ev for Si and hence neglected. Here, E_{VS0} is the classical value of valance band ceiling, 2nd term is the separation of topmost subband from valance band ceiling.

It is crucial that the parity of wave functions at source and channel must harmonize for effective BTBT. Thus, to obtain threshold voltage (i.e., at onset of BTBT), we analysed transition of electron from first source valance-subband to first channel conduction-subband. Principally, quantum mechanical shift for each and every parameter is derived by substituting quantum values from (11), (12), (13) in (7), (8), (9), (10).

4 Results

Figure 2 plots the variation of minimum tunnel length $[L_{\min} = (x_1)]$ with V_{ds} for different ε_{ox}. After quantization, the highest subband of M_1 does not coincide with the top of valance band and similarly, bottom of conduction band of M_2 does not coincide with bottom of conduction band. This phenomena increases the effective quantum bandgap $(E_g{}^q)$ and thus, higher values of tunneling barrier are obtained.

However, tunneling barrier $[L_{\min} = (x_1)]$ is decreased when the gate oxide is replaced by high-k dielectric material (vide Fig. 3). Such high-k dielectrics are capable to improve capacitive control of gate over the tunneling path. Same trend is observed in Fig. 3 where $L_{\min}$ exhibits quasi-exponential dependence on gate voltage. This result corroborate our analytical modeling with quantum theory.

Figure 4 depicts quantum threshold voltage $(V_{th}{}^q)$ and classical threshold voltage (V_{th}) while varying the drain voltage (V_{ds}) whereas V_{th} exhibit significant QMEs shift

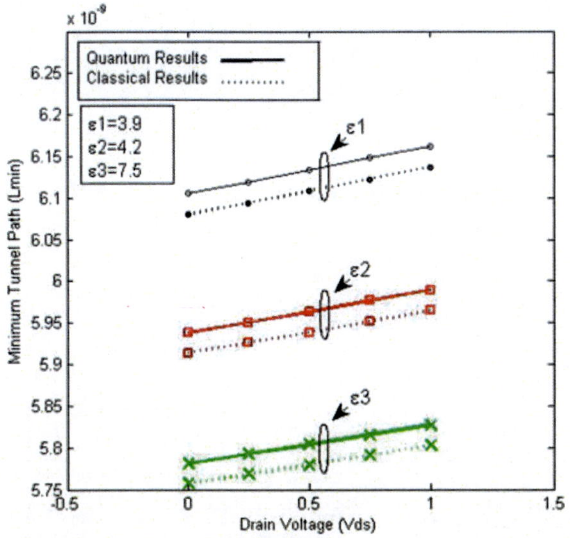

Fig. 2 Variation of minimum tunnel path (x_1) with drain voltage (V_{ds}) for different ε (permittivity of insulating layer). *Lines* Analytical results. *Symbols* Simulation results

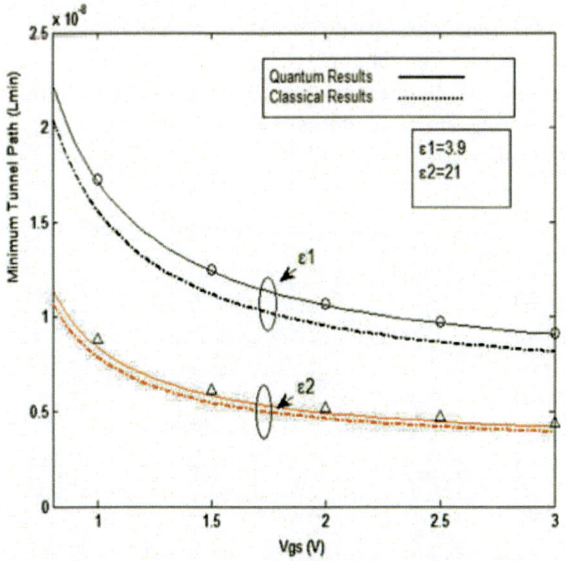

Fig. 3 Variation of minimum tunnel path (x_1) with gate voltage (V_{gs}) for different ε (permittivity of insulating layer). *Lines* Analytical results. *Symbols* Simulation results

as explained earlier. V_{th} increases with V_{ds} owing to the fact that at higher drain voltage, more gate voltage is required to control 'onset of tunneling.' Whereas the pattern improves when SiO_2 is replaced with high-*k* dielectric. Thus, high-*k* materials are confirmed to be better and even indispensable at low-dimensional, low-power applications.

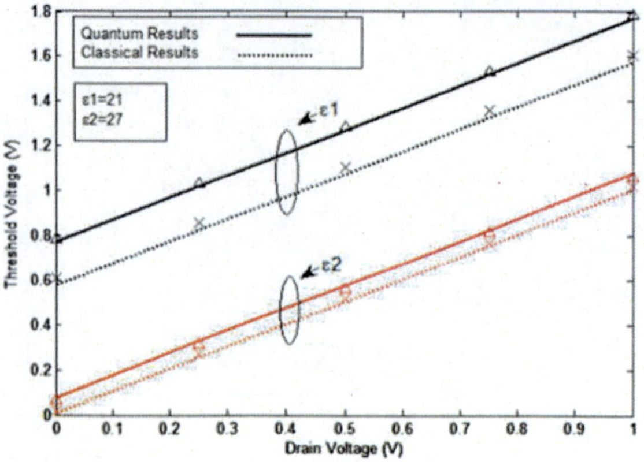

Fig. 4 Variation of threshold voltage (V_{th}) with drain voltage (V_{ds}) for different ε (permittivity of insulating layer). *Lines* Analytical results. *Symbols* Simulation results

The effect of dimensional scaling on threshold is further studied in Fig. 5. We observe quantum threshold voltage ($V_{th}{}^q$) and classical threshold voltage (V_{th}) differ more at lower dimensions, whereas, they tend to amalgamate at higher dimension. This plot indicates the limit of dimensional downscaling beyond which classical model loses its applicability in low dimensions.

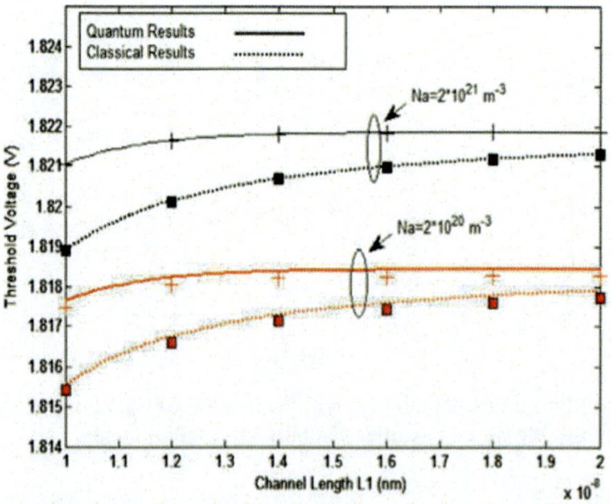

Fig. 5 Variation of threshold voltage (V_{th}) with channel length for different concentration (Na). *Lines* Analytical results. *Symbols* Simulation results

5 Conclusion

In this paper, we develop novel and completely analytical expressions of threshold voltage in TFET based on surface potential model. We observe that the threshold parameters are shifted significantly when QMEs are incorporated in the modeling. Our work also emphasis on the methods to mitigate quantum mechanical effects in such low-dimensional devices.

References

1. Padilla, J.L., Gámiz, F., Godoy, A.: The effect of quantum confinement on tunneling field-effect transistors with high-*k* gate dielectric. Appl. Phys. Lett. **103**, 112105 (2013)
2. Boucart, K., Ionesu, A.M.: Threshold Voltage in Tunnel FETs: Physical Definition, Extraction, Scaling and Impact on IC Design. IEEE 1-4244-1124-6/07 (2007)
3. Safa, S., Noor, S.L., Khan, M.Z.R.: Physics-based generalized threshold voltage model of multiple material gate tunneling FET structure. IEEE Trans. Electron Devices **64**(4) (2017)
4. Shee, S., Bhattacharyya, G., Sarkar, S.K.: Quantum analytical modeling for device parameters and *I*–*V* characteristics of nanoscale dual-material double-gate silicon-on-nothing MOSFET. IEEE Trans. Electron Devices **61**(8), 2697–2704 (2014)
5. Bagga, N., Sarkar, S.K.: An analytical model for tunnel barrier modulation in triple metal double gate TFET. IEEE Trans. Electron Devices **62**(7), 2136–2142 (2015)

Characterization and TCAD Simulation Studies of Single-Crystal Diamond Detectors

S. Mohapatra[1(✉)], P. K. Sahu[1], and N. V. L. Narasimha Murty[2]

[1] School of Electrical Sciences, Indian Institute of Technology Bhubaneswar, Arugul, Jatni 752050, India
sm43@iitbbs.ac.in

[2] Department of Electrical Engineering, Indian Institute of Technology Tirupati, Tirupati, Andhra Pradesh 517506, India

Abstract. The excellent electronic properties of single crystal (sc) ultra-high pure (UHP) diamond, such as wide band gap, high carrier mobility and high displacement energy of atoms make it the current material of choice for radiation detection applications. This paper presents the suitability of commercially available free standing single crystal diamond plate from Soni CVD Diamonds, India, for fabrication, optical, and electrical characterization of diamond-based bulk photodetectors intended for use in radiation detection applications. The optical characterization includes Raman spectroscopy and the electrical characterizations include leakage current, capacitance–voltage, and ultra-violet response current measurements. Numerical simulations using SYNOPSYS© Sentaurus TCAD have also been carried out to determine the optimum thickness and doping density of a boron-doped diamond film intended to be grown on a type-IIa diamond substrate, for its potential use in alpha-particle spectroscopy.

Keywords: Diamond detectors · Ultra-violet (UV) current response · TCAD modelling

1 Introduction

Diamond grown by chemical vapour deposition (CVD) has wide band gap, good thermal conductivity, high radiation hardness, and fast charge collection owing to high electron and hole mobility. These unique properties make it an attractive material for harsh environments, high temperature, and radiation detection applications [1–3]. Earlier studies of fabrication and characterization of scCVD photodetectors were carried out using a variety of diamond tiles manufactured by commercial vendors such as Element Six Ltd. (UK), Diamond Detectors Ltd. (UK), and IIa Technologies (Singapore) [1, 2, 4–6]. Recently, a number of commercial manufacturers are able to grow UHP grade scCVD diamond. The reason for this can be attributed to the improvement in CVD growth technology. This work presents the fabrication, optical, and electrical characterization of a metal–semiconductor-metal (MSM) diamond photodetector from commercially available free standing UHP scCVD diamond plate supplied by *SONI CVD DIAMONDS,*

© Springer Nature Singapore Pte Ltd. 2021
N. R. Das and S. Sarkar (eds.), *Computers and Devices for Communication*,
Lecture Notes in Networks and Systems 147,
https://doi.org/10.1007/978-981-15-8366-7_61

India. Moreover, it is found from the studies [7, 8] that suitable boron-doped alpha-particle diamond detector grown on scCVD type-IIa diamonds were carried out only a limited number of times in the past. The reason for this can be attributed to the difficulty in optimization of certain parameters such as doping density and thickness of the boron-doped layer on intrinsic type-IIa diamond. In this work, TCAD simulations have been carried out for deciding the optimum thickness and boron defect density of the boron-doped layer on type-IIa scCVD diamond. This may be useful in predicting the important parameter such as depletion layer width, which plays an important role in deciding the charge collection efficiency in diamond solid state radiation detectors.

2 Experimental Procedure

UHP grade scCVD diamond plate of dimensions 3 × 3 × 0.1 mm^3 from *SONI CVD DIAMONDS* is employed for the study.

2.1 Optical Characterization, Device Fabrication, and Electrical Characterization

Raman spectroscopy is carried out prior to detector fabrication. Argon-ion laser source of wavelength 514 nm is used for the excitation. The diamond plate is first cleaned by boiling it in a saturated solution of CrO_3 in H_2SO_4 and thereafter the diamond surfaces are exposed to oxygen plasma inorder to make the surfaces oxygen-terminated. This is followed by RF sputtering to deposit Cr/Au (50 nm/200 nm) contacts on both sides of the diamond plate. The fabrication process is completed by post-metallization annealing at 350 °C in argon environment. Leakage current is measured using Keithley 6517B electrometer. The capacitance–voltage (*C*–*V*) measurements are carried out using Agilent Technologies B1500A Semiconductor Device Analyzer. The UV response current of the detector is measured using a 20 W Xenon Flash Lamp Module with the repetition rate set to 100 Hz.

3 Sentaurus TCAD Simulation Set-Up

SYNOPSYS© Sentaurus TCAD is employed to find out the optimized boron defect density (or doping concentration) and the thickness of the boron-doped layer on intrinsic type-IIa scCVD diamond. The diamond material is created in the TCAD software. To model the physical processes, temperature dependent mobility and Shockley–Read–Hall (SRH) recombination models are used. The simulated device structure consists of a simple geometry, 3000 μm in length, 3000 μm wide, and 300 μm in thickness of intrinsic diamond with a boron-doped layer created on the top of the intrinsic diamond. Since the modeling of diamond is still a topic of debate, hence, diamond bulk (both intrinsic and boron doped) is modelled by the incorporation of electrically active defect centers [9] instead of specifying any external doping concentration. The defect level in the simulator is represented by four parameters: the trap energy level E_t (eV), the trap concentration N_t (cm^{-3}), and the capture cross section of electrons and holes σ_n/σ_p (cm^2), respectively. Three different trap concentrations 1×10^{15} cm^{-3}, 5×10^{17} cm^{-3}, and 1×10^{19} cm^{-3} are employed for the boron-doped layer inorder to decide its optimum doping concentration. The boron defect level is located at an energy level of 0.38 eV above the valence band [10] and the capture cross-sections for the same are included according to the studies [11]. Two different thicknesses of 3 μm and 5 μm are considered for the simulations [7]. The intrinsic diamond bulk is represented by trap levels having concentrations of the order of ~1×10^{13} cm^{-3} and the electron and hole capture cross sections are included from [12, 13] into the simulator. Metal contacts having work function of 4.3 eV are employed for the anode and cathode terminals. The anode is defined on the boron-doped layer and the cathode is defined on the intrinsic diamond substrate.

4 Results and Discussions

Figure 1 shows the Raman spectrum of the sample. The prominent Raman line at 1332 cm^{-1} corresponding to the T_{2g} first-order Raman mode of the scCVD diamond is observed.

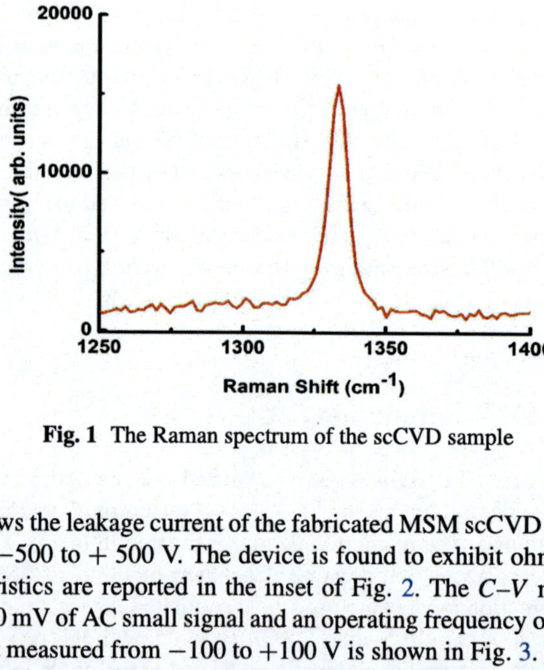

Fig. 1 The Raman spectrum of the scCVD sample

Figure 2 shows the leakage current of the fabricated MSM scCVD diamond detector measured from −500 to + 500 V. The device is found to exhibit ohmic behavior. The (*C*–*V*) characteristics are reported in the inset of Fig. 2. The *C*–*V* measurements are carried out for 30 mV of AC small signal and an operating frequency of 1 MHz. The UV response current measured from −100 to +100 V is shown in Fig. 3. The UV response current indicates higher orders of current magnitude compared to the leakage current because of the creation of electron–hole pairs due to the xenon flash lamp module. Alpha particles have a lower penetrating power and are stopped after few micrometers (~14 μm for ^{241}Am) [14] in CVD diamond. A robust solid-state-doped alpha-particle detector is intended to satisfy two necessary criteria (a) the charge collection distance of generated excess carriers should consists of less trapping events and (b) the doping concentration of the detector material should also be low so as to prevent more numbers of scattering events [14, 15]. The anode current for the simulated device structure is reported in Fig. 4 for a trap density of 5×10^{17} cm^{-3} of the boron layer. The current curve exhibits schottky nature.

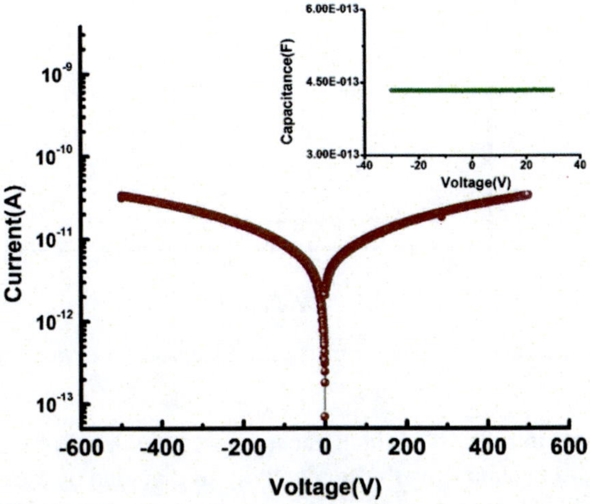

Fig. 2 Leakage current of the UHP scCVD photodetector. Inset shows the *C–V* characteristics

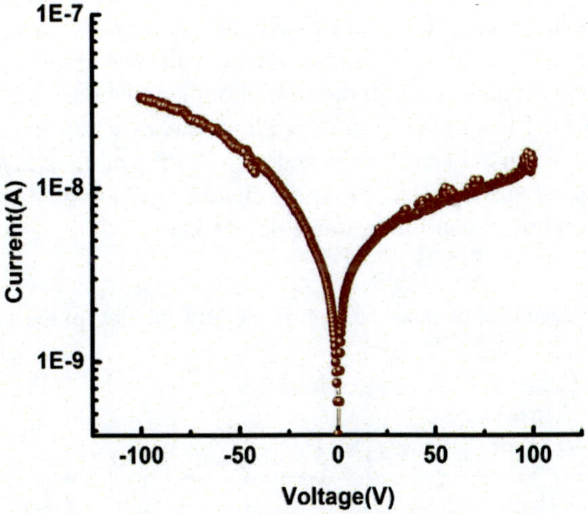

Fig. 3 UV response current of the UHP SC photo detector

The maximum anode current values for the forward-bias region has been reported in Table 1 for three different trap densities of 1×10^{15} cm^{-3}, 5×10^{17} cm^{-3}, and 1×10^{19} cm^{-3} and two different thicknesses of 3 and 5 μm for the boron layer. From the table, it can be noted that for a particular value of thickness of the boron layer, the maximum current value increases with the increase in trap density. This confirms the shallow nature

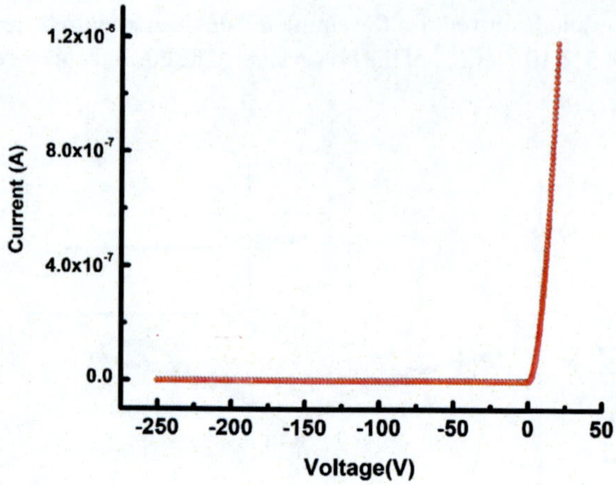

Fig. 4 Anode current of the simulated diamond detector device structure

of the boron dopants in diamond, which occupy an energy level of 0.38 eV above the valence band and aid in enhancing the conductivity of diamond by the creation of holes in the valence band [14]. Moreover, the maximum current value also increases with the decrease in thickness. This can be explained due the decrease in resistance across the cross-sectional region of the device when the thickness is decreased. The defect density of the boron-doped layer should be sufficient enough so that the generated depletion layer width will be more than that of the alpha particle penetration depth in diamond. Hence, there needs to be a trade-off between the doping concentration and the thickness so as to optimize the device design. In the case of our simulations, this criteria is found to best satisfied by the 5 μm thick boron-doped layer having a trap density (of the boron dopants) of 5×10^{17} cm^{-3}. The resistance due to the 5 μm thick layer will be more than that of the 3 μm thick layer and the space charge region will be increased more in the case of the 5 μm thick layer for the same defect density.

Table 1 Maximum magnitude of anode current (in the forward-bias) of the simulated diamond device

Thickness (μm)	Trap density		
	1×10^{15} cm^{-3} (μA)	5×10^{17} cm^{-3} (μA)	1×10^{19} cm^{-3} (μA)
3	1.16	1.19	1.2
5	1.15	1.18	1.19

The boron trap density of 5×10^{17} cm^{-3} is found to be optimum so as to cause less number of scattering events. The depletion layer due to the 5 μm thick boron layer extends to 20 μm from the top at a reverse bias voltage of 250 V as estimated from the space charge picture in TCAD simulations.

5 Conclusions

UHP grade scCVD diamond detector is fabricated and both optical and electrical characterizations are carried out. The Raman spectroscopy reveals the characteristic 1332 cm^{-1} peak of diamond. The electrical characterizations include leakage current, C–V characteristics, and UV response current measurements. Four orders of increase in the magnitude of current are observed from leakage current to UV response current. Sentaurus TCAD simulations are carried out to decide the optimum thickness (in μm) and doping concentration (cm^{-3}) of a boron-doped layer on an intrinsic-type *IIa* diamond to decide the typical structure of an efficient diamond-based alpha-particle detector. It is concluded from the simulation results that a thickness of 5 μm and a moderate boron doping of 2–3 ppm is found to be the optimum solution.

Acknowledgements. The fabrication and capacitance–voltage measurements are performed using facilities at CeNSE, funded by Ministry of Electronics and Information Technology (MeitY), Govt. of India, and located at the Indian Institute of Science, Bengaluru. We are grateful to Dr. Pratima Kumari Mishra, Chief Scientist, CSIR-IMMT for the Raman Measurements.

References

1. Tsubota, M., et al.: High-temperature characteristics of charge collection efficiency using single CVD diamond detectors. Nucl. Inst. Methods Phys. Res. A **789**, 50 (2015)
2. Schirru, F., et al.: Thin single crystal diamond detectors for alpha particle detection. Diam. Relat. Mater. **49**, 96 (2014)
3. Caiffi, B., et al.: Characterisation of scCVD diamond detectors with γ sources. Nucl. Inst. Methods Phys. Res. A **754**, 24 (2014)
4. Galbiati, A., et al.: Performance of monocrystalline diamond radiation detectors fabricated using TiW, Cr/Au and a novel ohmic DLC/Pt/Au electrical contact. IEEE Trans Electron. Dev. **56**, 1863 (2009)
5. Angelone, M., et al.: Performances of monocrystalline artificial diamond detectors operated at high temperature. IEEE Trans Electron. Dev. **59**, 2416 (2012)
6. Kumar, A., et al.: Prototyping and performance study of a single crystal diamond detector for operation at high temperatures. Nucl. Inst. Methods Phys. Res. A **858**, 12 (2017)
7. Kaneko, J., et al.: Development of a synthetic diamond radiation detector with a boron doped CVD diamond contact. Nucl. Inst. Methods Phys. Res. A **422**, 211 (1999)
8. De Feudis, M., et al.: Ohmic graphite-metal contacts on oxygen-terminated lightly boron-doped CVD monocrystalline diamond. Diamond Relat. Mater. **92**, 18 (2019)
9. Morozzi, A., et al.: Polycrystalline CVD diamond device level modelling for particle detection applications. JINST **11**, C12043 (2016)
10. Balducci, A., et al.: Trapping-detrapping defects in single crystal diamond films grown by chemical vapor deposition. Appl. Phys. Lett. **87**, 222101 (2005)
11. Bruzzi, M., et al.: Photo-induced deep level analysis in undoped CVD diamond films. Diam. Relat. Mater. **9**, 1081 (2000)
12. Forneris, J., et al.: Electrical control of deep NV centers in diamond by means of subsuperficial graphitic micro-electrodes. Carbon **113**, 76 (2017)
13. Edmonds, A.M., et al.: Production of oriented nitrogen-vacancy color centers in synthetic diamond. Phys. Rev. B **86**, 035201 (2012)

14. Sussman, R.: CVD diamond for electronic devices and sensors. In: Wiley Series in Materials for Electronics and Optoelectronics Application (2008)
15. Lutz, G.: Semiconductor Radiation Detectors: Device Physics. Springer Optical and Electronic Materials

Design of 8-Stage RF-to-DC Converter for Energy Harvesting Applications

Amena Najeeb[1(✉)], Mohammed Arifuddin Sohel[2], and Qudsia Masood[2]

[1] ECE Department, NSAKCET, Hyderabad, India
amena.najeeb123@gmail.com
[2] ECE Department, MJCET, Hyderabad, India

Abstract. The conversion of freely available energy to electrical energy is termed as energy harvesting. The sources of energy can be solar, light, vibration, thermal, and electromagnetic signals. Scavenging energy from these sources will reduce the use of batteries which need to be replaced regularly. RF energy is opted in this work due its high energy density and reliability. This paper presents an 8-stage RF-to-DC converter to convert 1800 MHz input radio frequency signal to a constant output voltage of 3.5 V for a load of 5 MΩ.

Keywords: Energy harvesting · RF energy harvesting · Energy scavenging · Power management unit (PMU) · RF-to-DC converter

1 Introduction

The process that derives energy from external sources to power electronic devices is called energy scavenging [1]. It is a remarkable way to power an electronic device by harvesting energy from the sources readily available in the environment. The two main design goals are to choose very low power electronics and to harvest and store energy efficiently.

The growing technologies need continuous power to function well. There are numerous ways to generate energy; solar energy is the one which is used widely. The only disadvantage being, it does not work in regions where there is less or no light.

Energy harvesters provide a feeble amount of power, sufficient to power the low energy electronic devices. The various forms of energy could be solar, mechanical, thermal, electromagnetic, and kinetic energy.

Power management unit is made of two blocks that are used for two different functions [2]; the energy shaping block helps to extract energy from the source and the power conversion block to supply the power when needed by the application.

The harvested energy is stored in any one of the following three options. One is using traditional batteries, where energy can be stored for long term. Second being capacitors, that are used for short-time storage and the best option is using super capacitors. Super capacitors helps in efficient charging and has high capacity. They are compact and best alternatives to batteries.

© Springer Nature Singapore Pte Ltd. 2021
N. R. Das and S. Sarkar (eds.), *Computers and Devices for Communication*,
Lecture Notes in Networks and Systems 147,
https://doi.org/10.1007/978-981-15-8366-7_62

2 Literature Survey

Thermoelectric generators work by converting the heat difference of a body to electrical energy. The photovoltaic cells help in converting light energy into electrical energy. Solar cells are best examples for photovoltaic cells.

Kinetic energy involves the conversion of vibrational energy into electrical energy. This includes the piezoelectric form of energy [3]. Piezoelectric energy converts mechanical strain to electric current and produces power in the order of milli Watts. It is useful for small applications like handheld devices, light bulbs, human motion, acoustic noise, vibrations, and pressure.

Radio frequency energy harvesting involves generation of DC output from incoming radio frequency signals that are in the form of AC input.

The other forms of energy involve the conversion of chemical energy to electrical energy. It includes the use of MFCs that produce energy directly from biodegradable substrates [4].

This paper deals with the conversion of radio signals present in AC form to a constant DC form energy. The available radio frequency signal is converted to DC using an 8-stage RF-to-DC converter.

3 RF Energy Harvesting

This paper focusses on conversion of radio frequency or RF energy. The range of radio frequency signals is 3 kHz to 300 GHz [5, 6]. The range can be categorized to low frequency or high frequency or Ultra high frequency depending on the desired area of application.

3.1 Block Diagram

The basic block diagram for radio frequency energy harvesting system consists of antenna, impedance matching circuit, RF-to-DC converter and energy storage device or load (Fig. 1).

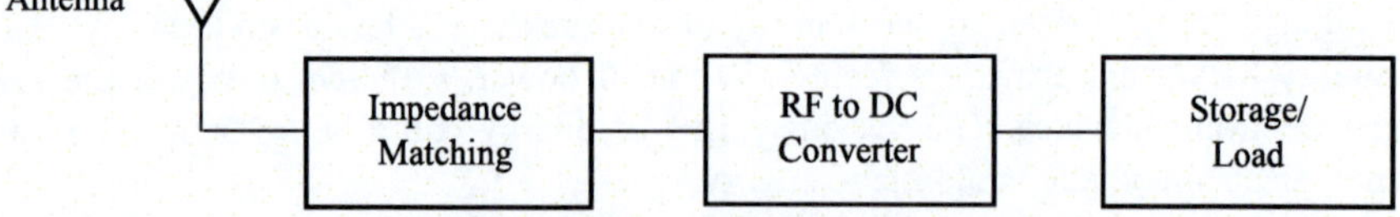

Fig. 1 Block diagram of RF energy harvesting system

3.1.1 Antenna

The antenna helps in absorbing the ambient RF energy present in the environment. As the transmitter–receiver distance increases, strength of the RF signal decreases. This requires an impedance matching circuit that will help in increasing the power received from the antenna.

3.1.2 Impedance Matching

Impedance matching circuit helps in transferring the power that is received at the receiving antenna to the next block. This happens by maximizing the received power and is crucial for optimization of an RF energy harvesting circuit. It is used to avoid a mismatch between antenna and rectifier.

3.1.3 RF-to-DC Converter

Full-wave rectifier acts as a voltage doubler, where two half-wave rectifiers are connected in cascade.

The RF-to-DC converter is a cascaded rectifier that helps in efficient conversion of AC form radio frequency signal to convert into DC form. The conversion should be such that output is usable and efficiency should be high.

The rectifier can take the form of Cockcroft-Walton [5, 6], Greinacher [7], Dickson Multiplier [8].

3.2 Multistage Voltage Doubler

Several voltage doublers can be used in a cascaded form to generate high output [9]. This increases the expected resultant voltage of the designed rectifier, but the efficiency diminishes as the number of stages increase [10]. The number of stages has to be chosen such that we get a usable output along with high efficiency.

3.3 Energy Storage

For energy storage, we can use battery, capacitor, or supercapacitor. Supercapacitors are found to be the most reliable energy storage option.

4 Design of the 8-Stage RF-to-DC Converter

The design is carried out on Cadence Virtuoso using 180 nm UMC technology. The output resistance was fixed to 5 MΩ. The circuit consists of an 8-stage RF-to-DC converter. The first stage consists of an NMOS with grounded body terminal whereas the rest of the stages have PMOS transistors help in rectification.

A connection is made between gate terminal of NMOS (stage 1) to source terminal of the following PMOS transistor (stage 2) and the gate terminal of this transistor is in turn attached to the drain terminal of NMOS transistor. The gate terminals of PMOS transistors (stage 3–7) are connected to the source terminals of the preceding transistor. The PMOS in the final stage is intentionally left uncompensated.

Pumping capacitance of 1 pF is used. The implemented 8-stage RF-to-DC converter is shown in Fig. 2.

The even stages of the rectifier are connected to the ground terminal. Input is provided through the odd stages of the rectifier.

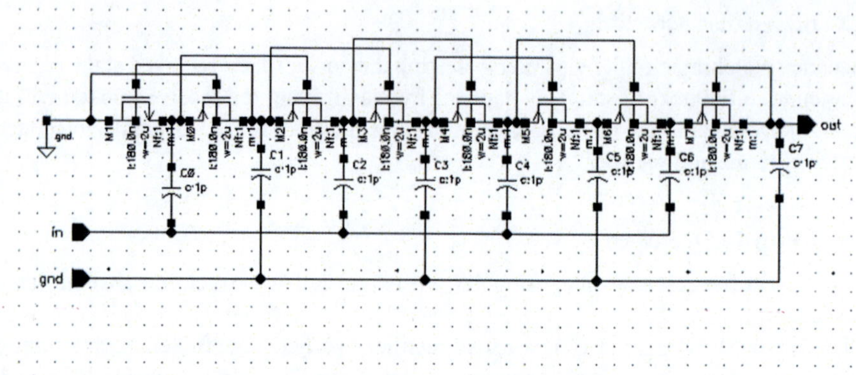

Fig. 2 Schematic of the 8-stage RF-to-DC converter

5 Experimental Results

Figure 3 indicates the voltage and current at the output, that are found to be 3.5 V and 715 nA, respectively. The settling time is found to be 230 ns.

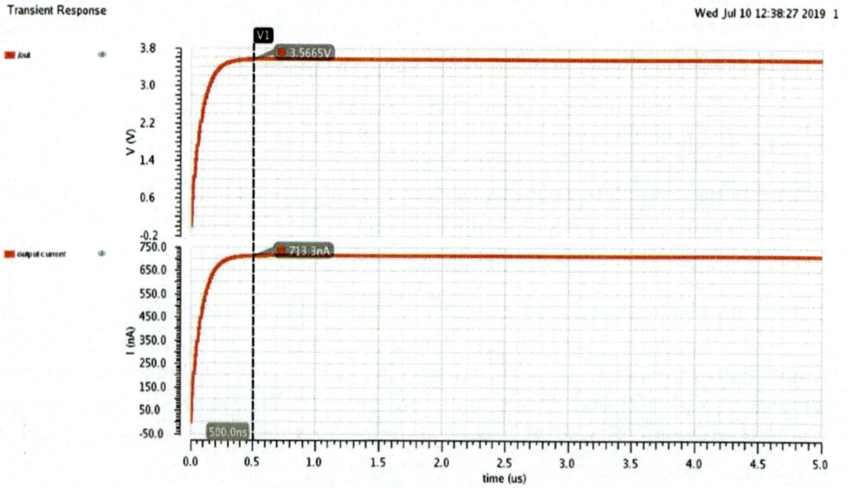

Fig. 3 Transient analysis of output voltage and current

Figure 4 demonstrates the output of corner analysis for an input signal having voltage of 500 mV and frequency 1800 MHz.

6 Conclusions

An 8-stage RF-to-DC converter has been designed in 180 nm CMOS technology. It converts input radio frequency signal to a constant output voltage of 3.5 V. NMOS is

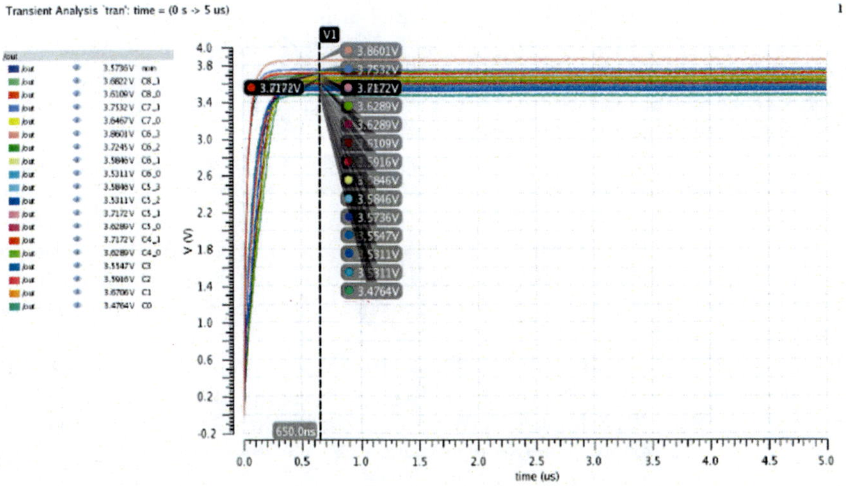

Fig. 4 Result of corner analysis

used at the first stage with a grounded body terminal followed by PMOS transistors. PMOS transistors are used for rectification (Table 1).

Table 1 Comparison with previous work

Parameters	This work	[10]	[2]	[11]
Technology	180 nm	130 nm	250 nm	180 nm
Frequency	1.8 GHz	902–928 MHz	906 MHz	1 GHz
Input power	−24 dBm	−15 dBm	−22.6 dBm	−18 dBm
V_{out}	3.5 V	3.2 V	2.1 V	1 V
Load	5 MΩ	1 MΩ	1.32 MΩ	100 kΩ
Number of stages	8	12	36	1
PCE	70%	32%	60%	65%

References

1. Khera, S., Turk, N., Kaur, N.: Energy harvesting aspects of wireless sensor networks: a review. Int. J. Recent Innov. Trends Comput. Commun. **5**(5), 875–878 (2017)
2. Le, T., Mayaram, K., Fiez, T.: Efficient far-field radio frequency energy harvesting for passively powered sensor networks. IEEE J. Solid-State Circ. **43**, 1287–1302 (2008)
3. Hao, Z., Wang, G., Li, W., Zhang, J., Kan J.: Effects of electrode material on the voltage of a tree-based energy generator. J. Renew. Sustain. Energy **10**, 043101 (2018)
4. Logan, B.E., Hamelers, B., Rozendal, R., Schroder, U., Keller, J., Freguia, S., et al.: Microbial fuel cells: methodology and technology. Env. Sci. Technol. **40**, 5181–5192 (2006)

5. bin Othman. M.A.: Waste of radio frequency signal analysis for wireless energy harvester. In: 2010 6th International Colloquium on Signal Processing & its Applications, Mallaca City, pp. 1–3 (2010)
6. Xue, L.: High voltage switched-mode step-up DC–DC converters in standard CMOS process. PhD Thesis, University of Florida, Gainesville, Florida (2013)
7. Chow, E.Y., Chlebowski, A.L., Chakraborty, S., Chappell, W.J., Irazoqui, P.P.: Fully wireless implantable cardiovascular pressure monitors integrated with a medical stent. IEEE Trans. Biomed. Eng. **57**(6), 1487–1496 (2010)
8. Dickson, J.F.: On-chip high-voltage generation in MNOS integrated circuits using an improved voltage multiplier technique. IEEE J. Solid-State Circ. **11**(3), 374–378 (1976)
9. Favrat, P., Deval, P., Declercq, M.: A high-efficiency CMOS voltage doubler. IEEE J. Solid-State Circ. **33**(3), 410–416 (1998)
10. Hameed, Z.: A 3.2 V–15 dBm Adaptive threshold-voltage compensated RF energy harvester in 130 nm CMOS. IEEE Trans. Circ. Syst. I Reg. Pap. **62**(4) 948–956 (2015)
11. Ouda, M.H., Khalil, W., Salama, K.N.: Wide-range adaptive RF-to-DC power converter for UHF RFIDs. IEEE Microw. Wirel. Compon. Lett. **26**(8), 634–636 (2016)

Performance and Circuit Analysis of Independent Gate FinFET

Ankush Chattopadhyay[1(✉)], Chayanika Bose[2], and K. Sarkar Chandan[2]

[1] ECE Department, St. Thomas' College of Engineering and Technology, Kolkata, West Bengal, India
ankush.chattopadhyay@gmail.com
[2] ETCE Department, Jadavpur University, Kolkata, West Bengal 700032, India
chayanikab@yahoo.com, phyhod@yahoo.co.in

Abstract. In this article, we examine the influence of separate gate biasing on independent gate FinFET and related circuit through RF performance, gain and harmonic distortion analysis. Non-quasi static channel approach is considered in the small-signal modelling of 4 T-IG-FinFET. Intrinsic parameters such as C_{gs}, C_{gd}, R_{gd}, τ_m, f_t, $f_{\max}$ are investigated over an wide range of frequency (10–100 GHz). The device is then employed to simulate a single stage cascode amplifier, which offers flexibility in controlling its noise margins, gain and HDs through the applied back gate bias.

Keywords: IG-FinFET · Back gate bias · Cascode amplifier

1 Introduction

Scaling down of device dimensions opened up a new arena of multi-gate MOSFET architectures viz. double-gate MOSFET [1], gate all around MOSFET [2], FinFET [3], cylindrical nano-wires FETs [4, 5]. Such architectures reduce the effect of floating body [5] and offer excellent gate control, thereby improving the channel potential. Increasing an equivalent number of gates reduces the drain current transient tail, as well as bipolar gain [6]. Various structural modifications, such as FinFETs with underlap [7], junction-less FET [8] and tunnel FET [9], exhibit unique features improved with high-K spacer. In FinFET architecture, the channel is wrapped by the gate from three sides, which allows better electrical control. But, the major drawback of FinFET is its fixed threshold voltage, which limits its low power circuit applications. Thus, independent gate (IG) FinFET comes up with gates placed at the front and back of the Fin. In IG-FinFET configuration, the applied gate bias induces shift in inversion charge-centroid [10] and in turn reduces the SCEs. Operating IG-FinFET in synchronized driving mode offers an alternative approach to achieve flexible threshold voltage (V_{T}), which makes it suitable for RF mixer design [10]. In this paper, n-channel IG-FinFET of channel length 22 nm will be simulated using Sentaurus TCAD. Influence of independent gate biasing on the device performance will be reviewed in terms of current conducting capability as well as

© Springer Nature Singapore Pte Ltd. 2021
N. R. Das and S. Sarkar (eds.), *Computers and Devices for Communication*,
Lecture Notes in Networks and Systems 147,
https://doi.org/10.1007/978-981-15-8366-7_63

OFF-current. A non-quasi static channel approach will be employed to determine its RF performance, which involves charging and discharging of the associated parasitic capacitances. Finally, influence of the back gate bias on a high voltage gain with high output impedance amplifier such as cascode amplifier circuit will be investigated to establish importance of the separated gate FinFET configuration.

2 Device Structure and Simulation Framework

The 3D view of an IG-FinFET with drain/source symmetric underlap and spacers in gate sidewalls is depicted in Fig. 1a. Figure 1b depicts the top view and Fig. 1c the simulated structure of FinFET with extended spacers. Molybdenum is used as gate metal of length 22 nm and height 8 nm. Other parameters used for device simulation are chosen from the ITRS roadmap for RF and mixed signal analysis [11]. Device simulator Sentaurus TCAD [12] is calibrated with the standard modelled data [13], and is used to carry out all the simulations. Lombardi mobility model [14] is employed here. Carrier transport phenomenon is governed by the drift–diffusion model. The back gate bias (V_{bg}) is varied from −0.4 to +0.4 V, maintaining a step size of 0.2 V, and analog and RF performance of the device are analyzed and reported in this literature.

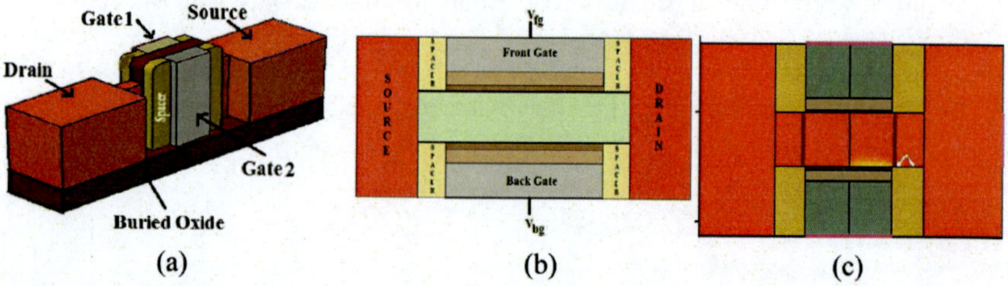

Fig. 1 An IG-FinFET: **a** 3D view, and **b** top view and **c** simulated structure (with extended spacers)

3 FinFET with Back Gate Bias

Drain currents in the IG-FinFET with variation in V_{fg} for different V_{bg} are shown in Fig. 2a, both in linear and logarithm (to have a better view of the subthreshold region) scales. The figure reveals that application of negative V_{bg} suppresses the I_{OFF}. Figure 2b exhibits the electron density along the bottom channel, as obtained from Svisual, for different values of V_{bg}. The figure indicates weak inversion for negative V_{bg}. Figure 3a presents the variations of I_{ON}/I_{OFF} ratio and threshold voltage with V_{bg}. The I_{ON}/I_{OFF} ratio increases rapidly for negative V_{bg}, whereas V_T falls smoothly with increase in V_{bg} over the entire range considered here. A device with higher V_T offers smaller OFF-current and therefore results in less prominent sub-threshold leakages. The figure clearly indicates that the V_T can be varied with the application of V_{bg}, thus making the device more immune to the SCEs. Plot of transconductance (g_m) with respect to front gate bias at different V_{bg} is illustrated in Fig. 3b. Sharp increase in g_m, especially in weak-inversion region, appears at V_{fg} below 0.8 V.

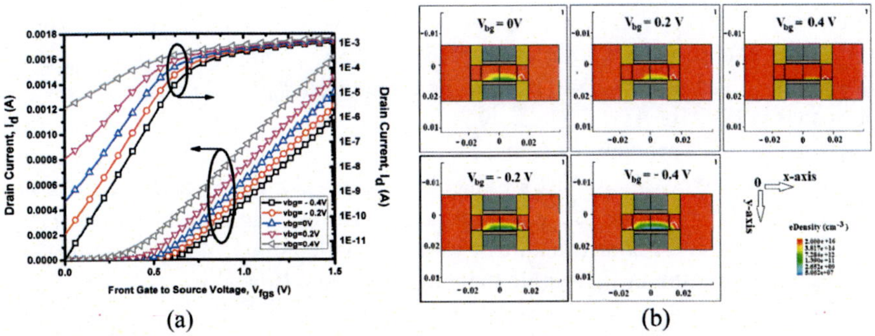

Fig. 2 **a** Transfer characteristics of IG-FinFET. **b** Electron density at the bottom channel is plotted along the channel length, for different V_{bg} at $V_{ds} = 0.82$ V

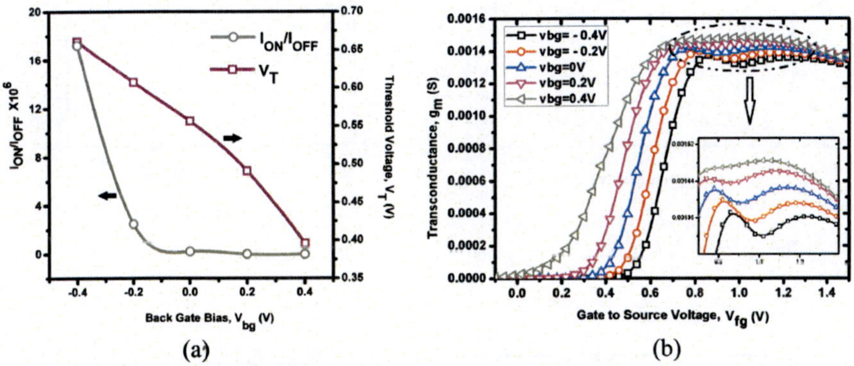

Fig. 3 **a** Variation of I_{ON}/I_{OFF} ratio and threshold voltage (V_T) with V_{bg}. **b** Transconductance versus front gate bias plot for different V_{bg} at $V_{ds} = 0.82$ V

4 RF Analysis of IG-FinFET with Back Gate Bias

The applicability of the proposed structure in high-frequency domain is studied in this section. In order to have accurate results, RF analysis is performed considering the non-quasi static effect [15] in gate and channel regions of the AC small-signal equivalent circuit, as shown in Fig. 4.

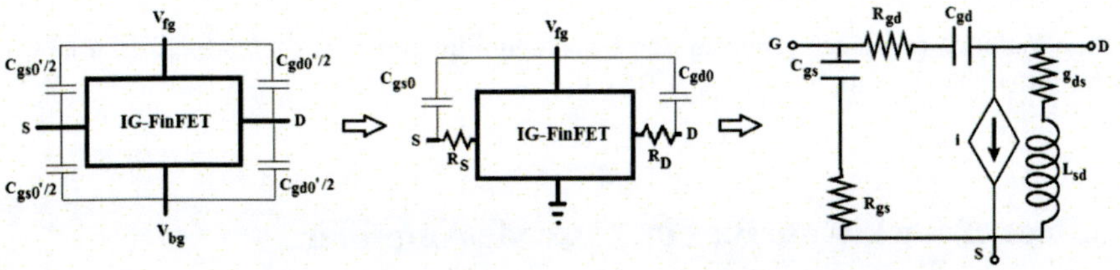

Fig. 4 AC small-signal equivalent circuit of IG-FinFET

Influence of V_{bg} on intrinsic capacitances of IG-FinFET is depicted in Fig. 5a. The intrinsic gate-to-source/drain capacitances (C_{gs}; C_{gd}) are found to increase with increasing V_{bg}. The effect of V_{bg} on intrinsic resistance (R_{gd}) and transport delay (τ_m) is presented in Fig. 5b. As τ_m is mainly influenced by the ratio of C_{gd} and g_m, it also follows their trend. Variations of f_t and f_{max} with V_{fg} for different V_{bg} are resented in Fig. 6a, b, respectively. f_t is mostly influenced by gm. As V_{bg} is increased, the peak value of g_m gradually increases and appears at smaller V_{fg} (Fig. 3b). In Fig. 6b, f_{max} is also found to exhibit the similar nature of variation, with around 14.4% increment in its peak value for variation in V_{bg} within the range under consideration.

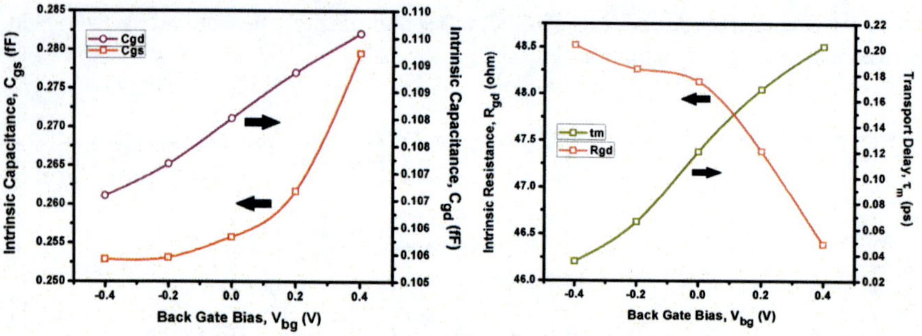

(a) (b)

Fig. 5 **a** Intrinsic capacitance (C_{gs}; C_{gd}). **b** Transport delay and intrinsic resistance versus back gate bias at $V_{fg} = 0.5$ V and $V_{ds} = 0.82$ V

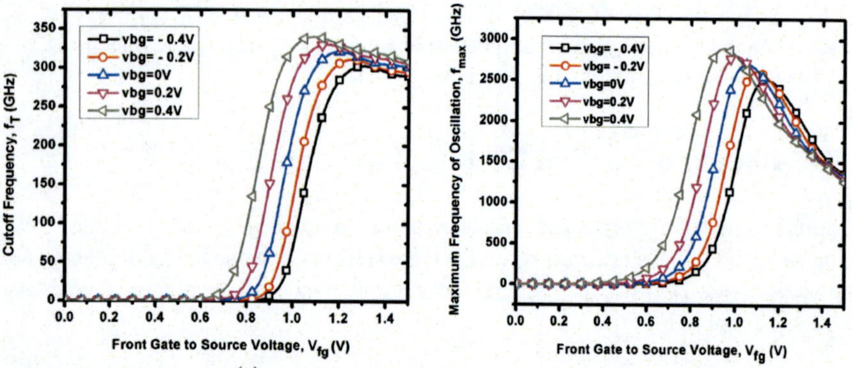

(a) (b)

Fig. 6 **a** Variation of cut-off frequency and **b** maximum frequency of oscillation with front gate bias for different back gate bias at $V_{ds} = 0.82$ V

5 Effect of Back Gate Bias on Cascode Amplifier

A single stage cascode amplifier designed using IG-FinFET is shown in Fig. 7. Both the gates of the transistor M_1 are shorted, while gates of the driving transistor M_2 are kept separated. Input voltage (V_{in}) is applied at the forward gate of M_2 and V_{bias} is 0.8 V. The load resistance (R) of 30 KΩ is connected to the drain of M_1, and 0.82 V of dc voltage (V_{dd}) is applied to it. The transfer characteristic of the configuration is investigated for different V_{bg} applied at M_2. Increase in V_{bg} lowers the threshold voltage; as a result, the current driving capability of the transistor increases. This causes an early fall of the output voltage, as observed in Fig. 7b. As the input voltage V_{in} increases, M_2 turns ON and conducts current. For positive V_{bg}, the current conducting capability of the transistor increases (due to Fig. 2b), and faster discharge of the output node takes place. The noise margins (NM_L and NM_H) calculated from the output characteristics are listed in Table 1. Gain versus frequency plot for the amplifier is illustrated in Fig. 7c, which indicates the highest gain for the grounded V_{bg}. In HD analysis, the minima is observed at V_{bg} = 0 V, and it shifts upward for negative V_{bg}. Early occurrence of minima is observed due to the application of positive V_{bg}. The nature of the HD3 plot is also quite similar to that of HD2 as far as the occurrences of minima are concerned. Grounded back gate produces ultimate minima in HD3 as shown in Fig. 7d–f.

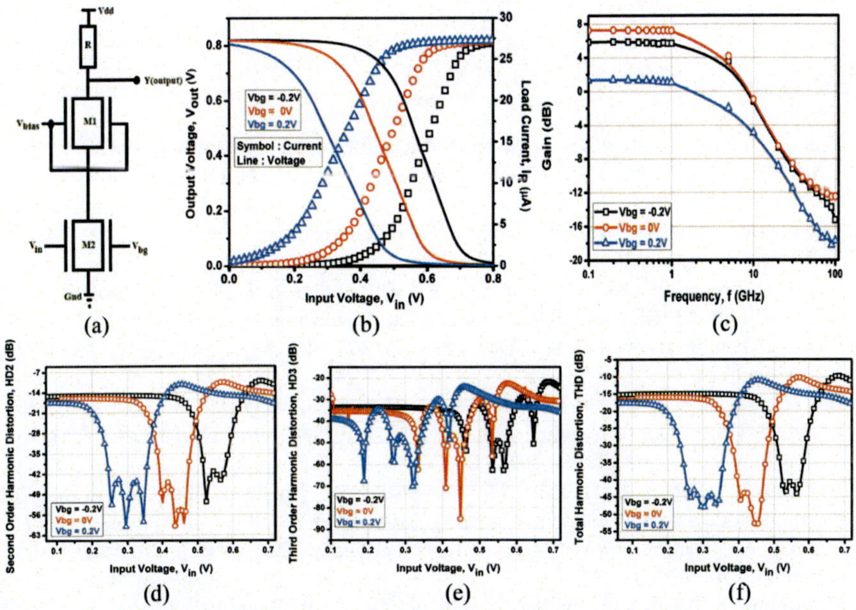

Fig. 7 Cascode amplifier realized with using IG-FinFETs: **a** Circuit. **b** Output characteristics and **c** gain versus frequency plot **d** HD2 **e** HD3 **f** THD plots for different V_{bg} at V_{ds} = 0.82 V

6 Conclusion

In this paper, influence of back gate bias on the performance of an IG-FinFET has been investigated. Significant improvement in I_{ON} is noticed for positive voltage applied at the back gate; I_{ON} getting doubled for V_{bg} varied over the entire range (−0.4 to +

Table 1 Data extracted from cascode amplifier circuit

Back gate voltages (V)	V_{IL} (V)	V_{IH} (V)	V_{OL} (V)	V_{OH} (V)	NM_L (V)	NM_H (V)
−0.2	0.44	0.709	0.006	0.815	0.434	0.106
0	0.33	0.603	0.006	0.815	0.324	0.212
0.2	0.175	0.485	0.006	0.805	0.169	0.32

0.4 V). C_{gs}, C_{gd}, $f_{\max}$ and f_t undergo an increment of 10.7%, 2.8%, 14.4% and 12.3%, respectively. Application of suitable back gate bias also controls the NMs of the cascode amplifier circuit. Both the gain and bandwidth of the circuit are enhanced for grounded V_{bg}. Thus, appropriately chosen back gate bias can make an IG-FinFET extremely suitable for numerous circuit applications.

Acknowledgements. The financial help provided by the UGC for "University with Potential for Excellence scheme" (UPE-II) is gratefully acknowledged.

References

1. Paul, B., Bansal, A., Roy, K.: Underlap DGMOS for digital sub-threshold operation. IEEE Trans. Electron Devices **53**(4), 910–913 (2006). /https://doi.org/10.1109/DRC.2005.1553132
2. Colinge, J.P., Gao, M.H., Romano-Rodríguez, A., Maes, H., Claeys, C.: Silicon-on-insulator gate-all-around device. In: Proceedings of the International Electron Devices Meeting Technical Digest, Dec 1990, pp. 595–598. https://doi.org/10.1109/SOSSOI.1990.145749
3. Choi, Y., Lindert, N., Xuan, P., Tang, S., Ha, D., Anderson, E., King, T., Bokor, J., Hu, C.: Sub-20 nm CMOS FinFET technologies. In: Proceedings of the International Electron Devices Meeting Techical Digest, Dec 2001, pp. 421–424. https://doi.org/10.1109/IEDM.2001.979526
4. Jiménez, D., Iñíguez, B., Suñé, J., Marsal. L.F, Pallarès, J., Roig, J., Flores, D.: Continuous analytic *I*–*V* model for surrounding-gate MOS-FETs. IEEE Electron Device Lett. **25**, 571–573 (2004). https://doi.org/10.1109/LED.2004.831902
5. Bescond, M., Nehari, K., Autran, J.L., Cavassilas, N., Munteanu, D., Lannoo, M.: 3-D quantum-modeling and simulation of multi-gate nanowire MOSFETs. In: Proceedings of the International Electron Devices Meeting Technical Digest, Dec 2004, pp. 617–620. https://doi.org/10.1109/IEDM.2004.1419237
6. Munteanu, D., Autran, J.L., Ferlet-Cavrois, V., Paillet, P., Baggio, J., Castellani, K.: 3-D quantumnumerical simulation of single-event transients in multiple-gate nanowire MOSFETs. IEEE Trans. Nucl. Sci. **54**(4), 994–1001 (2007). https://doi.org/10.1109/TNS.2007.892284
7. Sachid, A.B., Manoj, C.R., Sharma, D.K., Rao, V.R.: Gate fringe induced barrier lowering in underlap FinFET structures and its optimization. IEEE Electron Dev. Lett. **29**(1), 128–130 (2008). https://doi.org/10.1109/LED.2007.911974
8. Gundapaneni, S., Ganguly, S., Kottantharayil, A.: Enhanced electrostatic integrity of short-channel junctionless transistor with high-k spacers. IEEE Electron Dev. Lett. **32**(10), 1325–1327 (2011). https://doi.org/10.1109/LED.2011.2162309
9. Virani, H.G., Adari, R.B.R., Kottantharayil, A.: Dual-k spacer device architecture for the improvement of performance of silicon n-channel tunnel FETs. IEEE Trans. Electron Dev. **57**(10), 2410–2417 (2010). https://doi.org/10.1109/TED.2010.2057195

10. Zhang, W., Fossum, J.G., Mathew, L., Du, Y.: Physical insights regarding design and performance of independent-gate FinFETs. IEEE Trans. Electron Devices **52**(10) (2005). https://doi.org/10.1109/TED.2005.856184
11. International Technology Roadmap for Semiconductor (2012)
12. Sentaurus Device User Guide, Synopsys, Inc.: Mountain View, CA, USA (2010)
13. Kundu, A., Syamal, B., Koley, K., Sarkar, C.K., Mohankumar, N.: RF parameter extraction of Bulk FinFET: a non quasi static approach. In: IEEE International Conference of Electron Devices and Solid-State Circuits (EDSSC) (2010). https://doi.org/10.1109/EDSSC.2010.5713679
14. Lombardi, C., Manzini, S., Saporito, A., Vanzi, M.: A physically based mobility model for numerical simulation of nonplanar devices. IEEE Trans. Comput. Aid. Des. **7**(11), 1164–1171 (1988). https://doi.org/10.1109/43.9186
15. Kang, I.M., Shin, H.: Non-quasi-static small-signal modeling and analytical parameter extraction of SOI FinFETs. IEEE Trans. Nanotechnol. **5**(3), 205–210 (2006). https://doi.org/10.1109/TNANO.2006.869946

Impact of Trap Charges and High Temperature on Reliability of GaAs/Al_2O_3-Based Junctionless FinFET

Neha Garg[1], Yogesh Pratap[1], Mridula Gupta[2], and Sneha Kabra[1](✉)

[1] Department of Instrumentation, Shaheed Rajguru College of Applied Sciences for Women, University of Delhi, New Delhi, India
gneha148@gmail.com, yogi.pratap87@gmail.com, snehakabra1@gmail.com

[2] Semiconductor Device Research Laboratory, Department of Electronic Science, University of Delhi, South Campus, New Delhi, Delhi 110021, India
mridula@south.du.ac.in

Abstract. In the present work, the reliability issues of GaAs/Al_2O_3 Junctionless FinFET have been investigated by considering interface trap charges at semiconductor/oxide interface. RF/Analog performance of GaAs/Al_2O_3 Junctionless FinFET has been studied by evaluating different figures of merit such as drain current, I_{on}/I_{off} ratio, transconductance, output conductance, capacitance (gate to source) and cut-off frequency. To analyze the effect of temperature on trap charges, the simulation study has been done at 300, 400 and 500 K temperature. In addition to this, a comparative analysis between GaAs/Al_2O_3 and Si/SiO_2 Junctionless FinFET has also been carried out using a 3D device simulator (ATLAS). The results express that GaAs/Al_2O_3 Junctionless FinFET shows better performance in terms of the I_{on}/I_{off} ratio and gives better immunity to trap charges as compared to Si/SiO_2 Junctionless FinFET.

Keywords: Junctionless FinFET (JL FinFET) · Short channel effects (SCEs) · Trap charges

1 Introduction

As technology is advancing, the device size is reducing at a very fast rate, which causes various unwanted SCEs [1]. To suppress these effects, manufacturers of semiconductors devices are compelled to develop novel device architectures. Many architectures are proposed for nanoscale devices. One of such devices is junctionless FinFET. FinFET provides better gate controllability because of the presence of the gate at three sides and it also reduces various SCEs [2]. Junctionless devices eliminate the problems related to the complex fabrication process of junction-based device and large S/D contact resistance [3]. As junctionless devices do not contain any junction, they offer numerous advantages like better subthreshold slope, high on-current and they also provide better

© Springer Nature Singapore Pte Ltd. 2021
N. R. Das and S. Sarkar (eds.), *Computers and Devices for Communication*,
Lecture Notes in Networks and Systems 147,
https://doi.org/10.1007/978-981-15-8366-7_64

immunity to various SCEs [4, 5]. During device manufacturing, various types of charges are incorporated in the device which affects the device adversely, and hence, they need to be addressed correctly. These interface trap charges are generated at the semiconductor/insulator interface due to lattice mismatch [6]. Despite the excellent potential of JL FinFET, their performance may be limited by various effects that occur due to the presence of trap charges. The presence of trap charges not only changes the gain but also shifts the bias point of the device which leads to difficulty in device designing. Therefore, it is very crucial to study the impact of trap charges on long-term reliability of the device [7].

A lot of investigation in the previous years has been carried out to study the impact of trap charges on various advanced semiconductor devices like Double Gate (DG) MOSFET [8], FinFET [9], Tunnel FET [10], GAA MOSFET [11]. Chabbra et al. [12] analyzed GaAs-based JL FinFET for high-performance analog application and compared with Si and Ge. GaAs JL FinFET shows better performance in terms of switching and various other factors. However, reliability issues of GaAs-based JL FinFET are still not addressed. Therefore in this work, a comprehensive analysis of various reliability issues that occur due to trap charges present at the interface in GaAs/Al_2O_3 JL FinFET for channel length of 20 nm has been carried out. Moreover, the impact of temperature on GaAs/Al_2O_3 JL FinFET has also been studied to predict the high-temperature performance of the device. Also, a comparative analysis in terms of performance degradation is done between GaAs/Al_2O_3 and Si/SiO_2 JL FinFET. More degradation is observed in Si/SiO_2 JL FinFET as compared to GaAs/Al_2O_3 JL FinFET due to trap charges present at the interface.

The research work in this paper is organized as follows: Structure of device and simulation methodology is included in Sect. 2. Different models used in the simulation of the device are also described in this section. In Sect. 3, results of Analog/RF performance of GaAs/Al_2O_3 JL FinFET have been included by evaluating various figure of merits like drain current, I_{on}/I_{off} ratio, transconductance, output conductance, gate capacitance and cut-off frequency. Finally, in Sect. 4, conclusion of the proposed work is drawn.

2 Structure and Simulation of JL FinFET

Figure 1a shows the structure of conventional FinFET [13] and Figure 1b shows a 2D structure of GaAs/Al_2O_3-based junctionless FinFET. The device under consideration has gates on three sides and hence provides better gate controllability over the channel, thus reducing various unwanted SCEs. The device is doped uniformly throughout from source to drain with a doping concentration of 5×10^{18} cm^{-3} and no junctions are considered. To study the impact of trap charges, uniform localized charges are considered at the semiconductor/oxide interface. During the simulation oxide thickness (t_{ox}), 1.5 nm and metal work function (Φ_m) 4.6 eV have been taken under consideration. During simulation, various models which have been considered include Shockley Read Hall (SRH) for thermal generation and minority recombination, band gap narrowing (BGN) model for high channel concentration, field dependent mobility (FLDMOB) model to account for velocity saturation effect and CVT mobility model for simulating temperature variation. Carrier–carrier scattering mobility (CCSMOB) model and concentration dependent mobility (CONMOB) models are also considered for evaluating simulation results presented in the next section [14].

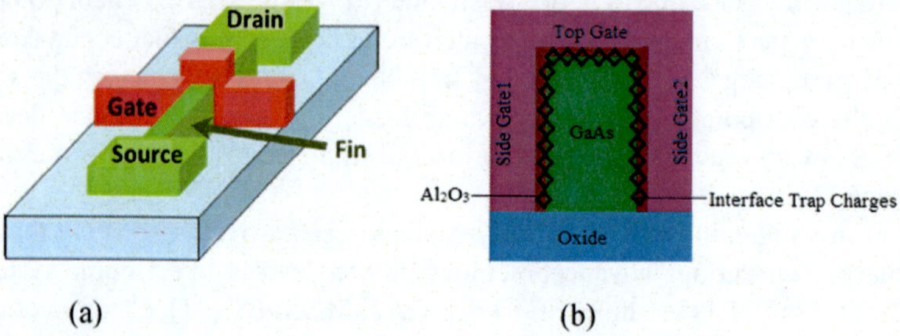

Fig. 1 **a** Structure of conventional FinFET [13] and **b** 2D view of $GaAs/Al_2O_3$ JL FinFET

3 Results and Discussion

Figure 2 depicts the change in drain current with variation in gate voltage of $GaAs/Al_2O_3$ and Si/SiO_2 JL FinFET. The lower value of off current (I_{off}) and hence better I_{on}/I_{off} ratio is observed in $GaAs/Al_2O_3$ JL FinFET as compared to Si/SiO_2 JL FinFET. This is because the dielectric constant of Al_2O_3 is high which leads to better dipole alignment that produces a higher vertical electric field and hence provides better gate controllability over the channel. Also, it can be analyzed from Table1 that the trap charges have more impact on drain current (I_{off}) of Si/SiO_2 Junctionless FinFET.

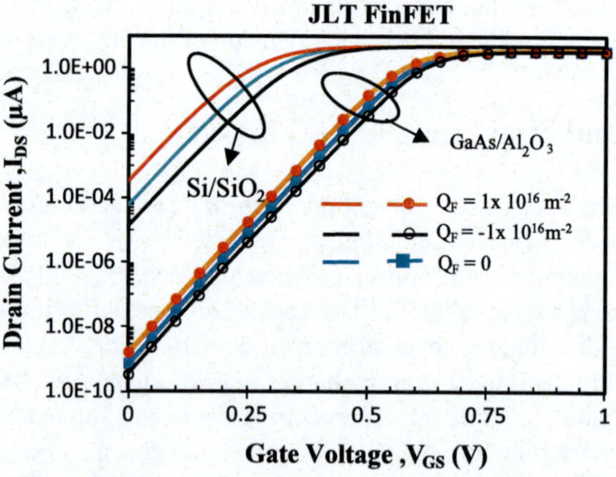

Fig. 2 Impact of trap charges on drain current for $GaAs/Al_2O_3$ and Si/SiO_2 JL FinFET

Figure 3a depicts the change in drain current with the variation in gate voltage for GaAs JL FinFET in the presence of trap charges when the device is operated at a high

Table 1 I_{on}/I_{off} ratio for $GaAs/Al_2O_3$ and Si/SiO_2 JL FinFET

Device	$Q_F = 0$	$Q_F = 1 \times 10^{16}\ m^{-2}$	$Q_F = -1 \times 10^{16}\ m^{-2}$
$GaAs/Al_2O_3$	4.71×10^9	2.16×10^9	10.0×10^9
Si/SiO_2 JL FinFET	8.65×10^4	1.60×10^2	3.77×10^5

temperature ranging from 300 to 500 K. As depicted from the figure, the change in drain current is more noticeable in the subthreshold region. The drain current increases with the inclusion of positive trap charges. It can also be noticed in the figure that as the temperature decreases, the effect of trap charges is enhanced. Change in I_{off} with the inclusion of positive trap charges is 117%, 79% and 59% at a temperature of 300 K, 400 K and 500 K, respectively. The transconductance of any device is a very significant factor as it decides the gain of the device, and for high cut-off frequency, it is desirable to be as high as possible [15]. Figure 3b shows the transconductance variation with gate voltage for $GaAs/Al_2O_3$ and Si/SiO_2 JL FinFET. Although the value of transconductance is high for Si/SiO_2 JL FinFET compared to $GaAs/Al_2O_3$, degradation due to trap charges is more in it. The bias point shift as can be observed from Fig. 3b is 8.3% for $GaAs/Al_2O_3$ and 14% for Si/SiO_2 JL FinFET.

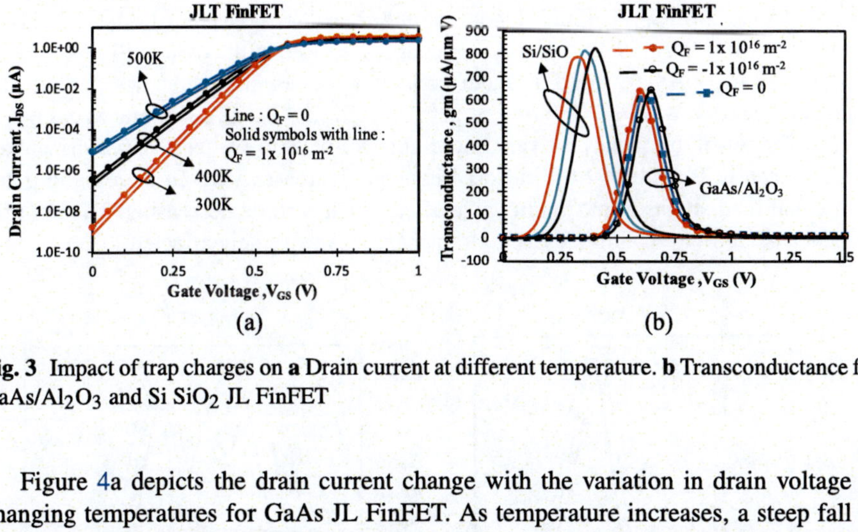

Fig. 3 Impact of trap charges on **a** Drain current at different temperature. **b** Transconductance for $GaAs/Al_2O_3$ and Si SiO_2 JL FinFET

Figure 4a depicts the drain current change with the variation in drain voltage at changing temperatures for GaAs JL FinFET. As temperature increases, a steep fall in the drain current can be observed. It can be noticed that the value of drain saturation current also decreases with the increase in temperature. The drain current saturation value, $I_{DSat} = 3.73\ \mu A$ at 300 k, $I_{DSat} = 3.14\ \mu A$ at 400 k, $I_{DSat} = 2.55\ \mu A$ at 500 k. Figure 4b depicts the change in output conductance with the variation in drain voltage at different temperatures for GaAs JL FinFET. The maximum value of output conductance at 300 K, 400 K and 500 K are 49.7 µS, 34.3 µS, 24.4 µS, respectively. The high value of g_d at 300 K as compared to 400 K and 500 K is observed because the drain current is also higher at room temperature. With the increase in temperature, output conductance

(g_d) decreases which in turn increases the output resistance and leads to improvement in device driving capability.

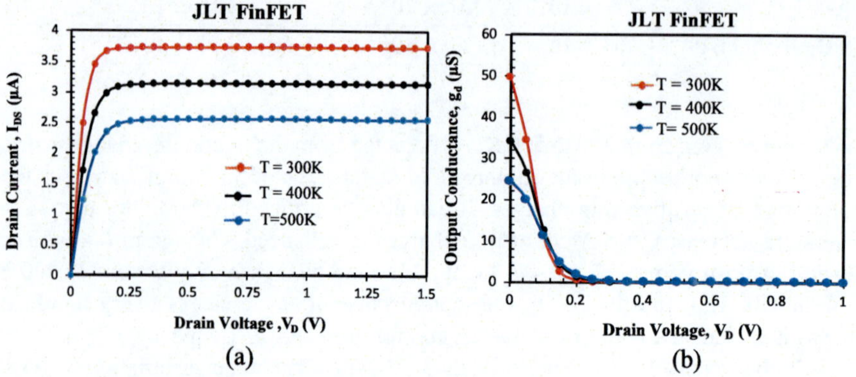

Fig. 4 **a** Drain current and **b** output conductance as drain voltage function for different temperatures for $GaAs/Al_2O_3$ JL FinFET

Figure 5a shows the variation of gate to source capacitance with the variation of gate voltage for $GaAs/Al_2O_3$ and Si/SiO_2 JL FinFET. High C_{gs} is observed for $GaAs/Al_2O_3$ as compared to Si/SiO_2 JL FinFET because the dielectric constant of Al_2O_3 is high compared to SiO_2. Degradation due to trap charges is more in the case of Si JL FinFET as compared to GaAs JL FinFET under consideration. Figure 5b shows the cut-off frequency variation with the variation of gate voltage for $GaAs/Al_2O_3$ and Si/SiO_2 JL FinFET. The cut-off frequency is a very important parameter to evaluate RF performance. It can be seen in Fig. 5b that the cut-off frequency increases with an increase in gate voltage and then attains a maximum value after which it starts decreasing. The cut-off frequency of Si/SiO_2 JL FinFET due to trap charges is extremely affected.

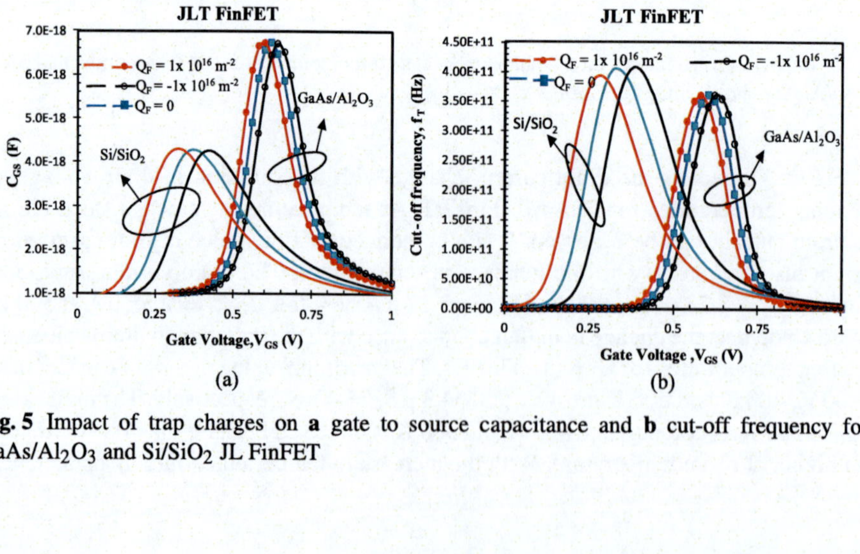

Fig. 5 Impact of trap charges on **a** gate to source capacitance and **b** cut-off frequency for $GaAs/Al_2O_3$ and Si/SiO_2 JL FinFET

4 Conclusion

Analog/RF reliability performance of GaAs/Al_2O_3 JL FinFET has been investigated by including the effect of trap charges at the interface. The impact of high temperature on device performance has also been carried out for temperature ranging from 300 to 500 K. Interface charges cause more degradation in the device performance of Si/SiO_2 JL FinFET than GaAs/Al_2O_3 JL FinFET. Trap charges not only change the gain of the device but also shift the bias point of the device which affects the reliability of circuits. It has been shown that the damage due to trap charges is more in the subthreshold region. Results also illustrate that the impact of trap charges is more pronounced at room temperature. Also, GaAs/Al_2O_3 JL FinFET are more immune to trap charges as compared to Si/SiO_2 JL FinFET. Thus, GaAs/Al_2O_3 JL FinFET is a more reliable device.

Acknowledgements. The authors are grateful to Shaheed Rajguru College of Applied Sciences for Women, University of Delhi and Science and Engineering Research Board, Department of Science and Technology (SERB-DST project "ECR/2017/000576"), Government of India, for providing the opportunity for carrying out this work.

References

1. Sarvari, H., Ghayour, A.H., Chen, Z., Ghayour, R.: Analyses of short channel effects of single-gate and double-gate graphene nanoribbon field effect transistors. J. Mater. **16**, 8 (2016). Article ID 8242469
2. Kaundal, S., Rana, A.K.: Design and structural optimization of junctionless FinFET with Gaussian-doped channel. J. Comput. Electron. **17**(2), 637–645 (2018)
3. Singh, J., Ciavatti, J., Sundaram, K., Wong, J.S., Bandyopadhyay, A., Zhang, X., Li, S., Bellaouar, A., Watts, J., Lee, J.G., Samavedam, S.B.: 14-nm FinFET technology for analog and RF applications. IEEE Trans. Electron Devices **65**(1) (2018)
4. Colinge Colinge, J.-P., Lee, C.-W., Afzalian, A., Akhavan, N.D., Yan, R., Ferain, I., Razavi, P., O'Neill, B., Blake, A., White, M., Kelleher, A.-M., McCarthy, B., Murphy, R.: Nanowire transistors without junctions. Nat. Nanotechnol. **5**, 225–229 (2010)
5. Wang, Y., Tang, Y., Sun, L., Cao, F.: High performance of junctionless MOSFET with asymmetric gate. Proc. Comput. Sci. **125**, 825–831 (2018)
6. Garg, N., Pratap, Y., Gupta, M., Kabra, S.: Impact of different localized trap charge profiles on the short channel double gate junctionless nanowire transistor based inverter and ring oscillator circuit. AEUE Int. J. Electron Commun. **108**, 251–261 (2019)
7. Yu, L.C., Dunne, G.T., Matocha, K.S., Cheung, K.P., Suehle, J.S., Sheng, K.: Reliability issues of SiC MOSFETs: a technology for high-temperature environments, IEEE Trans. Dev. Mater. Reliabil. **10**(4) (2010)
8. Garg, N., Pratap, Y., Gupta, M., Kabra, S.: Analysis of interface trap charges of double gate junctionless nanowire transistor (DG-JNT) for digital circuit applications. In: IEEE Electron Device Kolkata Conference. IEEE EDKCON (2018)
9. Singh, J., Ciavatti, J., Sundaram, K., Wong, J.S., Bandyopadhyay, A., Zhang, X., Li, S., Bellaouar, A., Watts, J., Lee, J.G., Samavedam, S.B.: 14-nm FinFET technology for analog and RF applications. IEEE Trans. Electron Dev. **65**(1), 31–37 (2018)

10. Rahimian, M., Fathipour, M.: Junctionless nanowire TFET with built-in N-P-N bipolar action: physics and operational principle. J. Appl. Phys. **120**, 225702 (2016). https://doi.org/10.1063/1.4971345
11. Nagy, D., Indalecio, G., García-Loureiro, A.J., Elmessary, M.A., Kalna, K., Seoane, N.: FinFET versus gate-all-around nanowire FET: performance, scaling, and variability. IEEE J. Electron Dev. Soc. **6**, 332–340 (2018)
12. Kumar, A., Chhabra, A., Chaujar, R.: GaAs junctionless FinFET using high-dielectric for high-performance applications. In: 2018 IEEE 38th International Conference on Electronics and Nanotechnology (ELNANO)
13. https://en.wikipedia.org/wiki/Multigate_device
14. ATLAS Device Simulation Software. Silvaco Int., Santa Clara, CA, USA (2018)
15. Mohapatra, S.K., Pradhan, K.P., Sahu, P.K., Kumar, M.R.: The performance measure of GS-DG MOSFET: an impact of metal gate work function. Adv. Natl. Sci. Nanosci. Nanotechnol. **5**(2), 025002 (2014)

Power Analysis and Optimization Using Nonlinear Modeling of Memristor: A Design Case Study

Panthadeb Saha[1](✉) and Prasun Ghosal[2]

[1] B. P. Poddar Institute of Management and Technology, Kolkata, West Bengal 700052, India
panthadebinraphel@yahoo.com
[2] Indian Institute of Engineering Science and Technology, Shibpur, Howrah, West Bengal 711103, India
prasun@ieee.org

Abstract. Ever-increasing power consumption due to continuous increase in circuit complexity in nanoscale electronic circuits and systems design is playing a pivotal role today. The necessity of mitigating higher power consumption issues and exploring different types of power minimization techniques are gaining rising importance. Search and exploration with different emerging circuit design technologies are also of increasing importance. A memristor is one such viable candidate with its inherent ability of low power consumption. Proper modeling and intelligent use of memristors in hybrid mode with CMOS can show a bright direction for future design. Here we tried to focus the light on this issue by inspecting the problem with a specific case study of the Wien-bridge oscillator circuit. The basic nonlinear modeling approach of the memristor is followed. Different configurations of circuit replacement are simulated to bring heterogeneity. Thus, an effort is presented in this paper to identify the proper and judicious use of memristors for power optimization. Simulation results are quite encouraging to predict the appropriate configuration for power reduction properly.

Keywords: Memristor · Nonlinear ion drift · Wien-bridge oscillator · Power minimization

1 Introduction

Due to its pinched hysteresis and dynamic negative resistance memristor can find several important applications in nanoelectronic design [1]. As it can directly relate magnetic flux and charge, it is considered as the fourth fundamental passive circuit element besides resistor, inductor, and capacitor.

Several memristive applications and technology involving memristors are in proximity to reality. A little about that is mentioned below to give an idea about the present status in this field of research and development. They are

- Nonvolatile memory applications

© Springer Nature Singapore Pte Ltd. 2021
N. R. Das and S. Sarkar (eds.), *Computers and Devices for Communication*,
Lecture Notes in Networks and Systems 147,
https://doi.org/10.1007/978-981-15-8366-7_65

- Low power and remote sensing applications
- Crossbar latches as transistor replacements or augmentors
- Analog computation and circuit applications
- Circuits which mimic neuromorphic and biological systems (learning circuits)
- Programmable logic and signal processing.

2 Memristor Modeling: Challenges and Improvements

Memristors can be used in a wide range of applications. Hence, different characteristics from it are required for each application. Some of them are good scalability, less consumption of power, flexibility, and compatibility with CMOS. A model exhibiting these attributes is required to design and analyze memristive device-based circuits and applications. The major models of memristors are mentioned in this section.

A memristive device model to become effective needs to be sufficiently accurate and computationally efficient. It is desired to be simple, intuitive, and of closed-form. Again, it should be generalized to suit the different types of memristive devices. Different modeling approaches include A. Linear Ion Drift Model, B. Nonlinear Ion Drift Model, C. Simmons Tunnel Barrier Model, D. ThrEshold Adaptive Memristor Model (TEAM), E. VTEAM.

3 Related Research and Motivation

Due to the very limited reach to the experimental realization of the memristor researchers resort to the behavioral model of the memristor developed in HP company Laboratory in 2008 or emulate the memristor model using active circuitry [2, 3]. Several approaches were made to model the memristor using different computational tools namely SPICE, Verilog-A/AMS, MATLAB, and Simulink/Simscape.

From experiments, it is found that the fabricated memristive devices are highly nonlinear [4, 5] and act significantly different from the linear ion drift model. An extensive study is needed for the choice of effective memristive model and optimization of the incorporation of memristors replacing existing circuit components in the practical circuits to fulfill the aim of minimization of power consumption.

Keeping this in mind in this work, we have followed the approaches of [6, 7] and revisited those same case studies but considering the nonlinear ion drift model to model a memristor and take a look on the extent of this modeling approach in minimizing power consumption in Wien-bridge oscillator circuit incorporating memristors. Relative comparison is then made, and the possibility of the replacement of existing circuit components in Wien-bridge oscillator by memristors is analyzed.

4 Proposed Work

In this paper, the modeling in Simulink/Simscape environment is done following the nonlinear ion drift model as the mathematical model to express the behavior of a memristor with a suitable choice of values of the different fitting parameters. None of the available window functions is used here primarily for the modeling purpose, but they can easily be incorporated in the process.

4.1 Detailed Experimental Framework and Methodology

In the beginning, we have created our component block to represent a memristor and work as a circuit element for the simulation in MATLAB Simulink/Simscape environment as no such block is available there readily in the directory. Then we have created different MATLAB model Circuits/MATLAB models to represent the different configurations of the Wien-bridge oscillator circuit and then simulated by adjusting different parameters to achieve the perfect MATLAB model.

Here Wien-bridge oscillator is used as a specific case study circuit. The schematic representation is shown in Fig. 1. A Wien-bridge oscillator tank circuit consists of two capacitors and four resistors. The instability achieves sustained oscillation (range 20 Hz–20 kHz) that occurred by a combination of + ve and −ve feedback without any external source.

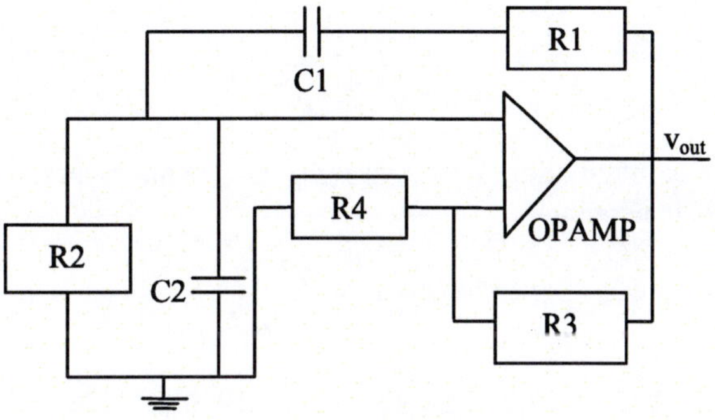

Fig. 1 Schematic of Wien-bridge oscillator

Equation (1) presents the condition for sustained oscillation, and Eq. (2) presents the frequency expression.

$$(C_2/C_1) + (R_1/R_2) \geq 2 \tag{1}$$

and

$$f = 1/\sqrt{(R_1 R_2 C_1 C_2)} \tag{2}$$

By controlling memristance during replacing resistors by memristors, the oscillation frequency can be changed to the desired value (by Eq. 2). Five different configurations are studied (following Table 1) which are the same as therein [6, 7].

In configuration 1, memristor M_1 whose resistance is labeled as R_{m1} replaces R_1. In configuration 2, R_2 is replaced with memristor M_2 whose resistance is R_{m2}, and so on (as per Table 1).

(1) *One Resistor Replaced with Memristor*: When only one resistor (R_1/R_2) is replaced with memristor (M_1/M_2), then the required balancing property is according to

Table 1 Variations of Wien oscillators

Configuration	Independent modifications	Dependent modifications
Configuration 1	$R_1 \rightarrow M_1$	$R_3 \rightarrow R_4(1 + (R_{m1}/R_2))$
Configuration 2	$R_2 \rightarrow M_2$	$R_3 \rightarrow R_4(1 + (R_1/R_{m2}))$
Configuration 3	$R_3 \rightarrow M_3, R_4 \rightarrow M_4$	$R_3 \rightarrow 2R_{m4}, R_1 = R_2$
Configuration 4	$R_2 \rightarrow M_2, R_4 \rightarrow M_4$	$R_3 \rightarrow R_{m2} + R_1$
Configuration 5	Resistors $\rightarrow$ Memristors	$R_3 \rightarrow 2R_{m4}$, Given that $R_{m1} = R_{m2}$

Eq. (1). Here both configurations 1 and 2 are exhibiting this example. In configuration 1, R_3 needs to be changed to achieve sustained oscillation. So here, R_3 is designed to be

$$\left(R_4\left(1 + \frac{R_{m1}}{R_2}\right)\right).$$

Similarly, for configuration 2, R_2 was replaced with a memristor to achieve oscillation. R_3 is wired to be

$$\left(R_4\left(1 + \frac{R_1}{R_{m2}}\right)\right).$$

(2) *Two Resistors Replaced with Memristors*: Two different configurations are studied with two resistors replacement. In configuration 3, R_3, and R_4 are replaced with memristors. Again, to fulfill the condition of oscillation, different rating memristors were used. Given that $R_4 = R_{m4}$ and $R_1 = R_2$, R_3 is calculated to be $2 \times R_{m4}$. In configuration 4, R_2 and R_4 are replaced with memristors. Given that $R_{m2} = R_{m4}$, R_3 is replaced with a memristor-resistor combination expressed as $R_{m2} + R_1$.

(3) *All Resistors Replaced with Memristors*: Lastly, all four resistors are replaced with memristors in configuration 5 as shown in Fig. 2. To perceive the sustained oscillation condition, R_3 was replaced with $2 \times R_{m4}$. The power variation of the simulated circuit is shown in Fig. 3 (Table 2).

5 Result and Discussions

A comparative study is made in Table 3 for the consumption of power by the Wien-bridge oscillator circuits incorporating the memristor for the two different memristor modeling approaches one nonlinear and the other linear as detailed in [7]. Here the resulting power ratio of the nonlinear model to a linear model (P_{nl}/P_l) is calculated in a logarithmic scale for ease of representation. So, a more negative value indicates lesser the resulting power in the case of the nonlinear model of memristor compared to that of the linear model. The following comment can be made by studying Table 3.

For configuration 1 and configuration 3, the resulting power is found large compared to the simulation results obtained considering the linear ion drift model.

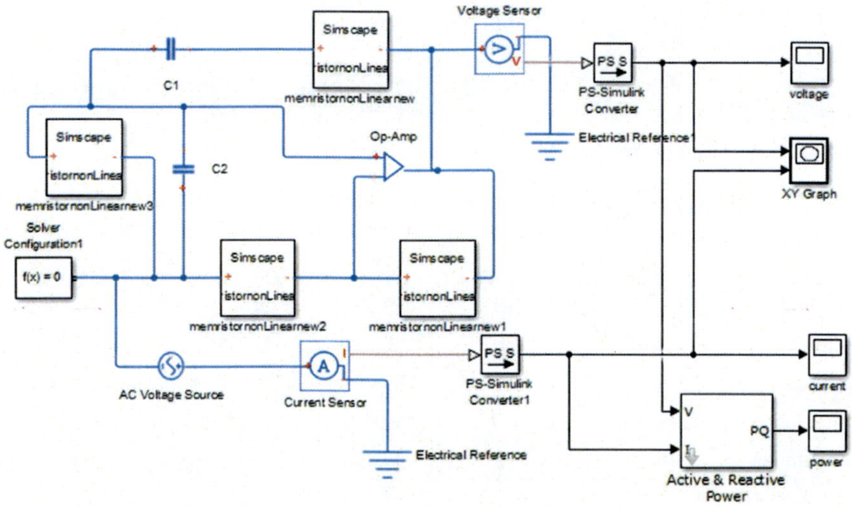

Fig. 2 Configuration 5: simscape/simulink model replacing all resistor with memristor

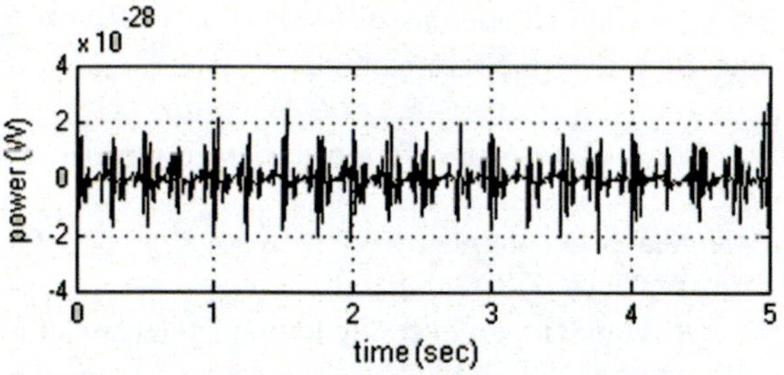

Fig. 3 Study of power for configuration 5

Table 2 Variations of power in Wien oscillators

Configuration	Maximum linear power (L_{pow}) (W) [$L_{pow} = V_m \times I_m$]	Maximum resulting power (R_{pow}) (W)	Reduction factor (R_f) [$R_f = R_{pow}/L_{pow}$]
Configuration 1	1.05×10^{-18}	1.67×10^{-19}	0.16
Configuration 2	3.50×10^{-21}	3.23×10^{-23}	0.009
Configuration 3	3.52×10^{-12}	9.4×10^{-14}	0.027
Configuration 4	1.45×10^{-22}	1.82×10^{-24}	0.012
Configuration 5	7.6×10^{-27}	3×10^{-28}	0.04

Table 3 Comparison of variations of power in Wien oscillators for different models

Memristor model	Nonlinear	Linear [7]	Percentage reduction of resulting power considering the nonlinear model	The ratio P_{nl}/P_l	$\log(P_{nl}/P_l)$
Configuration	Maximum resulting power (P_{nl}) (W)	Maximum resulting power (P_l) (W)			
Configuration 1	1.67×10^{-19}	8×10^{-24}	−99.995	20,875	4.31962
Configuration 2	3.23×10^{-23}	3×10^{-22}	89.23	0.10767	−0.96790
Configuration 3	9.4×10^{-14}	4×10^{-22}	−99.99	235,000,000	8.37106
Configuration 4	1.82×10^{-24}	1.5×10^{-19}	99.99	0.00001213	4.91613
Configuration 5	3×10^{-28}	2×10^{-25}	99.85	0.0015	2.82390

5.1 Power Variation Analysis

The peak power variation, from all configurations, is shown in Table 2. The following comments are obvious from the table information.

- In configuration 5, where all resistors are replaced with memristors, the consumed power is the lowest.
- Consumed power is greater in configuration 3 than configuration 1, where a number of resistors are being replaced with memristors.
- The reduction of power is maximum in configuration 2 than any other configuration.
- The design and configuration of the circuit have a great impact on the power reduction, and it is not simply dependent on a number of resistors to be replaced.

So, it is obvious that power reduction can be achieved for maximum cases by replacing resistors with memristors.

Observation 1: *Power reduction by replacing resistors with memristors does not solely dependent on a number of resistors replaced by memristors. It also depends on the configuration of the designed circuit. It cannot be achieved by simply replacing more and more resistors with a memristor.*

Observation 2: *The nature of variation of power and its reduction is similar to that with the linear model.*

In Observation 1, an analogy between the variable power reduction factors for different configurations is drawn.

Observation 3: *Replacement of resistors mainly in the feedback path of the circuit results consumption of larger power in simulations considering nonlinear ion drift model except the al memristor configuration.*

6 Conclusion

From experiments, it is found that the fabricated memristive devices are highly nonlinear. So, we have followed the approaches of [7] and revisited those same case studies but considering the nonlinear ion drift model and take a look at the extent of this modeling approach in minimizing power consumption in Wien-bridge oscillator.

This work may be extended using a generalized memristor model, which is based on experimentally published data, to study the effect on power minimization by replacement of resistors by memristors in different analog electronic circuits.

References

1. Mohanty, S.P.: Nanoelectronic Mixed-Signal System Design. McGraw-Hill Education (2015)
2. Pershin, V., Di Ventra, M.: Practical approach to programmable analog circuits with memristors. IEEE Trans. Circ. Syst. I **57**(8), 1857–1864 (2010)
3. Pershin, V., Di Ventra, M.: Memristive Circuits Simulate memcapacitors and meminductors. Electron. Lett. **46**(7), 517–518 (2010)
4. Yang, J.J., Pickett, M.D., Li, X., Ohlberg, D.A., Stewart, D.R., Williams, R.S.: Nat. Nanotechnol. **3**(7), 429–433 (2008)
5. Strukov, D.V., Williams, R.S.: "Exponential ionic drift: fast switching and low volatility of thin-film memristors. Appl. Phys. A Mater. Sci. Process. **94**(3), 515–519 (2009)
6. Agu, E., Mohanty, S.P., Kougianos, E., Gautam, M.: Simscape design flow for memristor based programmable oscillators. In: Proceedings of the 24th Edition of the GLSVLSI, pp. 223–224. ACM (2014)
7. Ghosal, P., Mohanty, S.P.: Power minimization of a memristor-based Wien bridge oscillator through a Simscape framework. IEEE iNIS 2015, pp. 83–88

Check for updates

Study of High-Frequency Performance in GeSn-Based QWIP

Soumava Ghosh, Swagata Dey(✉), Bratati Mukhopadhyay, and Gopa Sen

Institute of Radio Physics and Electronics, University of Calcutta, 92, Acharya Prafulla Chandra Road, Kolkata, West Bengal 700009, India
swagatadey2009@gmail.com

Abstract. QWIP using group IV elements are of great research attention for its potential application in optical communication and in optical interconnects. The high-frequency performance of GeSn–SiGeSn QWIP has been studied considering the intersubband transition and transit time effect of electrons. The band structure of GeSn–SiGeSn QWIP and the analytical results of responsivity are also presented in this paper.

Keywords: QWIP · Intersubband transition · MQW · Transit time

1 Introduction

Photonic devices using group IV semiconductors grown on Ge or Si substrate are an attractive research area due to their compatibility with CMOS circuits. However, over past two decades, realization of such devices, especially light emitters and modulators, etc., remained a great challenge due to the indirect bandgap nature in the materials [1–3]. Though Ge-based photodetectors do not have such limitation, the direct growth of GeSn on Ge substrate poses problems due to high degree of lattice mismatch. Recently, an alloy of GeSiSn directly grown on Si substrates has been made possible with practical chemical vapor deposition (CVD) technique [4, 5]. A noteworthy feature is that the direct Γ-valley conduction band decreases more rapidly than the indirect L-valley band by applying proper tensile strain and as a result, an indirect to direct crossover is possible. Direct gap type I heterostructures have been realized in tensile strained Ge using GeSiSn barriers [6, 7].

However, Chang et al. [8] proposed and analyzed a strain balanced GeSn–SiGeSn multiple quantum-well (MQW) laser. Chakraborty et al. [9] recently theoretically modeled the loss mechanisms of the infrared laser using SiGeSn barrier-well barrier structure. Over the last few years, the intersubband transition is utilized from the bound state in quantum wells (QW) to the continuum state above the barrier in infrared technology on the basis of quantum well-infrared photodetectors (QWIPs) [10]. In this work, the multiple quantum well (MQW) structure is analyzed theoretically by the method of Ryzhii et al. [11] and the frequency-dependent responsivity of GeSn–SiGeSn QWIP is determined analytically.

© Springer Nature Singapore Pte Ltd. 2021
N. R. Das and S. Sarkar (eds.), *Computers and Devices for Communication*,
Lecture Notes in Networks and Systems 147,
https://doi.org/10.1007/978-981-15-8366-7_66

2 Theory

We consider $Ge_{0.92}Sn_{0.08}/Si_{0.11}$ $Ge_{0.7}$ $Sn_{0.19}$ QWIPs consisting of a large number of donor-doped QWs separated by relatively thick barriers. An electric field is applied perpendicularly to the plane of MQW structure which is sandwiched between the emitter and the extreme barriers followed by two contact layers. Direct bandgap and type I nature of $Ge_{0.92}Sn_{0.08}$ (5.717 Å) QW material are encouraged us for analytical exploration of this model [12]. The concentrations of well and barrier material are chosen in such a way that the well has a tensile strain effect with respect to the virtual substrate $Si_{0.11}Ge_{0.7}Sn_{0.19}$ (5.785 Å), and the strain of the barrier is balanced with respect to the virtual substrate. All the layers are grown on a Si substrate.

The band structure calculation helps us to investigate the intersubband transition effect of this proposed structure. The bandgaps in different materials and the barrier height of QWs are calculated by the model solid theory suggested by Van de Walle [13, 14] which has been used by Menendez et al. [6]. The bowing parameters [8] are considered during our calculation.

$$\begin{aligned} E_{v\Gamma}(\text{activelayer}) = & -\frac{\Delta_0\left(\text{Si}_x\text{Ge}_{1-x-y}\text{Sn}_y\right)}{3} - \Delta E_{v,\text{ave}}\left(\text{Si}_x\text{Ge}_{1-x-y}\text{Sn}_y\right) \\ & - \frac{\Delta_0(\text{activelayer})}{6} + \delta E_h^v + \frac{1}{4}\delta E_{001} \\ & + \frac{1}{2}\sqrt{\left((\Delta_0(\text{activelayer}) + \frac{1}{2}\delta E'_{001})^2 + 2\left(\delta E''_{001}\right)^2\right)} \end{aligned} \quad (1)$$

$$E_{C\Gamma}(\text{activelayer}) = E_{v\Gamma}(\text{activelayer}) + E_0(\text{activelayer}) + \delta E_h^{C\Gamma} \quad (2)$$

$$E_{\text{CL}}(\text{activelayer}) = E_{v\Gamma}(\text{activelayer}) + E_{\text{ind}}(\text{activelayer}) + \delta E_h^{\text{CL}} \quad (3)$$

The strain e_{11} is given by,

$$e_{11} = \frac{a(\text{substrate}) - a(\text{activelayer})}{a(\text{activelayer})} \quad (4)$$

where a is the lattice constant of the constituent material.

So, the barrier height can be expressed as,

$$E_0 = E_{C\Gamma}\left(\text{Si}_x\text{Ge}_{1-x-y}\text{Sn}_y\right) - E_{C\Gamma}(\text{activelayer}) \quad (5)$$

Now, the responsivity can be calculated by the method given by Ryzhii [11]. The frequency-dependent normalized responsivity of a QWIP can be expressed as,

$$\begin{aligned} R_f = & R\frac{2}{2j\pi ft(n+1)}\frac{\exp(2j\pi f) - 1}{\left[1 - \delta\exp(2j\pi ft)\right]^2} \\ & \times \left[\left\{\delta^{(n+1)}\exp(2j\pi ft(n+1))\right\} - \{\delta(n+1)\exp(2j\pi ft(n+1))\} + n\right] \end{aligned} \quad (6)$$

where δ is the capture parameter and the electron transit time t is given by $t = \left(n + 1 - \frac{K}{V_s}\right)L_w$.

Here, n is the total number of QW's, Vs is the saturation velocity, $K = 1, 2, 3 \ldots n$ and Lw is the period of the QW structure. Also,

$$R = \frac{q\xi n_s}{2\Sigma_i} \tag{7}$$

The electron excitation energy Σ_i is estimated from the bottom of the lowest subband in QWs.ξ is the photoexcitation cross section, n_s is the donor sheet concentration in each QW. q is the electronic charge.

3 Results and Discussion

The values of different parameters are taken as reported by Dey et al. [15]. The normalized responsivity versus the modulation frequency is shown in Fig. 1 for a fixed value of capture parameter (0.98) of the QWIP. The responsivity increases slightly with increasing frequency for different number of QWs. In Fig. 2, the similar results are shown with different capture parameter for a fixed value of quantum well ($n = 16$). It is seen that for a higher value of capturing probability, the responsivity is large. But, the value of responsivity is very appreciable for a small value of capture parameter since the responsivity is less than unity for the lower value of capture parameter. This result is more significant than the relevant QWIP structure made of group III–V material reported by Ryzhii et al. [11].

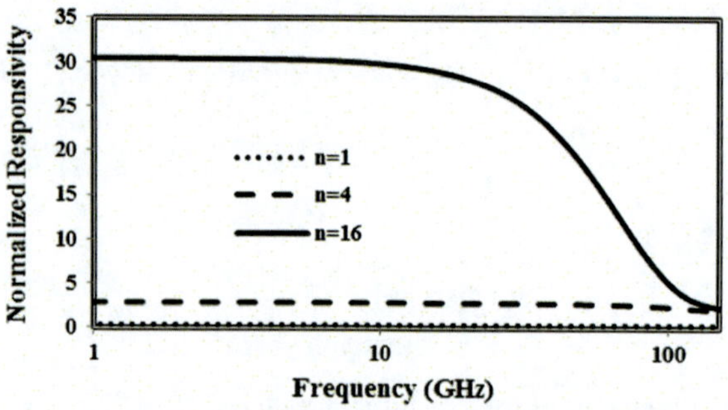

Fig. 1 Normalized responsivity as a function of modulation frequency for the QWIP's with different number of quantum well

4 Conclusion

We have studied the responsivity of $Ge_{0.92}Sn_{0.08}/Si_{0.11}\ Ge_{0.7}\ Sn_{0.19}$ QWIPs using the developed analytical model as functions of the modulation frequency, the number of QWs, and the different capture parameters. The analytical results are more appreciable in our study due to the transit time effects of photoexcited electrons from the QWs. So, it can be concluded that the performance of QWIP can be operated in infrared to mid-infrared region of electromagnetic spectrum.

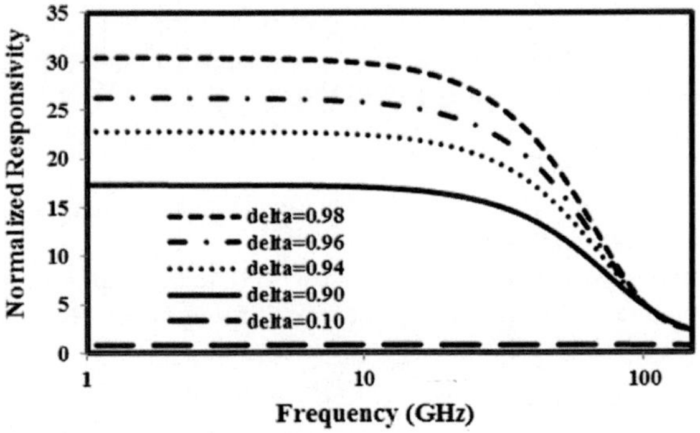

Fig. 2 Normalized responsivity as a function of modulation frequency for the QWIP's with different capture parameter

Acknowledgements. The first author (SG) acknowledges the joint Indo-Taiwan Research project high responsivity GeSn short-wave infrared phototransistor. The second author (SD) acknowledges support by TEQIP-Phase III under University College of Technology-Calcutta University (UCT-CU) through the award of a fellowship.

References

1. Basu, P.K.: Theory of Optical Processes in Semiconductors. Oxford University Press, Oxford, UK (2003)
2. Deen, M.J., Basu, P.K.: Silicon Phonics: Fundamentals and Devices. Wiley, Chichester, UK (2012)
3. Basu, P.K., Mukhopadhyay, B., Basu, R.: Semiconductor Laser Theory. CRC Press, Taylor & Francis Group
4. Bauer, M., Taraci, J., Tolle, J., Chizmeshya, A.V.G., Zollner, S., Smith, D.J., Menendez, J., Hu, C., Kouvetakis, J.: Ge–Sn semiconductors for bandgap and lattice engineering. Appl. Phys. Lett. **81**(1–3), 2992 (2002)
5. D'Costa, V.R., Cook, C.S., Birdwell, A.G., Littler, C.L., Canonico, M., Zollner, S., Kouvetakis, J., Menendez, J.: Optical critical points of thin-film $Ge_{1-y}Sn_y$ alloys: a comparative $Ge_{1-y}Sn_y/Ge_{1-x}Si_x$ study. Phys. Rev. B. **73**(1–16), 125–207 (2006)
6. Menendez, J., Kouvetakis, J.: Type-I Ge/GeSiSn strained layer heterostructures with a direct Ge band gap. Appl. Phys. Lett. **85**, 1175–1178 (2004)
7. Chakraborty, V., Mukhopadhyay, B., Basu, P.K.: Group IV heterojunction laser structure based on S–Ge–Sn–C around 1550 nm: determination of gain coefficient. CODEC 2015. IEEE, Swissotel, Kolkata, Dec 2015. ISBN: 978-1-4673-9511-3
8. Chang, G.E., Chang, S.W., Chuang, S.L.: Strain-balanced Ge_zSn_{1-z}–$Si_xGe_ySn_{1-x-y}$ multiple quantum-well lasers. IEEE J. Quantum Electron. **46**(12), 1813–1820 (2010)
9. Chakraborty, V., Mukhopadhyay, B., Basu, P.K.: Effect of different loss mechanisms in SiGeSn based Mid_Infrared laser. Semiconductors **49**(6), 836–842 (2015). ISSN 1063_7826
10. Levine, B.F.: Quantum-well infrared photodetectors. J. Appl. Phys. **74**, R1 (1993)
11. Ryzhii, V., Khmyrova, I., Ryzhii, M.: Impact of transit-time and capture effects on high-frequency performance of multiple quantum-well infrared photodetectors. IEEE Trans. Electron Devices **45**(1), 293–298 (1998)

12. Ghosh, S., Mukhopadhyay, B., Sen, G., Basu, P.K.: Study of Si–Ge–Sn based heterobipolar phototransistor (HPT) exploiting quantum confined stark effect and Franz Keldysh effect with and without resonant cavity. Phys. E **106**, 62 (2018)
13. Van De Walle, C.G.: Band lineups and deformation potentials in the model-solid theory. Phys. Rev. B **39**, 1871–1883 (1989)
14. Van De Walle, C.G., Martin, R.M.: Theoretical calculations of heterojunction discontinuities in the Si/Ge system. Phys. Rev. B **57**, 6493 (1998)
15. Dey, S., Chakraborty, V., Mukhopadhyay, B., Sen, G.: Modeling of tunneling current density of GeC based double barrier multiple quantum well resonant tunneling diode. J. Semicond. **39**(10), 104003(1–5) (2018)

An Asymmetric π - Gate MOSHEMT Architecture for High Frequency Applications

Khushwant Sehra[1], Vandana Kumari[2], Mridula Gupta[1], Meena Mishra[3], D. S. Rawal[3], and Manoj Saxena[4(✉)]

[1] Department of Electronic Science, Semiconductor Device Research Laboratory, University of Delhi, South Campus, New Delhi, Delhi 110021, India
[2] Department of Electronics, Maharaja Agrasen College, University of Delhi, New Delhi, Delhi 110096, India
[3] Solid State Physics Laboratory, Defence Research and Development Organization, New Delhi, Delhi 110054, India
[4] Department of Electronics, Deen Dayal Upadhyaya College, University of Delhi, New Delhi, Delhi 110078, India

saxenamanoj77@gmail.com

Abstract. This paper evaluates the RF performance of an Asymmetric π - Gate MOSHEMT and HEMT architecture for high frequency applications through extensive TCAD simulations. To ensure that the simulation results conform with the actual device, simulation results have been calibrated with respect to experimental data. The calibrated device then acts as a primer for realizing the π - Gate HEMT and its Asymmetric MOSHEMT architecture. The performance has been investigated by introducing MOSHEMT architecture under both legs of the π - structure. Comparisons demonstrate an improvement in terms of current gain cut - off frequency by 20% (under right leg) — 25% (under left leg) when compared with the original π - Gate HEMT.

Keywords: HEMTs · MOSHEMT · 2DEG · Cut - Off Frequency · Hot Electron Effect · Parasitic Capacitance · Threshold Voltage · Transconductance

1 Introduction

High Electron Mobility Transistors (HEMTs) have emerged as a promising candidate for high power, high frequency applications pertaining to wireless communications and satellite links from past few decades [1]. AlGaN/GaN HEMTs are capable to operate at high voltages and also enable high frequency operation (>100 GHz) [2–4] due to high sheet carrier densities and high electron mobilities in the 2D Electron Gas (2DEG) channel which accounts for high frequency operation [5–7]. Wider bandgap energy coupled with higher breakdown electric fields enables such devices to operate at high supply voltages and to sustain higher orders of electric field strengths [2, 3]. Over the past few years, new architectures such as the MOSHEMTs [8–10], MISHEMTs [11–13],

© Springer Nature Singapore Pte Ltd. 2021
N. R. Das and S. Sarkar (eds.), *Computers and Devices for Communication*,
Lecture Notes in Networks and Systems 147,
https://doi.org/10.1007/978-981-15-8366-7_67

along with novel gate architectures such as the circular [14], Pi(π) [15] are being studied in detail.

In this paper, investigation of Asymmetric π - Gate MOSHEMT architecture for RF performance has been presented and compared with π - Gate HEMT. The resulting architectures are then compared on the basis of device simulation [16] results for Current Gain, Maximum Transducer Gain and Unilateral Power Gain metrics. In Sect. 2, the device structure and simulation models used are described. In Sect. 3, the RF analysis results are presented followed by conclusion in Sect. 4.

2 Device Architecture: π-Gate HEMT

Schematic view of the π - Gate HEMT architecture analyzed in this paper is shown in Fig. 1a. The π - Gate structure is realized by first calibrating the T - Gate HEMT with experimental data [15] through exhaustive TCAD simulations [16]. For device simulations, Shockley - Read - Hall (SRH) recombination along with Albrecht model for low field mobility in GaN has been invoked. To account for velocity saturation at high fields, inbuilt GANSAT model for high field mobility has also been invoked. Interface charges at the AlN/GaN interface and polarization models have been defined to match the sheet carrier density and current levels.

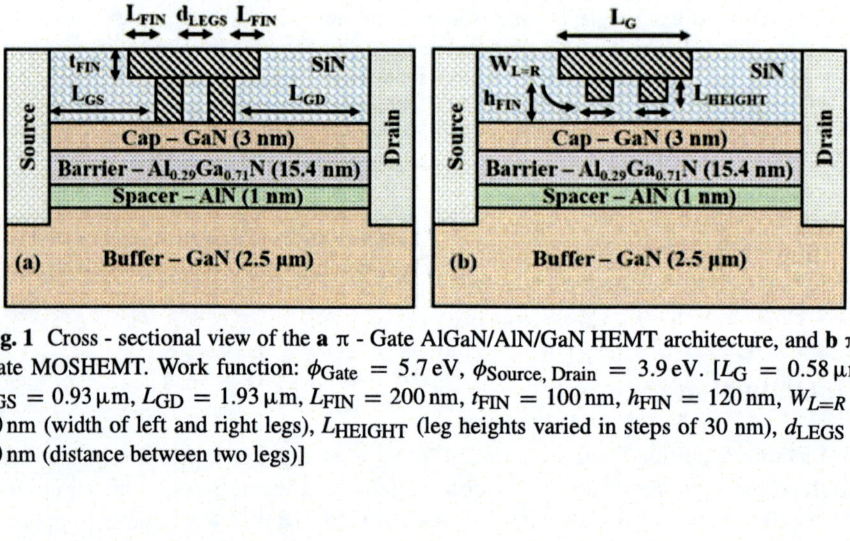

Fig. 1 Cross - sectional view of the **a** π - Gate AlGaN/AlN/GaN HEMT architecture, and **b** π - Gate MOSHEMT. Work function: $\phi_{Gate} = 5.7\,eV$, $\phi_{Source, Drain} = 3.9\,eV$. [$L_G = 0.58\,\mu m$, $L_{GS} = 0.93\,\mu m$, $L_{GD} = 1.93\,\mu m$, $L_{FIN} = 200\,nm$, $t_{FIN} = 100\,nm$, $h_{FIN} = 120\,nm$, $W_{L=R} = 70\,nm$ (width of left and right legs), L_{HEIGHT} (leg heights varied in steps of 30 nm), $d_{LEGS} = 40\,nm$ (distance between two legs)]

3 Results and Discussions

The RF performance of the π - Gate HEMT and MOS-HEMT is shown in Fig. 2a–d. The π - Gate HEMT architecture exhibits a cut - off frequency (f_T) of 64 GHz and maximum frequency of oscillation (f_{max}) of 208 GHz. The cut - off frequency of π - Gate structure gets reduced in comparison to T - Gate (~10%) due to the introduction of fringing capacitance between the legs of π - Gate structure. On introducing the MOS architecture under the left leg of π - structure (L_{MOS}), with $L_{HEIGHT} = 0.5 * h_{FIN} = 60$ nm, an 25% enhancement in f_T is observed. As a consequence of this modification, deterioration of f_{max} by 25% is recorded in the same case. For the MOS architecture under right leg (R_{MOS}), 20% enhancement in f_T is coupled with a 35% deterioration in f_{max}. The enhancement of f_T is directly related to increased drain current ($I_{DS(max)}$) as a result of reduced gate action, as well as reduction in parasitic gate-to-drain (C_{GD}) and gate-to-source (C_{GS}) capacitances as depicted in Fig. 3a, b. Additionally, since f_{max} depends on gate resistance (R_G) and gate charging resistance (R_i, mainly because of electrons captured by traps situated beneath the gate region) [17], the extra MOS architecture manifests itself as increased R_G/R_i, thereby deteriorating the device f_{max}.

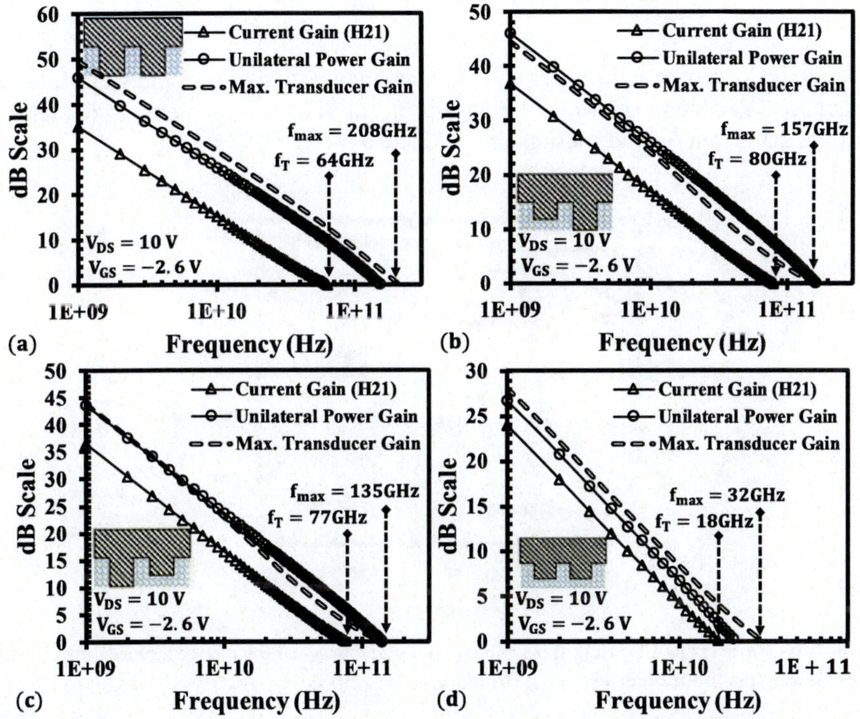

Fig. 2 RF performance of **a** π - Gate, **b** L_{MOS} π - Gate, **c** R_{MOS} π - Gate, and **d** LR_{MOS} π - Gate HEMT at $V_{DS} = 10$ V and $V_{GS} = -2.6$ V. [MOS structure introduced under left leg (L_{MOS}), right leg (R_{MOS}), and both legs (LR_{MOS}). Leg height (L_{HEIGHT}) = 60 nm]

When MOS architecture is introduced under the entire π - structure, the threshold voltage (V_{TH}) of the structure shifts to –18 V, courtesy of the 60 nm nitride layer between the gate electrode and GaN cap layer, that increases the potential barrier at the gate region thereby causing a negative shift in the threshold voltage of the device. The resulting device though exhibits a lower parasitic capacitance C_{GD} and C_{GS}, the negative shift of V_{TH} results in a lower transconductance ($g_{m(max)}$) value at the biasing conditions so taken ($V_{DS} = 10$ V and $V_{GS} = -2.6$ V) for comparison of various device architectures

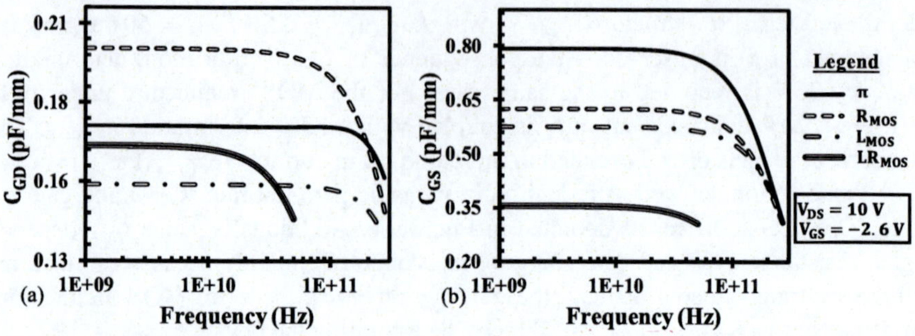

Fig. 3 Comparison of **a** C_{GD} and **b** C_{GS} as a function of frequency for the device architectures under study. Biasing conditions are: $V_{DS} = 10$ V and $V_{GS} = -2.6$ V

under study. Moreover, a device exhibiting a highly negative V_{TH} would serve little to no purpose for RF applications, wherein a higher gate voltage is required just to bias the device at maximum f_T (and consequently increased power dissipation).

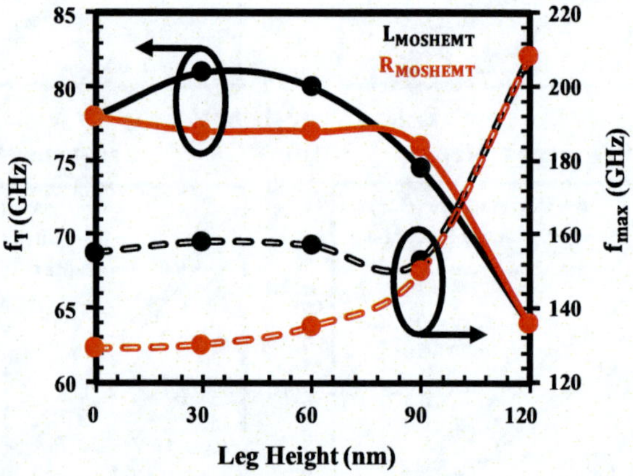

Fig. 4 Effect of π - Gate Leg Height (L_{HEIGHT}) on f_T and f_{max} for the device architectures under study. Biasing conditions are: $V_{DS} = 10$ V and $V_{GS} = -2.6$ V

For optimization purposes, we investigate the impact of π - Gate Leg Heights (L_{HEIGHT}) on the RF performance of the device as shown in Fig. 4. On increasing L_{HEIGHT}, there is a transition of gate architecture from field plated device ($L_{HEIGHT} = 0$ nm) to MOS architecture ($L_{HEIGHT} = 30$ nm), which accounts for the increased f_T and f_{max} due to field plate device exhibiting larger parasitic capacitances (and thus lower f_T) in comparison to the original structure. As the L_{HEIGHT} is being increased, the effective gate resistances R_G/R_i are being reduced which results in the enhancement of f_{max}. At the same time, the maximum drain current ($I_{D(max)}$) gets reduced due to stronger gate coupling with channel, thereby accounting for the reduced f_T.

4 Conclusions

The RF performance of π - Gate HEMT architecture has been investigated in this paper through extensive TCAD simulations and compared with π - Gate MOSHEMT architecture. Asymmetric π - Gate MOSHEMT architecture results in the enhancement of cut - off frequency by 25% due to reduction in parasitic C_{GD} and C_{GS}. The deterioration in f_{max} is because of increased gate access and gate charging resistances (R_G/R_i). The LR_{MOS} architecture has an effect of shifting the V_{TH} to a much higher negative value, thereby losing its suitability for concerned RF applications due to much power being dissipated just to bias the device at maximum f_T. Moreover, since the π - Gate architecture, as reported is capable in reducing hot electron generation, the presented modifications to π - Gate in terms of MOSHEMT validate its suitability for various RF applications.

Acknowledgements. The authors wish to acknowledge DBT Star College Laboratory at Deen Dayal Upadhyaya College, University of Delhi; Semiconductor Device Research Laboratory (SDRL) at Department of Electronic Science, University of Delhi South Campus; and CARS Project No.: 1115/CARS-73/TS/SPL/18 funded by Solid State Physics Laboratory (SSPL), Defence Research and Development Organization (DRDO), Government of India, for providing necessary tools and financial assistance for completion of this work.

References

1. Mimura, T., Hiyamizu, S., Fujii, T., Nanbu, K.: A new field-effect transistor with selectively doped GaAs/n-$Al_xGa_{1-x}As$ heterojunctions. Jpn J Appl Phys **19**(5), L.225–L.227 (1980)
2. Islam, S.S., Anwar, A.F.M.: Design of GaN/AlGaN HEMT class-E power amplifier considering trapping and thermal effects. In: IEEE Lester Eastman Conference on High Performance Devices, pp. 155–163 (2003)
3. Sahoo, D.K., Lal, R.K., Kim, H., Tilak, V., Eastman, L.F.: High-field effects in silicon nitride passivated GaN MODFets. IEEE Trans. Electron Devices **50**(5), 1163–1170 (2003)
4. Nakarmi, M.L., Nepal, N., Lini, J.Y., Jianga, H.X.: Unintentionally doped n-type $Al_{0.67}Ga_{0.33}N$ epilayers. Appl. Phys. Lett. **86**(26), 261902-1-261902–3 (2005)
5. Subramani, N.K., Couvidat, J., Hajjar, A.A., Nallatamby, J.C., Sommet, R., Quéré, R.: Identification of GaN buffer traps in microwave power AlGaN/GaN HEMTs through low frequency S-parameters measurements and TCAD-based physical device simulations. IEEE J. Electron Dev. Soc. **5**(3), 175–181 (2017)
6. Ibbetson, J.P., Fini, P.T., Ness, K.D., DenBaars, S.P., Speck, J.S., Mishra, U.K.: Polarization effects, surface states, and the source of electrons in AlGaN/GaN heterostructure field effect transistors. Appl. Phys. Lett. **77**(2), 250–252 (2000)
7. Fletcher, A.S.A., Nirmal, D.: A survey of gallium nitride HEMT for RF and high power application. Superlattices Microstruct. **109**, 519–537 (2017) (Elsevier)
8. Medjdoub, F., Sarazin, N., Tordjman, M., Magis, M., Forte-Poisson, M.A.D., Knez, M., Delos, E., Gaquiere, C., Delage, S.L., Kohn, E.: Characteristics of Al_2O_3/AlInN/GaN MOSHEMT. IEEE Electron. Lett. **43**(12), 691–692 (2007)
9. Swain, R., Lenka, T.R.: Normally-off $Al_{0.25}Ga_{0.75}N$/GaN MOSHEMT with stack gate dielectric structure. In: IEEE International Conference on Electron Devices and Solid-State Circuits, pp. 567–570 (2015)
10. Pal, P., Pratap, Y., Gupta, M., Kabra, S.: Modeling and simulation of AlGaN/GaN MOS-HEMT for biosensor applications. IEEE Sens. J. **19**(2), 587–593 (2019)

11. Chen, C., Liu, X., Tian, B., Shu, P., Chen, Y., Zhang, W., Jiang, H., Li, Y.: Fabrication of enhancement-mode AlGaN/GaN MISHEMTs by using fluorinated Al_2O_3 as gate dielectrics. IEEE Electron Dev. Lett. **32**(10), 1373–1375 (2011)
12. Zhou, Q., Chen, H., Zhou, C., Feng, Z.H., Cai, S.J., Chen, K.J.: Schottky source/drain InAlN/AlN/GaN MISHEMT with enhanced breakdown voltage. IEEE Electron Dev. Lett. **33**(1), 38–40 (2011)
13. Sun, W., Joh, J., Krishnan, S., Pendharkar, S., Jackson, C.M., Ringel, S.A., Arehart, A.R.: Investigation of trap-induced threshold voltage instability in GaN-on-Si MISHEMTs. IEEE Trans. Electron Devices **66**(2), 890–895 (2019)
14. Cai, Y., Gong, Y., Bai, J., Yu, X., Zhu, C., Esendag, V., Lee, K.B., Wang, T.: Controllable uniform green light emitters enabled by circular HEMT-LED devices. IEEE Photon. J. **10**(5) (2018)
15. Rey, A.D.L., Albrecht, J.D., Saraniti, M.: A π-shaped gate design for reducing hot-electron generation in GaN HEMTs. IEEE Trans. Electron Devices **65**(10), 4263–4270 (2018)
16. Silvaco Atlas TCAD Tool Version 5.24.1.R. User's Manual. Available at: https://www.silvaco.com
17. Chung, J.W., Hoke, W.E., Chumbes, E.M., Palacios, T.: AlGaN/GaN HEMT with 300-GHz f_{max}. IEEE Electron Device Lett. **31**(3), 195–197 (2010)

Gate Leakage Current Assessment of AlGaN/GaN HEMT with AlN Cap Layer

Shreyasi Das[1], Vandana Kumari[2], Mridula Gupta[3], and Manoj Saxena[4(✉)]

[1] Electronics and Communication Engineering, Institute of Radio Physics and Electronics, University of Calcutta, 92, Acharya Prafulla Chandra Road, Kolkata, West Bengal 700009, India
[2] Department of Electronics, Maharaja Agrasen College, University of Delhi, New Delhi, Delhi 110096, India
[3] Department of Electronics Science, University of Delhi, South Campus, New Delhi, Delhi 110021, India
[4] Department of Electronics, Deen Dayal Upadhyaya College, University of Delhi, New Delhi, Delhi 110078, India
saxena_manoj77@yahoo.co.in

Abstract. In this work, TCAD-based investigation has been performed to explore the influence of AlN cap layer on the reliability of AlGaN/GaN HEMT. Effect of cap layer thickness (AlN) and gate length on gate leakage current, drain current and off-current (I_{off}) has also been demonstrated in this study. Using AlN cap layer, higher drain current has been achieved with significantly lower gate leakage current. However, almost same off-state current is achieved by using AlN material instead of GaN-based cap layer. With the enhancement in drain voltage, significant enhancement in gate leakage current is observed which is significantly higher in the device having GaN cap layer. Also, the improvement in gate leakage current with positive gate bias is significantly higher than the negative gate bias.

Keywords: TCAD · HEMT · Gate leakage · AlN cap layer

1 Introduction

AlGaN/GaN-based HEMT shows significant potential for vide variety of applications such as: high frequency and high power because of its unique properties like: wide bandgap, higher breakdown voltage, high electron mobility, etc. [1–3]. For high-power application, device should have high saturation drain current along with the capability to withstand high voltage which is possible with AlGaN/GaN-based HEMT [4]. Various innovative architectures like Field Plate HEMT [5], the Schottky-source/drain [6] has been proposed previously to improve the device operating withstand voltage. However, the gate leakage current is still a major limiting factor in AlGaN/GaN-based HEMT which affect the reliability of the device [7]. To overcome this shortcoming, MOS-HEMT has been proposed in which insulator has been introduced between metal gate and AlGaN barrier [8]. In past few years, AlN has also been introduced in AlGaN/GaN

© Springer Nature Singapore Pte Ltd. 2021
N. R. Das and S. Sarkar (eds.), *Computers and Devices for Communication*, Lecture Notes in Networks and Systems 147,
https://doi.org/10.1007/978-981-15-8366-7_68

HEMT between gate and GaN cap layer to improve the dynamic on-state resistance [9, 10]. AlN has also been used as passivation layer between gate oxide and cap layer to improve the interface quality of AlGaN/GaN HEMT [11]. The thermal behaviour of AlN/GaN/AlGaN HEMT has also been explored previously using Synopsys TCAD based on both SiC and Sapphire substrate [12].

In this work, AlN layer (undoped) is used as cap layer between gate and AlGaN barrier. In order to check the reliability of the HEMT (AlGaN/GaN) architecture, gate leakage current at different gate length has been compared using GaN and AlN cap layer. Variation of cap layer thickness has also been done for estimating drain current and gate leakage current of AlGaN/GaN HEMT. All simulations have been performed using ATLAS TCAD simulation [13] tool at room temperature.

2 Results and Discussion

The schematic cross section of AlGaN/GaN HEMT having AlN cap layer has been plotted in Fig. 1a. For calibrating TCAD models, simulation results of GaN cap-based AlGaN/GaN HEMT have been matched with the previously reported experimental results [14]. Matched simulation and experimental results plotted in Fig. 1b in terms of transfer characteristics validate the various models used for further investigation of device behaviour. For estimating gate leakage current, electron tunnel (e. tunnel) model along with universal Schottky tunnelling (UST) [13] model has been used during simulation.

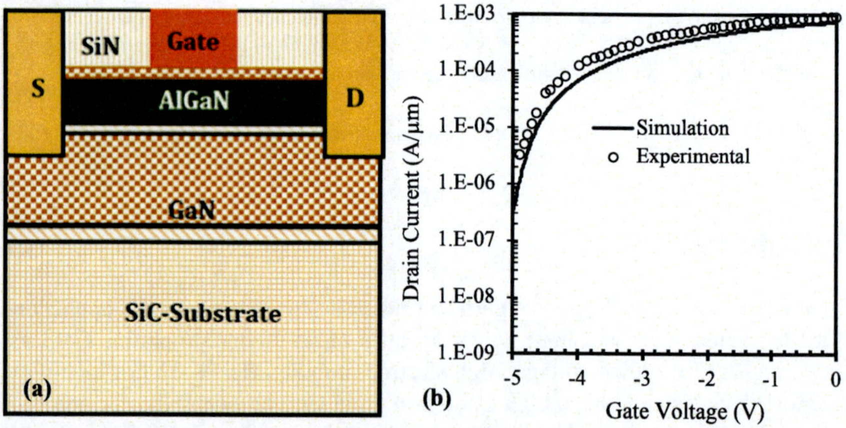

Fig. 1 **a** AlGaN/GaN HEMT architecture with AlN cap layer which is used for simulation. **b** Comparison of the experimental [14] and simulation results for AlGaN/GaN HEMT in terms of current

Figure 2a, b compares the variation of I_{ds} with V_{ds} for AlN and GaN cap layer at two gate length. At lower gate length, significantly higher drain current can be achieved from the device. Maximum drain current has increased from 1.1 to 1.3 mA/μm using AlN Cap layer instead of GaN in AlGaN/GaN HEMT and this improvement is significantly higher at lower channel length, i.e. 0.25 μm device. However, slight variation in I_{off} (off-state current) has been seen by changing the material of cap layer (from GaN to AlN) which results in further improvement of on-to-off ratio (I_{on}/I_{off}) because of higher I_{ds} (on-state current). With the enhancement in cap (AlN) layer thickness, significant improvement in I_{ds} has been observed from Fig. 2a, b. Subsequently, gate leakage current increases which is discussed in detail in next section.

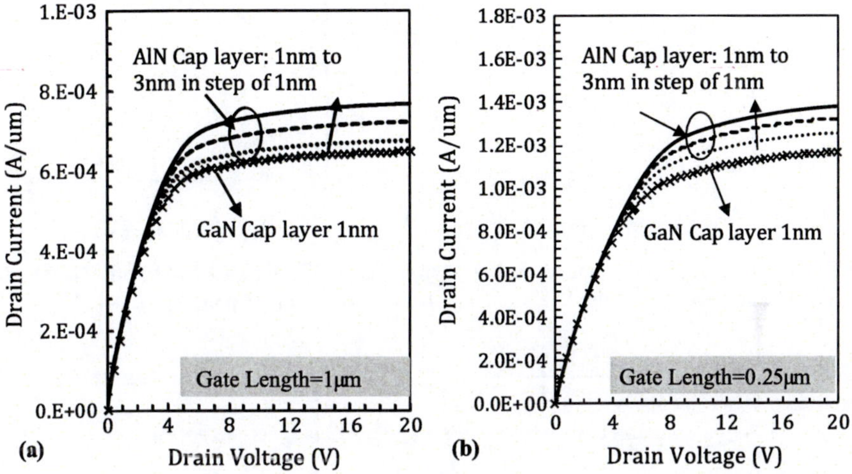

Fig. 2 I_{ds} (Drain current) comparison of AlGaN/GaN HEMT with AlN and GaN cap layer at **a** gate length = 1 μm and **b** gate length = 0.25 μm: $V_{gs} = 1$ V

Figure 3a, b investigates gate current variation with the drain bias for AlN and GaN cap layer at different gate length. With the reduction in gate length, marginal change (reduction) in gate current has been seen with GaN cap layer. However, using AlN cap layer, significant reduction in gate current has been observed from Fig. 3a, b. With the enhancement in AlN cap layer thickness, gate leakage current of the device increases. At 1 μm gate length, the influence of cap layer material is slightly lower than that 0.25 μm gate length. Also, the change in leakage current with AlN cap layer thickness is higher at shorter gate length.

Figure 4a, b compares the change in gate current with V_{gs} for different cap layer material at 1 μm and 0.25 μm gate length, respectively. Using AlN cap layer, significantly improved (lower) gate leakage current has been obtained from the device compared to GaN cap layer. At zero gate voltage, gate leakage current reduces the almost zero due to zero field from the gate terminal. However, enhancement in gate voltage (either negative or positive) leads to enhancement in gate leakage current and this enhancement is higher for positive gate bias voltage. With the enhancement in cap layer thickness, gate leakage current of the device also increases which is significantly higher at lower channel length.

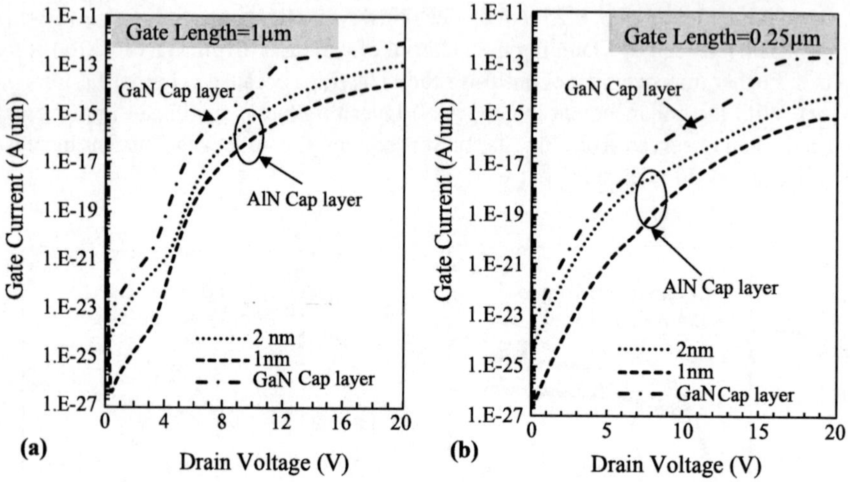

Fig. 3 Gate current (variation with drain bias) variation of AlGaN/GaN HEMT with V_{ds} for AlN and GaN cap layer at **a** gate length = 1 μm and **b** gate length = 0.25 μm: $V_{gs} = -1$ V

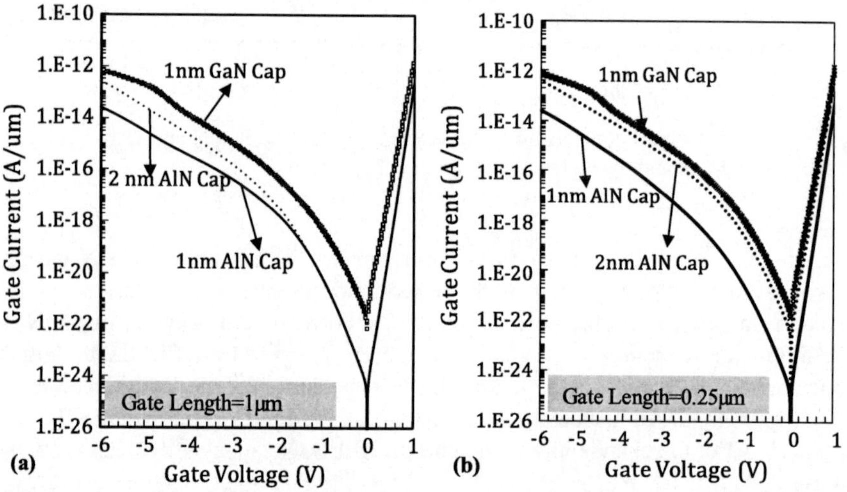

Fig. 4 Plot of gate current for AlGaN/GaN HEMT with V_{gs} (gate voltage) for AlN and GaN cap layer at **a** gate length = 1 μm and **b** gate length = 0.25 μm: at $V_{ds} = 5$ V

3 Conclusion

Present work mainly focused on the investigation of reliability for HEMT (AlGaN/GaN) architecture in terms of gate leakage current. Comparison has been drawn between the AlGaN/GaN HEMT having GaN and AlN-based cap layer. Significant improvement in terms of gate leakage current has been obtained by using AlN cap layer instead of GaN cap layer. From the results, it is observed that the lower gate length results in reduction in gate leakage current and the influence of AlN cap layer thickness is higher at shorter channel length. Instead of reducing gate leakage current, AlN cap layer-based HEMT also shows improved output characteristics in terms of drain current along with negligible change in off-state leakage current (I_{off}) and sub-threshold Slope. Thus, the I_{off}/I_{on} ratio of the device has also been improved by using AlN cap layer instead of GaN cap layer.

Acknowledgements. The authors wish to thank DBT Star College Laboratory at University of Delhi, DDU College, SDR Laboratory at Department of Electronic Science, UDSC; and SSPL-DRDO (for CARS Project No.: 1115/CARS-73/TS/SPL/18) for necessary TCAD tools used for carrying out the work. One of the authors, Shreyasi Das, would like to acknowledge IASc—INSA—NASI ("Science Academies") for providing fellowship under SRFP 2019 vide registration number ENGS3771.

References

1. Tang, Y., Shinohara, K., Regan, D., Corrion, A., Brown, D., Wong, J., Schmitz, A., Fung, H., Kim, S., Micovic, M.: Ultrahigh-speed GaN high electron-mobility transistors with f_T/f_{max} of 454/444 GHz. IEEE Electron Dev. Lett. **36**, 549–551 (2015)
2. Lu, B., Palacios, T.: T High Breakdown (>1500 V) AlGaN/GaN HEMTs by substrate-transfer technology. IEEE Electron Dev. Lett. **31**, 951–953 (2010)
3. Ibbetson, J.P., et al.: Polarization effects, surface states, and the source of electrons in AlGaN/GaN heterostructure field effect transistors. Appl. Phys. Lett. **77**, 250–252 (2000)
4. Saito, W., Omura, I., Ogura, T., Ohashi, H.: Theoretical limit estimation of lateral wide band-gap semiconductor power-switching device. Solid State Electron. **48**, 1555–1562 (2004)
5. Karmalkar, S., Mishra, U.K.: Enhancement of breakdown voltage in AlGaN/GaN high electron mobility transistors using a field plate. IEEE Trans. Electron Devices **48**, 1515–1521 (2001)
6. Zhou, Q., Chen, W., Liu, S., Zhang, B., Feng, Z., Cai, S., Chen, K.J.: Schottky-contact technology in InAlN/GaN HEMTs for breakdown voltage improvement. IEEE Trans. Electron Devices **60**, 1075–1081 (2013)
7. Dutta, G., Das, G.N., Das, G.A.: Gate leakage mechanisms in AlInN/GaN and AlGaN/GaN MIS-HEMTs and its modeling. IEEE Trans. Electron Devices **64**, 3609–3615 (2017)
8. Huang, S., Yang, S., Roberts, J., Chen, K.J.: Threshold voltage instability in Al_2O_3/GaN/AlGaN/GaN metal–insulator–semiconductor high-electron-mobility transistors. Jpn. J. Appl. Phys. **50**, 110202-1-110202–3 (2011)
9. Liu, S., Yang, S., Tang, Z., Jiang, Q., Liu, Q., Wang, M., Chen, K.J.: Al_2O_3/AlN/GaN MOS-channel-HEMTs with an AlN interfacial layer. IEEE Electron Dev. Lett. **35**, 723–725 (2014)
10. Huang, S., Jiang, Q., Yang, S., Zhou, C., Chen, K.J.: Effective passivation of AlGaN/GaN HEMTs by ALD-grown AlN thin film. IEEE Electron Dev. Lett. **33**, 516–518 (2012)
11. Huang, S., Jiang, Q., Yang, S., Tang, Z., Chen, K.J.: Mechanism of PEALD-grown AlN passivation for AlGaN/GaN HEMTs: compensation of interface traps by polarization charges. IEEE Electron Dev. Lett. **34**(2), 193–195 (2013)
12. Sahoo, A.K., Subramani, N.K., Nallatamby, J.-C., Sommet, R., Quéré, R., Rolland, N., Medjdoub, F.: Thermal analysis of AlN/GaN/AlGaN HEMTs grown on Si and SiC substrate through TCAD simulations and measurements. In: Proceedings of the 11th European Microwave Integrated Circuits Conference (2016). https://doi.org/10.1109/EuMIC.2016.7777511

13. Silvaco Atlas TCAD Tool Version 5.26.1.R. User's Manual. Available at: https://www.silvaco.com
14. Madhulikaa, Malik, A., Jain, N., Mishra, M., Kumar, S., Rawal, D.S., Singh, A.K.: Nanoscale structural parameters based analytical model for GaN HEMTs. Superlattices Microstruct. **130**, 267–276 (2019)

A Study on the Optimum Selection of Interpolation Factor for the Design of Narrow Transition Band FIR Filter Using IBM

Subhabrata Roy(✉) and Abhijit Chandra

Department of Instrumentation and Electronics Engineering, Jadavpur University, Kolkata, West Bengal 700106, India
subhabrata_ece@yahoo.com, abhijit922@yahoo.co.in

Abstract. Hardware efficient digital systems have drawn significant attention to the researchers throughout the world over the last few years. In order to achieve this objective, this paper presents the design scheme of sharp cut-off FIR filter using an optimal value of interpolation factor, that produces minimum hardware complexity without affecting the filter performance. Simulation results help to achieve the favourable value of interpolation factor from a set of values. Moreover, the designed filter has subsequently been synthesized using Altera's Cyclone IV FPGA board and the superiority of the design has been established by comparing its hardware cost with the other state-of-the-art design frameworks.

Keywords: FIR filter · Field programmable gate array (FPGA) · Interpolation factor · Narrow transition band

1 Introduction

Recent works on design of hardware efficient digital filters have drawn sufficient attention which aims to incorporate minimum hardware elements during its implementation and hence consumes less amount of power [1–4]. One of the most efficient ways of synthesizing linear-phase narrow transition width digital FIR filters with least computational complexity is frequency response masking (FRM) approach [5, 6]. A generalized two-stage FRM filter was proposed in [7] without any constraint on the subfilters or the interpolation factors. In [6], a brief review of FRM technique is discussed and a reconfigurable FRM approach is derived for the sake of realizing of maximum number of channel from similar model filter. The author in [5] estimated the filter complexities for each of the interpolation factors considered in the study and made a choice for the one which gives lower estimate. Chen et al. in [8] developed a target function in terms of interpolation factor, M and pass-band and stop-band edges, ω_p and ω_s of the resultant sharp cut-off FIR filter for minimizing the complexity measure of the filter. However, the authors neither developed any optimization framework for complexity measure nor solved the interpolation factor, pass-band and stop-band edges analytically that would

© Springer Nature Singapore Pte Ltd. 2021
N. R. Das and S. Sarkar (eds.), *Computers and Devices for Communication*,
Lecture Notes in Networks and Systems 147,
https://doi.org/10.1007/978-981-15-8366-7_69

cause minimization on the overall hardware complexity. The authors in [9] defined an objective function in terms of transition width of the overall narrow transition-band FIR filter along with the other subfilters to minimize the cost function, $C(\Delta)$. However, the authors neither solved any optimization problem to formulate any closed form expression of the interpolation factor, M nor implemented the designed FIR filter on hardware chip to achieve a significant saving in terms of multipliers count and accomplish inferior group delay. The solution obtained in complexity minimization problem addressed in [9] is sub-optimal. In [10], a novel two-stage FRM structure has been proposed with the help of a band-edge shaping filter having a non-periodical frequency response to achieve minimum complexity in terms of number of multipliers compared to the so called FRM design methods like interpolated finite impulse response frequency response masking (IFIR-FRM) [11], serial-masking based frequency response masking [12], etc. Recently, an interpolated bandpass method (IBM) has been introduced so as to design narrow transition width low-pass FIR filter [13] in which a fixed value of interpolation factor was considered in the entire study.

Motivated by the techniques as reported in [5, 8, 9], our proposition aims to select an optimal value of interpolation factor, which helps to generate a narrow transition-band FIR filter without compromising the hardware constraints. We have also extended our study towards the analysis of hardware cost of the IBM based FIR filter resulting from our design by implementing it on Altera's Cyclone IV FPGA chip with the aid of hardware description language (HDL). Power consumption of the resultant interpolated bandpass method based FIR filter has also been examined. The main contribution of our proposition can be outlined as below:

- An attempt has been made for the selection of most favourable interpolation factor, which essentially minimizes the hardware complexity of the designed FIR filter and that too with maintaining the transition bandwidth, pass-band ripple and stop-band attenuation at desired level.
- The proposed work is then extended with hardware implementation of the resultant FIR filter on Altera's Cyclone IV FPGA chip. Thus, finite word-length concept in the presence of powers-of-two representation is also considered while minimizing the requirement of explicit multipliers and thereby constructing the overall filter representation less complex as well as cost effective.

The rest of the paper is organized as follows: Sect. 2 represents the design formulation. Numerical results with detailed discussion on the selection of interpolation factor are presented in Sect. 3. Section 4 describes the detailed steps to implement the FIR filter in FPGA chip. Finally, the paper is concluded in Sect. 5.

2 Design Formulation

Let us consider the bandpass filter (BPF), $H_{bp}(z)$, as depicted in Fig. 1c with pass-band edges specified at ϕ_1 and ϕ_2, and stop-band edges specified at θ_1 and θ_2, respectively, are formulated in between the edge frequencies of $H_a(z)$ and $H_c(z)$ as displayed in Fig. 1a, b respectively. Figure 1d displays complementary BPF, $H_{cbp}(z)$ which is drafted as below:

$$H_{cbp}(z) = z^{\frac{-(N_b-1)}{2}} - H_{bp}(z) \tag{1}$$

where N_b represents length of BPF. Frequency response of the overall filter as shown in Fig. 1g is thus given by

$$H(z) = H_{bp}\left(z^M\right)H_{ma}(z) + \left[z^{-M(N_b-1)/2} - H_{bp}\left(z^M\right)\right]H_{mc}(z) \tag{2}$$

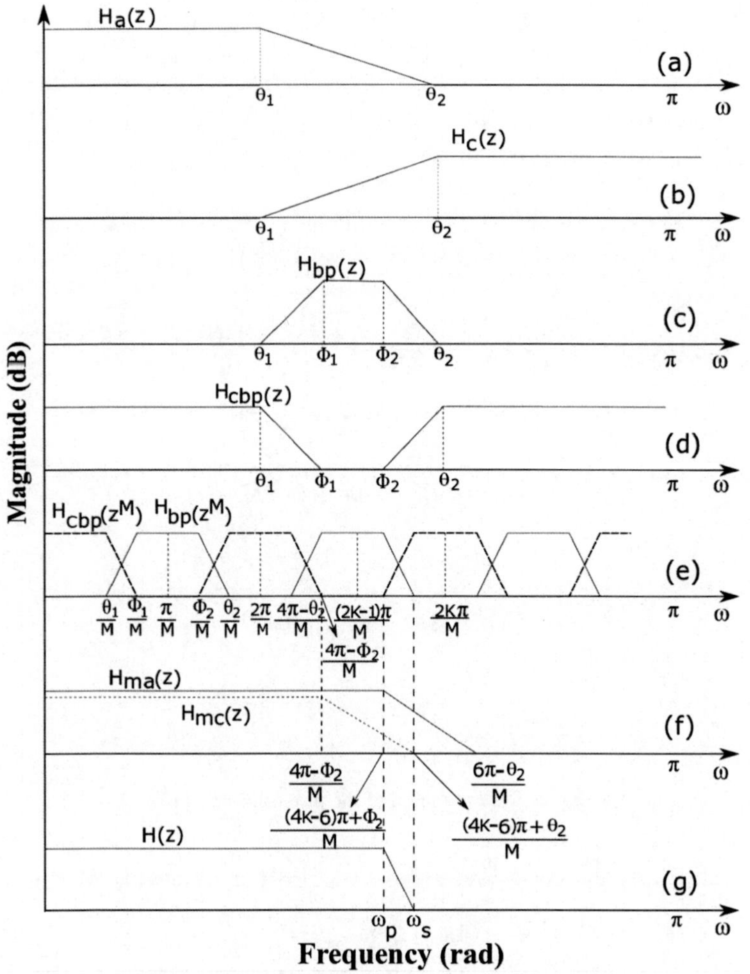

Fig. 1 Frequency response of different subfilters

Complexity of resultant filter, $H(z)$ essentially depends on the length of the subfilters. For the filter arrangement displayed in Fig. 1, it can be concluded that complexity must be a function of interpolation factor, M. Therefore, lengths of the three subfilters denote the total number of multiplication count. As a matter of that, total sum of the length of these subfilters can be used to measure the computational complexity, $C(M)$. Length of these subfilters can be calculated by observing the corresponding transition bandwidths as displayed in Fig. 1. Therefore, measure of complexity can be written as

$$C(M) \propto \left(\frac{1}{\Delta_{bp}}+\frac{1}{\Delta_{ma}}+\frac{1}{\Delta_{mc}}\right)=\beta\ \frac{1}{\Delta_{bp}}+\frac{1}{\Delta_{ma}}+\frac{1}{\Delta_{mc}}\Big\} \tag{3}$$

where β is a proportionality constant which essentially relies upon pass-band ripple and stop-band attenuation of the concerned filter [14]. Δ_{bp}, Δ_{ma} and Δ_{mc} represent the bandwidths of $H_{bp}(z)$, $H_{ma}(z)$ and $H_{mc}(z)$ respectively. Let θ_1, θ_2 and ϕ_1, ϕ_2 signify two stop-band and two pass-band edge frequencies of $H_{bp}(z)$, respectively. Looking at Fig. 1c, f, it can be concluded that total width of $H_{bp}(z)$ and transition width of $H_{ma}(z)$ and $H_{mc}(z)$ are determined to be $(\theta_2-\theta_1)$, $\left[\frac{4\pi(3-K)-(\theta_2+\phi_2)}{M}\right]$ and $\left[\frac{2\pi(2K-5)+(\theta_2+\phi_2)}{M}\right]$ respectively. Hence, complexity measure of $H(z)$ can be rewritten as

$$\begin{aligned} C(M) = \{1/\{(\theta_2-\theta_1)\} &+ M/\{4\pi(3-K)-(\theta_2+\phi_2)\} \\ &+ M/\{2\pi(2K-5)+(\theta_2+\phi_2)\}\} \end{aligned} \tag{4}$$

where $K=\left\lfloor \frac{\omega_p M}{2\pi} \right\rfloor$ signifies the total count of pass-band edge in the interpolated filter response. Moreover, ϕ_2 and θ_2 can be obtained from the following set of equations:

$$\omega_p = \{(4K-6)\pi+\phi_2\}/M \tag{5}$$

$$\omega_s = \{(4K-6)\pi+\theta_2\}/M \tag{6}$$

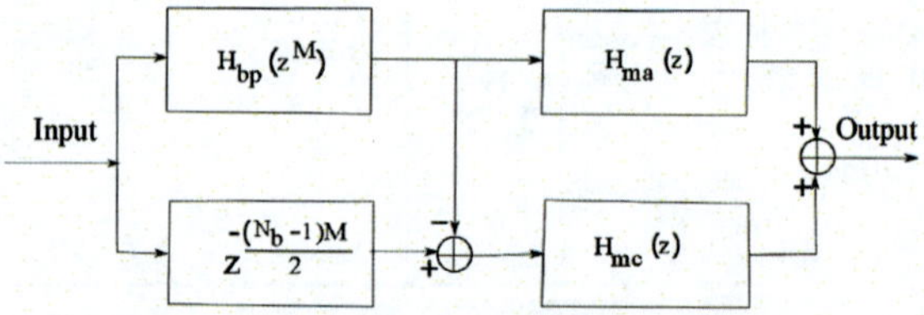

Fig. 2 System model of the proposed design

Substituting the values of ϕ_2 and θ_2 from Eqs. (5) and (6) into (4), we get

$$\begin{aligned} C(M) = \{1/\{\omega_s M-(4K-6)\pi-\theta_1\} \\ +M/\{4\pi(3-K)-(\omega_s+\omega_p)M-4\pi(2K-3)\} \\ +M/\ 2\pi(2K-5)+(\omega_s+\omega_p)M-4\pi(2K-3)\}\} \end{aligned} \tag{7}$$

In the literature, there is no well-known scientific expression of interpolation factor, M. However, a good choice of M can be investigated by predicting the filter complexity for each and every value of M maintaining the narrow-band performance of the filter and then selecting the value of M which corresponds to the lowest cost. We define a cost function $\kappa(M)$, to measure the overall complexity of the designed FIR filter, which can be drafted as below:

$$\kappa(M) = \mathrm{e}^{w_1 C(M)} + \mathrm{e}^{w_2 \Delta_{H(z)}} \tag{8}$$

where $\mathrm{e}^{C(M)}$ and $\mathrm{e}^{\Delta_{H(z)}}$ signify the dependency of hardware complexity and transition bandwidth on M respectively and w_1 and w_2 denotes the corresponding weight value assigned to each of the filters.

3 Results and Discussions

We will demonstrate this technique by considering the construction of a low-pass FIR filter for the filter specification as given below: edge frequencies of the low-pass model filter are specified at 0.3 rad/π and 0.7 rad/π respectively whereas $\theta_1 = 0.3$ rad/π, $\phi_1 = 0.4\,\text{rad}/\pi$, $\phi_2 = 0.6\,\text{rad}/\pi$ and $\theta_2 = 0.7$ rad/π with pass-band ripple (δ_p) and stop-band attenuation (δ_s) are considered as ± 0.01 dB and 60 dB respectively.

Figures 3 and 4 have outlined the variation of hardware complexity, $C(M)$ and transition bandwidth, $\Delta_{H(z)}$ respectively with interpolation factor M. From Fig. 3, it can be unambiguously observed that with the increase of M complexity of the designed filter, $C(M)$ increases. It is quite obvious because with the increase of M, length of both the masking filters, N_{ma} and N_{mc}, increases which in turn enhances the arithmetic complexity. On the other hand, it can be seen from Fig. 4 that transition bandwidth, $\Delta_{H(z)}$, decreases gradually with the increase of M. This can be validated in the sense that as M is increased from its initial value, length of both the masking filters, N_{ma} and N_{mc}, increases which essentially decreases the transition width of H_{ma} and H_{mc}, respectively and hence, the $\Delta_{H(z)}$ value decreases.

Based on the observations depicted in Figs. 3 and 4, let us consider three distinct cases such as

Case I: $w_1 = w_2$ (same priority to both transition bandwidth and measure of complexity)
Case II: $w_1 > w_2$ (priority of measure of complexity over transition bandwidth)
Case III: $w_1 < w_2$ (priority of transition bandwidth over measure of complexity) (Tables 1, 2, 3 and 4).

where $\tau(\mathrm{e}^{w_1 C(M)})$ and $\tau(\mathrm{e}^{w_2 \Delta_{H(z)}})$ signify the normalized value of $\mathrm{e}^{w_1 C(M)}$ and $\mathrm{e}^{w_2 \Delta_{H(z)}}$ respectively. As an attempt to select an optimum value for the interpolation factor M, proper trade-off has been carried out between the measure of complexity and the transition bandwidth. Experimental results have shown that $M = 4$ is a good choice when both the constraints of filter design are given equal importance. In order to validate the superiority of the proposed work, resultant filter response $H(z)$ is demonstrated in Fig. 5 using different values of M and that too with filter length of 50, 60 and 70 respectively.

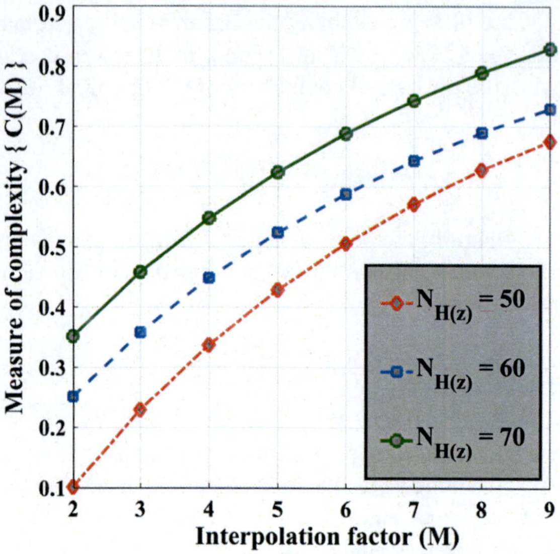

Fig. 3 Measure of complexity $\{C\ (M)\}$ versus M

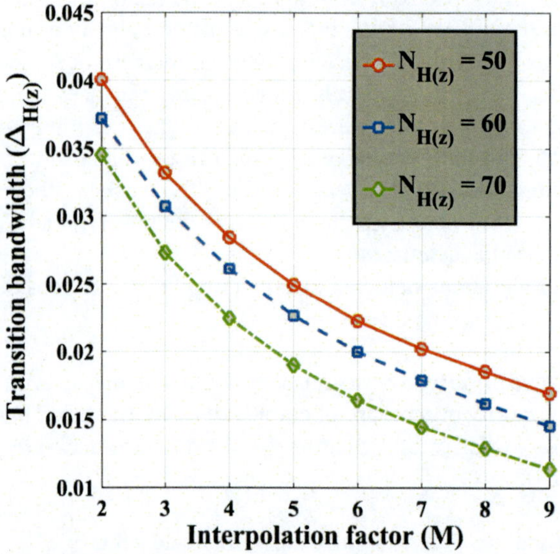

Fig. 4 Measure of transition bandwidth $\{\Delta_{H(z)}\}$ versus M

4 FPGA Implementation

This section illustrates the successive procedures towards the implementation of designed FIR filter on redesignable hardware setup. We start our synthesis cycle using the designed narrow-band FIR filter with double-precision floating-point coefficients. The synthesis

Table 1 Simulation parameter values

Case	Parameter	Values
Case I	w_1	1
	w_2	1
Case II	w_1	2
	w_2	1
Case III	w_1	1
	w_2	2

Table 2 Parameter values of $\tau(e^{w_1 C(M)})$ and $\tau(e^{w_2 \Delta_{H(z)}})$ for case I

Interpolation factor (M)	$\tau(e^{w_1 C(M)})$	$\tau(e^{w_2 \Delta_{H(z)}})$	$\kappa(M)$
2	0	1	1
3	0.176	0.6792	0.8732
4	**0.3428**	**0.4914**	**0.8342**
5	0.4985	0.3444	0.8429
6	0.6422	0.231	0.8732
7	0.7734	0.1386	0.912
8	0.8925	0.0672	0.9597
9	1	0	1

Bold indicates that M=4 is a good choice for the selection of optimal interpolation factor

tool applied in this proposition is Altera's Quartus II and the simulation is achieved using Altera Modelsim with VHDL test bench platform. Modelsim result of the proposed design is displayed in Fig. 6.

Total cost pertaining to the design of the FIR filter is investigated in terms of the total logic elements, combinational functions, registers, I/O pins and so on by recognizing it on Altera's Cyclone IV EP4CE15F17C8 FPGA chip. A correlative investigation amongst a set of narrow-band FIR filter with respect to synthesis results has been depicted in Table 5. The entries of Table 5 reveal the supremacy of the proposed filter arrangement over the other state-of-the-art design methods. It can be seen that proposed filter arrangement requires 5631 logic elements (LEs), corresponding to 37% of the FPGA capacity, 4521 combinational functions (29%), 45 pins (27%) and 1453 registers. From Table 5, it can be evaluated that there is a fluctuation of I/O thermal power dissipation while core static thermal power dissipation maintains its stability. Hence, the requirement of total hardware elements and their power dissipation with respect to the other filtering techniques as reported in [15–17] is summarized in Table 5 while implementing on Cyclone IV FPGA chip.

Table 3 Parameter values of $\tau(e^{w_1 C(M)})$ and $\tau(e^{w_2 \Delta_{H(z)}})$ for case II

Interpolation factor (M)	$\tau(e^{w_1 C(M)})$	$\tau(e^{w_2 \Delta_{H(z)}})$	$\kappa(M)$
2	0	1	1
3	0.1354	0.6972	0.8326
4	**0.2798**	**0.4914**	**0.7712**
5	0.4285	0.3444	0.7729
6	0.5779	0.231	0.8089
7	0.7244	0.1386	0.8630
8	0.8657	0.0672	0.9329
9	1	0	1

Bold indicates that M=4 is a good choice for the selection of optimal interpolation factor

Table 4 Parameter values of $\tau(e^{w_1 C(M)})$ and $\tau(e^{w_2 \Delta_{H(z)}})$ for case III

Interpolation factor (M)	$\tau(e^{w_1 C(M)})$	$\tau(e^{w_2 \Delta_{H(z)}})$	$\kappa(M)$
2	0	1	1
3	0.176	0.6967	0.8727
4	**0.3428**	**0.4898**	**0.8326**
5	0.4985	0.3381	0.8366
6	0.6422	0.2254	0.8676
7	0.7734	0.1373	0.9107
8	0.8925	0.0656	0.9581
9	1	0	1

Bold indicates that M=4 is a good choice for the selection of optimal interpolation factor

5 Conclusion and Scope of Future Work

In this proposition, selection of optimal value of interpolation factor M is investigated for the framing of linear-phase, narrow transition band, low-pass FIR filter with reduced hardware complexity. In addition to that, the designed FIR filter is synthesized in Altera's Cyclone IV FPGA board in order to analyse its performance on reconfigurable hardware setup. Power consumption resulted from the proposed design has also been inspected using power estimator tool.

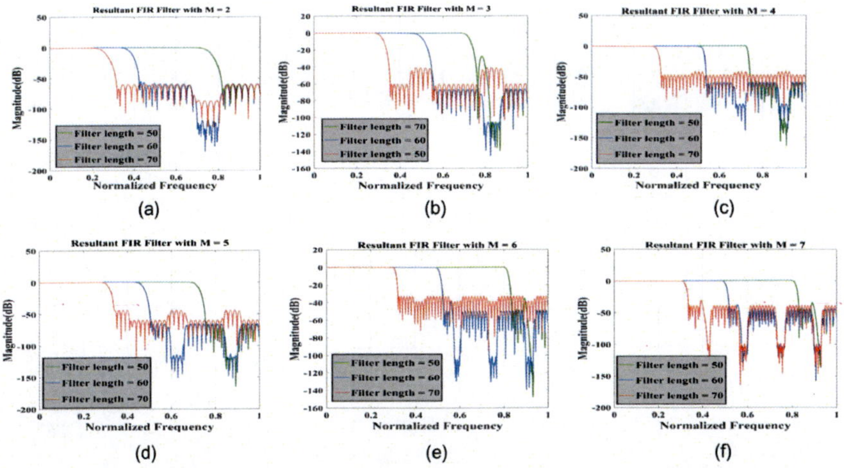

Fig. 5 Frequency response of the designed FIR filter $H(z)$ with different values of M

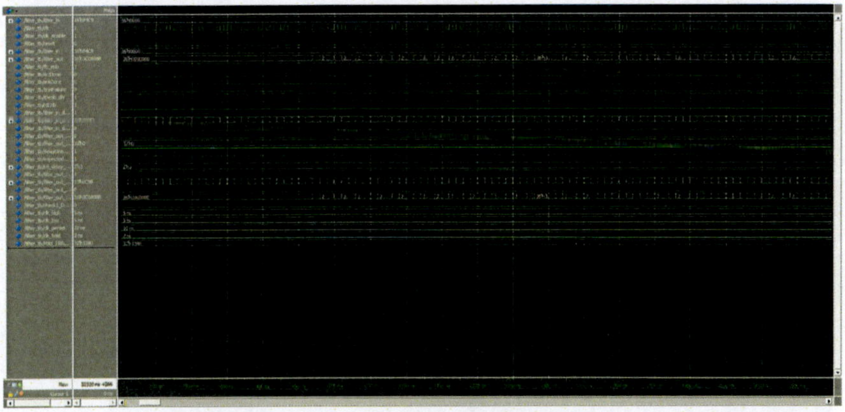

Fig. 6 Modelsim result of the proposed design

Table 5 Synthesis results for designed narrow-band FIR filter

Filter length	Method	Total logic elements	Total combinational functions	Total registers	Total pins	Core static thermal power dissipation (mW)	I/O thermal power dissipation (mW)	Total thermal power dissipation (mW)
50	Alam and Gustafsson [15]	6442	5940	1999	46	49.03	14.66	63.70
	Dhabu and Vinod [16]	6307	5913	1889	46	49.03	14.66	63.69
	Lehto et al. [17]	12,069	11,040	3584	47	49.11	14.72	63.83
	Proposed method	5631	4521	1453	45	46.71	11.72	58.43

Acknowledgements. This research work is funded by Science and Engineering Research Board (SERB), Department of Science and Technology, Government of India, vide sanction order no. ECR/2017/000440.

References

1. Lim, Y., Parker, S.: FIR filter design over a discrete powers-of-two coefficient space. IEEE Trans. Acoust. Speech Signal Process. **31**(3), 583–591 (1983)
2. Feng, Z.G., Teo, K.L.: A discrete filled function method for the design of FIR filters with signed-powers-of-two coefficients. IEEE Trans. Signal Process. **56**, 134–139 (2008)
3. Ito, R., Fujie, T., Suyama, K., Hirabayashi, R.: A powers-of-two term allocation algorithm for designing FIR filters with CSD coefficients in a min-max sense. In: 2004 12th European Signal Processing Conference, Sept 2004, pp. 987–990
4. Roy, S., Chandra, A.: A new design strategy of sharp cut-off FIR filter with powers-of-two coefficients. In: 2018 International Conference on Wireless Communications, Signal Processing and Networking (WiSPNET), Mar 2018, pp. 1–6
5. Lim, Y.: Frequency-response masking approach for the synthesis of sharp linear phase digital filters. IEEE Trans. Circ. Syst. **33**, 357–364 (1986)
6. Haridas, N., Elias, E.: Reconfigurable farrow structure-based FRM filters for wireless communication systems. Circ. Syst. Signal Process. **36**(1), 315–338 (2017)
7. Wei, Y., Huang, S., Ma, X.: A novel approach to design low-cost two-stage frequency-response masking filters. IEEE Trans. Circuits Syst. II Express Briefs **62**(10), 982–986 (2015)
8. Chen, C.-K., Lee, J.-H.: Design of sharp-cutoff FIR digital filters with prescribed constant group delay. IEEE Trans. Circuits Syst. II Analog Digital Signal Process. **43**(1), 1–13 (1996)
9. Yang, C.Z., Lian, Y.: Reduce the complexity of frequency-response masking filter using multiplication free filter. In: Proceedings of the 2003 International Symposium on Circuits and Systems, 2003. ISCAS'03, vol. 4, p. IV. IEEE (2003)
10. Wei, Y., Liu, D.: Improved design of frequency-response masking filters using band-edge shaping filter with non-periodical frequency response. IEEE Trans. Signal Process. **61**(13), 3269–3278 (2013)

11. Lian, Y., Zhang, L., Ko, C.C.: An improved frequency response masking approach for designing sharp FIR filters. Sig. Process. **81**(12), 2573–2581 (2001)
12. Wei, Y., Lian, Y.: Frequency-response masking filters based on serial masking schemes. Circ. Syst. Signal Process. **29**(1), 7–24 (2010)
13. Roy, S., Chandra, A.: Design of narrow transition band digital filter: an analytical approach. Integration **68**, 38–49 (2019)
14. Roy, S., Chandra, A.: On the order minimization of interpolated bandpass method based narrow transition band FIR filter design. IEEE Trans. Circuits Syst. I Regul. Pap. **66**, 4287–4295 (2019)
15. Alam, S.A., Gustafsson, O.: Implementation of narrow-band frequency-response masking for efficient narrow transition band FIR filters on FPGAs. In: NORCHIP, 2011, pp. 1–4. IEEE (2011)
16. Dhabu, S., Vinod, A.P.: Design and FPGA implementation of reconfigurable linear-phase digital filter with wide cutoff frequency range and narrow transition bandwidth. IEEE Trans. Circuits Syst. II Express Briefs **63**(2), 181–185 (2016)
17. Lehto, R., Taurén, T., Vainio, O.: Recursive FIR filter structures on FPGA. Microprocess. Microsyst. **35**(7), 595–602 (2011)

Design of Dynamic Threshold OTA-Based Transconductance-Capacitance Loop Filter for PLL Applications

Priti Gupta(✉) and Sanjay Kumar Jana

Department of Electronics and Communication Engineering, National Institute of Technology Sikkim, Sikkim, India
pritigupta@nitsikkim.ac.in

Abstract. With the increasing demand of low-power integrated circuit for the RF transreceiver, it is always desirable to design the low-power phase lock loop. Nowadays, the demand of ultra-low-power PLL is increasing rapidly. Loop filter is one of the power consuming building block that takes large amount of power. This paper deals with the designing of the low-power transconductance-capacitance based loop filter with the help of dynamic threshold MOS technique. The simulation results show transconductance-capacitance loop filter operating at −3 dB frequency of 39.9 MHz with the power consumption of 252.78 μW along with the supply voltage of 1 V.

Keywords: Dynamic threshold MOS · Phase lock loop · Transconductance-capacitance based filter · Current buffer compensation

1 Introduction

The recent trends in the VLSI technology deals with the high-speed, low-power, and less-area integrated circuit design [1–9]. With the decreasing size of modern CMOS processes, it is always required to decrease the supply voltage for the integrated circuit. Modern communication demands high-speed performance in data transmission with low power consumption, low noise, small size, and also low cost [2]. Phase lock loop (PLL) is the one of the important building blocks of the any transreceiver [1–3]. It generates frequency for the transreceiver. PLL is used in various applications such as clock generation, frequency synthesizer, and clock data recovery (CDR) in a serial data link [3–7]. There are four major functional blocks in PLL which are phase frequency detector, loop filter, voltage-controlled oscillator, and frequency divider as shown in Fig. 1. Loop filter is the one of the basic building blocks of the phase look loop. It plays vital role in determining PLL operating characteristics including PLL loop bandwidth, locking time, and output phase noise [8]. It provides the tuning voltage for the VCO [9–12].

It is also required to design the G_m-C-based loop filter to provide automatic tuning voltage to the VCO input that required to maintain stability of the filter with the help

© Springer Nature Singapore Pte Ltd. 2021
N. R. Das and S. Sarkar (eds.), *Computers and Devices for Communication*,
Lecture Notes in Networks and Systems 147,
https://doi.org/10.1007/978-981-15-8366-7_70

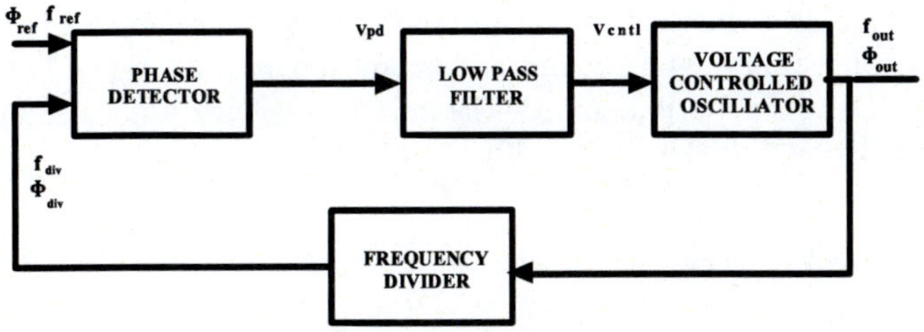

Fig. 1 PLL block diagram

of optimized operational transconductance amplifier (OTA). The OTA is preferred as a basic building block of the G_m-C filter because of automatic tuning facility of G_m (transconductance) with the help of bias current. The symbolic representation of the OTA is as shown in Fig. 2. The output of the OTA is given as

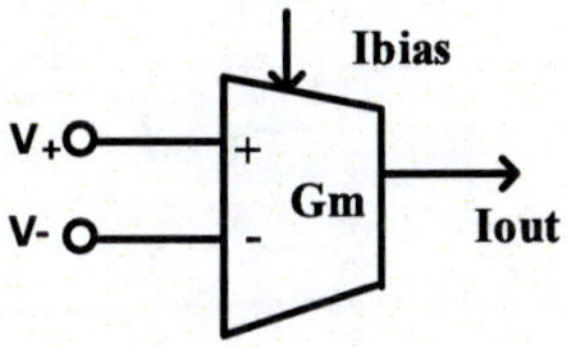

Fig. 2 OTA basic symbol

$$I_{out} = G_m(V_+ - V_-) \tag{1}$$

where I_{out} is the output current, V_+ and V_- is the voltage of the non-inverting and inverting amplifier. To avoid the stability issue in the close loop OTA, it is required to design OTAs with the frequency compensation techniques [13, 14]. So, there are many compensation strategies to develop the optimized OTA [15–19]. Current buffer (CB) approach is one of the better compensation approach that preserves the output swing [14, 15].

2 Dynamic Threshold MOS for Low-Power Design

To fulfill the demand of the low-power integrated circuit, there are many low-power design approaches are available such as bulk-driven technique [19], dynamic threshold MOS technique [20], quasi floating gate [19], and quasi floating bulk techniques [19]. Dynamic threshold (DT) technique is used to increase the gain and unity gain bandwidth of the operational transconductance amplifier [20]. Dynamic threshold MOS technique is that in which body (bulk) terminal is connected to the gate terminal that is a promising method for achieving enhanced performance without even modifying the existing structure of MOSFET. DT MOS-based OTA is designed that is basically used to improve the DC gain and transconductance of the overall circuit [19, 20]. This is to reduce the leakage current during off state and reducing the threshold voltage during on state to increase the overdrive voltage for digital circuits. For analog applications, the body terminal of the MOSFET acts as the fourth terminal. The threshold voltage is reduced by factor of 25% by manipulating the value of V_{bs} [20].

3 Design and Analysis

The proposed DTMOS based OTA as shown in Fig. 3 is simulated using 180 nm SCL technology with the help of cadence virtuoso IC 616. DTMOS OTA has reduced the power consumption of the circuit and improved the gain of the circuit.

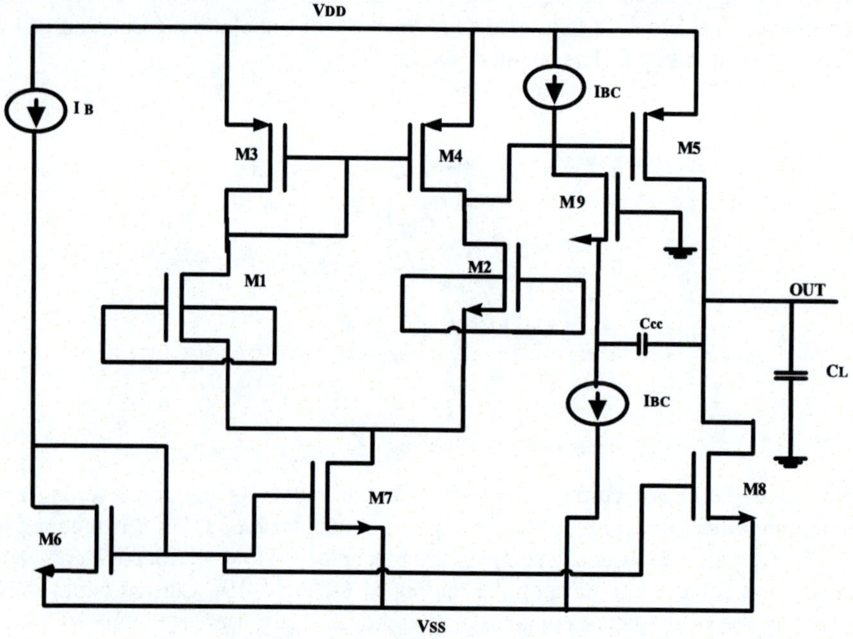

Fig. 3 Design of current buffer compensated DTMOS OTA

Frequency response of the DTMOS OTA shows the magnitude response and phase response is shown in Figs. 4 and 5. that provides the −3 dB cut of frequency is approximately 44.40 MHz and DC gain is approximately 71.11 dB. The gain cross-over frequency (w_{gc}) and phase cross-over frequency (w_{pc}) are 34.96 and 51.37 MHz that described that the filter is stable. From the Magnitude response and phase response graph, the phase margin is measured approximately 31°.

Table 1 shows the specifications of DT MOS-based current buffer compensated OTA, which is the basic building block that can be frequently used for the G_m-C filter design. DTMOS-based approach has improved the gain of the OTA circuit along with the reduction of the power consumption that can be used in the low-power applications.

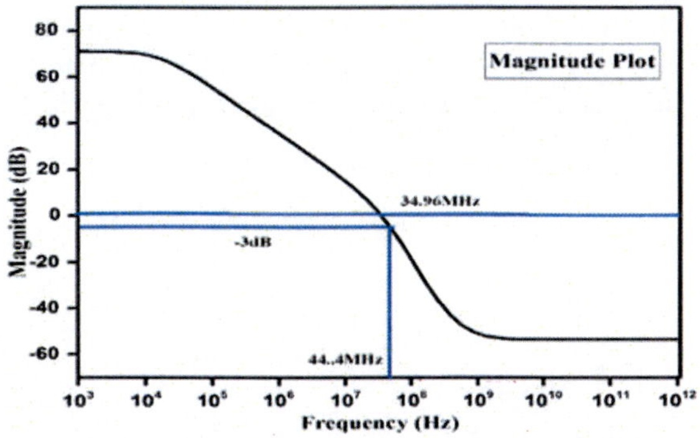

Fig. 4 DT MOS OTA magnitude response with frequency variation

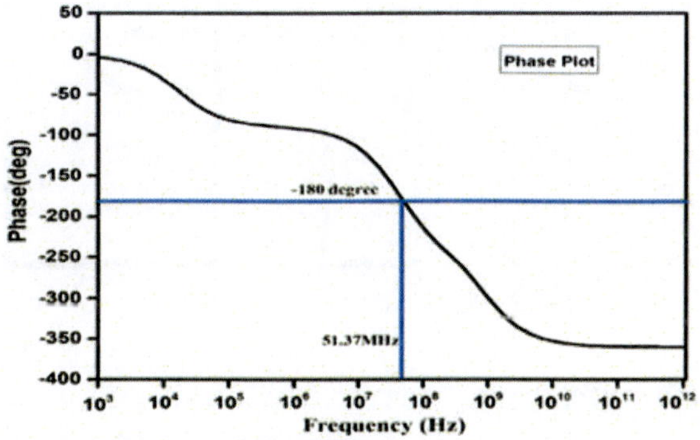

Fig. 5 DT MOS OTA phase response with frequency variation

Table 1 DT MOS-based OTA specifications

S. No.	Parameter	Dynamic threshold MOS-based OTA	[21]	[22]
1	Technology (μm)	0.18	0.18	0.18
2	Supply voltage (V)	1	0.5	1.8
3	UGB (MHz)	39.514	2.5	47.5
4	DC gain (dB)	71.11	52	72
5	Phase margin (degree)	20	65	50
6	Gain margin (dB)	5.901	–	85.5
7	Power consumption (mW)	0.128	0.110	11.9

4 Proposed DTMOS G_m-C Filter Design

The architecture of second-order transconductance -capacitor (G_m-C) voltage mode multifunction filter is shown in Fig. 6. The two OTA used are labelled as G_{m1} and

G_{m2} along with two external capacitors C_1 and C_2 to realize low pass (LP), band pass (BP), and high pass (HP) filters. When $V_{inBP} = V_{inHP} = 0$, the architecture performs second-order low pass filtering. The transfer function is given as

$$\frac{V_{out}}{V_{in}} = \frac{G_{m1}/C_1C_2}{s^2 + \left(\frac{G_{m2}}{C_2}\right)s + (G_{m1}G_{m1}/C_1C_2)} \tag{2}$$

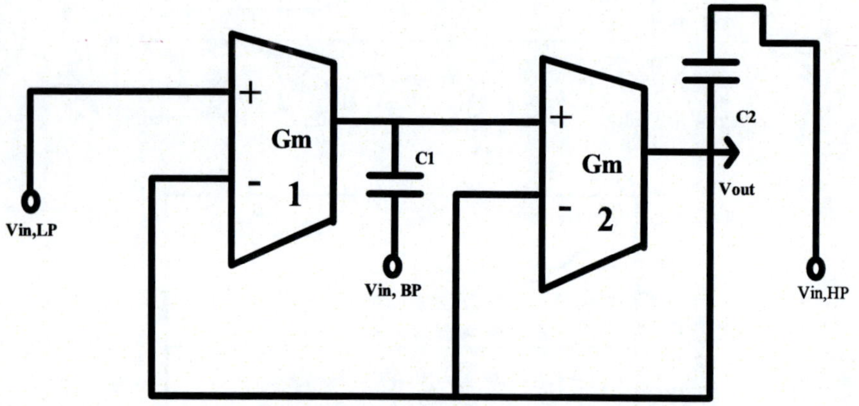

Fig. 6 Design of multifunction filter

The cut off frequency of the multifunction filter is given as

$$w_o = \sqrt{\frac{G_{m1}G_{m2}}{C_1C_2}} \tag{3}$$

From the −3 dB frequency, the quality factor is calculated

$$Q = \sqrt{\frac{G_{m1}C_2}{G_{m2}C_1}} \tag{4}$$

Second-order multifunction low pass filter is shown in Fig. 7 which has cut off frequency of 32.8 MHz. DTMOS-based multifunction filter provides cut off frequency of 39.91 MHz. as shown in Fig. 8.

Table 2 shows the proposed specifications for the G_m-C based loop filter specification for the PLL.

5 Conclusion

This paper deals with the designing of the low-power loop filter for PLL applications. From the comparison table, it is clear that DTMOS based loop filter consumes low power of order of 252.78 μW along with the cut off frequency 39.91 MHz with the supply voltage of 1 V. The proposed DTMOS-based loop filter can be further used in designing of 1–3 GHz PLL.

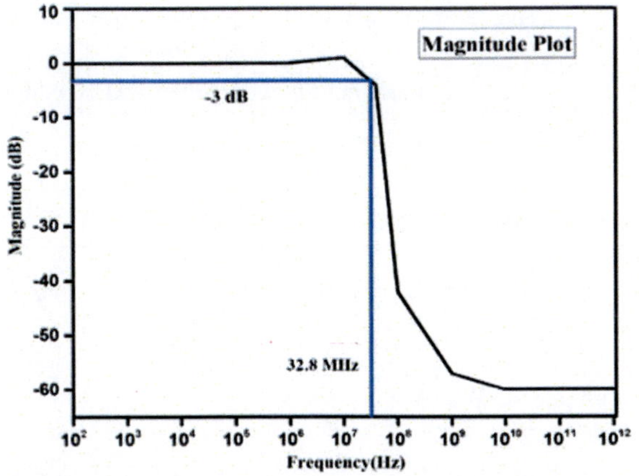

Fig. 7 Basic multifunction filter magnitude response with frequency variation

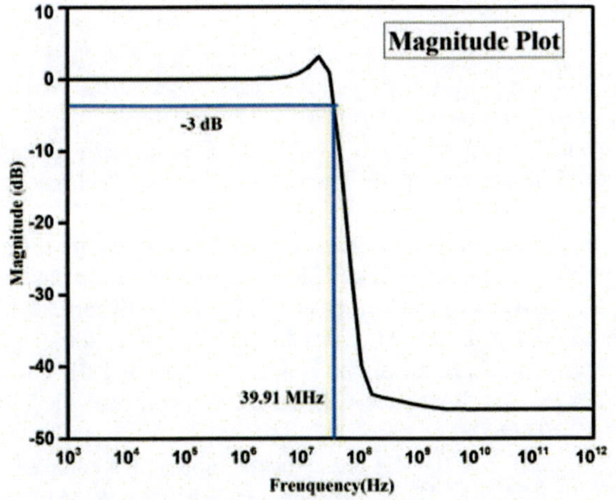

Fig. 8 DTMOS-based multifunction filter magnitude response with frequency variation

Table 2 Multifunction filter specifications

S. No.	Parameter	Basic multifunction filter	DTMOS multifunction filter
1	Technology (nm)	180	180
2	Supply voltage (V)	1.8	1
3	Cut off frequency (MHz)	32.81	39.91
4	C1 (pf)	1	1
5	C2 (pf)	1	1
6	Quality factor (Q)	1	1
7	Power consumption (μW)	314.7	252.78

Acknowledgements. Authors would like to thank, Ministry of Electronics and Information Technology (MeitY), Govt. of India for providing financial support under SMDP-C2SD Project.

References

1. Sumi, Y., Tsukutani, T., Higashimura, M., Fukui, Y.: Electronically Tunable low-pass filter in PLL frequency synthesizer. In: IEEE International Vehicle Electronics Conference, Sept 2001, pp. 55–59
2. Musch, T., Gerding, M., Schiek, B.: A phase-locked-loop concept for the generation of two RF-signals with a small frequency offset. IEEE Trans. Instrum. Meas. **54**(2), 709–712 (2005)
3. Cao, Z., Li, Y., Ya, S.: A 0.4 ps RMS-Jitter 1–3 GHz ring-oscillator PLL using phase-noise preamplification. IEEE J. Solid-State Circuits **43**(9), 2079–2089 (2008)
4. Oskooei, M.S., Masoumi, N., Kamarei, M.: A 5.2 mW 240–550 MHz continuous-time low-pass filter and VGA for a UWB receiver in 0.18 μm CMOS process. Analog Integr. Circ. Sig. Process **56**(3), 185–197 (2008)
5. Li, S., Hsieh, H., Lu, L.H.: A 10 GHz phase-locked loop with a compact low-pass filter in 0.18 μm CMOS. IEEE Microw. Wirel. Compon. Lett. **19**(10), 659–661 (2009)
6. Chiu, W., Huang, Y., Lin, T.: A dynamic phase error compensation technique for fast-locking phase-locked loops. IEEE J. Solid State Circ. **45**(6), 1137–1148 (2010)
7. Choi, K.C., Kim, S.G., Lee, S., Lee, B., Choi, W.: A 990 μW 1.6-GHz PLL based on a novel supply-regulated active-loop-filter VCO. IEEE Trans. Circ. Syst. II **60**(6), 311–315 (2013)
8. Jiang, B., Xia, T., Wang, G.: PLL Low pass filter design considering unified specification constraints. Analog Integr. Circ. Sig. Process **80**, 113–120 (2014)
9. Moon, J., Choi, K.C., Choi, W.Y.: A 0.4 V 90–350 MHz PLL with an active loop-filter charge pump. IEEE Trans. Circ. Syst. II **61**(5), 319–323 (2014)
10. Kim, S., Rhim, J., Kwon, D., Kim, M., Choi, W.Y.: A low-voltage PLL with a supply-noise compensated feedforward ring VCO. IEEE Trans. Circ. Syst. II **63**(6), 548–552 (2016)
11. Shu, K., Sinencio, E.S.: CMOS PLL synthesizers-analysis and design, pp. 321–330. Springer, New York (2005)
12. Kuang X.F., Wu, N.J.: A fast-settling PLL frequency synthesizer with direct frequency presetting. In: Proceedings of the IEEE ISSCC Digest of Technical Papers, Feb 2006, pp. 741–742
13. Scaumann, R., Valkenburg, M.: Design of Analog Filters, p. 628. Oxford University (2008)

14. Mahattanakul, J.: Design procedure for two-stage CMOS operational amplifiers employing current buffer. IEEE Trans. Circ. Syst. II **52**(11), 766–770 (2005)
15. Palmisano, G., Palumbo, G., Pennisi, S.: Design procedure for two-stage CMOS transconductance operational amplifiers: a tutorial. Analog Integr. Circ. Sig. Process **27**, 179–189 (2001)
16. Razavi, B.: Design of Analog CMOS Integrated Circuit, pp. 651–660. McGraw Hill, New York (2018)
17. Carusone, T.C., Johns, D., Martin, K.: Analog Integrated Circuit Design, pp. 737–746. Wiley, (2015)
18. Carlosena, A., Ugarte, M., Martin, A.: Loop filter approximation for PLLs. In: IEEE Symposium on Circuits and System, pp. 21–24 (2008)
19. Khateb, F.: Bulk-driven floating-gate and bulk-driven quasi-floating-gate techniques for low-voltage low-power analog circuits design. AEU Int. J. Electron. Commun. **68**(1), 64–72 (2014)
20. Kargaran, E., Sawan, M., Mafinezhad, K., Nabovati, H.: Design of 0.4 V, 386 nW OTA using DTMOS technique for biomedical applications. In: IEEE Symposium on Circuit and System, pp. 270–273
21. Chatterje, S., Tsividis, Y., Kinget, P.: A 0.5-V bulk-input fully differential operational transconductance amplifier. In: Proceedings of the ESSCIRC, pp. 147–150. Leuven, Belgium, 21–23 Sept 2004
22. Sutula, S., Dei, M., Teres, L., Serra-Graells, F.: Variable-mirror amplifier: a new family of process-independent class-AB single-stage OTAs for low-power SC circuits. IEEE Trans. Circ. Syst. I **63**(8), 1101–1110 (2016)

Performance Enhancement of InGaN/GaN Green QW LEDs with Different Interlayers and Doping in the Barriers

Apu Mistry(✉) and Dipankar Biswas

Institute of Radiophysics and Electronics, University of Calcutta, 92 A. P. C. Road, Kolkata 700009, India
apumistry@gmail.com

Abstract. The effect of introducing an InAlN or AlGaN IL (interlayer) in between the barrier and QW for the green emission has been studied. The tensile strain of AlGaN compensates the compressive strain of InGaN/GaN interface. The IL increases the barrier potential and reduces the carrier leakage from the QW. These increase the device efficiency. By changing the doping in the barrier, the optical output can be increased for green QW LEDs with and without interlayer. The best results are obtained for the AlGaN IL which increases the transition probability up to 2 times, as compared to the QW LEDs, without IL.

Keywords: InGaN QW LED · Interlayer · Transition energy · Transition probability

1 Introduction

Recently, III-V nitride-based quantum well (QW) light-emitting diodes (LEDs) are being widely used in solid-state lighting because of their high efficiency. RGB-based color mixed LEDs provide white light, with the use of red, green and blue LEDs. To emit the red and green light, higher indium (In) is required, which reduces the external quantum efficiencies of $In_xGa_{1-x}N$/GaN quantum well (QW) LEDs. For the higher In content defects, high dislocation density, In fluctuations and the strong polarization fields existing in InGaN materials in which deteriorates the optical performance of $In_xGa_{1-x}N$/GaN quantum well (QW) LEDs [1–4]. Several solutions have been proposed to reduce the effects and enhance the optical properties [5–15] of which InAlN interlayer (IL) and AlGaN IL seem to be important and worth studying.

Recent studies show, by applying an $Al_yGa_{1-y}N$ or $In_yAl_{1-y}N$ IL [16] in the active region, improve the efficiency of longer wavelength-based $In_xGa_{1-x}N$ QW LEDs [16, 17]. Sun et al. experimentally show that 10-Å-thick $Al_{0.42}Ga_{0.58}$ N provides the best optical output for green LEDs. Recently, it has also been shown that use of 10-Å-thick InAlN IL in the InGaN/GaN QW can enhance the green emission [16]. Although doping is a highly important parameter during the growth of the QW structures, there has not

© Springer Nature Singapore Pte Ltd. 2021
N. R. Das and S. Sarkar (eds.), *Computers and Devices for Communication*,
Lecture Notes in Networks and Systems 147,
https://doi.org/10.1007/978-981-15-8366-7_71

been any rigorous study that can describe its effects on the TE and the square of the overlap of the electron and hole wave functions (M_{eh}^2), i.e., TP with and without IL LEDs for a different doping in the barriers.

In this paper, we have maximized the efficiencies of green LEDs with ILs, by the variation of doping in the barriers. The interesting results reveal that, for certain doping in the barriers, the $M_{\text{eh}}^2 \text{M}_{\text{eh}}^2$ can be maximized and the maximum M_{eh}^2 is observed in the InGaN QW with AlGaN ILs.

2 Theoretical Details

In our study, we use three different single QW structures, with and without $\text{Al}_y\text{Ga}_{1-y}\text{N}$ and $\text{In}_y\text{Al}_{1-y}\text{N}$ IL. The structures consist of A: GaN/$\text{In}_x\text{Ga}_{1-x}\text{N}$/GaN QW LED, B: GaN/$\text{In}_x\text{Ga}_{1-x}\text{N}$/$\text{In}_y\text{Al}_{1-y}\text{N}$/GaN QW LED and C: GaN/$\text{In}_x\text{Ga}_{1-x}\text{N}$/$\text{Al}_y\text{Ga}_{1-y}\text{N}$/GaN QW LED. Here, GaN and $\text{In}_x\text{Ga}_{1-x}\text{N}$ layers are considered as the usual barrier and QW of the LED, respectively. $\text{Al}_y\text{Ga}_{1-y}\text{N}$ and $\text{In}_y\text{Al}_{1-y}\text{N}$ act as IL between the QW and barrier. The width of the QW is considered to be 30 Å. The n-type and p-type doping concentrations in the GaN barriers are varied, where both the QW and ILs are considered as undoped.

The nonlinear variation of the energy bandgaps of $\text{In}_x\text{Ga}_{1-x}\text{N}$, $\text{Al}_y\text{Ga}_{1-y}\text{N}$ and $\text{In}_y\text{Al}_{1-y}\text{N}$ is obtained by linear interpolation of InN, AlN and GaN materials. The bandgap of $\text{In}_x\text{Ga}_{1-x}\text{N}$, $\text{Al}_y\text{Ga}_{1-y}\text{N}$ and $\text{In}_y\text{Al}_{1-y}\text{N}$ is obtained as

$$E_g(\text{In}_x\text{Ga}_{1-x}N) = x.E_{g,\text{In}N} + (1-x)E_{g,\text{Ga}N} - b1.x(1-x) \tag{1}$$

$$E_g\left(\text{Al}_y\text{Ga}_{1-y}N\right) = y.E_{g,\text{Al}N} + (1-y)E_{g,\text{Ga}N} - b2.y(1-y) \tag{2}$$

$$E_g\left(\text{In}_y\text{Al}_{1-y}N\right) = y.E_{g,\text{In}N} + (1-y)E_{g,\text{Al}N} - b3.y(1-y) \tag{3}$$

where $E_{g,InN}$,$E_{g,AlN}$ and $E_{g,GaN}$ are the bandgaps of InN, AlN and GaN. b_1, b_2 and b_3 are the bowing parameters which are 1.4, 0.6 and 5.0, respectively [18].

The detailed calculations of piezoelectric and spontaneous polarization and the direction of fields between GaN, InGaN and ILs are taken from Ambacher et al. and Fiorentini et al. [19, 20].

The electrostatic field, the electron–hole concentration, the energy states, M_{eh}^2 are found out as stated in detail in our references [6, 21]. The computations have been carried out through the self-consistent solution of Schrödinger and Poisson equations using MATLAB. The Schrödinger and Poison equations are solved using the finite difference method with non-uniform mesh of discretization, and the discretized equation is solved using Newton–Raphson method [22].

3 Results and Discussion

Figure 1 shows the energy band diagram of structures A, B and C. From the energy band diagram, we have seen (1) the high compressive strain in InGaN/GaN QW is compensated by the tensile strained of AlGaN IL, as well as InAlN IL reduces the strain between the InGaN and GaN layers significantly. (2) The ILs reduce the electron and hole leakage from the QW. (3) The ILs act as a cap layer, reducing the out-diffusion of In from the QW. By changing the composition and width of the IL, better optical properties are obtained [16]. The experimental spectral characteristic of the studied green LEDs is shown in [17]. For green LEDs, a series of our simulation results show that the 10-Å-thick $In_{0.2}Al_{0.8}N$ IL and 10-Å-thick $Al_{0.8}Ga_{0.2}N$ IL provide the maximum M_{eh}^2, when the ILs are used. Figure 2 shows the variation TE with respect to doping concentrations in the structures A, B, and C. The doping levels in the p-side and the n-side are kept equal. The results show that the TE increases significantly when the ILs are used and it is maximum for the AlGaN IL. The emission covers the entire green range. It is also seen that the doping dependence of the TE for the AlGaN IL is smaller than the rectangular QW LEDs without IL.

The TP with respect to doping concentrations for the structures A, B, and C is shown in Fig. 3.

The results show that the TP is significantly increased with the inclusion of ILs. Initially, the TP of AlGaN IL LEDs is more than 2 times higher than the conventional LEDs without IL. It is also seen form the result that the TP is significantly increased with the increment of doping in the barriers. This information should be very important for the optoelectronic designer.

4 Conclusions

In conclusion, we have seen that inclusion of an IL compensates the compressive strain in the $In_xGa_{1-x}N$/GaN interface. Use of AlGaN or InAlN ILs between the QW and barriers increases the barrier potential height which potentially suppresses the electron and hole leakage from the QW. These increase the M_{eh}^2 of the green LEDs. By changing the doping in the barrier, we could modify the M_{eh}^2 and obtain better optical output. This should convey important information to the optoelectronics designer.

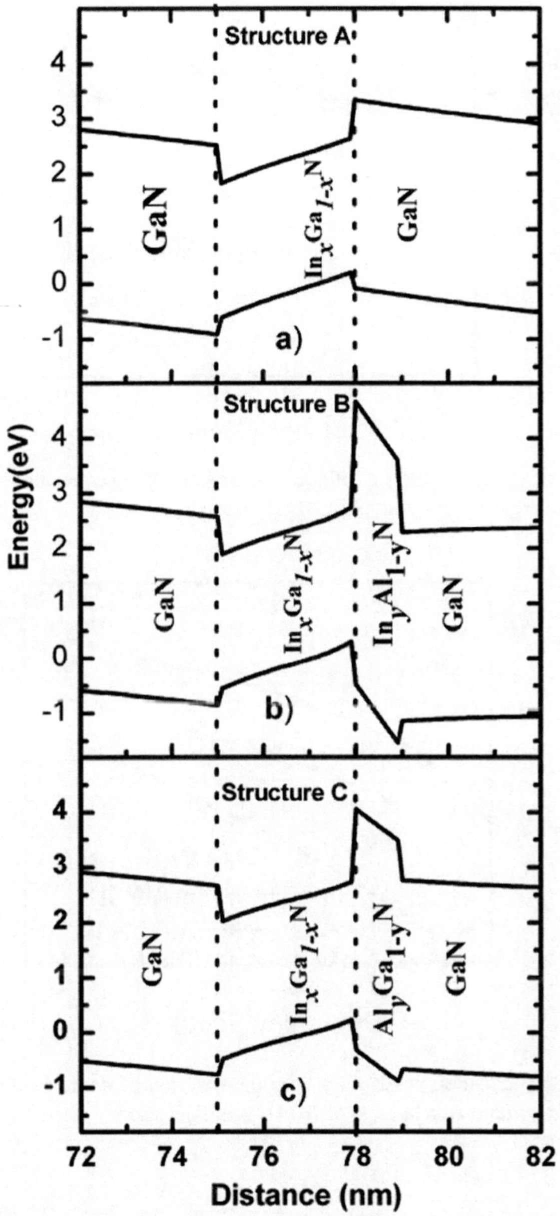

Fig. 1 Schematic band diagram of InGaN/GaN QW structure **a** A: Without interlayer, **b** B: with $In_yAl_{1-y}N$ IL and **c** C: with $In_yAl_{1-y}N$ IL

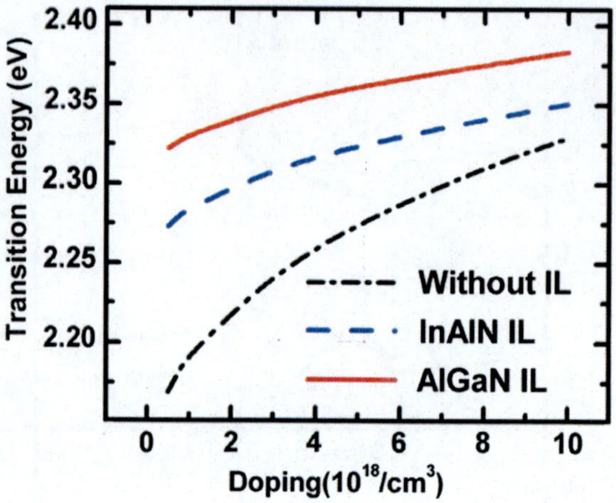

Fig. 2 Variation of the transition energy with the function of doping concentration in the barriers when with or without $Al_yGa_{1-y}N$ or $In_yAl_{1-y}N$ IL is used

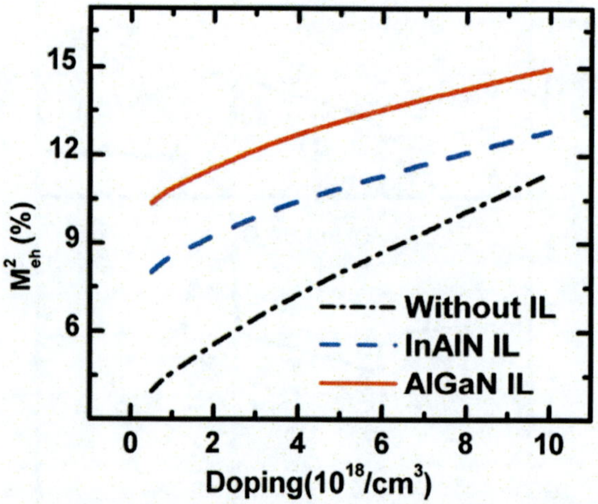

Fig. 3 Variation of the transition probability with the function of doping concentration in the barriers when with or without $Al_yGa_{1-y}N$ or $In_yAl_{1-y}N$ IL is used

References

1. Cho, H., Lee, J., Yang, G., Kim, C.: Formation mechanism of V defects in the InGaN/GaN multiple quantum wells grown on GaN layers with low threading dislocation density. Appl. Phys. Lett. **79**, 215–217 (2001)
2. Shi, J.J., Gan, Z.Z.: Effects of piezoelectricity and spontaneous polarization on localized excitons in self-formed InGaN quantum dots. J. Appl. Phys. **94**, 407–415 (2003)

3. Yang, T.J., Shivaraman, R., Speck, J.S., Wu, Y.R.: The influence of random indium alloy fluctuations in indium gallium nitride quantum wells on the device behavior. J. Appl. Phys. **116**, 113104 (2014)
4. Watson-Parris, D., Godfrey, M., Dawson, P., Oliver, R., Galtrey, M., Kappers, M., et al.: Carrier localization mechanisms in In x Ga 1–x N/GaN quantum wells. Phys. Rev. B **83**, 115321 (2011)
5. Zhao, H., Jiao, X., Tansu, N.: Analysis of interdiffused InGaN quantum wells for visible light-emitting diodes. J Display Technol **9**, 199–205 (2013)
6. Biswas, D., Mistry, A., Gorai, A.: Constructive and comprehensive studies on the advantages of using staggered InxGa1-xN/InyGa1-yN QWs in LEDs. Opt. Mater. **66**, 367–373 (2017)
7. Kwon, S.Y., Kim, H.J., Yoon, E., Jang, Y., Yee, K.J., Lee, D., et al.: Optical and microstructural studies of atomically flat ultrathin In-rich In Ga N/ Ga N multiple quantum wells. J. Appl. Phys. **103**, 063509 (2008)
8. Gorai, A., Mistry, A., Panda, S., Biswas, D.: Inclusion of Indium, with doping in the barriers of InxGa1-xN/InyGa1-yN quantum wells reveals striking modifications of the emission properties with current for better operation of LEDs. Photon. Nanostruct. Fundam. Appl. **28**, 70–74 (2018)
9. Park, S.H., Park, J., Yoon, E.: Optical gain in In Ga N/Ga N quantum well structures with embedded AlGaN δ layer. Appl. Phys. Lett. **90**, 023508 (2007)
10. Park, S.H., Ahn, D., Koo, B.H., Kim, J.W.: Dip-shaped InGaN/GaN quantum-well light-emitting diodes with high efficiency. Appl. Phys. Lett. **95**, 063507 (2009)
11. Zhao, H., Arif, R.A., Ee, Y.K., Tansu, N.: Self-consistent analysis of strain-compensated InGaN–AlGaN quantum wells for lasers and light-emitting diodes. IEEE J. Quantum Electron. **45**, 66–78 (2008)
12. Zhao, H., Liu, G., Tansu, N.: Analysis of InGaN-delta-InN quantum wells for light-emitting diodes. Appl. Phys. Lett. **97**, 131114 (2010)
13. Zhang, J., Tansu, N.: Improvement in spontaneous emission rates for InGaN quantum wells on ternary InGaN substrate for light-emitting diodes. J. Appl. Phys. **110**, 113110 (2011)
14. Saito, S., Hashimoto, R., Hwang, J., Nunoue, S.: InGaN light-emitting diodes on c-face sapphire substrates in green gap spectral range. Appl. Phys. Expr. **6**, 111004 (2013)
15. Koleske, D., Fischer, A., Bryant, B., Kotula, P., Wierer, J.: On the increased efficiency in InGaN-based multiple quantum wells emitting at 530–590 nm with AlGaN interlayers. J. Cryst. Growth **415**, 57–64 (2015)
16. Sun, W., Al Muyeed, S.A., Song, R., Wierer Jr, J.J., Tansu, N.: Integrating AlInN interlayers into InGaN/GaN multiple quantum wells for enhanced green emission. Appl. Phys. Lett. **112**, 201106 (2018)
17. Alhassan, A.I., Young, N.G., Farrell, R.M., Pynn, C., Wu, F., Alyamani, A.Y., et al.: Development of high performance green c-plane III-nitride light-emitting diodes. Opt. Expr **26**, 5591–5601 (2018)
18. Wu, J.: When group-III nitrides go infrared: new properties and perspectives. J. Appl. Phys. **106**, 5 (2009)
19. Ambacher, O., Smart, J., Shealy, J., Weimann, N., Chu, K., Murphy, M., et al.: Two-dimensional electron gases induced by spontaneous and piezoelectric polarization charges in N-and Ga-face AlGaN/GaN heterostructures. J. Appl. Phys. **85**, 3222–3233 (1999)
20. Fiorentini, V., Bernardini, F., Ambacher, O.: Evidence for nonlinear macroscopic polarization in III–V nitride alloy heterostructures. Appl. Phys. Lett. **80**, 1204–1206 (2002)
21. Panda, S., Biswas, D.: Effects of doping concentration on the transition energy of InGaN/GaN quantum well diodes. Solid State Commun. **168**, 60–63 (2013)
22. Tan, I.H., Snider, G., Chang, L., Hu, E.: A self-consistent solution of Schrödinger-Poisson equations using a nonuniform mesh. J. Appl. Phys. **68**, 4071–4076 (1990)

Design of a Novel High-Q Active Inductor at 2.5 GHz in CMOS 180-nm Technology

Moumita Das[1,2](✉), Shrabanti Das[2], Swarup Dandapat[2], and Sayan Chattearjee[2]

[1] STA, WBSETCL, Kolkata, West Bengal, India
moumitad50524@gmail.com

[2] Electronics and Telecommunication Engineering, Jadavpur University, Kolkata 70032, West Bengal, India

Abstract. This paper represents a novel approach of an inductor design having a high-quality factor using cascode topology. The proposed inductor consists of gyrator-C-based active inductor and parallel resonance circuit which are basically comprised of low value of spiral inductor and capacitor. This is validated in Cadence Virtuoso Tool using TSMC 180 nanometer technology CMOS process with power supply of 1.8 V. The designed inductor represents the inductance of above 40 nH with quality factor of over 880 around 2.45 GHz.

Keywords: Gyrator-C · Parallel resonance circuit · Feedback resistor · CMOS · Spiral inductor

1 Introduction

As the ultra-wideband spectrum dragged into consideration of various scientists, especially in wireless communication systems, the designers achieve this inductance in the form of on-chip spiral passive inductors. These structures consume very large surface areas and account to a very challenging factor to parameterize. Moreover, designers hardly produce inductance of having quality factors greater than five to ten at microwave signal frequencies. One of the most important arguments associated with standard CMOS technology is very low resistivity which results in low-quality factor for passive spiral inductors. Designers adopt various methods to increase the passive inductor quality factor, but these compensations enhance power dissipation and it also degrades the noise figure and limits the dynamic range [1, 2].

Active inductors could potentially alleviate many challenges in design of the analog circuits. The reactance value of the active inductors has efficiently changed either in uninterrupted manner or in a spontaneous step. This contributes flexibility in tuned matching circuits. Easier layout floor-planning, higher accuracy and absence of magnetic coupling are the alternative advantages of active inductors correlated to their passive inductors.

Nowadays, a lot of attempts have been adopted to restore passive inductors with active inductors [3, 4]. In designing an active inductor, the noise and resistive loss which

© Springer Nature Singapore Pte Ltd. 2021
N. R. Das and S. Sarkar (eds.), *Computers and Devices for Communication*,
Lecture Notes in Networks and Systems 147,
https://doi.org/10.1007/978-981-15-8366-7_72

eminently influence the performance of the device should be taken into account. Due to reputation of gyrator approach in designing of active inductors for high-frequency applications [1, 5], it has been adopted for design of proposed active inductor. The gyrator can be composed of operational trans-conductance amplifier, which is being constructed accurately by conventional CMOS technology [6, 7]. Moreover, gyrator can be modified with the applied bias but the gyrator-based active inductor has a limitation to enhance the frequency of tuning range and quality factor.

In this paper, a model active inductor has been designed for low noise and low loss applications and it has been implemented by adopting the gyrator-C approach. The loss and noise have been decreased by employing feed-forward method of design topology. In addition to it, all the MOS applied in the proposed design is considered to no body effect which is very compelling in submicron devices. In this paper, the value of inductance is as high as 40 nH with a very high-quality factor of about 880 at the 2.45 GHz.

2 Design Methodology

Passive inductors offer constant inductance value resulting in narrowband operation, and also it requires larger area in chip design. Traditionally, inductor has better quality factor than capacitors and from Leeson's formula it is depicted that the phase noise is inversely proportion to quality factor. Active inductors provide large tuning range as well as higher-quality factor. That is why it has been intended to design various circuits using active inductors in order to overcome passive inductor's limitations [8].

The gyrator-C structure is used to realize the conventional active inductor. This gyrator-C consists of two MOSs which generate the inductive reactance. However, these structures are having the limitation of enhancing quality factor and tuning range. To overcome these limitations, some external circuit (feedback resistor, resonating circuit, cascoded CMOS) has been added which improves its performance (Fig. 1).

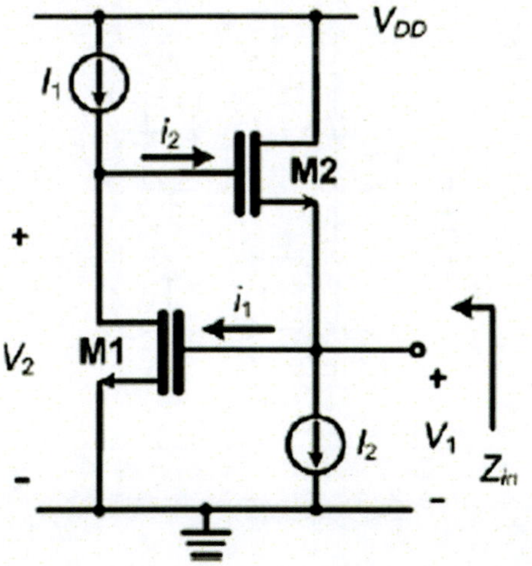

Fig. 1 Gyrator-C-based conventional active inductor

The simulation result of gyrator active inductor is to be highlighted that the quality factor can be increased by increasing r_0 of the MOS M_1 and the quality factor is also proportional to g_{m1}. Moreover, the quality factor can be deteriorated by enhancing the value of C_{gs2}. From that, it can be concluded the quality factor has changed dynamically by the influence of parasitic capacitance of M_2 at high frequency [9, 10]. An alternative method to improve the quality factor is to include a prototype inductor, which is a spiral inductor (Lf) in feedback loop connected between the source of M0 and the gate of M1 of the conventional gyrator active inductor. The compensation of the degradation of the quality factor by the parasitic capacitances can be achieved by applying a spiral inductor to the source terminal of M0. The value of the feedback spiral inductor can further be decreased by replacing the inductor with a LC resonate circuit at the source of the M0.

3 Active Inductor Design

In this paper, cascode topology has been adopted to reduce the output conductance which in turn enhances tuning range. Cascode topology accounts to series connection of components in which output of a component serves as input to the other component that results in reduction of output conduction to the proposed design. In addition to it, a feedback resistance has been applied that in turn degrades the series resistance component which eventually increases the quality factor. The quality factor of a tuned circuit is often used for measurement of its performance in a resonance circuit. It also provides an insight to indicate resonance's bandwidth relative to its center frequency.

To reduce further spiral inductance value, LC parallel resonance circuit has been used in the proposed design (Fig. 2).

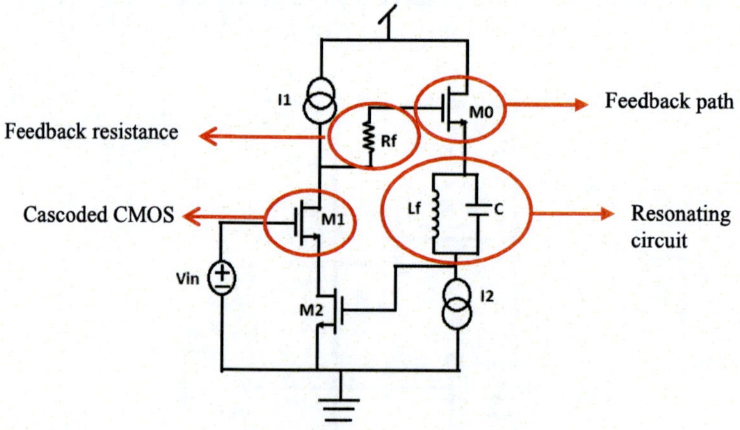

Fig. 2 Schematic of active inductor

4 Simulation Result

Since the spiral inductors have large amount of parasitic, due to the parasitic component the circuit performance degrades. In the present paper, an approach has been validated to avoid the parasitic of inductors.

To demonstrate the achievement of the proposed active inductor circuit, simulation results are given using Cadence Virtuoso Tool using TSMC 180 nanometer technology CMOS process. Figure 3 depicts the simulation results of Smith chart of high-quality factor using parallel feedback resonance circuit. The maximum quality factor at 2.45 GHZ is 850. From the calculation associated with Cadence Virtuoso, it has been found that the power dissipation is as less as 3.7 mW. Table 1 summarizes some important properties of the designed AI.

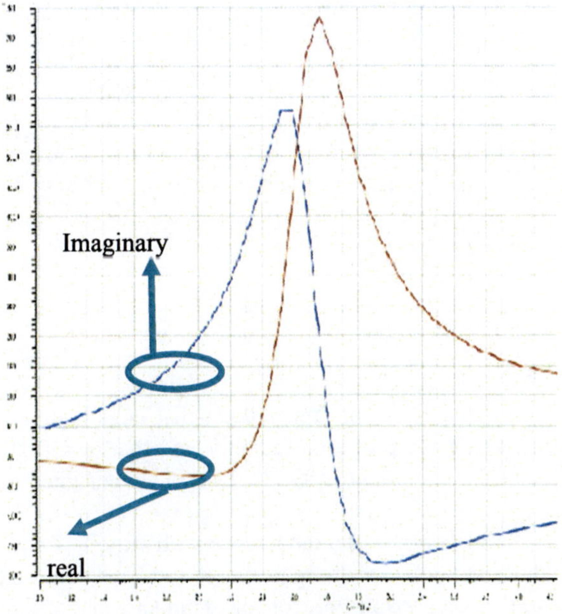

Fig. 3 Real and imaginary part of Z_{in}

Table 1 Comparison table

Ref @ 2.45 GHz	Technology (μm)	Inductance (nH)	Quality factor	Tuning range (GHZ)	Power dissipation (mW)
11	0.18	45	250	–	12
12	0.35	15	500	0.3–11.2	1.3
This work	0.18	40	880	1.5–10	3.7

Figure 3 represents the real and imaginary part of the proposed design circuit where Fig. 4 represents the Smith chart of the inductance. In Fig. 4, the inductance has been laid beyond the circle that is because of the active inductor. This means the active inductor performs amplification; i.e., it provides gain.

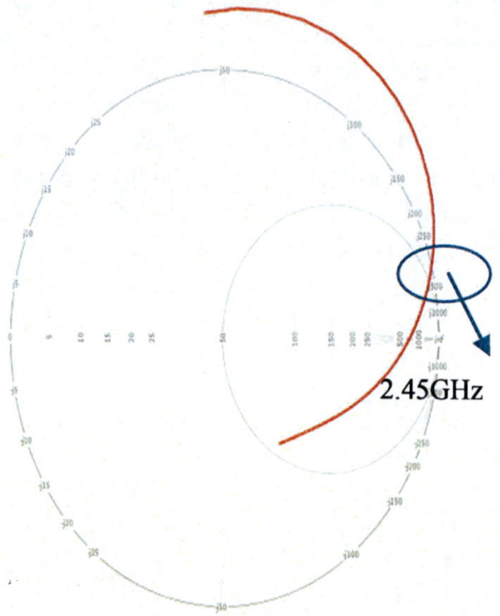

Fig. 4 Smith chart of high-quality factor inductor

Figure 5 depicts the magnitude of the inductor that is as high as 40nH. Figure 6 depicts the quality factor of the proposed inductor and that is 880 at 2.45 GHz. The resonance frequency of the LC circuit is 890 MHz, and the designed circuit is tuned at 2.45 GHz.

Figure 7 depicts the variation of the inductance with varying frequency, and Fig. 8 depicts the variation of quality factor of the proposed inductor with varying frequency.

Table 1 represents the comparison table for active inductor at 2.45 GHz using 180-nm technology.

5 Conclusion

In the present paper, the authors have presented a novel high-quality factor inductor using gyrator-C-based active inductor with feedback parallel resonance circuit and cascode technology. The proposed active inductor has the quality factor above 880 at 2.5 GHz with the maximum inductance of about 40 nH at the same frequency. In this paper, a novel approach of introduction to high-quality factor inductor provides a very high-quality factor and also a very large inductance using a small value of spiral inductor.

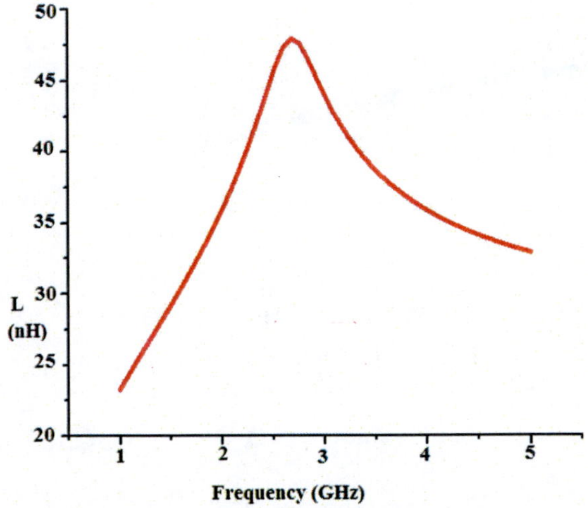

Fig. 5 Magnitude of the inductor

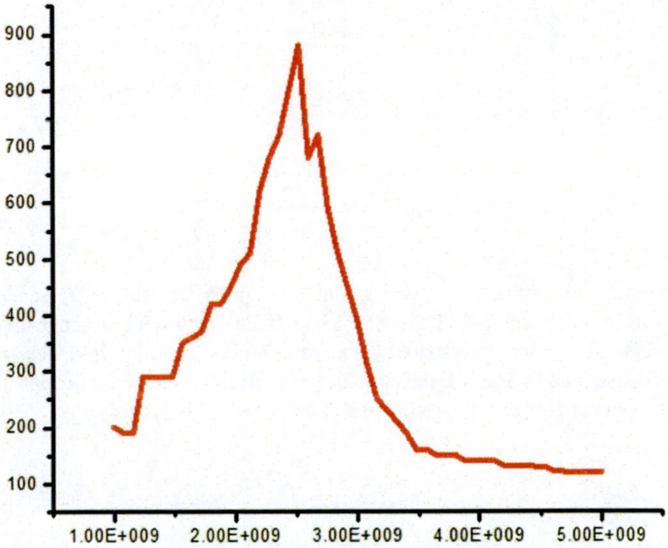

Fig. 6 Magnitude of the quality factor

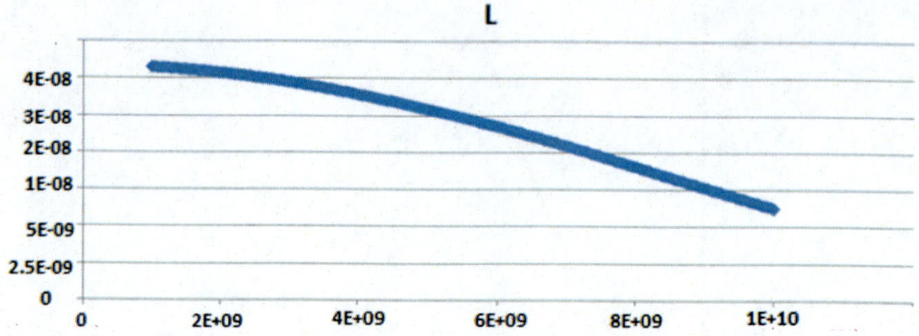

Fig. 7 Inductance versus frequency

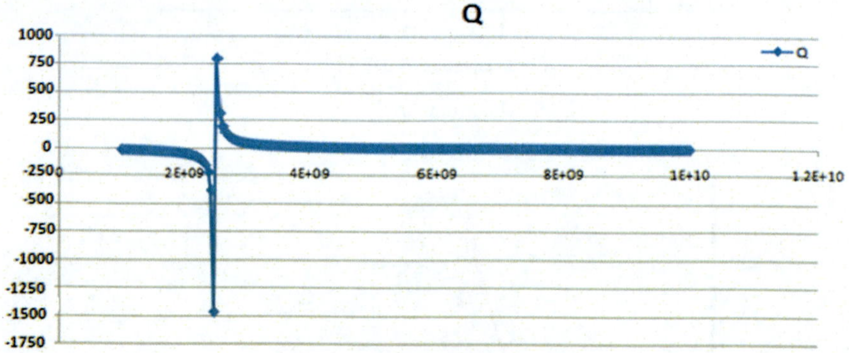

Fig. 8 Quality factor versus frequency

Acknowledgements. The authors are extremely thankful to SMDP-III(C2SD) (funded by Deity, under Ministry of Communication & IT, Government of India) for providing them with the required laboratory facilities and software to carry on the research works, and they also extend their sincere gratitude toward Integrated Circuit Centre of Electronics and Telecommunication Engineering Department of Jadavpur University for allowing them to use the laboratory.

References

1. Ghasemzadeh H., Yazgi, M., Kopru, R.: A low loss, low voltage and high Q active inductor with multi-regulated cascode stage for RF applications. IEEE conference, ICECS 2015 (In press)
2. Rafei, M., Mosavi, M.R.: A new 0.25–12.5 GHz high quality factor low-power active inductor using local RC feedback to cancel series-loss resistance. Arab. J. Sci. Eng. (2013)
3. Reja, M., Filanovsky, I., Moez,K.: A CMOS 2.0–11.2 GHz UWB LNA using active inductor circuit. IEEE International Symposium on Circuits and Systems, 2008. ISCAS 2008, Seattle, WA, pp. 2266–2269 (2008)
4. Nair, M.U., Zheng, Y.J., Lian, Y.: 1 V, 0.18 μm-area and power efficient UWB LNA utilising active inductors. Electron. Lett. **44**(19), 1127–1129 (2008)

5. Saberkari, A., Kazemi, Sh., Shirmohammadli, V., Yagoub, M.C.E.: Gmboosted flat gain UWB low noise amplifier with active inductor-based input matching network. Integr. VLSI J. (2015)
6. Uyanik, H.U., Tarim, N.: Compact low voltage high-Q CMOS active inductor suitable for RF applications. Analog. Integr. Circ. Sig. Process, Springer, Berlin (2007)
7. Kia, H.B.: Adaptive CMOS LNA using highly tunable active inductor. 2014 22nd Iranian Conference on Electrical Engineering (ICEE), vol. 1, no. 6, pp. 20–22 (May, 2014)
8. Momen, H.G., Yazgi, M., Kopru, R.: Designing a new high Q fully CMOS tunable floating active inductor based on modified tunable grounded active inductor. 2015 9th International Conference on Electrical and Electronics Engineering (ELECO), pp. 1–5, Bursa (2015)
9. Thanacbayanont, A., Payne, A.: VHF CMOS Integrated Active Inductor. Electron. Lett. **32**(11), 999–1000 (1996)
10. Hayashi, H., Muraguchi, M.: A high-q broad-band active inductor and its application to a low-loss analog phase shifter. IEEE Trans. Microw. Theor. Tech. **44**(12), 2369–2314 (1996)

Design of a Low-Power Linear Down-Conversion Mixer at 2.45 GHz CMOS 180-nm Technology

Swarup Dandapat(✉), Shrabanti Das, Moumita Das, and Sayan Chatterjee

Electronics and Telecommunication Engineering, Jadavpur University, Kolkata 70032, West Bengal, India

swarupdandapat92@gmail.com

Abstract. This paper explores the working principles of mixers, which is an integral part of superheterodyne transceiver system. Special emphasis has been led on the design of subharmonic mixers, which is a subclass of the mixer system by using 180-nm CMOS technology. The present paper highlights the design implementation of a 2× subharmonic mixer. From simulation results, it has been observed that the mixer attains 19 dBm IIP3, 12 dB conversion gain, 10 dB noise figure at 2–2.4 GHz RF range and 300–500 MHz of IF range. From 1.8 V supply, total power dissipation has been measured to be 3.8 mW.

Keywords: Mixer · Low power · Down-conversion · CMOS

1 Introduction

In a radio transceiver, the bandwidth of the valuable data is generally in the sort of a few KHz or MHz with carrier signal in the range of GHz [1]. For example, a transceiver of 802.11 wirelesses LAN sends and receives signals at 5–6 GHz with only 20 MHz bandwidth. Accordingly, transmitter requires up-conversion mixers to transmit signals using higher-frequency carriers and a receiver with down-conversion mixers to take out the information from carrier signals. Subsequently, mixer is one of the key blocks in a transceiver system and its dynamic range limits the entire receiver's or transmitter's dynamic range. Mixers change the high RF frequency to a low IF frequency in receivers and vice versa in transmitters. The corresponding circuits are referred to as down-conversion and up-conversion mixers, respectively [2].

A new CMOS subharmonic mixing procedure based on threshold voltage modulation with high LO-to-RF separation has been offered in [3] for 2.1 GHz. And this novel circuit design increases the second harmonic LO-to-RF isolation to above 67 dB and hence can mitigate LO leakage issues in wireless receivers. Accordingly in [4], a novel short power 5.6 GHz doubly balanced subharmonic mixer IF receiver in 0.25-μm CMOS technology has been proposed for Industrial Scientific Medical (ISM) Band which attains input compression almost of –12 dBm and 5.96 dB double sideband noise figure. In 2002, Youngwook Kim described a predistortion technique to decrease the intermodulation distortion (IMD) produced from the conversion method of a mixer [5]. Accordingly,

© Springer Nature Singapore Pte Ltd. 2021
N. R. Das and S. Sarkar (eds.), *Computers and Devices for Communication*,
Lecture Notes in Networks and Systems 147,
https://doi.org/10.1007/978-981-15-8366-7_73

the authors claim an improvement of about 16 dB of IMD3 at −18 dBm IF output power 10 MHz frequency band and increase of about 3.5 dB of P1 dB of the mixer. Consequently in [6], a CMOS down-conversion folded mixer circuit structure for the radio frequency purposes has been proposed and fabricated in TSMC 0.18 μm 1P6M CMOS process with an input of 1.9 GHz RF signals with −40 dBm of determined power at 0.948 GHz local LO signals. The proposed design attains third-order input intercept point (IIP3), and the power consumption including the buffer stage has been -3dBm and 6.4mW, respectively. In [7], a novel design of 4× subharmonic mixer circuit is presented using CMOS 0.18-μm technology and the conversion gain of approximately 6 dB, 1-dB compression point of −12 dBm, IIP3 of 2 dBm and IIP2 of 17 dBm. A low-voltage low-power high linearity quadrature mixer for software-defined radio applications in a 90-nm CMOS technology has been proposed in [8]. This proposed design consumes a DC power of 3.8 mW under 1 V supply. The conversion gain with 10 samples is 3.6–7.2 dB in the frequency range of 0.3–6 GHz. The IIP3 is 7.9–12.3 dBm 0.3–6 GHz, whereas the single-sideband noise figure (SSB NF) is 11.1–14.7 dB. Also in [9], the operational details of the mixer and the design of a mixer using CMOS technology have been presented and the design has been simulated to obtain the necessary data. Both the 2× and the 4× subharmonic mixers have been presented in this report to achieve conversion gain of 8 dB using a 2.1 GHz RF signal and conversion gain of 6 dB using 12 GHz Ku-band, respectively.

In the present paper, a pseudo-differential pair has been included to achieve better performance of the mixer in terms of linearity. The present paper highlights the design realization of a 2× subharmonic mixer. Special emphasis has been given on the design of subharmonic mixers, which is subclass of the mixer system by using 180-nm CMOS technology.

2 Subharmonic Mixer Design Mythology

2.1 Third-Order Intermodulation

The g_m'' (second derivation of transconductance) can be considered as inversely proportional to the IIP3. And it has been established to calculate the small-signal nonlinearity in mixer circuits [10]. The diagram of an anticipated quadrature mixer has been depicted in Fig. 1. The transconductance of input unit has been designed using two PDTs—M1-M4. MI and M2 have been biased to operate in saturation state by Vb1, while the M3 and M4 have been biased in the subthreshold state by Vb2.

By neglecting the channel length modulation consequence, the drain current (I_D) of M1 and M2 versus voltage can be as follows $I_D = (\mu_n C_{OX/2}) \times (W/L) \times (V_{GS} - V_{TH})$. For small-signal linearity reflection, once $V_{in} = 0$, the g_m'' can be given by:

$$g_m''|\Delta\ v_{in} = \frac{-3}{4\sqrt{I_{SS}}}(\mu_n C_{OX} W/L)^{3/2} \qquad (1)$$

The voltages for M3 and M4 are $V_{b2} \pm \Delta V_{in}/2$, respectively, where V_{b2} is bias voltage and V_{in} is small-signal voltage, while the current of MOSFET in subthreshold region can be $I_D = I_0 e(v_{GS}/\xi V_T)$, where $\xi > 1$ is a non-ideal factor, I_0 is a factor related

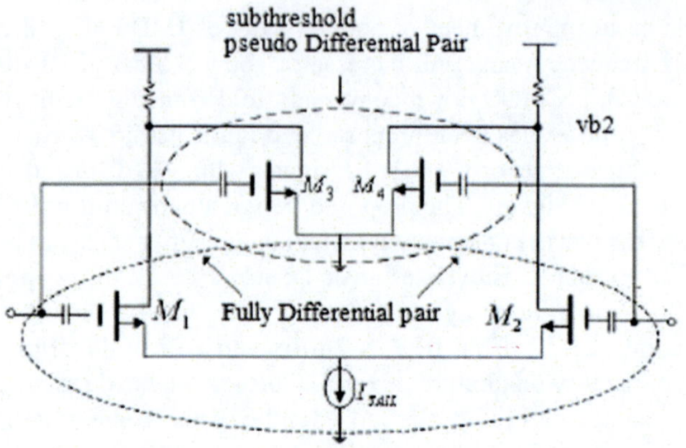

Fig. 1 Pseudo-differential transistor

to method and V_T is thermal voltage. For small-signal linearity consideration, while $V_{\text{in}} = 0$, the g_m'' can be given by:

$$g_m''|\Delta\, v_{\text{in}} = \frac{I_0}{4(\zeta V_T)^3} e^{\frac{V_b}{\zeta V_T}} > 0 \tag{2}$$

According to Eqs. (1) and (2), the g_m'' can be set to nearly zero by adjusting W/L and bias voltages Vb1 and Vb2 in PDT. Therefore, the highest IIP3 can be achieved. Figure 2 shows the simulated results for g_m'' versus degree of difference input voltage in the POTs. In small-signal action state, the –ve value of g_m'' which reduce the linearity in saturation PDT can be rewarded with positive value of g_m'' in subthreshold PDT. Seeing as the performance of mixer mostly depends on saturation PDT, the linearity is very much improved by this way without sacrificing other presentation like NF, gain and power consumption [8].

3 Mixer Design and Circuit Descriptions

Depending on the above study, a highly linear and low-power mixer circuit has been presented using the TSMC 180-nm RF CMOS technology, as shown in Fig. 3. This mixer be made up of a transconductance phase and switching couple. It utilizes a folded configuration, which offers adequate amount of voltage headroom for the transconductance phase even with a low supply (1.8 V). The use of a folded configuration also helps self-governing control of the bias current in the transconductance and switching couple, to attain appropriate conversion gain and noise figure. LC tanks offer driving current without voltage drop and show high impedance at the preferred frequency.

The input matching network comprises a DC block capacitance and a series-connected inductance with it. The mixer transconductance phase consists of a entirely differential pair M1, M2 and pseudo-differential pair M3, M4. To achieve the best possible nonlinearity compensation, analysis and simulation suggest the fully and pseudo-differential pair size has been chosen as 60/0.18 and 160/0.18, and the bias current is 2.1 mA.

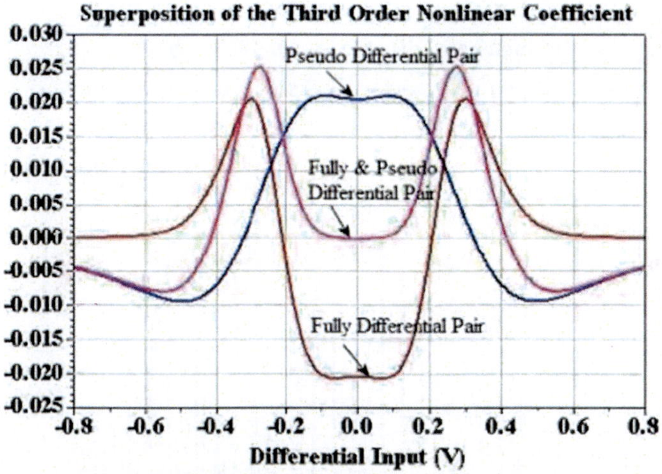

Fig. 2 g_m'' versus differential input voltage in the PDT

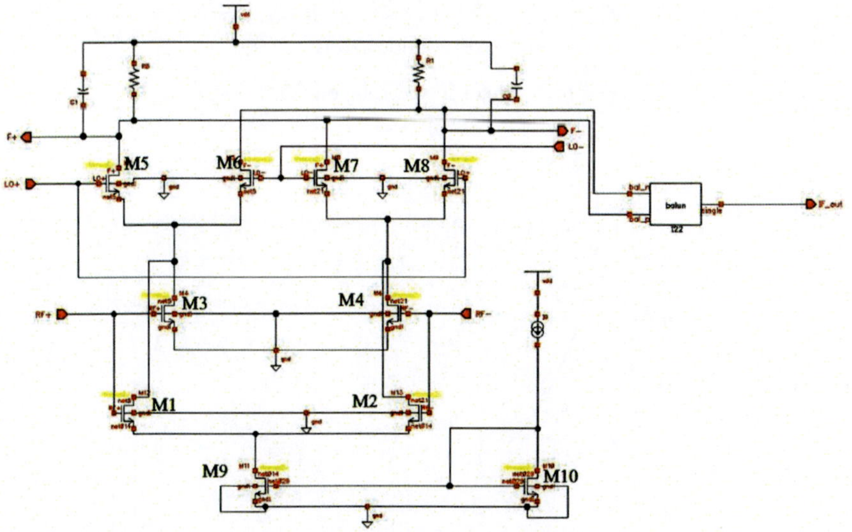

Fig. 3 Mixer schematic

4 Simulation Results

The improvement of the mixer performance has been shown by the following simulation results given below. Figure 4 represents the spectrum output of the proposed design, and it illustrates the design in advance design system simulation software. Figure 5 depicts the noise figure of the mixer. It shows that the value of the noise figure for the proposed design is 10 dB. The linearity of the mixer is represented by the IIP3 simulation, and the result has been depicted in Fig. 6. Figure 7 represents the conversion gain of the proposed design; it is as high as 12 dB (Fig. 8 and Table 1).

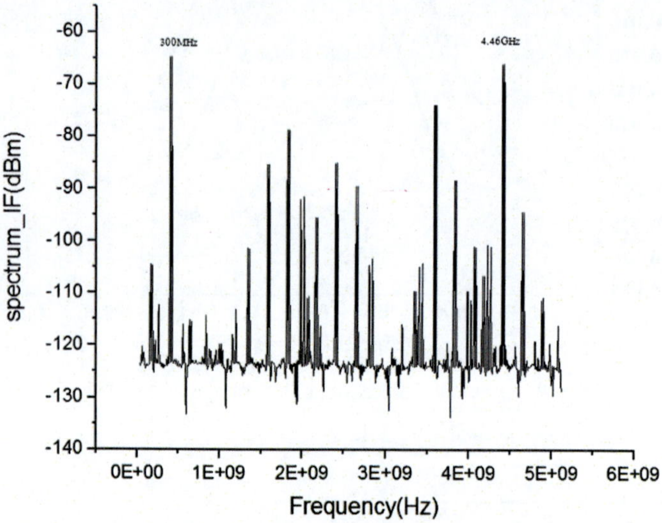

Fig. 4 Mixer IF spectrum output

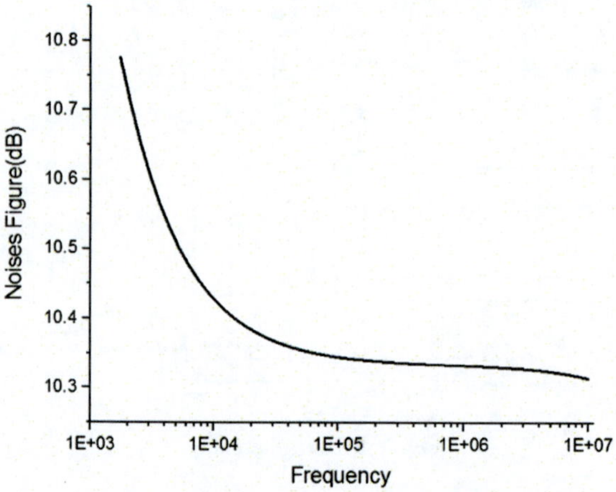

Fig. 5 Noise figure

5 Conclusion

While performing the analysis of the circuit, initially there were some discrepancies in the analyzed data. In the beginning, the circuit showed a negative conversion gain,

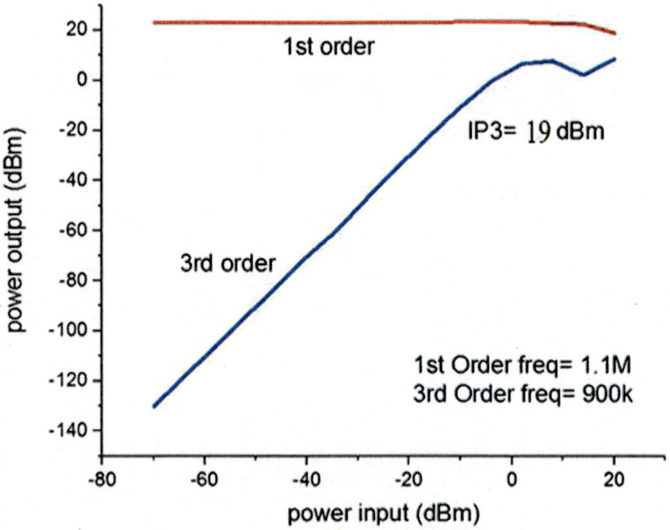

Fig. 6 IIP3

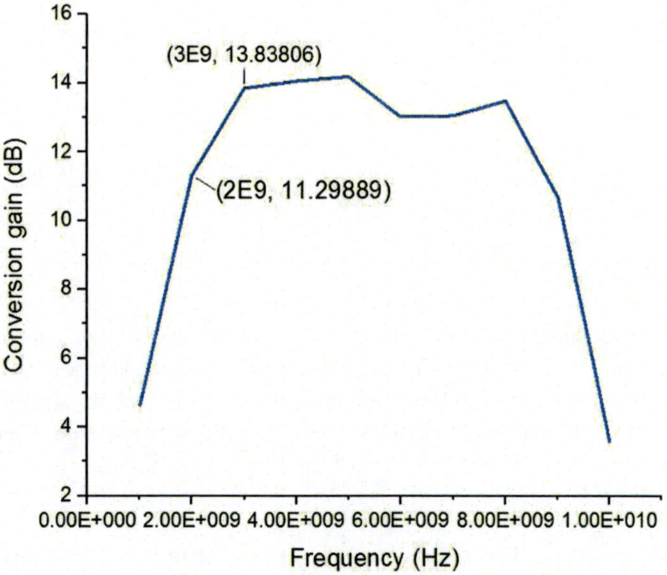

Fig. 7 Conversion gain of mixer

which was then rectified by modulating the MOSFET parameters (length only as width was fixed at 180 nm). Also over the design, the value of the drain resistor was increased accordingly to suit our requirements. The bias voltage was adjusted so that the circuit worked as per its requirements. To improve the linearity, an inductive degeneration was carried out, but it adversely affected the gain. So it was discarded in the final circuit. The

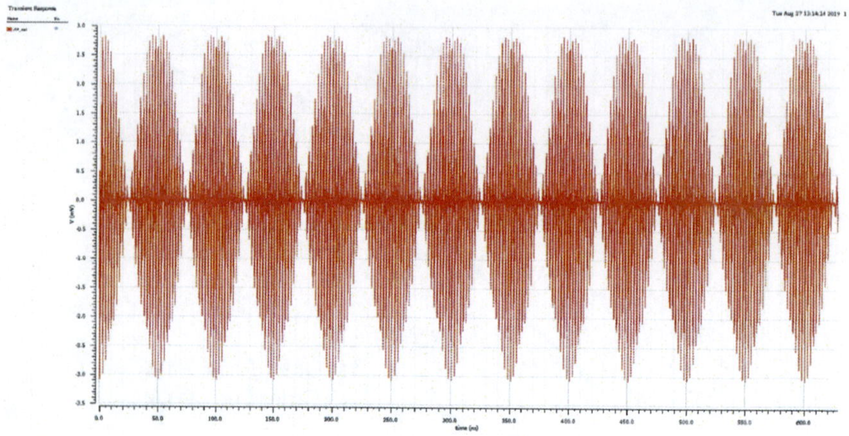

Fig. 8 Time domain output

Table 1 Literature review

Ref. no.	V_{dd}	Power (mW)	Conversion Gain(dB)	IIP3 (dBm)	Noise figure(dB)
[3]	2.5	12.5	10.5	−3.5	17.7
[4]	3	5.25	8.01	−6.5	5.96
[6]	1.2	6.4	2.2	−3	11.3
[7]	2.75	5	5.8	−2	18
[11]	1.2	6	15.8	−6.72	16.33
[12]	1.2	7.2	4.5	0	11
This work	1.8	3.8	12	19	10

subharmonic mixer illustrated in this report is a 2× mixer, and it uses a quadrature LO signaling to achieve its purpose. The circuit is basically a modified Gilbert cell topology. But there are multiple avenues of improvement in this circuit, and thus it can enable us to even design systems that are 4× in nature. Lastly, a direct-conversion receiver could be designed around this 2× mixer to identify the peak performance that is attainable with these mixers.

Acknowledgements. The authors are extremely thankful to SMDP-III(C2SD)(funded by Deity, under Ministry of Communication & IT, Government of India) for providing them with the required laboratory facilities and software to carry on the research works, and they also extend their sincere gratitude toward Integrated Circuit Centre of Electronics and Telecommunication Engineering Department of Jadavpur University for allowing them to use the laboratory.

References

1. Ding Y., Harjani, R.: High-Linearity CMOS Front-End Circuits. Springer, Berlin

2. Ellinger, F.: Radio Frequency Integrated Circuits and Technologies, 2nd edn. Springer, Berlin
3. Perumana, B.G., Lee, C.-H., Laskar, J., Charkraborty, S.: A subharmonic CMOS mixer based on threshold voltage modulation. IEEE MTT-S Int. Dig., pp 33–36 (2005)
4. Upadhyaya, P., Rajashekharaiah, M., Heo, D.: A 5.6 GHz CMOS doubly balanced sub-harmonic mixer for direct conversion zero IF receiver. IEEE Workshop on Microelectronics and Electron Devices Symposium, pp. 129–130 (2004)
5. Kim, Y., Kim, Y., Lee, S.: Linearized mixer using predistortion technique. IEEE Microw. Wirel. Compon. Lett. **12**, 204–205 (2002)
6. Weng, R.M., Wang, J.C., Li, S.Y., Wei, H.C.: A low power folded mixer using even harmonic technology. IEEE International Workshop Radio-Frequency Integration Technology Symposium, pp. 247–249 (2007)
7. Jackson, B., Saavedra, C.: A CMOS Ku-band 4x subharmonic mixer. IEEE J. Solid-State Circuits **43**(6), 1351–1359 (2008)
8. Wangl, K., Mal, K., Yel, W., Yeol, K.S., Zhang, B., Wang, Z.: A Low Voltage Low Power Highly Linear CMOS Quadrature Mixer Using Transconductance Cancellation Technique. 978-1-4673-1088-8/12/$31.00 © 2012 IEEE
9. Jackson, B.R.: Subharmonic Mixers in CMOS Microwave Integrated Circuits, PhD thesis, Department of Electrical & Computer Engineering, Queen's University, Canada (2009)
10. Kim, T., Kim, B., Lee, K.: Highly Linear Receiver Front-end Adopting MOSFET Transconductance Linearization by Multiple Gated Transistors. IEEE J. Solid-State Circuits **39**, 223–229 (2004)
11. Chong, W.K., Ramiah, H., Vitee, N., Tan, G.H.: Design of Inductorless, Low Power, High Conversion Gain CMOS Subharmonic Mixer for 2.4 GHz Application. In: IEEE Int. Microwave and RF Conf., pp. 274–277 (2014)
12. Jen, H.C., Rose, S.C., Meyer, R.G.: A 2.2 GHz sub-harmonic mixer for direct-conversion receivers in 0.13 μm CMOS. IEEE International Solid-State Circuits Symposium Digest of Technical Papers, pp. 1840–1849 (2006)

Check for updates

Comparative Study of AlGaN/GaN HEMT and MOS-HEMT Under Positive Gate Bias-Induced Stress

Amrutamayee Nayak[1], Vandana Kumari[2], Mridula Gupta[3], and Manoj Saxena[4(✉)]

[1] Electrical Engineering Department, National Institute of Science and Technology, Berhampur, India
[2] Department of Electronics, Maharaja Agrasen College, University of Delhi, New Delhi, India
[3] Department of Electronics Science, University of Delhi, South Campus, New Delhi, India
[4] Department of Electronics, Deen Dayal Upadhyaya College, University of Delhi, Delhi, New Delhi, India
msaxena@ddu.du.ac.in

Abstract. The present work compares the reliability of AlGaN/GaN HEMT and MOS-HEMT using ATLAS TCAD software in terms of positive gate bias-induced stress. The applied positive gate bias stress modulates the 2DEG density in the channel region, thereby resulting in change in off-state current and threshold voltage with marginal change in on-state current. The work presented in this paper also analyzes the influence of barrier thickness, passivation permittivity and oxide permittivity on the parameters such as: V_{th} (threshold voltage), I_{off} current and electron concentration. Maximum variation in I_{ds} (drain current) and V_{th} has been seen with Al_2O_3-based AlGaN/GaN MOS-HEMT compared to SiN-based MOS-HEMT. The observed change in device performance with positive gate bias-induced stress has been higher in AlGaN/GaN MOS-HEMT as compared to HEMT. The effect of oxide thickness and operating temperature has also been investigated, and it has been seen that the change in device performance is tremendously high in MOS-HEMT compared to AlGaN/GaN-based HEMT.

Keywords: TCAD · HEMT · MOS-HEMT · Bias-induced stress

1 Introduction

From past few decades, GaN-based HEMTs are widely used for applications like high power [1]. Various assets such as high critical breakdown, wide bandgap and improved saturation velocity allow GaN-based devices to operate at high operating voltages and temperatures [2]. However, high gate leakage current is still a critical issue in AlGaN/GaN-based HEMT due to Schottky gate contact [3]. To cope up with this, AlGaN/GaN-based MOS-HEMT has been explored previously [4]. The conduction in AlGaN/GaN-based HEMT/MOS-HEMT is mainly because of the polar nature of AlGaN/GaN heterostructure which creates 2DEG at the interface without using external

© Springer Nature Singapore Pte Ltd. 2021
N. R. Das and S. Sarkar (eds.), *Computers and Devices for Communication*,
Lecture Notes in Networks and Systems 147,
https://doi.org/10.1007/978-981-15-8366-7_74

doping. Subsequently, the associated traps (inside the device) also deteriorate the device performance due to various unavoidable effects such as: Kink effect and gate lag [5–7].

Several studies have been reported previously [8–12] related to the critical issue of AlGaN/GaN-based MOSHEMT and HEMT [13–16]; i.e., the modulation in the threshold voltage due to gate bias-induced stress which results in the deterioration of device current. The reason behind the modulation (i.e., positive shift) in threshold voltage is the interface defects mainly present at AlGaN/GaN interface [17–19]. In 2014, stress recovery experiment has been performed to study the complex dynamics of AlGaN/GaN-based MIS-HEMT [20]. In past, effect of both positive (because of traps mainly present under the gate region and access region between gate to drain [21]) and negative (due to electric field [15]) gate bias-induced stresses on HEMT has been assessed. However, the comparative analysis of MOS-HEMT with HEMT (AlGaN/GaN-based) has not been discussed yet under different operating conditions.

Thus, in this work, investigation of HEMT with MOS-HEMT (AlGaN/GaN) has been performed using simulation software, i.e., ATLAS TCAD [22] simulation for analyzing the device-related degradation due to positive gate bias-induced stress. Device parameter optimized in the paper is mainly: barrier thickness, oxide permittivity (in MOS-HEMT), passivation layer permittivity and operating temperature. Simulations are mainly performed at ambient temperature with 3 ms of integration time (mentioned elsewhere).

2 Results and Discussion

Cross-sectional view of AlGaN/GaN HEMT and MOS-HEMT has been provided, respectively, in Fig. 1a, b. Models, traps and polarization parameters were calibrated with the help of previously reported experimental results [23] of AlGaN/GaN HEMT. For analyzing positive gate bias-induced stress, the device has been ramped from negative gate bias (i.e., −5 V) to positive gate bias (i.e., +1 V). At +1 Vgate bias, same voltage has been continuously applied for 15 s and then the device was again ramped from positive +1 V to negative (−5 V) voltage.

The comparison of drain current and electron concentration before and after applying positive gate bias-induced stress has been plotted, respectively, in Fig. 2a, b for conventional HEMT and MOS-HEMT architecture. By applying positive voltage at the gate terminal for 15s, drain current through the device has been modified in terms of off-state leakage current, threshold voltage and trans-conductance. The said modifications in the device performance are because of the change in electron concentration (after applying stress), clearly visible from Fig. 2b. Since the concentration of electron reduces (after the induced stress) as depicted in Fig. 2b, the resultant drain current plotted in Fig. 2a also deteriorates slightly (at $V_{gs} = 1$ V, i.e., the voltage at which stress has been applied) along with the tremendously improved off-state leakage current (i.e., reduces). The noted change in off-state current is around one order of magnitude in AlGaN/GaN HEMT which is 50% higher than the corresponding change in MOS-HEMT. Subsequently, the V_{th} (threshold voltage) has been shifted toward the positive side which is higher in AlGaN/GaN-based HEMT (i.e., 0.3 V) as compared to MOS-HEMT (0.5 V). The effect of oxide thickness variation on the drain current of AlGaN/GaN MOS-HEMT

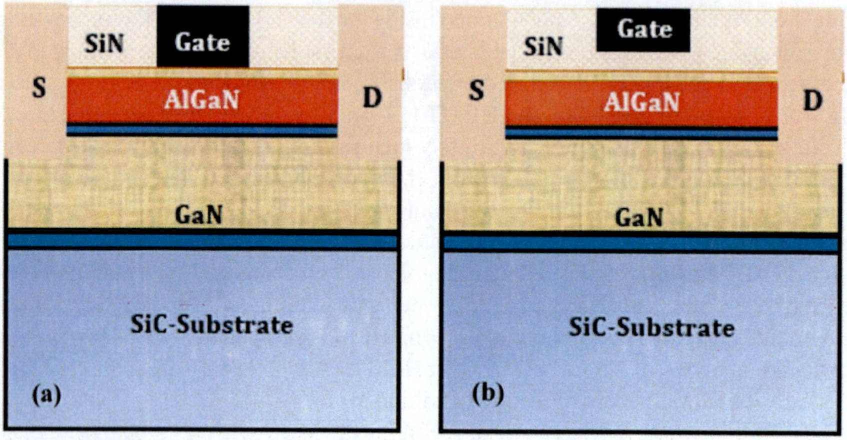

Fig. 1 Architecture of **a** AlGaN/GaN HEMT and AlGaN/GaN MOS-HEMT, T_{Barrier} = Barrier thickness and T_{ox} = Oxide thickness in MOS-HEMT

has also been plotted in Fig. 2a, and it has been observed that the shift in drain current characteristics after positive gate bias-induced stress is higher at $T_{\text{ox}} = 8$ nm due to higher electron concentration after de-trapping of the electrons as clearly visible from Fig. 2b. By using high-k dielectric material as gate oxide and passivation, the shift in the threshold voltage increases tremendously in AlGaN/GaN MOS-HEMT as shown in Fig. 3a compared to Fig. 2a. But the change in device performance with increase in passivation permittivity is negligible in AlGaN/GaN HEMT as clearly visible from Figs. 3a and 2a. The maximum shift in the device performance observed from the results plotted in Fig. 3 is in MOS-HEMT having Al_2O_3 as gate oxide and passivation which is because of the higher trap density in the device due to the presence of oxide layer between gate and cap layer.

Figure 4a, b compares the influence of positive gate bias-induced stress on AlGaN/GaN-based HEMT and MOS-HEMT, respectively, at a different barrier thickness and operating temperature. As depicted from the figures, lower barrier thickness results in improved threshold voltage (i.e., more positive) which is slightly higher in AlGaN/GaN-based MOS-HEMT. Also, the impact of positive V_{gs} (gate bias)-induced stress on the threshold voltage is lower at 15-nm barrier thickness. The change in off-state current is also lower at lower barrier thickness, i.e., less than one order of magnitude. The enhancement in operating temperature shows lower on-state current and higher I_{off} (off-state current) in Fig. 4 a and b. The corresponding change in I_{off} (i.e., before and after stress) is also tremendously high at 400 K instead of 300 K in both AlGaN/GaN HEMT and MOS-HEMT.

3 Conclusion

Work presented in this paper assessed the reliability of AlGaN/GaN-based HEMT and MOS-HEMT architectures. Positive gate bias-induced stress has been applied at the gate terminal for 15s, and drain current has been analyzed before and after gate bias-induced

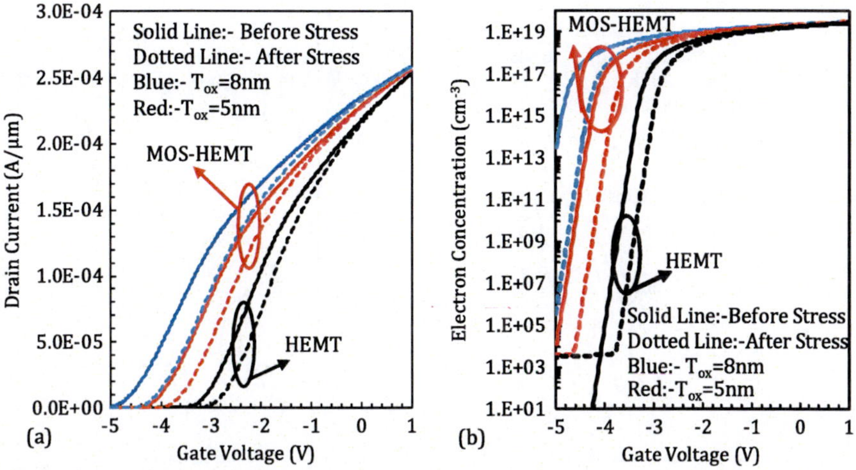

Fig. 2 **a** Drain current variation and **b** electron concentration variation of AlGaN/GaN HEMT and MOS-HEMT before and after positive gate bias-induced stress, $V_{ds} = 1$ V, and passivation and oxide material is silicon nitride

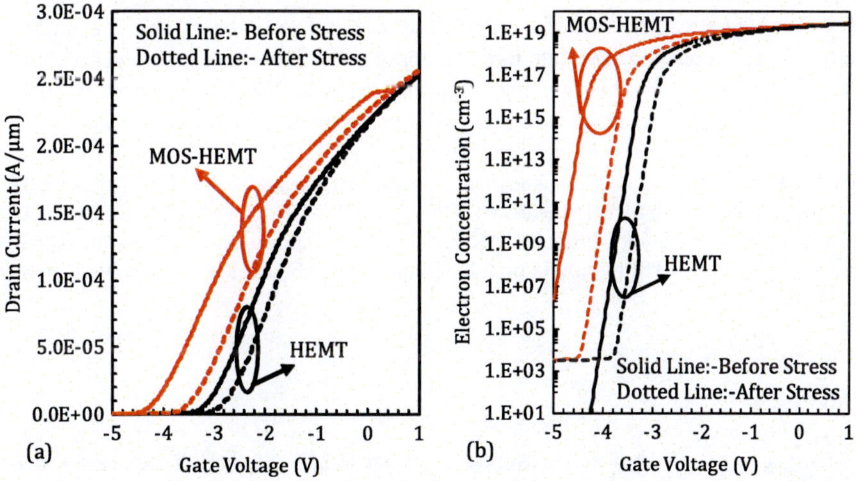

Fig. 3 **a** Drain current variation and **b** electron concentration variation of AlGaN/GaN HEMT and MOS-HEMT before and after positive gate bias induced stress, $V_{ds} = 1$ V, and passivation and oxide material is Al_2O_3, $T_{Barrier} = 25$ nm, $T_{ox} = 5$ nm

stress. Results presented in the paper shows that the change in device performance with gate bias-induced stress is higher in AlGaN/GaN-based MOS-HEMT as compared to conventional HEMT. Effect of oxide thickness, barrier thickness and operating temperature on the device reliability has also been assessed. The enhancement in both oxide thickness and temperature leads to higher shift in drain current after applying gate bias-induced stress. High-k gate dielectric (Al_2O_3) is also responsible for the enhancement in drain current variation with applied gate bias induced stress. Since the modulation in

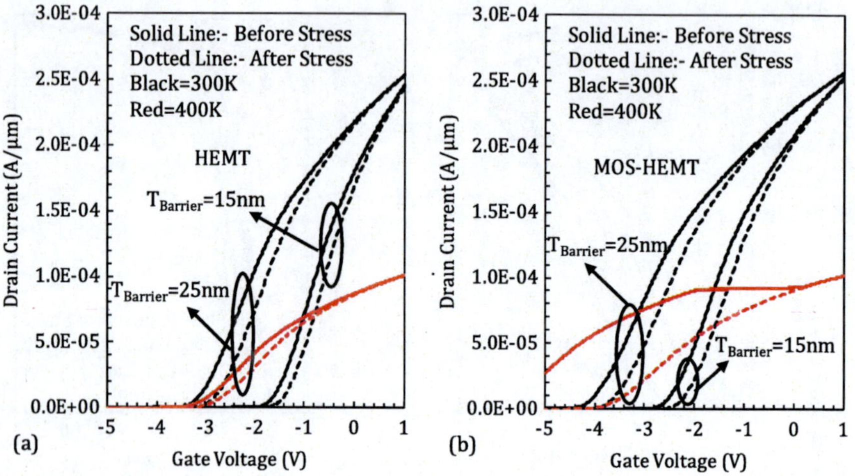

Fig. 4 Drain current variation with gate voltage before and after positive gate bias-induced stress at different barrier thicknesses (T_{Barrier}) for **a** AlGaN/GaN HEMT, **b** AlGaN/GaN MOS-HEMT, $V_{\text{ds}} = 1$ V, and passivation and oxide material is silicon nitride, $T_{\text{ox}} = 5$ nm

device characteristics (in terms of I_{off}, and threshold voltage) is because of the enhancement in electron concentration due to de-trapping of electrons from the level of trap, it can be concluded that the traps are more significant in AlGaN/GAN-based MOS-HEMT which can be further increased at higher operating temperature and oxide thickness.

Acknowledgements. The authors acknowledge the support provided by DDU College (for DBT Star College Laboratory); SDR Laboratory (SDRL) at DOES-UDSC; and SSPL-DRDO (CARS Project No.: 1115/CARS-73/TS/SPL/18). One of the authors (Amrutamayee Nayak) would like to thank IASc–INSA–NASI for providing fellowship under SRFP 2019 vide registration number ENGS4992.

References

1. Meneghesso, G., Rampazzo, F., Kordos, P., Verzellesi, G., Zanoni, E.: Current collapse and high-electric-field reliability of unpassivated GaN/AlGaN/GaN HEMTs. IEEE Trans. Electron. Devices **53**, 2932–2941 (2006)
2. Ambacher, O., Foutz, B., Smart, J., Shealy, J. R., Weimann, N. G., Chu, K., Murphy, M., Sierakowski, A. J., Schaff, W. J., Eastman, L. F., Dimitrov, R., Mitchell, A., Stutzmann, M.: Two dimensional electron gases induced by spontaneous and piezoelectric polarization undoped and doped AlGaN/GaN heterostructures, J. Appl. Phys.: **87**, 334–344 (2000)
3. Miller, E.J., Dang, X.Z., Yu, E.T.: Gate leakage current mechanisms in AlGaN/GaN heterostructure field-effect transistors. J. Appl. Phys. **88**, 5951–5958 (2000)
4. Khan, M.A., Simin, G., Yang, J., Zhang, J., Koudymov, A., Shur, M.S., et al.: Insulating gate III-N heterostructure field-effect transistors for high-power microwave and switching applications. IEEE Trans. Microw. Theory Technol. **51**, 624–632 (2003)
5. Vetury, R. Zhang, N Q. Keller, S., Mishra, U. K. The impact of surface states on the DC and RF characteristics of AlGaN/GaN HFETs. IEEE Trans Electron Dev, 48, 560–6, (2001).

6. Mishra, U.K., Parikh, P., Wu, Y.: AlGaN/GaN HEMTs—an overview of device operation and applications. Proc. IEEE **90**, 1022–1031 (2002)
7. Wang, M., Chen, K.J.: Kink effect in AlGaN/GaN HEMTs induced by drain and gate pumping. IEEE Electron. Dev. Lett. **32**, 482–484 (2011)
8. Meneghini, M., Rossetto, I., Bisi, D., Ruzzarin, M., Hove, M.V., Stoffels, S., Wu, T.L., Marcon, D., Decoutere, S., Meneghesso, G., Zanoni, E.: Negative bias-induced threshold voltage instability in GaN-on-Si power HEMTs. IEEE Electron. Device Lett. **37**, 474–477 (2016)
9. Huang, S., Yang, S., Roberts, J., Chen, K.J.: Threshold voltage instability in Al2O3/GaN/AlGaN/GaN metal-insulator–semiconductor high-electron mobility transistors Jpn. J. Appl. Phys.: **50**, 110202 (2011)
10. Lansbergen, G.P., Wong, K.Y., Lin, Y.S., Yu, J.L., Yang, F.J,. Tsai, C.L., Oates, A.S.: Threshold voltage drift (PBTI) in GaN D-MODE MISHEMTs: Characterization of fast trapping components. In: IEEE International Reliability Physics Symposium, 6C.4.1–6C.4.6 (2014)
11. Lagger, P., Ostermaier, C., Pobegen, G., Pogany, D.: Towards understanding the origin of threshold voltage instability of AlGaN/GaN MIS-HEMTs in IEDM Tech. Dig. pp. 13.1.1–13.1.4 (2012)
12. Johnson, D.W., Lee, R.T.P., Hill, R.J.W., Wong, M.H., Bersuker, G., Piner, E.L., Kirsch, P.D., Harris, H.R.: Threshold voltage shift due to charge trapping in dielectric-gated AlGaN/GaN high electron mobility transistors examined in Au-free technology. IEEE Trans. Electron Devices **60**(10), 3197–3203 (2013). https://doi.org/10.1109/TED.2013.2278677
13. Mizutani, T., Ohno, Y., Akita, M., Kishimoto, S., Maezawa, K.: A Study on Current Collapse in AlGaN/GaN HEMTs Induced by Bias Stress. IEEE Trans. Electron Devices **50**, 2015–2020 (2003)
14. Wu, T.L., Marcon, D., Zahid, M.B., Hove, M.V., Decoutere, S., Groeseneken, G.: Comprehensive investigation of on-state stress on DMode AlGaN/GaN MIS-HEMTs. In: IEEE International Reliability Physics Symposium (IRPS), pp. 3C.5.1–3C.5.7 (2013)
15. Malik, A., Sharma, C., Laishram, R., Bag, R.K., Rawal, D.S., Vinayak, S., Sharma, R.K.: Role of AlGaN/GaN interface traps on negative threshold voltage shift in AlGaN/GaN HEMT. Solid State Electron. **142**, 8–13 (2018)
16. Gupta, S.D., Sun, M., Armstrong, A., Kaplar, R.J., Marinella, M.J., Stanley, J.B., Atcitty, S., Palacios, T.: Slow detrapping transients due to gate and drain bias stress in high breakdown voltage AlGaN/GaN HEMTs. IEEE Trans. Electron Devices **59**, 2115–2122 (2012)
17. Imada, T., Motoyoshi, K., Kanamura, M., Kikkawa, T.: Reliability analysis of enhancement-mode GaN MIS-HEMT with gate-recess structure for power supplies. In: Proceedings of IIRW, pp. 38–41 (2011)
18. Liu, X., Chin, H., Tan, L., Yeo, Y.: In situ surface passivation of gallium nitride for metal organic chemical vapor deposition of high permittivity gate dielectric IEEE Trans. Electron Device **58**, 95–102 (2011)
19. Liu, H.Y., Lee, C.S., Hsu, W.C., Tseng, L.Y., Chou, B.Y., Ho, C.S., et al.: Investigation of AlGaN/AlN/GaN MOS-HEMTs on Si substrate by ozone water oxidation method. Trans. Electron Devices **60**, 2231–2237 (2013)
20. Lagger, P., Reiner, M., Pogany, D., Ostermaier, C.: Comprehensive study of the complex dynamics of forward bias-induced threshold voltage drifts in GaN based MIS-HEMTs by stress/recovery experiments. IEEE Trans. Electron Devices **61**, 1022–1030 (2014)
21. Chang, C.Y., Anderson, T., Hite, J., Lu, L., Lo, C.F., Chu, B.H., Cheney, D.J., Douglas, E.A., Gila, B.P., Ren, F., Via, G.D., Whiting, P., Holzworth, R., Jones, K.S., Jang, S., Pearton, S.J.: Reverse gate bias-induced degradation of AlGaN/GaN high electron mobility transistors. J. Vac. Sci. Technol. B **28**, 1044–1047 (2010)

22. Silvaco Atlas TCAD Tool Ver. 5.26.1.R. User's Manual Available at: https://www.silvaco.com
23. Karumuri, N., et al.: A Compact Model of Drain Current for GaN HEMTs Based on 2-DEG Charge Linearization. IEEE Trans. Electron Devices **63**, 4226–4232 (2016)

Novel Low-Power Nonvolatile High-K Memristor FET with Programmable SET/RESET for Synaptic Learning

Debashis Panda(✉) and Alaka Pradhan

Department of Physics, National Institute of Science and Technology (Autonomous), Palur Hills, Berhampur 761008, Orissa, India
dpanda@nist.edu

Abstract. Memristor field-effect transistor based on TiO_x layer has been fabricated; 15-nm active switching layer and position of the different gates are confirmed from the scanning electron microscope structure. Forming voltage of ~7 V is required to initiate the switching process. The set/reset voltages are strongly influenced by the gate voltage. As the gate voltage increases, the set/reset voltages also increase. The effect of gate on switching mechanism is also briefly discussed.

Keywords: Titanium oxide · Nonvolatile memory · Memristor · FET

1 Introduction

Technology and consumers have demanded for high package density and low-power devices. Memristors are crossbar devices with two distinct states of high and low resistance (HRS and LRS). The voltage applied between their two terminals alters their state that should ideally be stable otherwise. These nonvolatile memory devices can be realized using solid electrolytes, metal–insulator–metal (MIM) with many different metal oxides [1–7]. Many studies in the past were carried out to understand the underlying mechanisms of conduction and programming in memristors.

Memristors based on TiO_2 are very attractive after commercial development by Stanley Williams of HP. On the other hand, field-effect transistor (FET) based on high-k is almost established for scaled CMOS applications. The memristors based on TiO_2 are very attractive for the synaptic learning.

Here, we use a memristor device based on TiO_2 as an active switching layer with a gate electrode using HfO_2 as a gate dielectric that can be used to independently modify the conduction state of the device (Fig. 1) to gain insight into its underlying conduction and programming mechanisms.

2 Experimental

Memristor FET devices are fabricated by sputtering of 10-nm Ti on 50-nm/5-nm Pt/Ti on oxidized p-Si that was patterned to form the 3-μm-wide source electrode. The top

© Springer Nature Singapore Pte Ltd. 2021
N. R. Das and S. Sarkar (eds.), *Computers and Devices for Communication*,
Lecture Notes in Networks and Systems 147,
https://doi.org/10.1007/978-981-15-8366-7_75

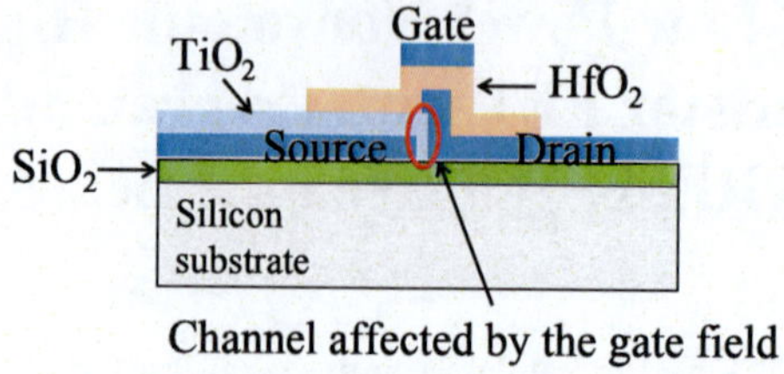

Fig. 1 Cross-sectional schematic of the memristor FET, showing the device structure

Ti layer was then oxidized at 500 °C for 10 min in oxygen ambient to form 10–15-nm TiO_x switching layer. Subsequently, 5-nm Ti followed by 50-nm Pt was sputtered and patterned to form the drain electrode that was intentionally overlapped (ideally by 0.1 μm) with the source; 100-nm Pt is sputtered again and patterned to achieve mechanically robust electrical connection. Finally, a 50-nm-thick HfO_2 layer is deposited by ALD and patterned it as the gate dielectric that was followed by 100-nm Pt that formed the gate electrode (Figs. 2 and 3). Source–drain electrodes are 3 μm wide and 10-μm long. Length of the active switching TiO_x layer is about 15 nm. Optical microscope and scanning electron microscope (SEM) are used to see the trigated structure.

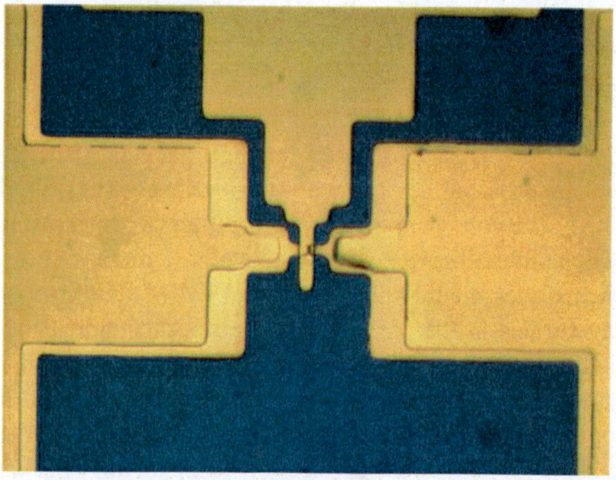

Fig. 2 Optical microscope image of the gated memristor FET

3 Results and Discussion

Figure 4 shows typical forming (the initial SET operation) and corresponding RESET (erasing the conductive state) current–voltage measurement of one of our devices. Higher SET voltage was required in the forming operation for the initial alignment of oxygen vacancies to form conducting filaments. There is an oxygen vacancy gradient that naturally occurs after the titanium oxidation. The oxygen vacancies initially have lower

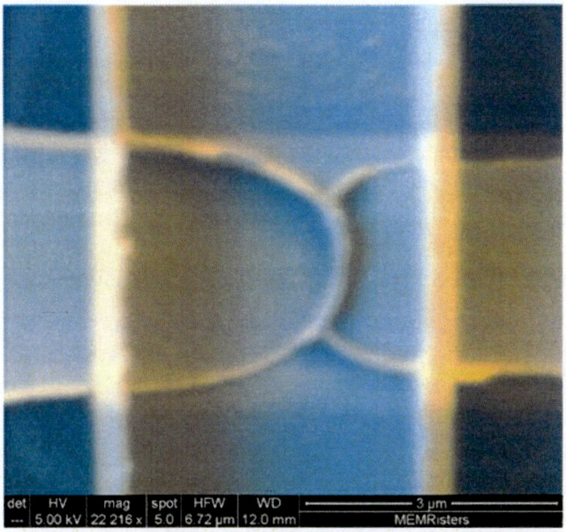

Fig. 3 SEM image of the gated memristor FET

densities near the top surface of the switching active layer under the gate oxide and under the overlapping drain electrode. However, oxygen vacancies have higher concentrations near the bottom interface over the source electrode that is titanium-rich. We anticipate that the gate field effect rearranges the oxygen vacancies in the channel between the drain and source and also can impede or enhance the conductive filament formation.

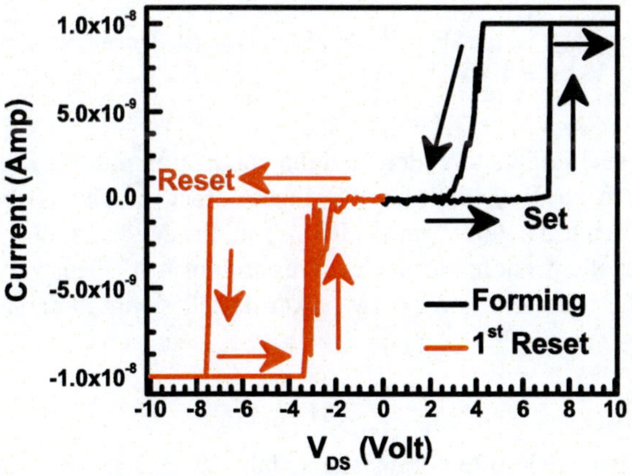

Fig. 4 Forming and first reset of the gated memristor FET ($V_f = 7$ V, $I_f = 1.8$ pA)

A negative gate ($V_g = -2$ V) reduced the SET voltage, while a positive gate ($V_g = 2$ V) increased it in our devices as shown in Fig. 5. The change in the SET voltage seems to be due to the attraction of positively charged ions/vacancies to the surface of the channel at $V_g < 0$ and their repulsion from the channel at $V_g > 0$. Additionally, we note that once the conducting filament was formed (i.e., the device is SET), the gate voltage can rupture it reducing the RESET voltage (Fig. 6) for $V_g > 0$. But the gate field effect cannot "form" the conducting filament; it can only enhance its formation by attracting the oxygen vacancies to the surface. A negative gate voltage increased the RESET voltage by attracting positive ions/vacancies to the channel surface where the filament is formed. The combined observation of field effects in SET and RESET operation voltages of our memristors (schematically shown in Fig. 7) consistently indicates that the conducting regions and filaments in these devices are composed of positively charged ions/vacancies. By integrating the current curve, we estimated that ~5×10^{-10} C positive ions/vacancies were attracted to the channel surface at $V_g = -2$ V.

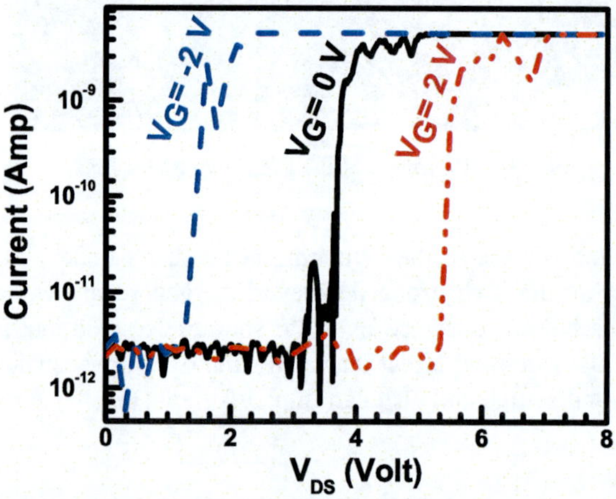

Fig. 5 Effect of gate voltage on the SET process of a typical gated memristor FET. I_S ~ pA; $V_{SET} = 3.6$ V ($V_G = 0$ V); $V_{SET} = 1.1$ V ($V_G = -2$ V) $V_{SET} = 5.2$ V ($V_G = +2$ V)

In addition to enabling us to understand the conduction process in memristors, the gate field effect can be used to isolate the input/output circuits of these nonvolatile memories. Or, we can use a global gate to RESET all the devices at once. Given that the subthreshold slope of current in our devices was around 1 mV/decade, we can also use the three-terminal memristors as very low-power switches with integrated memory.

4 Conclusion

TiO_x-based memristor field-effect transistor is fabricated. Thickness of the switching layer of 15 nm and the position of the gate electrode on active layer are confirmed from the SEM micrograph. About 7 V forming voltage is necessary to initiate the switching process. The gate effectively controlled the set/reset process. The set/reset voltages increase with the increase of gate voltage. The switching mechanism of the FET memristor is also briefly explained. The device can be effectively used for synaptic learning.

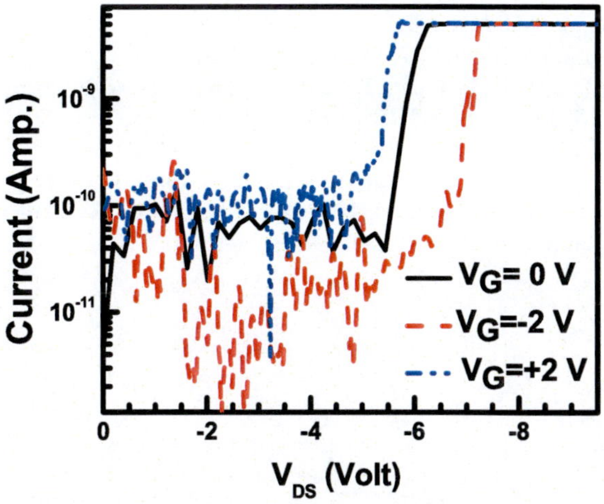

Fig. 6 Effect of gate voltage on the RESET process of a typical gated memristor FET. $V_{RESET} = -6.3$ V ($V_G = 0$ V); $V_{RESET} = -7.2$ V ($V_G = -2$ V), $V_{RESET} = -5.8$ V ($V_G = +2$ V)

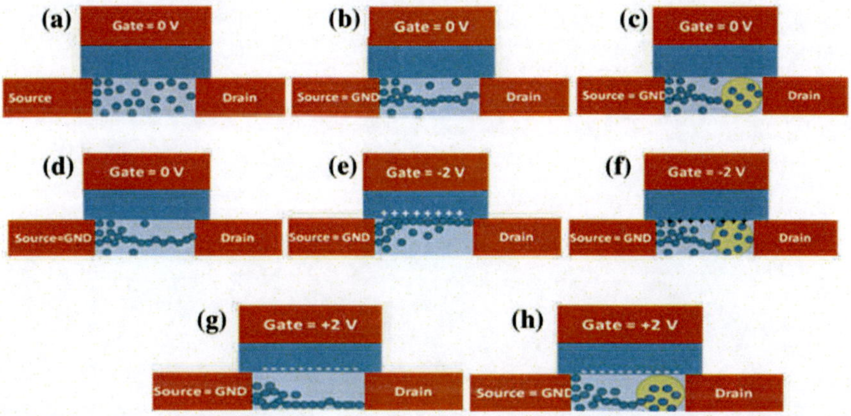

Fig. 7 Schematic model of the effect of gate voltage on the switching process of the gated memristor FET. **a** Device is in initial state, without applying any voltage; **b** first formation of conducting filament by applying +ve voltage between drain and source (forming process); **c** first reset process; **d** SET process without applying any gate voltage; **e** SET process applying negative gate voltage; **f** RESET process with negative gate voltage; **g** SET process after applying positive gate voltage; and **f** RESET process with positive gate voltage

Acknowledgements. This work was supported by the DST-SERB project grant (Grant no: SRG/2019/000129), Government of India. Some experimental help from Utah Nanofab, University of Utah, USA, is greatly acknowledged.

References

1. Tabib-Azar, M., Xie, Y.: ECS Trans **3**(10), 281–289 (2006)

2. Waser, R., Aono, M.: Nat. Mater. **6**, 833–840 (2007)
3. Panda, D., Tseng, T.Y.: Thin Solid Films **531**, 1–20 (2013)
4. Panda, D., Simanjuntak, F.M., Tseng, T.-Y.: AIP Adv. **6**, 0753141–7 (2016)
5. Chandrasekaran, S., Simanjuntak, F., Saminathan, R., Panda, D., Tseng, T.Y.: Nanotechnology **30**, 445205 (2019)
6. Panda, D., Sahu, P.P., Tseng, T.-Y.: Nanoscale Res. Lett. **13**(8), 1 (2018)
7. Chandrasekaran, S., Simanjuntak, F.M., Panda, D., Tseng, T.-Y.: IEEE Trans. Electron Device **66**(11), 4722–4726 (2019)

Numerical Investigation of Gate Field Plate AlGaN/GaN HEMT with Multi-recessed Buffer

Neha[1], Vandana Kumari[2], Mridula Gupta[3], and Manoj Saxena[1](✉)

[1] Department of Electronics, Deen Dayal Upadhyaya College, University of Delhi, New Delhi 110078, India
msaxena@ddu.du.ac.in

[2] Department of Electronics, Maharaja Agrasen College, University of Delhi, New Delhi, India

[3] Department of Electronics Science, University of Delhi, South campus, New Delhi, India

Abstract. Multi-recessed Buffer Gate Field Plate AlGaN/GaN HEMT is considered in this paper for improving various device characteristics such as: breakdown voltage, threshold voltage. Recessed buffer layer spreads the electric field along the gate to drain region resulting in high breakdown voltage. In the present work, 119.5 V of breakdown voltage has been obtained at 0.25 μm gate length, when both gate to source and gate to drain buffer regions are recessed. Positive shift in threshold voltage (−5.8 V (for conventional GFP HEMT) to −0.1 V (for present structure) has also been observed from the GFP-AlGaN/GaN HEMT with multi-recessed buffer. Thus, by further optimizing the device parameter, enhancement mode behavior can be achieved from the device along with superior breakdown voltage. The high breakdown voltage is because of the lower ion-generation rate below the gate electrode as compared to conventional GFP-AlGaN/GaN HEMT.

Keywords: GFP HEMT · Ion-generation · Breakdown voltage · Threshold voltage · Multi-recessed buffer

1 Introduction

Wide band gap materials such as diamond, silicon carbide, and gallium nitride (GaN)-based power electronics devices offer higher resistance to high voltages and lower on-resistance as compared to silicon-based devices because of their attractive properties like high critical breakdown electric field strength, carrier drift velocity, high thermal conductivity, and large carrier mobility [1–3]. GaN HEMT becomes popular in various fields like civil communication, petroleum exploration, aerospace, and many more because of high frequency, high power, high withstand voltage, and high efficiency [4]. Recently, tremendous research in the field of GaN HEMTs is based on peripheral circuits for regulation and compensation of transistors to achieve better output characteristics. However, this will also lead to certain shortcomings like poor transistor withstand voltage, large parasitic capacitance, and narrow transconductance saturation region, which degrade output characteristics of the device [5]. Various techniques were used to overcome these shortcomings like field plate engineering, etc. [6]. In 2019, Zhu et al. also demonstrates a

© Springer Nature Singapore Pte Ltd. 2021
N. R. Das and S. Sarkar (eds.), *Computers and Devices for Communication*,
Lecture Notes in Networks and Systems 147,
https://doi.org/10.1007/978-981-15-8366-7_76

High Gate and Multi-recessed buffer (HG-MRB) AlGaN/GaN HEMT to enhance device performance in terms of breakdown voltage [7]. Since at shorter channel length (nearly 0.75 μm), improved cut-off frequency comes at the cost of inferior breakdown voltage of the device. Thus, in present work, a high gate is used in gate field plate AlGaN/GaN HEMT and along with the multi-recessed buffer near source to gate side and gate to drain side region to improve device breakdown voltage at shorter device dimensions. Parameter explored in this work is: breakdown voltage, threshold voltage, and impact ionization rate. Results have also been compared with the conventional Gate Field Plate GFP-AlGaN/GaN HEMT.

2 Results and Discussion

The device structure of multi-recessed buffer GFP-AlGaN/GaN HEMT used for present investigation is shown in Fig. 1. ATLAS TCAD device simulator is used for investigating AlGaN/GaN HEMT [8]. Models used during simulation are: (a) parallel field mobility model, (b) Shockley-Read–Hall recombination model and (c) Selberherr impact ionization models (selb) for breakdown calculation [8]. The electron mobility used at room temperature is 1800 cm^2/V_s [6]. Interface charges are also used at the buffer–barrier interface.

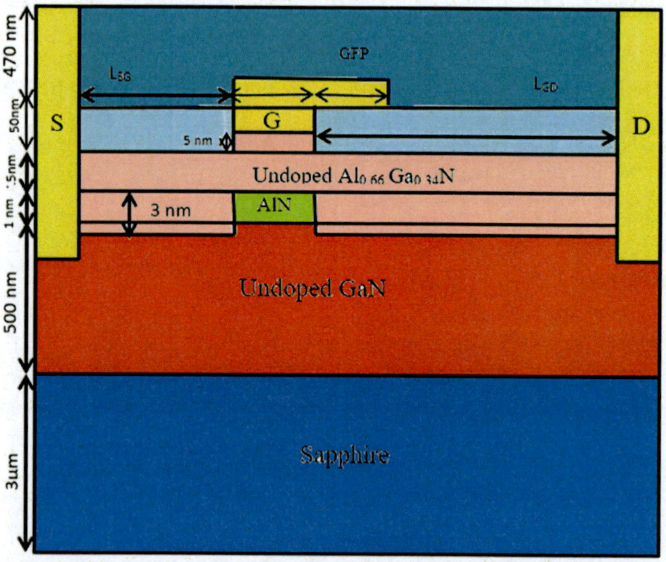

Fig. 1 Cross-sectional view of Gate Field plate AlGaN/GaN HEMT multi-recessed buffer regions. $L_G = 0.25$ μm; $L_{GFP} = 0.5$ μm; $L_{GS} = 0.5$ μm; $L_{GD} = 1.5$ μm

In Fig. 2, electric field along the channel region has been plotted for conventional Gate Field Plate AlGaN/GaN HEMT and multi-recessed buffer Gate Field Plate HEMT. Two different cases of multi-recessed buffer has been used for consideration: (1) only drain side buffer region has been recessed and (2) both drain and source side buffer region have been recessed. In comparison to GFP AlGaN/GaN HEMT, electric field peak reduced with the inclusion multi-recessed buffer in the gate to drain region. By further using recessed architecture along the source side, electric field along the source gate region reduces.

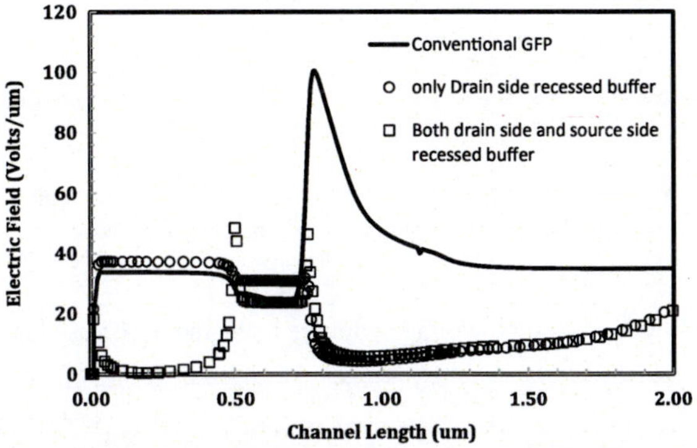

Fig. 2 Comparison of electric field distribution along channel length for recessed buffer structure with conventional GFP AlGaN/GaN HEMT

Impact of the variation in length of recess region in buffer on the breakdown voltage has been plotted in Fig. 3a, b. At first, only drain side buffer region has been recessed, and its impact on the breakdown voltage has been plotted in Fig. 3a. Results show that with the increase in the length recess region toward drain side, breakdown voltage of the device enhanced significantly. As compared to conventional GFP AlGaN/GaN HEMT (without recessed buffer region), breakdown voltage increased by 89.1%, when complete gate to drain buffer region is recessed. Since the higher BV has been achieved from the device having 1.5 μm of recess region, further simulation has been done at that specific recess length. As shown in Fig. 3b, breakdown voltage further increases, when complete source side buffer region is recessed.

In Table 1, threshold voltage has been shown for conventional GFP HEMT and multi-recessed buffer GFP-HEMT. Positive shift in threshold voltage has been observed from the results when buffer regions are recessed as compared to conventional GFP HEMT structure as shown in Table 1. Thus, using proper optimization, enhancement mode operation can be achieved from multi-recessed buffer GFP AlGaN/GaN HEMT.

Impact ionization rate for different structures has also been studied in Fig. 4(a–c) at $V_{DS} = 20$ V in the form of contour plot using ATLAS TCAD. For GFP architecture, the ion-generation is near the drain side edge of the gate electrode. However, using recessed buffer, the region has been shifted toward the drain electrode which results in the improvement in breakdown voltage of the device. Also, the observed ion-generation rate is lower for multi-recessed buffer GFP HEMT at $V_{DS} = 20$ V, thereby improving the voltage handling capability of the device as compared to conventional FP-HEMT.

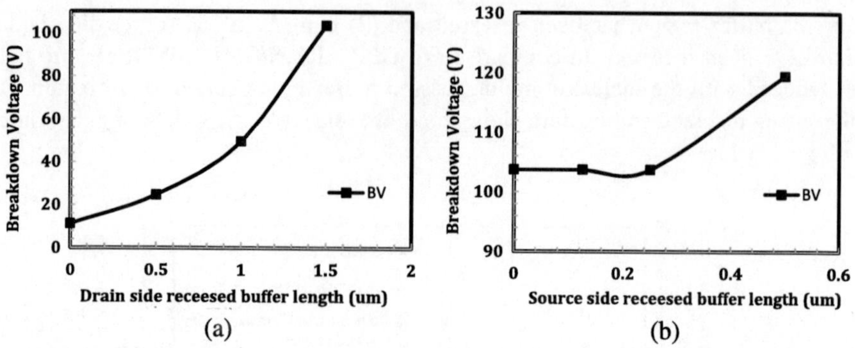

Fig. 3 Variation in breakdown voltage with variation in (**a**) drain side recessed buffer region length only, (**b**) source side recessed buffer region length with drain side recessed buffer = 1.5 μm

Table 1 Variation in threshold voltage with the inclusion of recessed buffer region

Structure	Threshold voltage(V)
Conventional GFP HEMT	−5.8
Only Drain side recessed buffer region = 1.5 μm	−0.3
Both drain side recessed buffer region = 1.5 μm and source side = 0.5 μm	−0.1

3 Conclusion

Multi-recessed buffer technique has been used onto the conventional gate field plate AlGaN/GaN HEMT in this work for investigated various output characteristics using TCAD ATLAS simulator. Enhancement in breakdown voltage is observed when buffer region is recessed toward gate to drain region and gate to source region. The enhancement in device voltage handling capability is because of the shift (and reduction) in the ion-generation region from below the gate edge (toward drain electrode). About 89.1% increase in breakdown voltage is observed as compared to conventional GFP HEMT using multi-recessed buffer technique at 0.75 μm gate length. Observed threshold voltage of the device also shift toward the positive side by using multi-recessed buffer architecture.

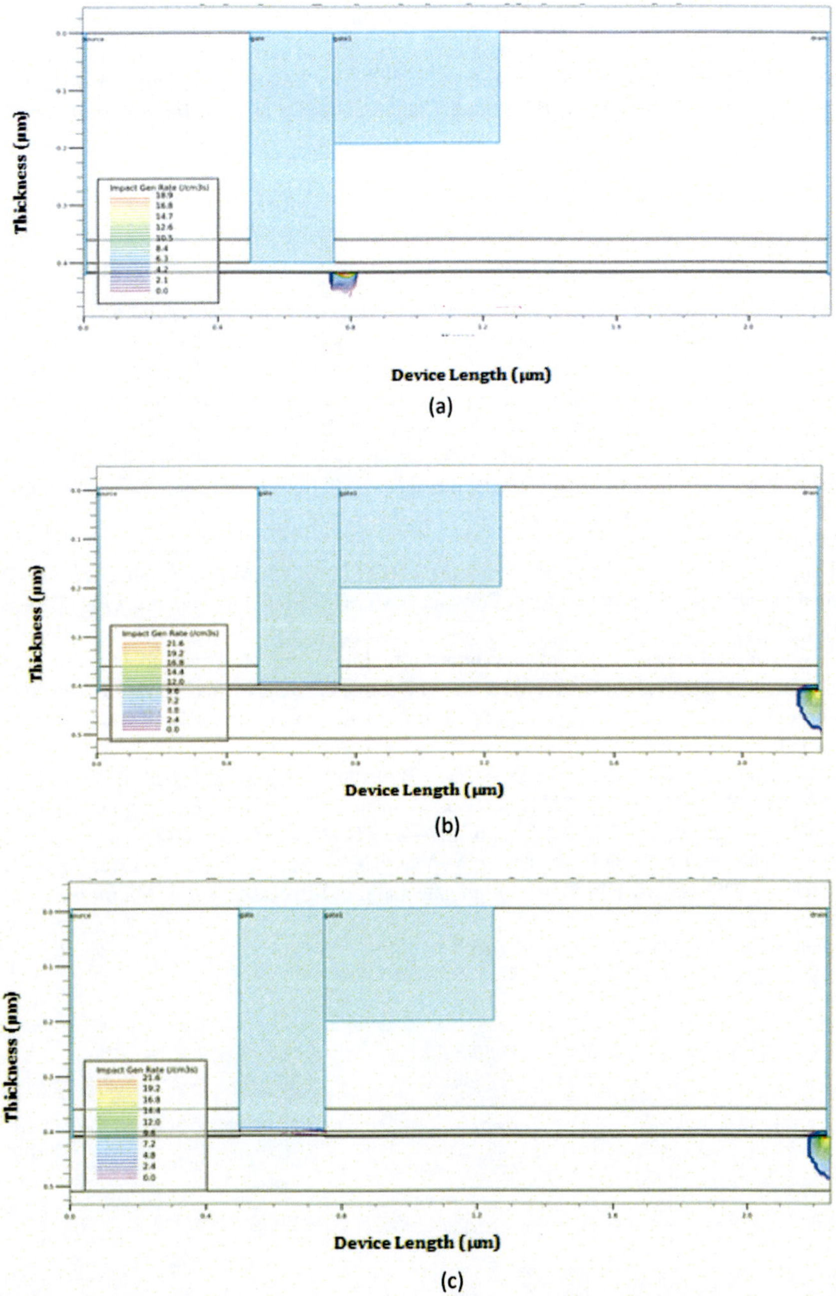

Fig. 4 Contour graphs showing impact ionization rate for (**a**) Conventional Gate Field Plate GFP-HEMT, (**b**) only drain side recessed region = 1.5 µm and (**c**) Both drain side recessed region = 0.5 µm and source side recessed region = 1.5 µm GFP-AlGaN/GaN HEMT at $V_{DS} = 20$ V and $V_{GS} = -7$ V

Acknowledgements. Authors would like to thank Solid State Physics Laboratory-SSPL-DRDO (1115/CARS-73/TS/SPL/18) for providing necessary help for carrying out this work, and one of the author (Neha) would like to thank Semiconductor Device Research Laboratory, Department of Electronic Science, University of Delhi, South Campus, New Delhi, India for providing necessary guidance.

References

1. Saremi, M., Hartwar, R., Dutta, M., Koeck, F.A., Nemanich, R.J., Chowdhury, S., Goodnick, S.M.: Analysis of reverse I-V characteristics of diamond- based PIN diodes. Appl. Phys. Letter **111**, 043507 (2017)
2. Jamali Mahabadi, S.E., Moghadam, H.A: Comprehensive study of 4H-SiC MES-MOSFET. Phys. E. low dimens. Syst. Nanostructure. **74**, 25–29 (2015)
3. Moghadam, H.A., Dimitrijev, S., Han, J., Haasmann, D., Aminbeidokhti, A.: Transient current method for measurement of active near interface oxide traps in 4H-Sic MOS capacitors and MOSFETs. IEEE Trans. Electron Devices **62**, 2670–2674 (2015)
4. Micovic, M., Brown, D.F., Regan, D., Wong, J.,Tang, Y., Herrault, F., Santos, D., Burnham, S.D., Tai, J., Khalaf, I.: High frequency GaN HEMTs for RF MMIC Applications. In: Proceedings of 2016 IEEE International Electron Devices Meeting, San Francisco, CA, USA, 3–7 December 2016
5. Benvegnu, A., Laurent, S., Jardel, O., Muraro, J.L., Meneghini, M., Barataud, D., Meneghesso, G., Zanoni, E., Quere, R.: Characterization of Defects in AlGaN/GaN HEMTs Based on non-linear microwave current transient spectroscopy. IEEE Trans. Electron Devices **64**, 2135–2141 (2017)
6. Neha, Kumari, V., Gupta, M., Saxena, M.: Breakdown voltage analysis of different field plate AlGaN/GaN HEMTs: TCAD based assessment. In: 2018 IEEE Electron Device Kolkata Conference (2018 IEEE EDKCON), pp. 407–412 (24–25 November 2018)
7. Zhu, S., Jia, H., Li, T., Tong, Y., Liang, Y., Wang, X., Zeng, T., Yang, Y.: Novel high energy efficiency AlGaN/GaN HEMT with high gate and multi- recessed buffer. Micromachines **10**, 444 (2019)
8. Silvaco Atlas TCAD tool ver.- 5.24.1.R.

Technology CAD for Dual-Bit Non-volatile Flash Memory to Enhance Storage Capability

B. Sachitra Kumar Patra, Aniket Padhy, E. Roshni, V. Ramya, Shrabani Mahata, Sandipan Mallik, and Satya Sopan Mahato(✉)

Electronics and Communication Engineering Department, National Institute of Science and Technology (Autonomous), Institute Park, Pallur Hills, Berhampur, India
{shrabani.mahata,ssmahato}@nist.edu

Abstract. A flash memory cell may store a single bit (single-level cell-SLC) or multi-bit binary data (multi-level cell-MLC). MLC devices are of less cost and allowed for higher storage density. Here, a dual-bit storage flash memory cell is presented which is capable to store two bits of information. It consists of two floating gates. One is present at the drain side, and the other one is present at the source side. The floating gates are electrically isolated from all the other electrodes by an inter-gate dielectric. The control gate and floating gate are separated by the control oxide layer, and the floating gate and substrate are separated by the tunnel oxide layer. Programming of the memory can be achieved by hot-carrier injection or F-N tunneling while erasing of the memory can be achieved by F-N tunneling.

Keywords: Floating gate · Hot-electron injection · F-N tunneling · Single-level cell · Multi-level cell

1 Introduction

The key device that the flash memory is based on is the floating gate transistor. A floating gate transistor is similar to a conventional MOS transistor, except for the additional electrode called the floating gate and an additional dielectric layer between the floating gate and control gate [1]. Substrate and the floating gate are separated by the tunnel oxide layer. The control gate and floating gate are separated by the control oxide layer, and the floating gate and substrate are separated by tunnel oxide layer. The substrate is of p-type, and source and drain are highly doped n+ type, or the substrate is of n-type, and source and drain are highly doped p+ type. The floating gate acts as the charge storage node capacitively coupled to the control gate and other nodes. Programming can be done by two methods. One by hot-electron injection [2] and second by F-N tunneling mechanism, and erasing can be performed by F-N tunneling [3] mechanism. Storage and removal of charge can be reflected into the transfer characteristics of the device; after write operation, negative charge is stored on the floating gate, the threshold voltage of the device increases, and after erasing negative charge tunnel back to the channel and the threshold voltage changes back to the original value. In a neutral state, no electron on

© Springer Nature Singapore Pte Ltd. 2021
N. R. Das and S. Sarkar (eds.), *Computers and Devices for Communication*,
Lecture Notes in Networks and Systems 147,
https://doi.org/10.1007/978-981-15-8366-7_77

the floating gate. Electron pushed from substrate to floating gate during write operation which increases the potential of the floating gate.

The floating gate memory element consists of an electrode that is completely encased in an insulator, and adjacent conductors are capacitively coupled to the floating gate. The potential on the floating gate can be sensed by its effect on the current, which alters the logical output by changing the charge state on the floating gate is referred to as "writing" to the device.

It is in the case of a single-level cell where a single bit is stored in the floating gate of the memory cell. Dual-bit storage flash memory cell is capable of storing two bits but in two floating gate transistors. And it is a multi-level cell (MLC). The structure of the dual bit storage is given in Fig. 1.

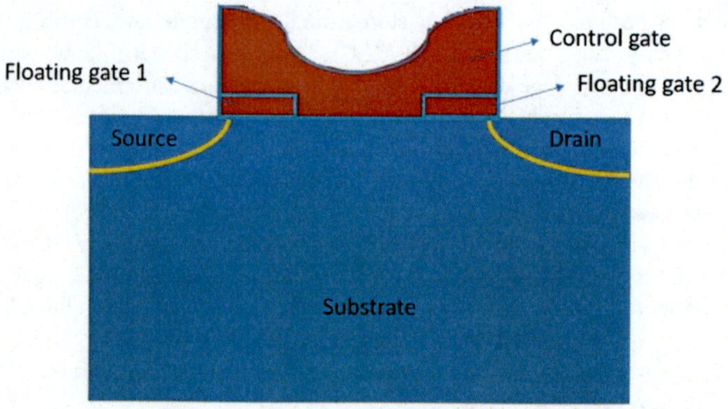

Fig. 1 Dual-bit storage flash memory cell

As shown in Fig. 1, both of the floating gates are electrically isolated and separated by control oxide layer as well as the control gate.

2 Operations

Programming of flash memory can be done by either Channel Hot-Electron Injection (CHEI) or Fowler–Nordheim (F–N) tunneling. Erase operation also performed by F–N tunneling. Some cells use F–N tunneling for both programming and erasing, while others engage CHE injection, for programming and F–N tunneling for erasure only.

1. Channel Hot-Electron Injection (CHEI):
   The wide-band gap of the insulator provides a potential barrier to prevent electrons from tunneling onto or from the floating gate. Hot carriers are injected when carriers are accelerated to the energy level to cross the barrier. Several mechanisms may accelerate the carriers, but the acceleration of electrons by lateral fields in the channel is the only hot-electron mechanism that is currently used by floating gate memories for writing. The channel hot electron must reach at the oxide interface barrier with high energy to cross barrier height and be stored in the floating gate.

2. F-N tunneling:
   F-N tunneling was introduced by Fowler and Nordheim in 1928 and was named after them. This method is realized by tunneling between an electrically isolated electrode and another conducting electrode. This effect can be employed for performing the floating gate programming or erasure. The concept of F-N tunneling is presented in Fig. 2. The energy discrepancy between the conduction band and valence bands in Si is about 1.1 eV while SiO2 is approximately 9 eV. When these two substances are merged together, the conduction band in SiO2 is higher than that in Si by 3.25 eV. The difference in the valence band and conduction band energies is even greater, i.e., more than 4 eV. Since the thermal energy of electron average only 0.025 eV at room temperature, the possibility of the electron in silicon gaining sufficient thermal energy to break through the barrier and enter the conduction band in SiO2 is extremely low.

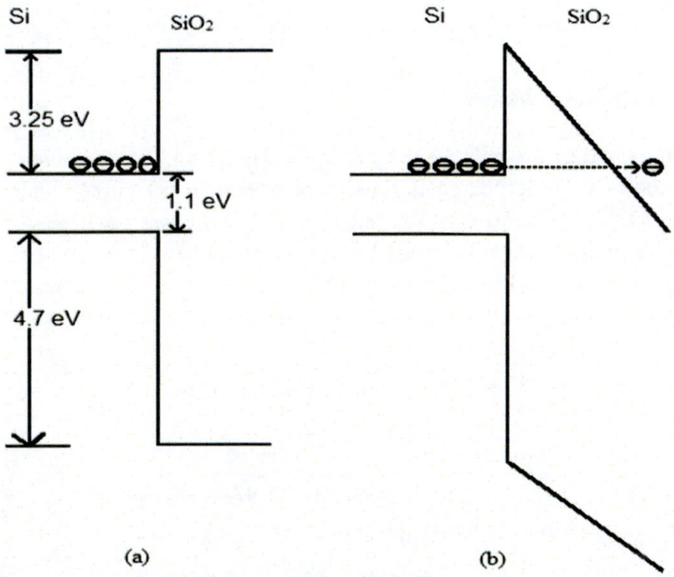

Fig. 2 Potential diagram of the Si/SiO2 interface, **a** with no electric field and **b** with a strong applied electric field [3]

3 Device Structure

Dual-bit storage non-volatile flash memory has been designed using the TCAD (Silvaco) tool. The continuous floating gate (in case of conventional flash memory) is now replaced by two floating gates that are isolated from each other by the control gate. Shockley–Read–Hall (SRH) mechanism was used during the simulation for modeling of trapping and recombination of the traps. The proposed structured is given in Fig. 3.

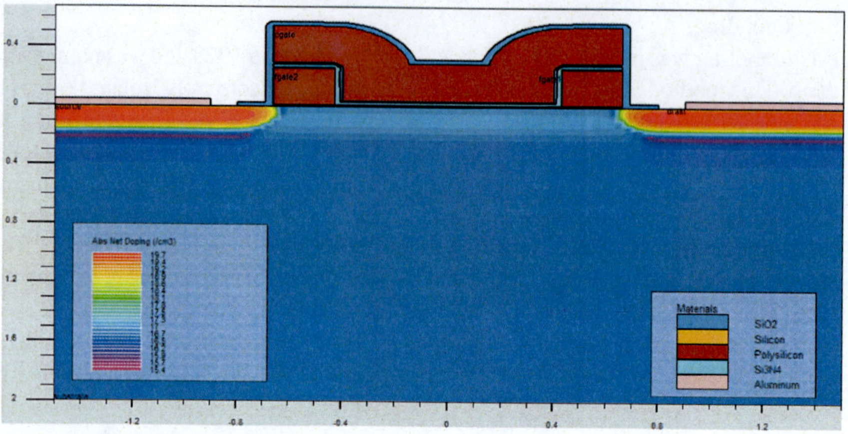

Fig. 3 Doping profile of dual-bit storage flash memory

4 Result and Discussion

Consider that q1 and q2 are charges present in the Fgate1 and Fgate2, respectively.

Table 1 shows that the charge is present or absent in the corresponding floating gates. As it is a dual-bit storage device, maximum of four cases possible. According to the charge values, the bit value is decided. For example, when there is no charge in both the floating gate, then the corresponding bit value is (1 1). Here, in all conditions, charge refers to negative charge.

Table 1 Possible cases of bit value

Bit value	Fgate1	Fgate2
1 1	Charge is absent	Charge is absent
1 0	Charge is absent	Charge is present
0 1	Charge is present	Charge is absent
0 0	Charge is present	Charge is present

a. Programming/Writing Operation:
   Programming can be done by carrier hot-electron injection and F-N tunneling. In programing mode, two types of hot-electron injection are present based on the location of charge injection. First one is source side programming, and second one is drain side programming.

- When a first positive voltage is applied to the cgate and second positive voltage is applied to the drain and the substrate and source are connected to ground, hot carrier injection takes place in drain side. Here, q1 (charge present on fgate1) is negative. It did not affect the q2, i.e., the charge present on fgate2, i.e., conversion of bit values from (1 1) to (1 0), (0 1) to (0 0) (Fig. 4).

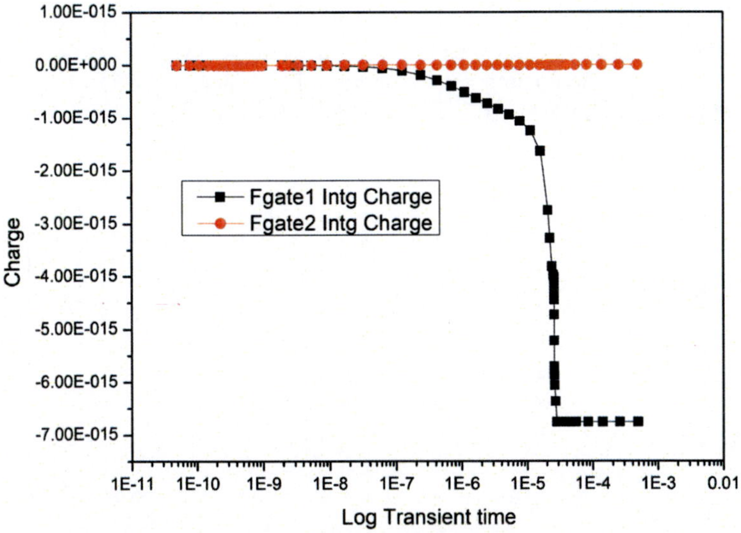

Fig. 4 Charge on the floating gates during drain side programming

- When a first positive voltage is applied to the cgate and second positive voltage is applied to the source and the substrate and drain are connected to ground, hot carrier injection takes place in source side. Here, q2 (charge present on fgate2) is negative. It did not affect the q1, i.e., the charge present on fgate1, i.e., conversion of bit values from (1 1) to (0 1), (1 0) to (0 0) (Fig. 5).

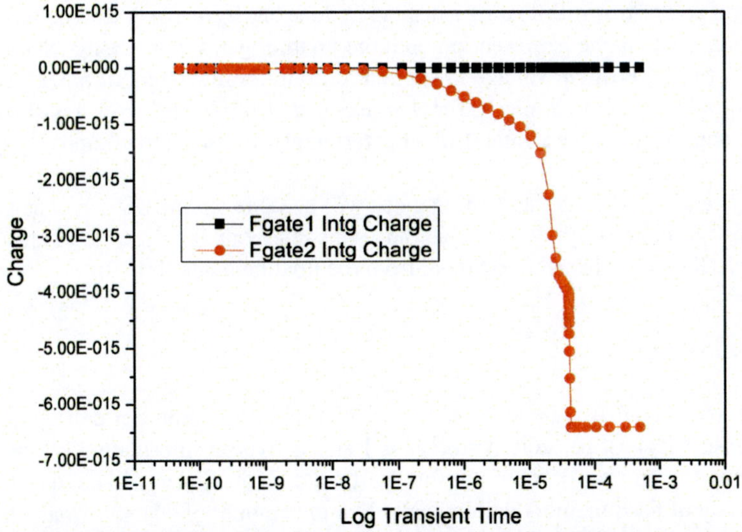

Fig. 5 Charge on the floating gates during source side programming

- When a high positive voltage is applied in cgate, both the charges present on the floating gates (q1 and q2) are changed due to F-N tunneling. Considering there is no charge present in the floating gates, charges will be increased after F-N tunneling, i.e., conversion of bit values from (1 1) to (0 0) (Fig. 6).

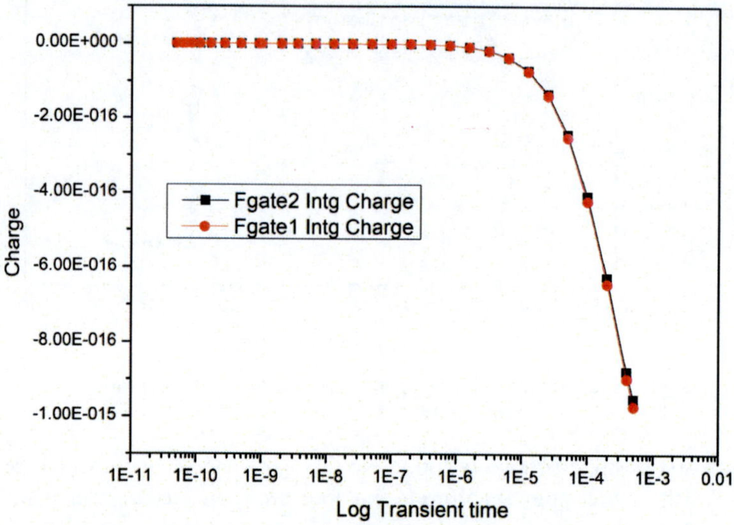

Fig. 6 Charge on the floating gates during F-N tunneling (write)

b. Erase operation:

- Erase of charge from the floating gates can be done one by one or both simultaneously.
- Consider that charge is present on both the floating gates. For erasing charge from Fgate1 (q1), a positive voltage is applied to drain terminal and all other terminals connected to ground. Similarly, for charge erasure from Fgate2 (q2), a positive voltage is applied to source terminal all other terminals connected to ground (Figs. 7 and 8).
- Charge erasure of both the floating gates will be done simultaneously when a high negative voltage is applied to cgate and substrate/body, and junctions are held at ground. It causes electrons to tunnel from the floating gates (Fig. 9).

5 Conclusions

The proposed model is capable to store two bits of information per cell. Due to the higher density, its cost per unit of storage is less. As two-bit information is stored in a single transistor, number of transistor for a particular memory size will be reduced. The present study of flash memory cell indicates that the amount of charge storage depends on cell structure and value and duration of read–write pulse applied.

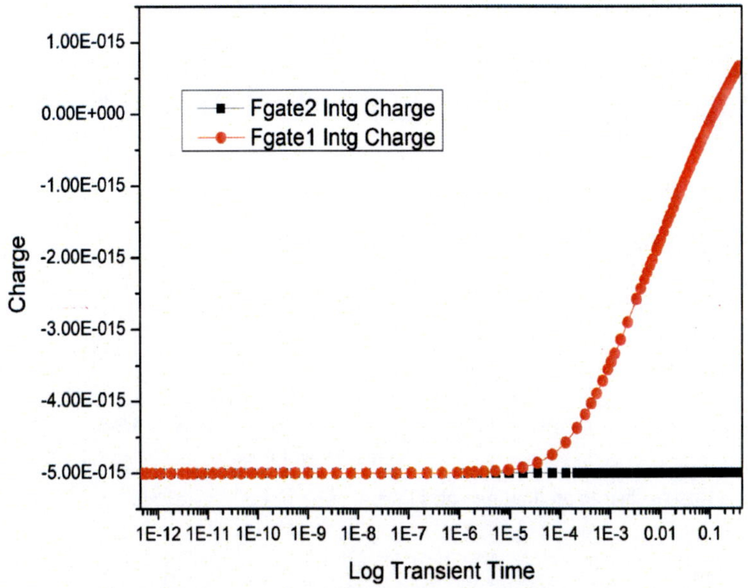

Fig. 7 Charge on floating gates during erase of fgate1

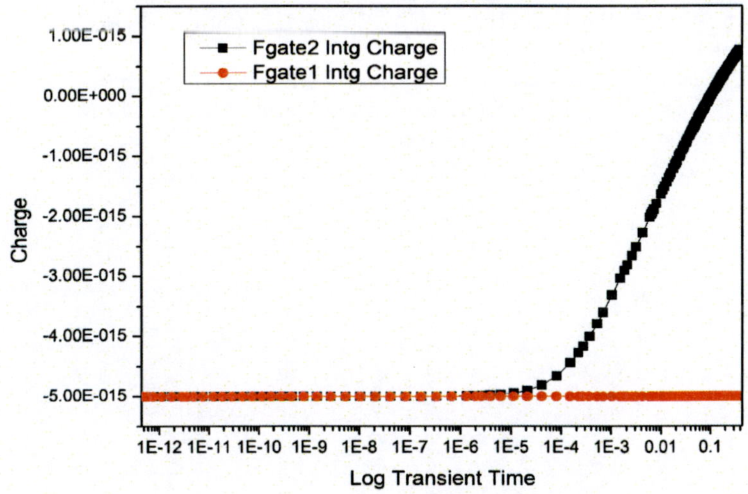

Fig. 8 Charge on floating gates during erase of fgate2

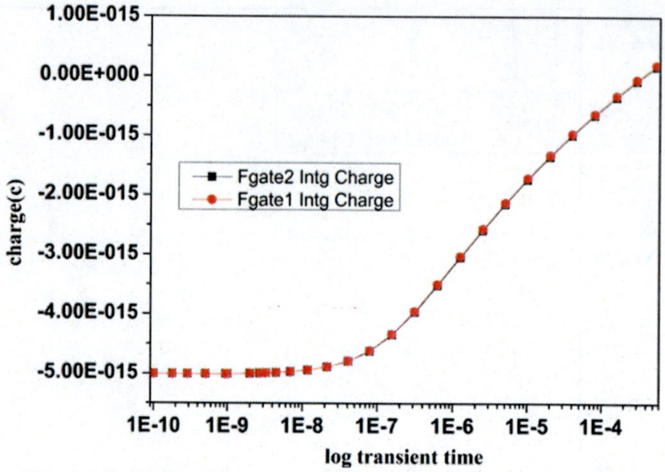

Fig. 9 Charge on floating gates during erase of both fgate1 and fgate2

References

1. Bez R, Camerlenghi E, Modelli A, Visconti A (2003) Introduction to flash memory. In: Proceedings of the IEEE, vol. 91, no. 4 (2003)
2. Tiwari, S., Rana, F., Hanafi, H., Hartstein, A., Crabbe, E.F., Chan, K.: Floating gate memory cell. Appl. Phys. Lett. **68**, 1377 (1996)
3. IEEE Standard Definitions and Characterization of Floating Gate Semiconductor Arrays Standards Committee of the IEEE Electron Devices Society (1998)

Performance Analysis of $Ga_{0.47}In_{0.53}Sb$-FinFET and Si-FinFET for RF and Low-Power Design Applications

Ankit Dixit, Dip Prakash Samajdar(✉), and Dheeraj Sharma

Nanoscale Device, Circuit and System Design Laboratory, Electronics and Communication Engineering Discipline, PDPM, Indian Institute of Information Technology, Jabalpur 482005, India

{ankit.dixit,dheeraj}@iiitdmj.ac.in, dipprakash010@gmail.com

Abstract. In this paper, a tri-gate N-channel $Ga_{0.47}In_{0.53}Sb$-based FinFET is compared with conventional Si-based device for high-frequency applications. TCAD simulation tool is used to investigate the DC behavior of device for different bias voltages. Capacitative analysis is utilized to investigate the RF parameters such as intrinsic delay, power and energy dissipation, transconductance generation factor, and high cut-off frequency of the device. Si-FinFET exhibits better DC performance whereas $Ga_{0.47}In_{0.53}Sb$-FinFET excels in terms of RF performance factors such as lower gate delay, lesser power consumption, and energy dissipation with higher cut-off frequency. All these findings conclude that InGaSb-based FinFETs can serve as suitable candidates for novel nanoelectronic devices operating in the high-frequency regime.

Keywords: III-V FinFET · DC analysis · e-mobility · Gate capacitance · RF analysis

1 Introduction

In the last few decades, non-silicon materials have been investigated to go beyond the Moore's law and mitigate the negative aspects of short channel effects for ultra nanoscale devices. Since the speed of any MOSFET is an important parameter, inclusion of III-V materials in channel region have gained importance in device research due to their higher carrier mobilities coupled with low band gap, which can serve as efficient factors for designing low-power and high-speed nanodevices [1, 2]. Modeling, optimization, and fabrication of heterostructures using III-V materials like InGaAs, InAs, InSb, etc., have been investigated by the researchers worldwide due to their potential advantages [2–4]. It is a well-established fact that FinFET structures provide better gate control over conventional MOSFET, though there few investigations on III-V FinFET structures [5–7]. For the first time, a tri-gate $Ga_{0.47}In_{0.53}Sb$-FinFET is simulated, and its DC and high frequency RF performance are compared with the conventional Si-FinFET. Exhaustive analysis has been done for the both the device for extraction of vital device parameters.

© Springer Nature Singapore Pte Ltd. 2021
N. R. Das and S. Sarkar (eds.), *Computers and Devices for Communication*, Lecture Notes in Networks and Systems 147,
https://doi.org/10.1007/978-981-15-8366-7_78

2 Device Structure and Simulation Approach

A three-dimensional structure of the FinFET device with gate length of 22 nm has been designed using 3D Sentaurus Structure Editor (SDE) [8], and electrical simulation is performed on S device TCAD [9]. Figures 1a, b show the cross-sectional view of the proposed device [10]. Silicon wafer of thickness 10 nm is used as a base of the structure over which 15 nm SiO_2 which acts as a Buried Oxide (BOX).

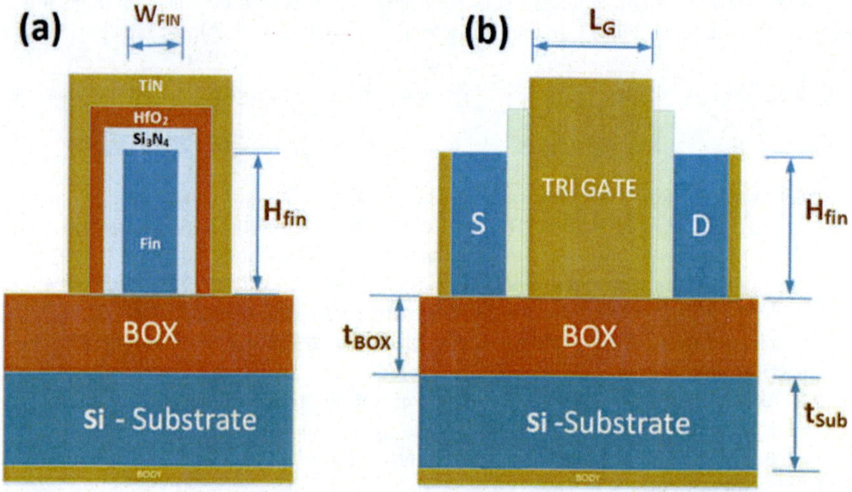

Fig. 1 **a** Front view, **b** Side view of the FinFET device [10]

$Ga_{0.47}In_{0.53}Sb/Si$ is used as Fin Material. The mole fraction used for GaInSb gives optimized performance over DC characteristics [11, 12]. Low-Doped Drain (LDD) regions are added to the structure to reduce the detrimental short channel effects (SCEs) like hot-carrier effect and DIBL [10]. High K-dielectric HfO_2 (K ~ 22) with oxide thickness of 5 nm is used as a gate oxide to obtain Effective Oxide Thickness (EOT) of 1 nm.

3 Results and Discussion

Figure 2a shows the $I_D - V_{GS}$ characteristic for drain voltages of 0.05 and 0.7 V while Fig. 2b illustrates the $I_D - V_{DS}$ characteristic for gate voltages of 0.5 and 0.7 V. From the nature of the graphs, it can be concluded that $Ga_{0.47}In_{0.53}Sb$-based FinFET has 12% less ON current, and transconductance is 30% less in comparison to the Si-based FinFET device. It is also evident that saturation current of $Ga_{0.47}In_{0.53}Sb$-based FinFET is 22% more for gate voltage of 0.7 V. Indeed, the current driving capability of $Ga_{0.47}In_{0.53}Sb$ material is better than Silicon.

Figure 3a illustrates the variation of Subthreshold Slope (SS) with respect to the gate bias. It is observed that for V_{DS} of 0.05 V, $Ga_{0.47}In_{0.53}Sb$-FinFET has only 9% higher

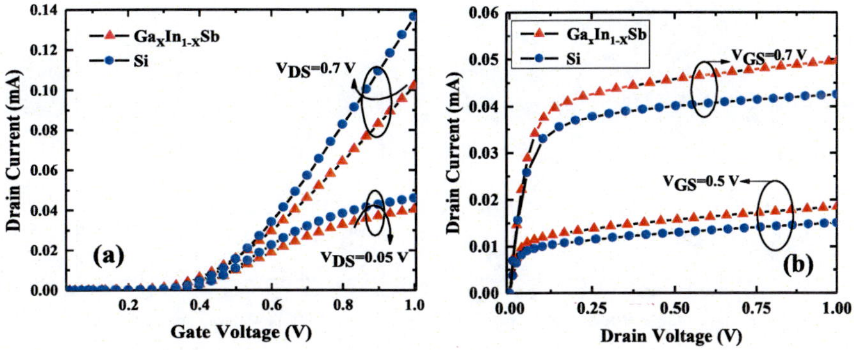

Fig. 2 **a** Comparison of $I_D - V_{GS}$ characteristic and $I_D - V_{DS}$ for $Ga_{0.47}In_{0.53}Sb$- and Si- based tri-gate FinFET for $L_G = 22$ nm and $H_{Fin} = 30$ nm

value of SS as compared to the Si-FinFET. Electron mobility along the channel length plotted in Fig. 3b exhibits superior performance of $Ga_{0.47}In_{0.53}Sb$ FinFET. However, most of the III-V material like GaInSb, InGaAs, InAs, and InSb possess lower band gap, smaller effective mass, and high dielectric constant which makes them more prone to SCEs [13].

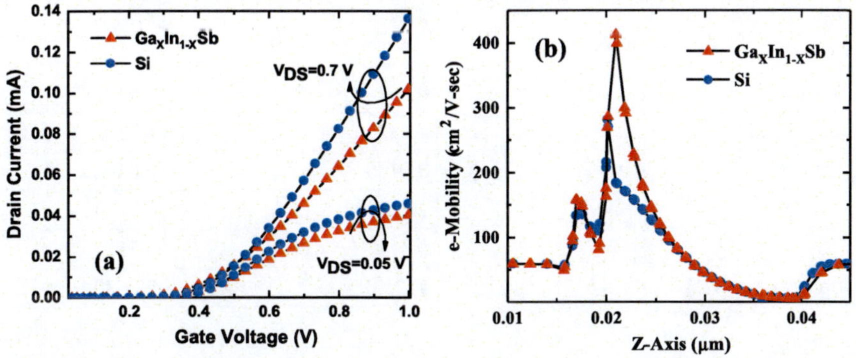

Fig. 3 Variation of **a** SS with respect to gate voltage and **b** electron mobility of $Ga_{0.47}In_{0.53}Sb$- and Si-based tri-gate FinFET for $L_G = 22$ nm and $H_{Fin} = 30$ nm

In Table 1, we have listed all the DC parameters extracted from the simulated curves. The threshold voltage shows significance improvement in $Ga_{0.47}In_{0.53}Sb$ and is approximately 17% lesser as compared to Si-FinFET. Lower output resistance is preferable for designing of a voltage amplifier circuit, and the reverse is true for current amplifier. So GaInSb-based FinFETs can act as efficient voltage amplifiers.

The impact of $Ga_{0.47}In_{0.53}Sb$ material on the analog performance of FinFET device in RF frequency regime is studied. From the circuit point of view, RF performance is measured in terms of RF Figure of Merits (FOMs) such as Intrinsic Delay, Intrinsic Gain, Dynamic Power Dissipation, Power Delay Product, Energy Dissipation, Energy Delay

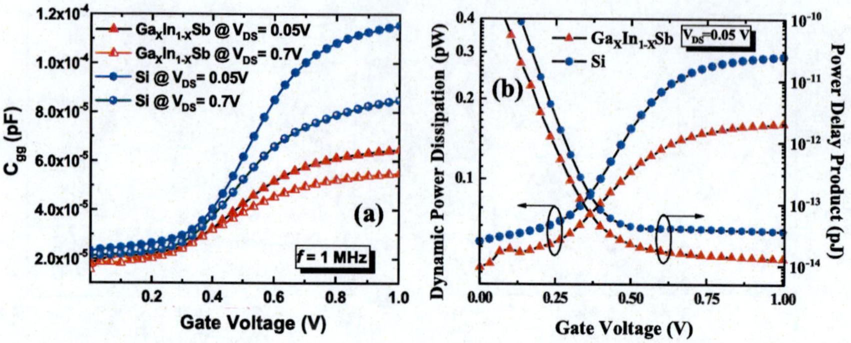

Fig. 4. **a** Gate capacitance. **b** Dynamic power dissipation and power delay product, variation with respect to gate bias

Table 1 Comparison of DC parameters for $Ga_{0.47}In_{0.53}Sb$-FinFET and Si-FinFET

Parameter	Remark	$Ga_{0.47}In_{0.53}Sb$-FinFET	Si-FinFET
V_{th} (V)	@$V_{DS} = 0.05$ V	0.24	0.29
I_{on} (mA)		0.0403	0.0458
I_{off} (A)		1.23×10^{-11}	1.03×10^{-11}
I_{on}/I_{off}		3.27×10^{6}	4.44×10^{6}
SS (mV/dec)		78	71
g_m (mS)		0.0913	0.119
DIBL (mV/V)		76.92	61.53
I_{Dsat} (mA)	@ $V_{GS} = 0.5$ V	0.0185	0.0150
R_{out} (KΩ)		207.73	278.83
g_o (μS)		4.81	3.58
e-Mobility (cm^2/V-s)		412.40	281.30

Product, Transconductance Generation Factor (TGF), Unit Gain Bandwidth (UGB), and Gain Bandwidth Product (GBP) [14–16].

For an ideal device, intrinsic delay, power and energy dissipation should be as low as possible, and intrinsic gain, UGB, GBP should be high. All these RF parameters are highly dependent on the device capacitance value [17]. RF analysis for $Ga_{0.47}In_{0.53}Sb$-FinFET and Si-FinFET is performed for gate length of 22 nm and operating frequency of 1 MHz at $V_{DS} = 0.05$ V. Variation of device gate capacitance with respect to gate bias for different drain voltages is plotted in Fig. 4a. It is interesting to note that for $Ga_{0.47}In_{0.53}Sb$-FinFET, strong inversion region occurs earlier in comparison to Si-FinFET due to lower threshold voltage of the device.

Dynamic power dissipation and power delay product are plotted in Fig. 4b, which clearly predicts better performance of GaInSb-FinFETs over Si-FinFET. In Fig. 5a, we have shown the variation of UGB and GBP as a function of gate bias. It is clear from the graph that for moderate inversion region, both the parameters increases with bias voltage but in the strong inversion region, opposite trend is observed. $Ga_{0.47}In_{0.53}Sb$-FinFET dominates over Si-FinFET irrespective of the gate voltage. The simulated intrinsic delay and intrinsic gain plotted in Fig. 5b shows that Si-FinFET gives better gain whereas $Ga_{0.47}In_{0.53}Sb$-FinFET exhibits lesser gate delay. This is because the intrinsic gain is a strong function of transconductance parameter, and intrinsic delay is a function of gate capacitance.

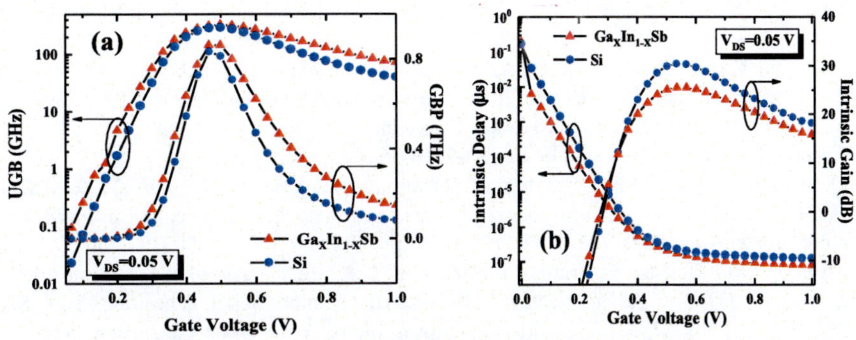

Fig. 5 RF FOMs performance, **a** UGB and GBP, **b** Intrinsic delay and intrinsic gain for $Ga_{0.47}In_{0.53}Sb$- and Si-based tri-gate FinFET for $L_G = 22$ nm

4 Conclusions

In this work, a comparative analysis of the DC and RF performance of $Ga_{0.47}In_{0.53}Sb$-FinFET and Si-FinFET is carried out. Si-FinFET is superior in terms of ON current, higher I_{on}/I_{off} ratio, SS and DIBL whereas $Ga_{0.47}In_{0.53}Sb$-FinFET proves to be a better candidate for higher frequency device applications due to its lower gate delay, lower power and energy dissipation, higher UGB and GBP. It can also be concluded that $Ga_{0.47}In_{0.53}Sb$-FinFET exhibits smaller value of device capacitance and reaches strong inversion region at lower threshold value as compared to Si-FinFET device.

Acknowledgements. The Science and Engineering Research Board, Department of Science and Technology, Government of India under Grant No. ECR/2016/000862 have supported this work.

References

1. ITRS: International Technology Roadmap for Semiconductors (ITRS): More Moore. pp. 1–52 (2015)
2. Johansson, S., Memisevic, E., Wernersson, L., Lind, E.: High-frequency gate-all-around vertical InAs nanowire MOSFETs on Si substrates. IEEE Electron Device Lett. **35**, 518–520 (2014). https://doi.org/10.1109/LED.2014.2310119

3. Tomioka, K., Yoshimura, M., Fukui, T.: A III-V nanowire channel on silicon for high-performance vertical transistors. Nature **488**, 189–192 (2012). https://doi.org/10.1038/nature11293
4. Moroz, V., Smith, L., Huang, J., Choi, M., Ma, T., Liu, J., Zhang, Y., Lin, X., Kawa, J., Saad, Y.: Modeling and optimization of group IV and III-V FinFETs and nano-wires. 2014 IEEE Int. Electron Devices Meet. 7.4.1–7.4.4 (2014). https://doi.org/10.1109/IEDM.2014.7047004
5. Chen, S.H., Liao, W.S., Yang, H.C., Wang, S.J., Liaw, Y.G., Wang, H., Gu, H., Wang, M.C.: High-performance III-V MOSFET with nano-stacked high-k gate dielectric and 3D fin-shaped structure. Nanoscale Res. Lett. **7**, 1 (2012). https://doi.org/10.1186/1556-276X-7-431
6. Lu, W., Roh, I.P., Geum, D., Kim, S., Song, J.D., Kong, L., Alamo, J.A.: 10-nm Fin-width InGaSb p-channel self-aligned FinFETs using antimonide-compatible digital etch. 433–436 (2017)
7. Zhao, X., Vardi, A., Del Alamo, J.A.: Excess off-state current in InGaAs FinFETs. IEEE Electron Device Lett. **39**, 476–479 (2018). https://doi.org/10.1109/LED.2018.2806559
8. Synopsys, T.: Sentaurus Structure Editor User Guide (2014)
9. Synopsys, T.: Sdevice manual. Release H-2014.06, Zurich, Switz
10. Dixit, A., Samajdar, D.P.: Extraction of performance parameters of nanoscale SOI LDD-FinFET using a semi-analytical model of capacitance and channel potential for low-power applications. Appl. Phys. A **126**, 782 (2020). https://doi.org/10.1007/s00339-020-03970-z
11. Moison, J.M., Houzay, F., Barthe, F., Sébenne, C.A.: First steps of the build-up of the Si3N4/Ga0.47In0.53 As interface by UV-induced chemical vapor deposition. Surf. Sci. **251–252**, 165–169 (1991). https://doi.org/https://doi.org/10.1016/0039-6028(91)90974-W
12. Tsang, W.T.: Al0.48In0.52As/Ga0.47In0.53As/Al0.48In0.52As double-heterostructure lasers grown by molecular-beam epitaxy with lasing wavelength at 1.65 μm. J. Appl. Phys. **52**, 3861–3864 (1981). https://doi.org/https://doi.org/10.1063/1.329852
13. Seung, H.P., Yang, L., Kharche, N., Jelodar, M.S., Klimeck, G., Lundstrom, M.S., Luisier, M.: Performance comparisons of III–V and strained-Si in planar FETs and nonplanar FinFETs at ultrashort gate length (12 nm). IEEE Trans. Electron Devices **59**, 2107–2114 (2012)
14. Singh, S., Kondekar, P., Dixit, A.: Digital and analog performance of gate inside p-type junctionless transistor (GI-JLT). In: 2013 Fifth International Conference on Computational Intelligence, Modelling and Simulation. pp. 394–397 (2013)
15. Mohapatra, S.K., Pradhan, K.P., Sahu, P.K.: Linearity and analog performance analysis in GSDG-MOSFET with gate and channel engineering. 11th IEEE India Conference on Emerging Trends Technology. INDICON 2014. pp. 0–4 (2015). https://doi.org/https://doi.org/10.1109/INDICON.2014.7030435
16. Bhaskar, A., Sunil, P., Nigam, K., Kondekar, P.N.: Effect of ITC's on linearity and distortion performance of junctionless tunnel field effect transistor. Superlattices Microstruct. **111**, 293–301 (2017). https://doi.org/.1037//0033-2909.I26.1.78.
17. Mohapatra, S.K., Pradhan, K.P., Singh, D., Sahu, P.K.: The role of geometry parameters and fin aspect ratio of sub-20nm SOI-FinFET: an analysis towards analog and RF circuit design. IEEE Trans. Nanotechnol. **14**, 546–554 (2015). https://doi.org/10.1109/TNANO.2015.2415555

Photo-Absorption Enhancement of Hybrid Solar Cells Through Metallic Nanoparticles Embedded with Nanopyramid Patterning

Sachchidanand and Dip Prakash Samajdar(✉)

Department of Electronics and Communication Engineering, PDPM Indian Institute of Information Technology, Design and Manufacturing, Jabalpur 482005, India
sachbade@gmail.com, dipprakash010@gmail.com

Abstract. The hybrid solar cells (HSCs) are the most abundant solar cell structures with higher charge collection efficiency, newer power conversion mechanisms, coupled with low cost. Recent studies have shown that metallic nanoparticles (MNPs) embedded with nanopyramid (NP) structures help to improve the photo-absorption over a wide range of wavelengths. Nanopyramidal structures, which consist of the tapered sidewalls, ensure gradual increment in the refractive index as light travels from air to the optical nerve of NP. In our study, we have used gold (Au) MNPs, which absorb and scatter the incident light into the substrate with the help of plasmonic resonance effects. In the proposed ITO/PEDOT:PSS/c-Si HSC structure, Au MNPs are located in the gap between nanopyramid arrays and the bottom of silicon (Si) substrate with molybdenum trioxide (MoO_3) coating. The optical simulation of HSCs is carried out using finite difference time domain (FDTD) module of Lumerical solutions software package.

Keywords: Hybrid solar cells · Nanopyramid · Metallic nanoparticles · Plasmonics · FDTD

1 Introduction

In order to solve the ever-growing challenge of energy crisis, solar energy is considered to be the most realistic renewable energy resource as harvesting of the endless energy from our sun can provide power equivalent to almost 10^4 times the mean consumption of humans worldwide [1]. To harness the maximum available solar energy, the researchers are striving to find out novel methods to optimize the power conversion efficiency (PCE) of solar cells (SCs), which have become a dominant research area in the field of photovoltaic research. Out of the available versions of SCs, hybrid solar cells (HSCs) can be considered as a potential candidates due to its reduced material consumption and manufacturing cost [2]. HSCs-based photovoltaic devices with higher optical absorption, low production cost, ease of fabrication and major pay back have emerged as a significant competitor in the contemporary photovoltaic market [3, 4]. There are numerous variety of nanostructures, like nanopyramids [5], metallic nanoparticles [6], nanowires [7],

© Springer Nature Singapore Pte Ltd. 2021
N. R. Das and S. Sarkar (eds.), *Computers and Devices for Communication*, Lecture Notes in Networks and Systems 147,
https://doi.org/10.1007/978-981-15-8366-7_79

nanorods [8], moth eye [9], nanoholes [10], truncated nanopyramids [11] and nanocylinders [12] which have contributed significantly in the improvement of the optoelectronic parameters of HSCs. The major role of each type of nanostructure is to improve the performance of HSCs by enhancing the efficiency and photo-absorption with increase in the light trapping mechanisms and reduction in the photo-reflection [13]. Moreover, considerable progress has been achieved in the improvement of PCE with conformal coating of conjugate polymer of poly (3, 4-ethylenedioxythiophene):poly (styrenesulfonate) (PEDOT:PSS) incorporated with transparent conductor indium tin oxide (ITO) [14].

In this article, we have considered the Au MNPs embedded with NPs into the proposed ITO/PEDOT:PSS/c-Si HSCs structure to utilize the broad range of power absorption spectrum through efficient light trapping. We have coated the Au MNPs with MoO_3 oxide layer for localized plasmonic surface resonance (LPSR) effect which can enhance the HSCs performance by generating more electron-hole pairs (EHP). We have also compared the optical performance of oxide coated/uncoated MNPs embedded with/without NPs HSC structures using FDTD Method.

2 Simulation Framework

A generalized 2D and 3D view of MNPs embedded with NPs in our proposed HSCs for optical simulation is shown in Figs. 1a, b. 3D FDTD method is used for the simulation of a unit cell of different HSC structures, and it is a widely used method for the analysis of nanostructure-based SCs [15]. In our analysis, we have assumed an un-polarized plane wave source with the wavelength range from 300 to 1200 nm. Period boundary condition along x- and y-direction and perfectly matched layer (PML) boundary condition along z-direction of the unit cell HSC structure are considered. Frequency domain power monitors placed at top of plane wave source and bottom of the Si substrate are used to compute the photo-reflection ($R(\lambda)$) and photo-transmission ($T(\lambda)$), respectively. The stored data in $R(\lambda)$ and $T(\lambda)$ monitors helps to calculate the photo-absorption ($A(\lambda)$) using the script command written as [16]:

$$A(\lambda) = 1 - R(\lambda) - T(\lambda) \tag{1}$$

In our design, we have considered c-Si thickness of 1.5 μm, the conformal coating thickness of PEDOT:PSS as 200 nm and the front transparent contact ITO of 110 nm and the back contact Al as 100 nm. The work function (ϕ) of PEDOT: PSS is chosen in such a way that it performs the role of the hole extraction layer (HEL) [17]. The geometrical parameters of nanostructures are depicted in Table 1. The MNPs are coated with MoO_3 with thickness of 13 nm. The optical database of the materials used in the HSCs is taken from the literature and software [18–20].

3 Results and Discussion

The optical short circuit current density, J_{sc} (optical), is evaluated by using the FDTD module for the proposed ITO/PEDOT:PSS/c-Si HSC with the assumption of 100%

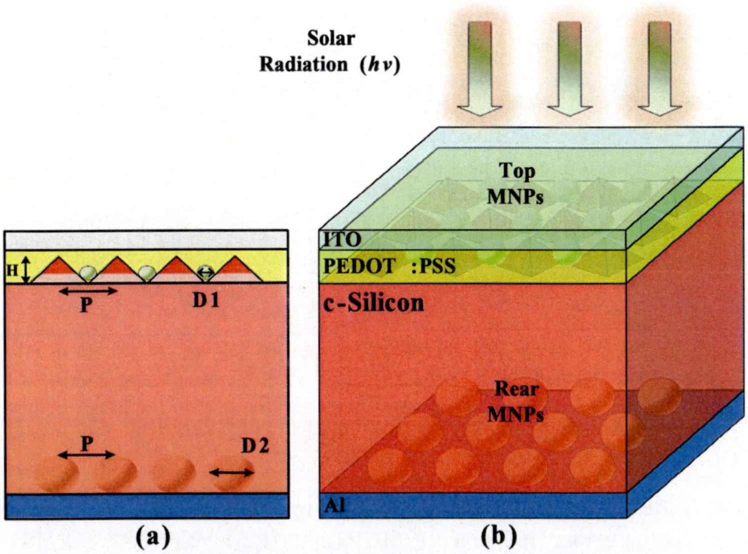

Fig. 1 **a** 2D, **b** 3D Schematic of MNPs embedded with NPs in ITO/PEDOT:PSS/c-Si HSC. P denotes the period of both nanostructures, H denotes the height of NPs, and D1 and D2 denote the top and rear diameters of Au MNPs, respectively. The numerical values of all the dimensions of these nanostructures are listed in Table 1

Table 1 Optimized geometrical parameter of nanostructures

Nanostructures	Geometrical parameters		
	Period, P (nm)	Height, H (nm) or diameter, D1,2 (nm)	Lattice constant, A (nm)
Top Au MNPs	60	25	–
Rear Au MNPs	60	45	–
Nanopyramid	60	150	60

internal quantum efficiency (IQE) which means that each incident photon will absorbed and generate a EHP [5]. The photo-absorption and photo-reflection spectra for the unit cell of HSCs are shown in Fig. 2a, b. From Fig. 2a, it is evident that the photo-absorption between 300 nm < λ < 500 nm is much higher for all nanostructures as compared to the planar counterpart due to reduced photo-reflection as depicted in Fig. 2b. However, for λ > 500 nm, the photo-absorption of planar and NP decreases sinusoidally and slight variation for MNPs embedded with NPs with/without oxide coating as compared to NP and Planar HSC which is shown in Figs. 2a.

Figures 3a–d depict that for λ = 463 nm, the E-field is higher at top end, moderate in the middle and low at the rear end of the all HSCs. In Figs. 3e–h, for λ = 709 nm, the E-field at the top MNP is low because of the plasmonic effect and moderate at the rear

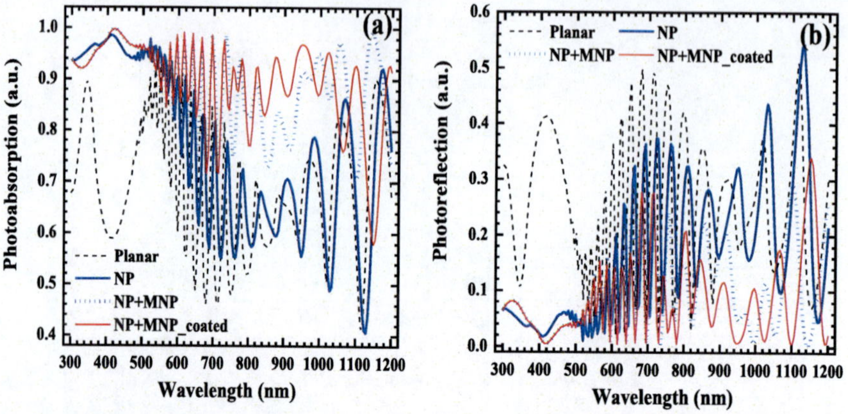

Fig. 2 **a** Photo-absorption, and **b** Photo-reflection spectra of ITO/PEDOT:PSS/c-Si HSCs

MNPs due to scattering [6]. However, for $\lambda = 1118$ nm, in Figs. 3i–l, the E-field is higher throughout the entire planar HSC and NP. MNPs embedded with NPs-based HSCs have a moderate E-field throughout the HSC and higher at the MNPs due to plasmonic and antireflection effect of NPs [21]. So, the overall E-field of HSC is boosted for higher generation of EHPs at $\lambda = 709$ nm and $\lambda = 1118$ nm as shown in Fig. 3e-l. Table 2 represents J_{sc} (optical), the photo-generation rate and maximum photo-absorbance of proposed HSC structures.

4 Conclusions

We have introduced some nanostructures like MNPs and NPs into the HSCs for improving the light trapping mechanisms and optical performance. The HSCs consisting of the Au MNPs embedded with Si NPs exhibit the highest J_{sc} (optical), photo-generation rate and absorbance. MNPs embedded with NPs with/without oxide coating achieve the highest J_{sc} (optical) of 40.92 mA/cm^2 and 40.33 mA/cm^2, which is ~31% and ~29% higher than planar counterpart. The absorbance of MNPs embedded with NPs with/without oxide coating HSCs attains 99.71 and 99.68% which is 6.29 and 6.26% higher than planar HSC.

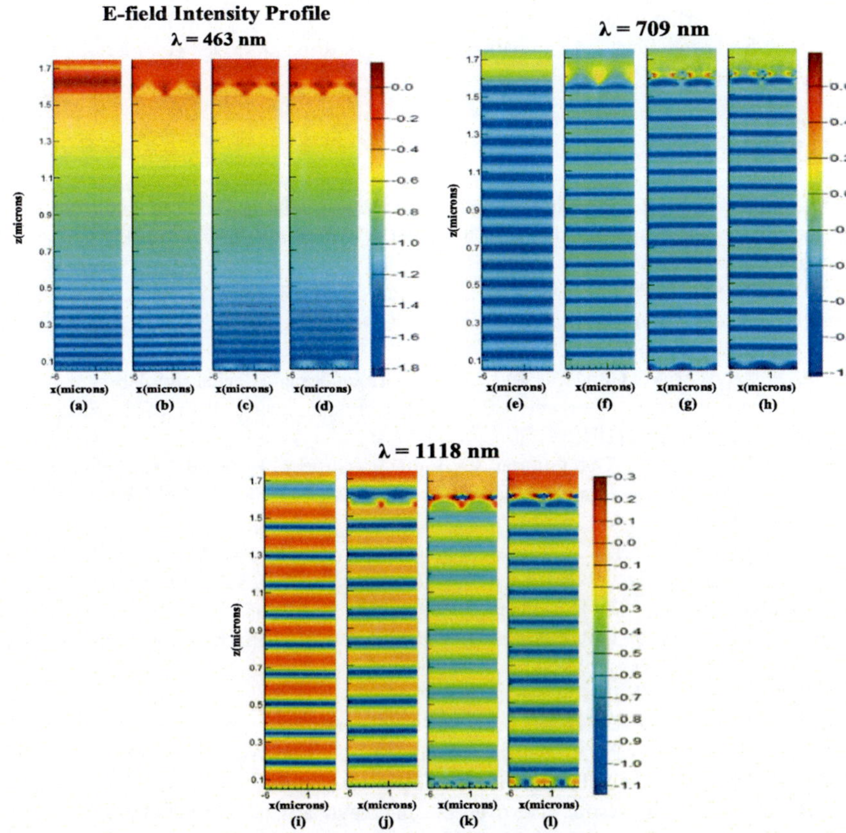

Fig. 3 E-field intensity profiles at 463 nm (**a–d**), 709 nm (**e–h**) and 1118 nm (**i–l**) for *Planar* (**a, e, i**), *NP* (**b, f, j**), *NP + Au_MNP* (**c, g, k**) and *NP + Au_MNP (coated)* (**d, h, l**) based HSCs

Table 2 J_{SC} (optical), maximum photo-generation rate and maximum photo-absorbance of all different HSCs

HSC Structure	J_{sc} (optical) (mA cm^{-2})	Max. photo-generation rate ($m^{-3}s^{-1}$)	Max. photo-absorbance (%)
Planar	31.24	2.31e+28	93.81
NP	35.01	2.64e+28	98.51
MNPs embedded with NPs	40.33	**4.78e+29**	99.68
MNPs (coated) embedded with NPs	**40.92**	2.62e+29	**99.71**

Acknowledgements. The author are thankful to the Science and Engineering Research Board, Department of Science and Technology, Government of India (ECR/2017/002369) (Established through an Act of Parliament) for providing financial support to carry out this work under the project "Analytical Modelling and Simulation of III–V nanostructure-based Hybrid Solar Cells"

References

1. Chen, W., Cabarrocas, P.R.: Rational design of nanowire solar cells : from single nanowire to nanowire arrays. Nanotechnology **30**, 194002 (2019). https://doi.org/10.1088/1361-6528/aaff8d
2. An, D.W.U., Ang, X.I.T., Ang, K.A.I.W., Ianqiang, X.L.I.: Effective coupled optoelectrical design method for fully infiltrated semiconductor nanowires based hybrid solar cells **24**, 1336–1348 (2016)
3. Jiang, W., Salvador, M., Ginger, D.S., Dunham, S.T.: Optics and device simulation of surface plasmonic enhancement of organic solar cell performance using silver nano-prisms. 17th SISPAD Conf. 245–248 (2012)
4. Moulé, A.J., Chang, L., Thambidurai, C., Vidu, R., Stroeve, P.: Hybrid solar cells: Basic principles and the role of ligands. J. Mater. Chem. **22**, 2351–2368 (2012). https://doi.org/10.1039/c1jm14829j
5. Sachchidanand, Samajdar, D.P.: Light-trapping strategy for PEDOT:PSS/ c-Si nanopyramid based hybrid solar cells embedded with metallic nanoparticles. Sol. Energy. **190**, 278–285 (2019). https://doi.org/10.1016/j.solener.2019.08.023
6. Srivastava, A., Samajdar, D.P., Sharma, D.: Plasmonic effect of different nanoarchitectures in the efficiency enhancement of polymer based solar cells: a review. Sol. Energy **173**, 905–919 (2018). https://doi.org/10.1016/j.solener.2018.08.028
7. Prashant, D.V, Samajdar, D.P., Sharma, D., Sachchidanand: Optical Simulation of III–V Semiconductor Nanowires/poly (3, 4-ethylenedioxythiophene) polystyrene sulfonate (PEDOT: PSS) based hybrid solar cells: significance of polymer coating thickness and geometrical parameters on light harvesting and over. 6th International Conference' Microelectronics, Circuits System Micro 2019. **I**, 67–71 (2019)
8. Hussain, S., Jacob, J., Usman, Z., Mahmood, K., Ali, A., Arshad, M.I., Khan, W.S., Farooq, Z., Farooq, M.U., Ashfaaq, A., Rehman, U.: Length dependent performance of Cu_2O/ZnO nanorods solar cells. Superlattices Microstruct. **126**, 181–185 (2019). https://doi.org/10.1016/j.spmi.2019.01.004
9. Luo, X., Lu, L., Yin, M., Fang, X., Chen, X., Li, D., Yang, L., Li, G., Ma, J.: Antireflective and self-cleaning glass with robust moth-eye surface nanostructures for photovoltaic utilization. Mater. Res. Bull. **109**, 183–189 (2019). https://doi.org/10.1016/j.materresbull.2018.09.029
10. Lu, R., Xu, L., Ge, Z., Li, R., Xu, J., Yu, L., Chen, K.: Improved efficiency of silicon nanoholes/gold nanoparticles/organic hybrid solar cells via localized surface plasmon resonance. Nanoscale Res. Lett. **11**, 0–6 (2016). https://doi.org/10.1186/s11671-016-1374-0
11. Yang, Z., Li, X., Wu, S., Gao, P., Ye, J.: High-efficiency photon capturing in ultrathin silicon solar cells with front nanobowl texture and truncated-nanopyramid reflector. Opt. Lett. **40**, 1077–1080 (2015). http://dx.doi.org/10.1364/OL.40.001077
12. Wang, W., Zhang, Y., Chen, M., Hao, Y., Ji, T., Zhu, F., Cui, Y.: Efficient light trapping in organic solar cell using a short-pitched hexagonal array of metallic nanocylinders. IEEE Photonics J. **8** (2016). https://doi.org/10.1109/JPHOT.2016.2614601
13. Pudasaini, P.R., Srivastava, S.K., Zhan, Y., Zepeda, F.R., Pandit, B.: Nanostructured solar cells. Int. J. Photoenergy. (2017)
14. Um, H., Choi, D., Choi, A., Seo, J.H., Seo, K.: Embedded metal electrode for organic—inorganic hybrid nanowire solar cells. ACS Nano. 6218–6224 (2017). https://doi.org/10.1021/acsnano.7b02322

15. High-Performance Nanophotonic Simulation Software-Lumerical. https://www.lumerical.com/
16. Tan, X., Tu, Y., Deng, C., Von Czarnowski, A., Yan, W., Ye, M., Yi, Y.: Enhancement of light trapping for ultrathin crystalline silicon solar cells. Opt. Commun. **426**, 584–588 (2018). https://doi.org/10.1016/j.optcom.2018.06.003
17. Li, J., Liu, J., Gao, C., Chen, G.: Nanocomposite Hole-extraction layers for organic solar cells. Int. J. Photoenergy (2011). https://doi.org/10.1155/2011/392832
18. Palik, E.D.: Handbook of Optical Constants by Palik, vol. 3 (1998)
19. Riaz, M., Earles, S.K., Kadhim, A., Azzahrani, A.: Computer analysis of microcrystalline silicon hetero-junction solar cell with lumerical FDTD/DEVICE **6**, 1–14 (2017). https://doi.org/10.1142/S2047684117500178
20. Arjmand, A., Mcguire, D.: Complete optoelectronic simulation of patterned silicon solar cells. Opt. Quantum Electron. 1379–1384 (2014). https://doi.org/10.1007/s11082-014-9898-y
21. Dai, H., Li, M., Li, Y., Yu, H., Bai, F., Ren, X.: Effective light trapping enhancement by plasmonic Ag nanoparticles on silicon pyramid surface. Opt. Express **20**, A502 (2012). https://doi.org/10.1364/oe.20.00a502

An Overview of Reactivity for Various Nano Zero Valent Iron Particles Towards Fenton's Oxidation

Avik De[1], Tanima Nandi[1], Santu Sarkar[2], and Sandip Haldar[1](✉)

[1] Department of Basic Science and Humanities, Asansol Engineering College, Asansol 713305, West Bengal, India
sand_ju@yahoo.com

[2] Department of Radio Physics, University College of Science and Technology, University of Calcutta, Kolkata 700010, India

Abstract. Historically Fenton's reaction, under darkness or under light conditions, is considered a homogeneous process. The most important characteristic of Fenton's reagent is related to its preparation due to it is an effective catalyst just using iron (II) ions, from specifically iron(II) sulphate salt ($FeSO_4$), and hydrogen peroxide (H_2O_2). But with advancement of technology researchers are shifted to nano dimension and the traditional Fenton's oxidation reactions are replaced by heterogeneous Fenton type reactions using various forms of nano zero-valent iron. It has been observed that there occurs a marked difference among various forms of nZVI, bare, ligand stabilized and composites towards the reaction with hydrogen peroxide. Primarily rate of hydroxyl radical production is greatly affected. The rate of pollutant's removal from synthetic or industrial wastewater was also influenced. In this review article, we have summarized the reason of varying reactivity and tried to find out the best possible solution towards water pollution abetment

Keywords: Homogenous Fenton's reaction · nZVI bare · nZVI modified · Heterogeneous Fenton's reaction

1 Introduction

It was the year 1894 when Fenton published [1] on the degradation of tartaric acid interacting with "certain oxidizing agents in the presence of $FeSO_4$. After that invention, Haber and Weiss established the presence of hydroxyl free radical (OH·) during Fenton's oxidation [1]. It has been an established fact that most of the reaction under Fenton's oxidation is homogenous in nature. It was detected the negative influence of chloride anions, from $FeCl_2$, due to they act as scavengers of ·OH radicals. In the case of $FeCl_3$ and $Fe_2(SO_4)_3$, these salts are not good precursors for Fenton's reaction due to the first reaction must be the reduction of Fe(III) ions, which delays the catalytic process. Moreover, the $FeCl_3$ increases the concentration of Cl^- scavengers and even though the $Fe_2(SO_4)_3$ increases the presence of complementary oxidants, $(SO_4)^{2-}$ anions, these features do not enhance the Fenton's reaction when it uses, exclusively, $FeSO_4$ [1, 7].

© Springer Nature Singapore Pte Ltd. 2021
N. R. Das and S. Sarkar (eds.), *Computers and Devices for Communication*,
Lecture Notes in Networks and Systems 147,
https://doi.org/10.1007/978-981-15-8366-7_80

Iron metal has traditionally been utilized in permeable reactive barriers to remediate contaminated subsurface sites. The application of iron in permeable reactive barrier gradually extinct due to its corrosive behaviour over time, which lead the introduction of micro size and nano size of elemental metallic iron into remediation technique [1, 2, 6].

The nano size zero-valent iron particles have advantages over micro size due to large specific surface area and high reactivity [2]. Various methods have been adopted to prepare nZVI of various kinds [6]. The most common one is the $NaBH_4$ reduction using suitable precursors. Apart from this, top down methods are also taken like precision milling or inverse micelle formation [2, 4] In last decade, nZVI was vigorously used as catalyst for heterogeneous Fenton's reaction [8].

The rate of hydroxyl radical production depends on the nature of the nZVI surface, i.e., whether bare, capped or composite. Many researchers [6] used nZVI as catalyst in Fenton/ultrasound process to decolorize dyes such as direct blue (DB15) and reactive black (RB5). The combination of ultrasound with the Fenton process (sono- Fenton) is one of the most promising advanced oxidation processes (AOPs) used to treat organic pollutants in wastewater.

2 General Discussion on Conventional Fenton Reactions

The oxidation processes utilizing activation of H_2O_2 by iron salts are referred to as Fenton's reagent. This reaction allows the generation of hydroxyl radicals as shown in reaction [5]. Main mechanism of hydroxyl radical formation can be represented as:

$$Fe^{2+} + H_2O_2 \rightarrow Fe^{3+} + OH^{\cdot} + OH^{-} \tag{1}$$

$$Fe^{3+} + H_2O_2 \rightarrow Fe^{2+} + HO_2. + H^{+} \tag{2}$$

If the concentrations of reactants are not limiting, the organic compound can be completely detoxified or mineralized by full conversion to CO_2, water and in the case of substituted organics, inorganic salts if the treatment is continued. The overall Fenton chemistry can be summarized as:

$$Fe^{2+} + H_2O_2 + H^{+} \rightarrow 2Fe^{3+} + 2H_2O$$

This reaction clearly depict the fact that Fenton reactions can best be carried out in acidic media, preferably near $pH = 3$ [5].

3 Introduction to Heterogeneous Fenton Reaction

Heterogeneous Fenton process includes nanoscale iron particles instead of macroscale iron along with hydrogen peroxide. Here the key role performed by the solid catalyst surface. The "nanoeffect" is the kinetic stabilization of metastable materials containing or representing the active sites. The extent of stabilization is size-dependent (surface free energy vs. cohesion energy).

Central problem in kinetic studies of catalytic reactions is a possible control by diffusion. Both model reactions to be considered here take place at the surface of the nanocatalyst. If the chemical turnover at the surface is much faster than the diffusion of the reactants, diffusion control sets in and the measured kinetic constants explicitly contain the diffusion constants of the reactants. For most thermally activated reactions, however, diffusion is fast and the measured constants refer to the rate-determining step only. The competition between mass transport and chemical reaction can be estimated by the second Damko¨hler number (DaII) which is the ratio of the rate of the reaction to the rate of diffusion [5]. The steps of heterogeneous Fenton process follow Haber–Weiss mechanism. In those studies, the bottom line is: aqueous H_2O_2 on the surface of hematite or any other iron oxide nanoparticles, decomposes to $H_2O + O_2$. This may either initiate via a true surface catalytic path of generation of intermediate HO· or $HO_2\cdot$ free radicals or, as we will discuss below, a noncatalytic, non-radical redox route by which solid Fe_2O_3 may actually release Fe ions to the solution via a reductive step that leads to aqueous Fe^{2+} ($+O_2$). The released aqueous Fe^{2+}, together with remaining H_2O_2 in the solution, could then fuel a continuing conventional homogeneous Fenton reaction step. The Fe_2O_3 dissolution reactions below will occur to a certain extent limited by pH and reach equilibrium with the specific predominant dissolved Fe species indicated for each pH range [2, 4]. The key reactions are listed below [1, 6]. They have considered that the mechanism of the heterogeneous Fenton process involves a complex series of reactions on the surface of the catalyst producing HO· and $HO_2\cdot$ radicals.

$$Fe^{3+} + H_2O_2 \rightarrow Fe^{2+} + OH\cdot + OH^- \tag{3}$$

$$Fe^{2+} + H_2O_2 \rightarrow Fe^{3+} + OH\cdot + OH^- \tag{4}$$

$$HO_2\cdot \longleftrightarrow Fe^{3+} + H + O_2\cdot \tag{5}$$

$$Fe^{3+} + HO_2 \rightarrow Fe^{2+} + H^+ + O_2 \tag{6}$$

In principle, it can be said that most of the reactions occur at the solid–liquid interfaces, either at the surface of the support or in the pores of the support, where the iron species remains substantially in the solid phase, and either as a mineral or as an adsorbed ion [2–4].

4 Relative Comparison of Reactivity

The relative reactivity depends on the ease of hydroxyl radical formed from H_2O_2 in presence of nZVI particles. It has been observed that ZVI can degrade and oxidize a series of organic compounds in the presence of dissolved oxygen (DO) since ZVI transfers two electrons to O_2 to produce H_2O_2. The produced can be reduced to water by another two-electron transfer from ZVI. Moreover, the combination of H_2O_2 and Fe^{2+} (known as Fenton reaction) can produce hydroxyl radicals (·OH) which possess strong oxidizing capability towards a variety of organic compounds.

Apart from radical generation, Fe^{2+} generation from nZVI surface is also a crucial part of the process. The released Fe^{2+}would trigger a rapid contaminant oxidation with the formation of hydroxyl radicals [5] reaction. Equations (1) to (3) causes the increase of pH due to the consumption of H^+ and generation of OH^-. With the increase of Fe^{2+} and OH^-concentration, the precipitation $Fe(OH)_2$ or $Fe(OH)_3$ happened. So, the release of dissolved iron is slower at higher pH values [2, 4, 5].

Due to some fundamental drawbacks, like lack of stability, difficult separation from the medium being purified, rapid passivation of the material and limited mobility of the particles due to the formation of agglomerates, nZVI surface modification has been done. This is associated with increased reactivity.

Compared to bare nZVI, supported nZVI like rectorite supported nZVI gives 100% efficiency towards orange II degradation (near pH = 3–5). Literature survey also revealed that green synthesized nZVI showed better performance as a Fenton catalyst towards monochloro benzene oxidation. It also showed reductive degradation of dye such as malachite green [6] or bromothymol blue.

Literature study also revealed that presence of nZVI is more effective in heterogeneous Fenton reaction compared to conventional Fenton's process towards achieving higher biodegradability [7]. The BOD5/COD ratio is increased by about 200%, indicating the suitability of this oxidation process to achieve a superior biodegradability enhancement.

4.1 Reaction Kinetics

The oxidation kinetics and mechanisms of heterogeneous Fenton-like reactions have been extensively studied. Many studies have found that the first-order linear relationship is obtained in the degradation of hazardous organic pollutants [4, 11]. However, existing literature claimed two-stage first-order degradation kinetic for the oxidation of 4-chlorophenol using zero-valent iron/H_2O_2 system [11]. Regarding the pathway of degradation mechanisms, there has been much doubt as to whether oxidation of organic pollutants in heterogeneous Fenton-like systems involves the hydroxyl radicals mechanism or Fe(IV)-based mechanism [2]. The formation of ferryl ion is found to enhance the rate of pollutant degradation reaction.

5 Conclusion

There is an increasing interest in the use of nZVI for the removal of contaminants from groundwater and wastewater. nZVI can be combined with other technologies such as ultrasonic irradiation, and an evident synergistic effect was often observed in such integrated processes. It has been also observed that compared to iron oxide-based nanomaterials nZVI acts as a better catalyst for heterogeneous Fenton type oxidation. Out of the all nZVI used supported nZVI is better performer in terms of better ground state stability, lesser agglomeration, higher surface dissolution (as Fe^{2+}) or hydroxyl radical generation.

References

1. Gaspar, O.M., Guzman, C.F., Payan-Martınez, L.F., Gonzalez-Reyes, L, Hernandez-Pérez, I, Garibay-Febles, V., Pérez-Orozco, J.P., Cabrera-Lara, L.I., Ramon-Garćıa, M.L., Galicia-Luis, L., Súarez-Parra, R.: Sizing the Fenton's Catalyst. J. Photochem. Photo Biol. doi: https://doi.org/10.1016/j.jphotochem.2017.12.022
2. Fenglian, F., Dionysios, D., Liu, H.: The use of zero-valent iron for groundwater remediation and wastewater treatment: a review. J. Hazard. Mater. **267**, 194–205 (2014)
3. Taha, R.M., Ibrahim, H.A.: Characterization of nano zero-valent iron (nZVI) and its application in sono-Fenton process to remove COD in palm oil mill effluent. J. Environ. Chem. Eng. **2**, 1–8 (2014)
4. Stefaniuk, M., Oleszczuk, P., Ok S.Y.: Review on nano zerovalent iron (nZVI): from synthesis to environmental applications
5. Aramyan, M.S., Moussavi, M.: Advances in fenton and fenton based oxidation processes for industrial effluent contaminants control—a review. Int. J. Environ. Sci. Nat. Resour. **2**(4) (2017), doi: https://doi.org/10.19080/IJESNR.2017.02.555594
6. Raman, D.C., Kanmani, S.: Textile dye degradation using nano zero valent iron: a review. J. Environ. Manage. **177**, 341–355 (2016)
7. Vilardi, G., Sebastiani, D., Miliziano, S., Verdone, N., Palma, L.D.: Heterogeneous nZVI-induced Fenton oxidation process to enhance biodegradability of excavation by-products. https://doi.org/10.1016/j.cej.2017.10.152
8. De, A., De K.A., Panda Sankar, G., Halder, S.: Synthesis of zero valent iron nanoparticle and its application as a dephenolization agent for coke oven plant wastewater situated in West Bengal, India. J. Environ. Progr. Sustain. Energy **36**(6), 1700–1708 (2017)
9. Freyria, F.S., Bonelli, B., Sethi, R., Armandi, M., Belluso, E., Garrone, E.: Reactions of acid orange 7 with iron nanoparticles in aqueous solutions. J. Phys. Chem. C, **115**, 24143–24152 (2011). https://doi.org/10.1021/jp204762u
10. Rahaman, N., Abedin, Z., Hossain, M.A.: Rapid degradation of azo dyes using nano-scale zero valent iron. Am. J. Environ. Sci. **10**(2), 157–163 (2014)
11. Xu, L., Wang, J.: A heterogeneous Fenton-like system with nanoparticulate zero-valent iron for removal of 4-chloro-3-methyl phenol. J. Hazardous Mater. **186**, 256–264 (2011)

Calculation of Intrinsic Carrier Density of $Ge_{1-x}Sn_x$ Alloy, Its Temperature Dependence Around Room Temperature and Its Effect on Maximum Electron Mobility

Shyamal Mukhopadhyay[1](✉), Bratati Mukhopadhyay[2], Gopa Sen[2], and P. K. Basu[2]

[1] ECE Department, Techno International Batanagar, Maheshtala, Kolkata 700141, India
shyamal.mukherjee1@gmail.com

[2] Institute of Radio Physics and Electronics, University of Calcutta, 92 Acharya Prafulla Chandra Road, Kolkata 700009, India

Abstract. Values of intrinsic carrier density n_i in $Ge_{1-x}Sn_x$ alloy ($0 \leq x \leq 0.2$) are calculated by including composition dependent effective masses in Γ and L valleys in the conduction band, in light hole (LH) and heavy hole (HH) valence bands, and composition-dependent band gaps. The maximum temperatures of operation of the alloys are also determined over a range of dopant densities and alloy compositions. The temperature dependence of intrinsic density around room temperature is obtained by using x-dependent α and β parameters in Varshney's equation. The maximum achievable electron mobility based on n_i is estimated.

Keywords: GeSn alloy · Intrinsic carrier density · Γ and L valleys · HH and LH bands · Extrinsic limit · Mobility

1 Introduction

Intrinsic carrier concentration, n_i, is an important physical parameter for any semiconductor [1]. In order to realize p or n-type layer, needed for any device, the doping density must exceed n_i. The maximum temperature of operation of an extrinsic semiconductor T_{max} is again determined by the values of n_i and doping density N_D [2, 3]. The maximum attainable mobility also has an upper limit due to increased impurity scattering, as $N_D > n_i$, and n_i increases with decreasing band gap.

$Ge_{1-x}Sn_x$, an important alloy of group IV materials, Ge and Sn, is a very promising electronic and photonic material [4–8], can be grown directly on Si substrate with high degree of strain and on Ge or GeSn or SiGeSn virtual substrates on Si in unstrained and strained conditions. A crossover from indirect to direct gap occurs for $x > 0.08$ [9] (Γ valley < L valleys] leading to very high mobility [10], high optical absorption coefficient [11], and even optical gain [12]. The present note reports for the first time the values of n_i and its temperature dependence for $0 < x < 0.2$. Maximum electron mobility for $N_D = n_i$ and above are also estimated.

© Springer Nature Singapore Pte Ltd. 2021
N. R. Das and S. Sarkar (eds.), *Computers and Devices for Communication*,
Lecture Notes in Networks and Systems 147,
https://doi.org/10.1007/978-981-15-8366-7_81

The modification of the basic equations [1], including both L and Γ conduction band valleys and light hole (LH) and heavy hole (HH) valence bands are given in Sect. 2. Section 3 gives the results for n_i versus T, and values of $T_{\max}$ for different n_i and dopant levels with relevant discussions. Section 4 gives the conclusion.

2 Theory

The electron concentration in a conduction band valley (c) is given by

$$n_j = N_{cj}\exp\left[-\left(E_{cj} - E_{fi}/k_B T\right)\right], j = L, \Gamma \tag{1}$$

The hole concentration in valence band (v) is expressed similarly as

$$p = N_v \exp\left[-\left(E_{fi} - E_v/k_B T\right)\right] \tag{2}$$

The effective density-of-states has the following general form

$$N_i = 2\left[\frac{2\pi k_B T m_{di}}{h^2}\right]^{3/2}, i = cL, c\Gamma, v \tag{3}$$

The DOS effective masses for different valleys and bands are expressed, respectively, as

$$m_{d\Gamma}(x), m_{dL}(x) = g_v^{3/2}\left[m_l(x)m_t^2(x)\right]^{1/3}, \text{ and } m_{dv}(x) = \left[m_{hh}(x)^{3/2} + m_{lh}(x)^{3/2}\right]^{2/3}$$

In Eqs. (1) and (2), E_{fi} is the intrinsic fermi level and other symbols have their usual meanings. Using the law of mass action $np = n_i^2$, and $n = n_L + n_\Gamma$, one obtains

$$n_i = \left[N_{cL}N_v\exp\{-E_{gL}(x)/k_B T\} + N_{c\Gamma}N_v\exp\{-E_{g\Gamma}(x)/k_B T\}\right]^{1/2} \tag{4}$$

The band gap related to the L and Γ valleys are given by

$$E_{cL} - E_v = E_{gL}(x) = (1-x)E_{gL(Ge)} + xE_{gL(Sn)} + b_L^{GeSn}x(1-x) \tag{5}$$

$$E_{c\Gamma} - E_v = E_{g\Gamma}(x) = (1-x)E_{g\Gamma(Ge)} + xE_{g\Gamma(Sn)} + b_\Gamma^{GeSn}x(1-x) \tag{6}$$

where b_L *and* b_Γ are the respective bowing parameters in L valley and Γ valley. The effective masses for different bands are obtained by linear interpolation.

The maximum temperature of operation for a given doping density N_{imp} can be calculated from Eq. (4) when $n_i = N_{\text{imp}}$.

The temperature dependence of the band gap is obtained from the well-known Varshni's equation [13] that reads

$$E_g(x, T) = E_g(x, 0) - \left[\alpha(x)T^2/(T + \beta(x))\right] \tag{7}$$

The values of mobility are obtained by using our own model developed in [10].

3 Results and Discussions

The values of parameters used in the calculation are entered in Table 1.

Parameter	Ge ($x = 0$)	Sn ($x = 1$)	Parameter	Ge ($x = 0$)	Sn ($x = 1$)
$E_{g\Gamma(300K)}$ (eV)	0.7985 [10]	−0.413 [10]	m_{tL}	0.082 [10]	0.075 [10]
$E_{gL(300K)}$ (eV)	0.664 [10]	0.092 [10]	$E_{g\Gamma}(0)$ (eV)	0.89 [14]	−0.2918
b_{Γ}^{GeSn} (eV)	= 2.15 [10]		$E_{gL}(0)$ (eV)	0.742 [14]	0.1939
b_{L}^{GeSn}(eV)	= 0.91 [10]		α_L(eV/K)	4.8×10^{-4} [14]	6×10^{-4} [15]
			$\beta_L(K)$	235 [14]	230 [15]
$m_{d\Gamma}$	0.037 [10]	0.058 [10]	α_Γ (eV/K)	5.82×10^{-4} [14]	7×10^{-4} [15]
m_{lL}	1.588 [10]	1.478 [10]	β_Γ (K)	296 [14]	220 [15]

Values of the effective masses in the LH and HH bands are taken from [16].

3.1 Intrinsic Carrier Density vs Composition

Figure 1 presents the variation of n_i with composition at 300 K. Starting with a value of 1.9×10^{19} m^{-3} for Ge ($x = 0$) which is close to the quoted value of 2.25×10^{19} m^{-3} [1], the values increase with increasing value of x.

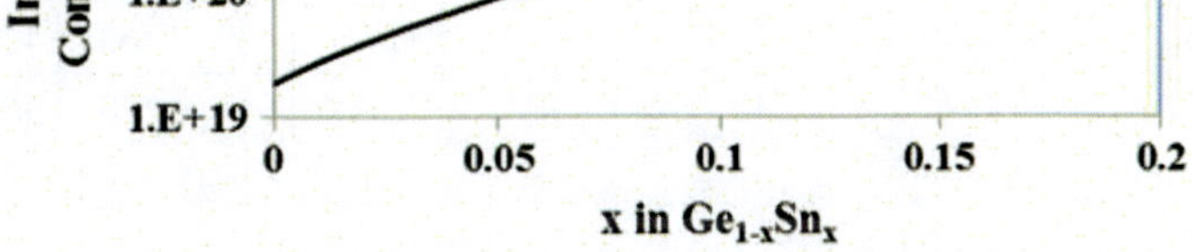

Fig. 1 Variation of n_i as a function of alloy composition (x)

3.2 Temperature Variation of n_i

The intrinsic carrier densities for different compositions x are plotted in Fig. 2 as a function of T in the range from 270 to 330 K. At any particular temperature T, the rate of increase of n_i with x is slower around the crossover point at $x = 0.08$.

3.3 Extrinsic Limit

The values of maximum temperature of operation of the alloy for different compositions x and doping density, calculated by using Eq. (7), are entered into Table 2. As expected, the maximum temperature of operation increases with doping density.

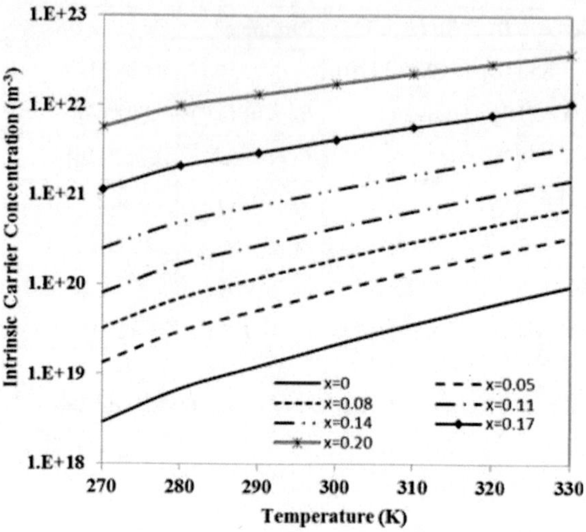

Fig. 2 Variation of n_i with temperature around 300 K for different alloy compositions

Table 2 T_{max} for different alloy composition (x) and dopant levels (N_D)

Dopant density in m^{-3}	Values of T_{max} for composition x					
N_D	0.05	0.08	0.11	0.14	0.17	0.20
10^{21}	300	320	300	280	240	200
10^{22}	360	340	320	300	260	220
10^{23}	420	400	380	360	320	280
10^{24}	520	500	480	460	420	380
10^{25}	680	700	660	620	560	520

3.4 Mobility

For extrinsic behavior, the donor density, N_D must exceed n_i. The mobility with N_D = 1×, 10×, 50× n_i are plotted against x in Fig. 3. A very high value ~10^5 cm^2/V.s is obtained when $N_D = n_i$. The value with 10 n_i is still large ~10^4. For highest N_D, the maximum mobility is reduced but is still higher than in pure Ge. The rise of mobility with x is due to increased separation between Γ and L valleys, leading to reduced contribution to mobility by electrons in lower mobility L valleys. The decrease of mobility for large x after the maxima is due to increased role of alloy scattering.

4 Conclusions

The values of intrinsic carrier density in $Ge_{1-x}Sn_x$ alloy at 300 K in the composition range $0 \leq x \leq 0.2$ have been presented for the first time taking into consideration both

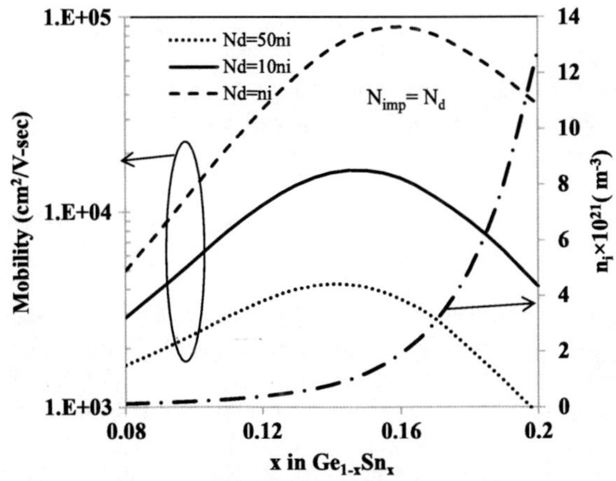

Fig. 3 Electron mobility versus composition x for three values of doping density

the Γ and L conduction band valleys and LH and HH valence bands. The value agrees with the value quoted in the literature for Ge ($x = 0$) and increases monotonically with increasing x as the band gap decreases. Using Varshni's equation for temperature dependence of band gap, the increase of n_i with T around 300 K for different composition x is also observed. The maximum temperatures of operation of the alloy material for different doping densities are ascertained. As expected, for operation as an extrinsic material above room temperature, the doping density must be higher for higher value of x for which n_i is larger. Mobility value as high as 10^5 cm^2/V s may be achieved at room temperature.

References

1. Streetman BG, Banerjee SK, Solid State Electronic Devices, 6th edn. PHI Learning Private Limited, New Delhi, Chapter 3 (2012)
2. Paul, S.K., Roy, J.B., Basu, P.K.: Empirical expressions for the alloy composition and temperature dependence of the band gap and intrinsic carrier density in $Ga_xIn_{1-x}As$. J. Appl. Phys. **69**, 827–829 (1991)
3. Paul, S.K., Basu, P.K: Use of empirically deduced composition and temperature dependent intrinsic carrier density in … and in calculating minimum capacitance in C-V plot of MIS structure. Solid-State Electron. **36**, 985–988 (1993)
4. Basu, P.K., Mukhopadhyay, B., Basu, R.: Ch. 14 in Semiconductor Laser Theory. CRC Press, Boca Raton (2015)
5. Wirths, S., Buca, D., Mantl, S.: Si–Ge–Sn alloys: from growth to applications. Prog. Cryst. Growth Charact. Mater. **62**, 1–39 (2016)
6. Gupta, S., Gong, X., Zhang, R., Yeo, Y.C., Takagi, S., Saraswat, K.C.: New materials for post-Si computing: Ge and GeSn devices. MRS Bull. **39**, 678–686 (2014)
7. Kouvetakis, J., Menendez, J., Chizmeshya, A.V.G.: Tin-based group IV semiconductors: new platforms for opto- and microelectronics on silicon. Annu. Rev. Mater. Res. **36**, 497–554 (2006)
8. Moontragoon, P., Soref, R.A., Ikonic, Z.: The direct and indirect bandgaps of unstrained SixGe1-x-ySny and their photonic device applications. J. Appl. Phys. **112**, 073106 (2012)

9. Dutt, B., Lin, H., Devanand, S., Sukhdeo, D.S., Vulovic, B.M., Gupta, S., Nam, D., Saraswat, K.C., Harris Jr., J.S.: Theoretical analysis of GeSn alloys as a gain medium for a Si-compatible laser. IEEE J. Sel Top Quantum Electron. **19**(1–6), 1502706 (2013)
10. Mukhopadhyay, B., Sen, G., Basu, R., Mukhopadhyay, S., Basu, P.K.: Prediction of large enhancement of electron mobility in direct gap Ge1-xSnx Alloy. Phys. Status Solidi B **254**, 1700244 (2017)
11. Chang, G.E., Basu, R., Mukhopadhyay, B., Basu, P.K.: Design and modeling of GeSn-based heterojunction phototransistors for communication applications. IEEE J. Sel Top in Quantum Electron 22 (2016)
12. Wirths, S., Geiger, R., von den Driesch, N., Mussler, G., Stoica, T., Mantl, S., Ikonic, Z., Luysberg, M., Chiussi, S., Hartmann, J.M., Sigg, H., Faist, J., Buca, D., Grützmacher, D.: Lasing in direct-bandgap GeSn alloy grown on Si. Nat. Photonics **9**, 88–92 (2015)
13. Varshni, Y.P.: Temperature dependence of the energy gap in semiconductors. Physica **34**, 149–154 (1967)
14. Dong-Liang, Z., Bu-Wen, C., Chun-Lai, X., Xu, Z., Hui, C., Zhi, L., Guang-Ze, Z., Qi-Ming, W: Theoretical study of the optical gain characteristics of a alloy for a short-wave infrared laser. Chin. Phys. B, **24**, 024211 (2015)
15. Vina, L., Hoechst, H., Cardona, M.: Dielectric function of Sn and its temperature dependence. Phys. Rev. B, **31** (1985)
16. Low, K.L., Yang, Y., Han, G., Fan, W., Yeo, Y.C.: Electronic band structure and effective mass parameters of Ge1-xSnx alloys. J. App. Phys. **112**, 103715 (2012)